The Universe of Conics

Georg Glaeser • Hellmuth Stachel
Boris Odehnal

The Universe of Conics

From the ancient Greeks to
21st century developments

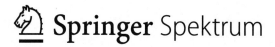
Springer Spektrum

Georg Glaeser
Department of Geometry
University of Applied Arts Vienna
Vienna, Austria

Boris Odehnal
Department of Geometry
University of Applied Arts Vienna
Vienna, Austria

Hellmuth Stachel
Institute of Discrete Mathematics and Geometry
Vienna University of Technology
Vienna, Austria

ISBN 978-3-662-56881-1   ISBN 978-3-662-45450-3 (eBook)
DOI 10.1007/978-3-662-45450-3

Springer Spektrum
© Springer-Verlag Berlin Heidelberg 2016
Softcover reprint of the hardcover 1st edition 2016

Printed on acid-free paper

This Springer Spektrum imprint is published by Springer Nature
The registered company is Springer-Verlag GmbH Berlin Heidelberg

Preface

This book attempts to cover the most important properties of conics. These are the simplest and yet so versatile planar curves, having been fascinating and useful to mathematicians and other scientists ever since their first description by the ancient Greeks.

The book has been written for people who love geometry and it is mainly based on figures and synthetic conclusions rather than on pure analytic calculations. In many proofs, illustrations help to explain ideas and to support the argumentation, and in a few cases, the picture can display a theorem at a glance together with its proof.

Large portions of the book can be understood by undergraduate students. Some sections, however, require a little more mathematical background and we hope that they will be a treasure trove even for professional mathematicians and geometers.

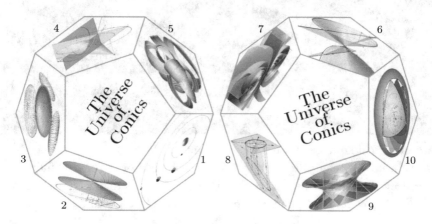

After a short historical summary (1), we explain the classical definitions of conics (2). Then, differential geometric properties are discussed (3), including the non-trivial fact that the orbits of planets are ellipses. Chapter 4 shows that conics can be planar intersections of surfaces (especially of quadratic cones). The framework of Projective Geometry is the proper setting for the study of conics (5 and 6). Still fitting into this context are polar systems and pencils of conics (7). After that we switch to the affine properties of conics (8). Chapter 9 deals with special problems. The last chapter is devoted to conics in various two-dimensional geometries: on the sphere, in Pseudo-Euclidean planes, and in hyperbolic planes.

When ISAAC NEWTON proved JOHANNES KEPLER's conjecture that the paths of the planets are ellipses, he utilized properties of ellipses which have in the mean-time almost been forgotten. The 23-years-old CARL FRIEDRICH GAUSS gained his reputation by rediscovering the dwarf planet Ceres which had only been observed on a very few occasions. He used the fact that an ellipse is uniquely determined by a single focal point (the Sun) and three additional points. GAUSS himself considered that this type of problem "commended itself to mathematicians by its difficulty and elegance". 215 years later, with Ceres still on its elliptic path, NASA's Dawn spacecraft has orbited the dwarf planet, sending stunning images of this icy remnant of our early solar system.

A geometrical homage to a famous problem:
Ceres in its icy beauty (©NASA_Dawn). The path of a planet is fully determined by three locations, due to fact that an ellipse is unambiguously defined when we know its focal point F and three points C_1, C_2, and C_3.

We wish all readers a pleasant journey into the universe of conics!

Vienna, September 2015

Acknowledgements
We say thanks to Bob Bix, David Brannan, Fritz Manhart, Andreas Rüdinger, and Norman Wildberger for assistence, suggestions, and proof reading.

Table of Contents

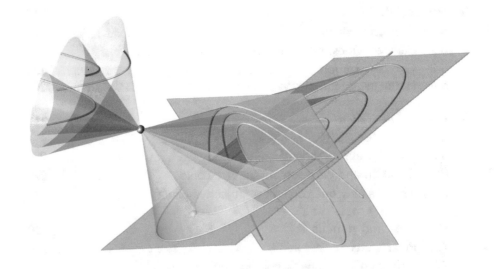

1 Introduction

Our Universe is full of conics, even if we cannot always see them – like the orbits of the planets. It needed very accurate observations to detect (JOHANNES KEPLER), and a great physicist (ISAAC NEWTON) to prove that.

Conics play a substantial role in the whole universe. On a large scale, the orbits of planets look like conics. Astronomers have recognized this fact and physicists have proved it. Many older and newer visualizations clearly show the ellipses as orbits of planets of our solar system with the Sun as the common focus.

At this point, we should mention the efforts of the ancient Greeks. Among them, especially APOLLONIUS OF PERGA (262 BC–190 BC) collected many results which were well known at his time. In his famous eight books, entitled *Konika*, he showed that all "four types" of conics can be obtained from the same cone. He even gave the names *ellipse*, *parabola*, and *hyperbola* to these curves. The first four books collect many elementary results which were known at that time.

FIGURE 1.1. A long history and a back and forth between languages: The 5th book of APOLLONIUS, here in a translation from Arabic into German.

The books five to seven present material which is probably due to APOLLONIUS. Therein, he discussed the problem of normals to a conic and prepared even the construction of the evolute of a conic. However, APOLLONIUS' books only survived in form of translations: first from Greek into Latin, then into Arabian, and back into Latin or other languages once again (cf. Figure 1.1, https://ia601408.us.archive.org/1/items/4737397/4737397.pdf). Unfortunately, the eighth book is lost.

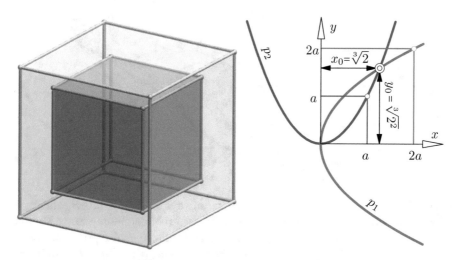

FIGURE 1.2. The Delian problem: The volumes of the two cubes on the left have ratio 2 : 1. How can one construct the scaling factor of the edge length?

The Greeks even used conics as auxiliary curves for the solution of mathematical problems: MENAECHMUS (380 BC–320 BC), *e.g.*, suggested to solve the classic "Delian problem" or "cube duplication" by means of parabolas (cf. Figure 1.2, for more details see p. 283).

Conics have a perfect shape, and with a bit of practice it is easy to judge whether a curve is a conic or just a "similar curve". This is why designers and architects use "true conics" instead of "just some freeform curve" when they deal with certain shapes.

Curved, but still entirely in a plane

There is only one type of algebraic curve of degree one: the straight line with an equation of the form

$$Ax + By + C = 0,$$

and zero curvature. A conic is an algebraic curve of degree two, described by the equation

$$Ax^2 + Bxy + Cy^2 + Dx + Ey + F = 0.$$

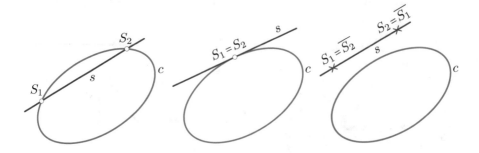

FIGURE 1.3. In the algebraic sense, a conic c always has two points S_1 and S_2 of intersection with a given straight line s. Left: two real common points. Middle: s is a tangent at the point $S_1 = S_2$. Right: The points S_1 and S_2 of intersection are complex conjugate. We cannot find them in the real plane. However, we sometimes indicate a pair of complex conjugate points by a pair of crosses on a line or curve, just in order to point out that there are two further points though we cannot see them.

Conics are always planar curves (there is no "true space curve" of degree two). One says that they have zero *torsion*. They are of *degree*[1] two, *i.e.*, they have two intersection points with a straight line – counted in the algebraic sense, where a point of contact counts twice (has multiplicity two) and complex conjugate solutions are also considered (Figure 1.3).

One possible classification of three types

Depending on the number of *real* intersection points with the line at infinity, we can distinguish three types of conics: Ellipses (no real intersection points), parabolas (the line of infinity touches the curve) and hyperbolas (two real intersection points).

Duality – points become straight lines and vice versa

Conics are also curves of *class* two which means that, from each point in the conic's plane, we have two tangents to the conic (in the algebraic sense). When the point lies on the conic, the tangents coincide (Figure 1.4). The set of all points, where there are no real tangents to the conic, can be called the *interior* of a conic. The points where the tangents are complex conjugate *and* perpendicular are called *focal points*.

---

[1]Sometimes and especially in the older literature the word *order* is used instead of *degree*. However, we prefer the term *degree*.

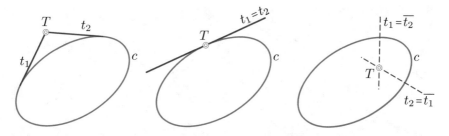

FIGURE 1.4. In the algebraic sense, a conic c always sends two tangents t_1 and t_2 through a given point T: For an exterior point T, we find two real tangents. Middle: At a point T on c, the two tangents coincide (one tangent with multiplicity two). Right: Through an interior point, there are two complex conjugate tangents. In order to make clear that there are such tangents, we sometimes indicate such tangents by dashed lines.

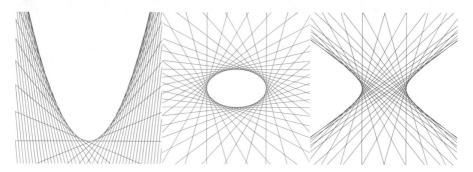

FIGURE 1.5. Dual curves of concis: the set of tangents.

Each theorem about conics that only contains the terms *point, straight line, to touch,* and *to intersect* remains, therefore, true if we interchange the terms point and straight line (and *vice versa*) but keep the relation of being incident. A conic considered as its set of *points and tangents* is a self-dual term. The dual of a conic is still a conic, represented as a set of lines, cf. Figure 1.5. We will explain this in more detail later on.

Ellipses and hyperbolas in nature

As we know, the Universe is "full of ellipses" (hyperbolas also occur sometimes, whereas parabolas are very unlikely), since all planets and comets have such trajectories (Figure 1.6).

As we mentioned earlier, conics usually appear as orbits, *i.e.*, they are not directly visible. If we push a stick with endpoint P into the ground

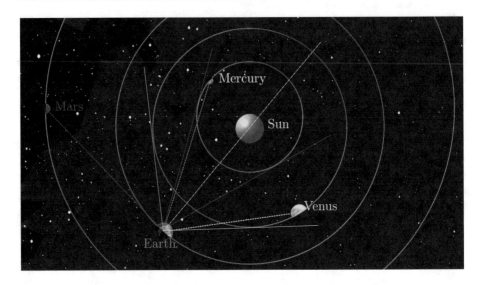

FIGURE 1.6. Ellipses as path curves of the inner planets. This image allows us to explain why one can see Mercury and Venus only close to the Sun (and thus around sunset and sunrise): We have two tangents to each of the inner orbits with comparatively small angle. To the orbit of Mars and all the other planets, there are no real tangents.

and mark the position of all the shadows P^s of P on the planar ground throughout a day, the set of all shadow points will almost exactly trace a hyperbola (or likewise an ellipse, if we are close to the poles). Only on two days of the year (the equinoctials), the conic "degenerates" into a straight line. The reason for this is the following: During one day, the Earth rotates only approximately 1° around the Sun. It fully spins, however, around the Earth axis during that time. Relatively speaking, the sun rays through a point P, therefore, form – in a good approximation – a cone of revolution. The intersection of the cone with the base plane results in a conic.

Since the inclination of the spin axis is the geographic latitude, the cone's axis is only steep enough to generate an ellipse when our location is in the neighborhood of the poles. Else we get a (branch of a) hyperbola. The angle between the oriented sun rays and the oriented rotation axis varies between $90° \pm \varepsilon$, where $\varepsilon \approx 23.44°$ is the Earth's axial tilt. Thus, the cone can also degenerate into a plane – which leads to a straight shadow orbit.

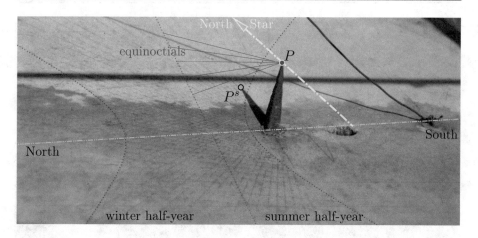

FIGURE 1.7. On each day of the year, the shadow of a point moves along a branch of a hyperbola. On the equinoctials, the curve degenerates into a straight line (Ottoman sundial Istanbul).

Figure 1.7 shows how this was used for sun dials and calenders: The photo was taken at the Topkapı-Palace in Istanbul[2].

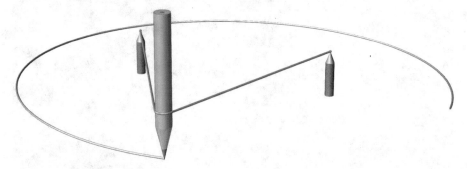

FIGURE 1.8. The "gardener's construction" of an ellipse.

THE ANCIENT GREEKS drew their geometric figures mainly in sand. Whether they also drew ellipses, we do not know. Maybe they already knew about the "gardener's construction" of the ellipse (see Figure 1.8).

For sure, they knew many facts about conic sections. We have good reasons to suppose that the painter of the scenes shown in Figure 1.9 (about

[2]http://www.muslimheritage.com/topics/default.cfm?ArticleID=942

FIGURE 1.9. High precision perspectives of circles produced 310 BC: Paintings in the tomb of PHILIPP III of Macedon. If this perspective wall painting was a few hundred years old, it would already be impressive. However, it is 2300 years old – and thus a little sensation.

310 BC) knew something about ellipses because otherwise he would not have been able to draw perspective views of circles of such a high precision. It is uncertain if MENAICHMOS (380 BC–320 BC) already knew that planar intersections of cones (and cylinders) of revolution were pro-

bably identical with the conics defined in the plane. ALBRECHT DÜRER (1471–1528) was probably aware of that, although his famous drawings in his book *Underweysung der Messung, mit dem Zirckel und Richtscheyt, in Linien, Ebenen unnd gantzen corporen,* from 1525 lack a little bit of symmetry (cf. Figure 1.10).

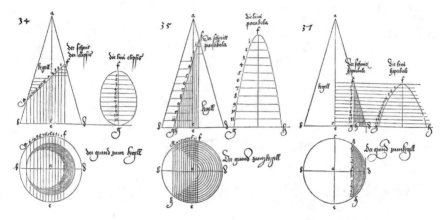

FIGURE 1.10. Dürer's constructions of conic sections.

It was the Belgian mathematician GERMINAL PIERRE DANDELIN 1794–1847 who gave the first (and very elegant) proof (see Figure 1.11).

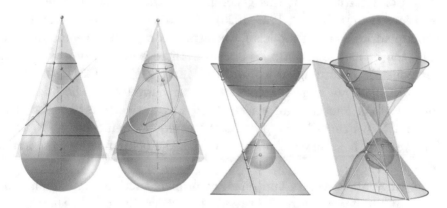

FIGURE 1.11. Dandelin spheres and focal points of conic sections.

Many elementary properties of conics have been discovered in the meantime and ever since. A fundamental theorem was discovered by BLAISE PASCAL (1623–1663) in 1639. His theorem, as illustrated in Figure 1.12, is a generalization of PAPPUS's theorem and relates conics and Projective Geometry, although this kind of geometry was introduced approximately 170 years later. The advent of Projective Geometry created a real "boom" in the theory of conics. Meanwhile, the chains of thought were extended to three-dimensional space which lead to the natural generalization of the conics, namely the quadrics.

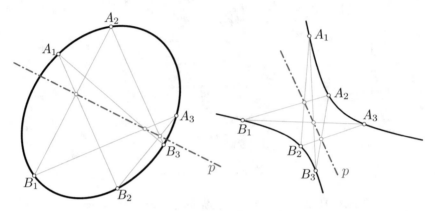

FIGURE 1.12. PASCAL's theorem: The three pairs of opposite sides of a hexagon inscribed in a conic section meet in three collinear points.

It was only natural that this generalization was extended to higher dimensions (FELIX KLEIN, 1849–1925). Nowadays, we quite often use such hyperquadrics as point models for arbitrary complicated geometries. Besides the geometric importance of conics, there is an additional "physical component": According to JOHANNES KEPLER (1571–1630), the orbits of the planets are ellipses (KEPLER's First Law, 1609). The final proof for that took another half century and was given by ISAAC NEWTON (1642–1727). Primarily the proof uses techniques from calculus. Newton, however, did it in an elementary way with the help of certain properties of conics (3.1).

Knowledge about conics and quadrics probably reached its culmination point at the beginning of the twentieth century. Since then there seems to be an increasing loss of knowledge. This book will help to sum up and preserve more or less known properties of these fascinating curves.

2 Euclidean plane

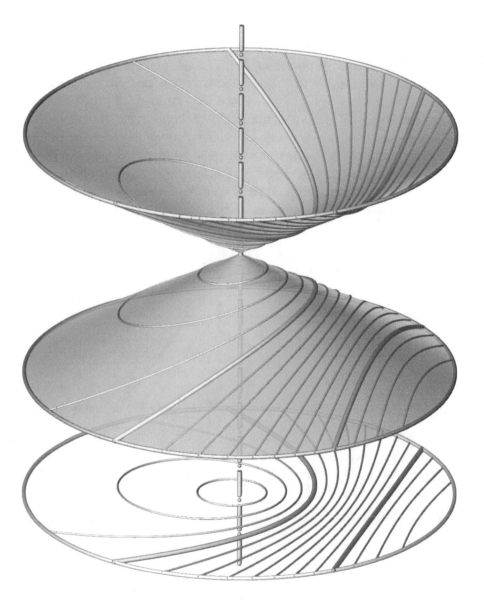

A pencil of planes meets a cone of revolution in a family of conics which maps
to a pencil of conics in the top view. These conics in the top view share a focal
point and the associated directrix.

2.1 Classical definitions

There are various definitions of conics. Depending on whether the underlying geometry is Euclidean, affine or projective, conics can be defined by their focal properties, as planar sections of cones of revolution, as perspective images of circles, by their quadratic equation, or from the viewpoint of projective geometry as the set of intersection points of projective pencils, or as the sets of self-conjugate points of a polarity in the plane. Also the ranges of the different definitions differ. Sometimes also the degenerate cases are included, sometimes only the regular ones, sometimes the circles are excluded. In the sequel, the term *conics* stands for ellipses (including circles), hyperbolas, and parabolas. Whenever we want to include degenerate cases like lines or pairs of lines, we will mention this.

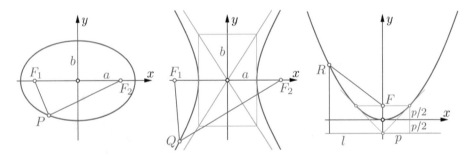

FIGURE 2.1. The standard definitions of ellipses, hyperbolas and parabolas.

We start with conics in the Euclidean plane $\mathbb{E}^2$, the plane of intuitive geometry and idealization of planes in our physical world. The standard definition of conics in $\mathbb{E}^2$ is based on the focal properties. There is a strict difference between ellipses, hyperbolas, and parabolas: ellipses are defined according to the "gardener's construction" (Figure 1.8) as the sets of points P in a plane with a constant sum of distances[1]

$$\overline{PF_1} + \overline{PF_2} = 2a \quad \text{with} \quad 2a > \overline{F_1F_2} \geq 0$$

---

[1]We use the overline to denote distances between two points or between a point and a line. $[P,Q]$ denotes the line connecting the points P and Q, while PQ is the segment terminated by P and Q.

from their *focal points* F_1 and F_2. Points Q of a hyperbola with focal points F_1 and F_2 are defined by a constant difference of distances

$$\left| \overline{QF_1} - \overline{QF_2} \right| = 2a > 0.$$

The points R of a parabola are characterized by equal distances to its focal point F and its *directrix l* provided $p = \overline{Rl} \neq 0$.

Not only the definitions of ellipses, hyperbolas, and parabolas are different but also their shapes as Figure 2.1 reveals. Hence, from this point of view it is surprising that there is a common metric definition for all three types, the definition due to APOLLONIUS OF PERGA. Later, when dealing with projective properties of conics, we will even notice that all conics are the projective transforms of one single conic (note Section 6.4).

Apollonian definition of conics

Our approach to conics in $\mathbb{E}^2$ is based on the Apollonian definition given below. In the sequel, we will learn step by step that this definition is equivalent to all other definitions of conics other than circles. We prefer this definition as it works for all three types of conics simultaneously. Also concerning proofs and graphical constructions, we prefer those which need no separation into different cases.

Definition 2.1.1 In the Euclidean plane $\mathbb{E}^2$, let F be a fixed point and l a line not passing through F. For any positive constant $\varepsilon > 0$, the set of points

$$c = \{ P \mid \overline{PF} = \varepsilon \cdot \overline{Pl} \}$$

is called a *conic* with *focal point* (or *focus*) F and associated *directrix l*. The constant ε is the *numerical eccentricity* of c; the distance $p := \varepsilon \cdot \overline{Fl}$ is the *parameter* of the conic c (Figure 2.2).

According to the eccentricity ε, we distinguish between *ellipses* with $\varepsilon < 1$, *parabolas* with $\varepsilon = 1$, and *hyperbolas* with $\varepsilon > 1$.

Remark 2.1.1 The notion 'parameter' is standard in the geometry of conics and should not be mixed with a 'parameter' used for the parametrization of curves.

The normal line s drawn from F to l is an axis of symmetry of the conic c. Figure 2.2 shows how on the axis s the point $A \in c$ between F and l

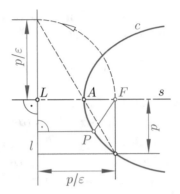

FIGURE 2.2. The Apollonian definition of the conic c with the focus F and the associated directrix l, with numerical eccentricity ε and parameter p.

can be obtained. The point A divides the segment FL on s between F and l in the ratio

$$\overline{AF} : \overline{AL} = \varepsilon : 1 = p : \frac{p}{\varepsilon}.$$

We call $A \in s$ a *vertex* of c. Figure 2.3 shows for a fixed F and l how the numerical eccentricity ε influences the shape of the conic.[2] The parameter p defines the size of c.

Lemma 2.1.1 *Ellipses and hyperbolas have a second axis of symmetry s_2, and therefore, a second focus F_2 with the associated directrix l_2. The corresponding numerical eccentricity ε and the parameter p remain unchanged.*

Proof: We are going to demonstrate that for $\varepsilon \neq 1$ on all lines parallel to the axis s there are two points of intersection with the conic c. To this end, we introduce a coordinate frame with the x-axis along s and the y-axis along the directrix l (Figure 2.4). Then, the focal point F has the coordinates $(p/\varepsilon, 0)$. For any point $P = (x, y)$, the condition $\overline{PF} = \varepsilon \cdot \overline{Pl}$ is equivalent to

$$x^2 - \frac{2p}{\varepsilon}x + \frac{p^2}{\varepsilon^2} + y^2 = \varepsilon^2 x^2,$$

hence

$$(1 - \varepsilon^2)x^2 - \frac{2p}{\varepsilon}x + \left(\frac{p^2}{\varepsilon^2} + y^2\right) = 0 \qquad (2.1)$$

which, in the case $\varepsilon \neq 1$ can be solved for x by

$$x = \frac{1}{1 - \varepsilon^2}\left[\frac{p}{\varepsilon} \pm \sqrt{p^2 + (\varepsilon^2 - 1)y^2}\right].$$

---

[2] Using the terminology of Section 7.3 below, we can state: The conics displayed in Figure 2.3 belong to a pencil of third kind with two complex conjugate base points and tangents (compare with Figure 7.21, right).

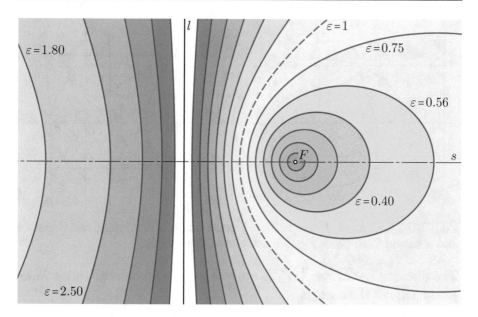

FIGURE 2.3. Conics sharing the focal point F and the associated directrix l.

In the hyperbolic case, we get for each $y \in \mathbb{R}$ a pair of real points $P, P_2 \in c$ sharing the y-coordinate. In the elliptic case, we have to take care of the limits $-\frac{p}{\sqrt{1-\varepsilon^2}} \le y \le \frac{p}{\sqrt{1-\varepsilon^2}}$. For $y = 0$ we obtain the two points

$$A = \left(\frac{p}{\varepsilon(1+\varepsilon)}, 0\right), \quad A_2 = \left(\frac{p}{\varepsilon(1-\varepsilon)}, 0\right). \tag{2.2}$$

The midpoints of all pairs (P, P_2) have the same x-coordinate

$$x_M = \frac{p}{\varepsilon(1-\varepsilon^2)}. \tag{2.3}$$

Hence, there is a second axis s_2 of symmetry, sometimes called *minor* or *secondary axis* in comparison with the *major* or *principal axis* s through F. Point $M = (x_M, 0)$ which is the crossing point of the two axes of symmetry, is called the *center* of the ellipse or hyperbola c.

By reflection in s_2, we obtain a second focal point F_2 and a second directrix l_2 (Figure 2.4). Let $P, P_2 \in c$ be symmetric with respect to s_2. Then,

$$\overline{P_2 F_2} = \overline{PF} = \varepsilon \, \overline{Pl} = \varepsilon \, \overline{P_2 l_2}.$$

This reveals that the same ellipse or hyperbola c can also be defined by the condition $\overline{PF_2} = \varepsilon \cdot \overline{Pl_2}$. Hence, the numerical eccentricity remains the same. The same holds for the parameter p, since $\overline{Fl} = p/\varepsilon = \overline{F_2 l_2}$. ■

The distances from l and F to the center M are as follows:

$$\overline{Ml} = |x_M|, \quad e := \overline{MF} = \left|x_M - \frac{p}{\varepsilon}\right| = \left|\frac{p\varepsilon}{1-\varepsilon^2}\right|.$$

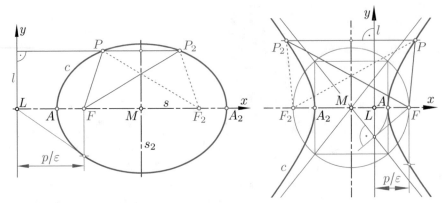

FIGURE 2.4. Conics with $\varepsilon \neq 1$ have a second axis s_2 of symmetry, and therefore, also a second focal point F_2 with associated directrix l_2.

The distance $e = \overline{MF} = \overline{MF_2}$ is a new type of eccentricity, called *linear eccentricity* of the conic c.

The vertex A as well as the second vertex A_2 given in (2.2) yield the same ratio

$$\overline{AF} : \overline{AL} = \overline{A_2F} : \overline{A_2L} = \varepsilon : 1$$

with $L = s \cap l$. The vertex A lies between L and F and A_2 outside (see Figure 2.4). We say, A_2 is the *harmonic conjugate* of A with respect to ('w.r.t.' by short) F and L.[3] In the elliptic case $\varepsilon < 1$ the two vertices A and A_2 lie on the same side of the directrix l, in the hyperbolic case $\varepsilon > 1$ on opposite sides. For a parabola, the fourth harmonic conjugate A_2 of A w.r.t. F and L lies at infinity.

Theorem 2.1.1 *The Apollonian definition of conics is equivalent to the union of the standard definitions of ellipses without circles, of hyperbolas and parabolas.*

Proof: Here we need to separate the different types of conics.

(1) Obviously, in the case $\varepsilon = 1$ the Apollonian definition is identical with the standard definition of a parabola.

(2) In the *elliptic* case (Figure 2.4, left), we conclude from the existence of the second axis of symmetry, by virtue of (2.3),

$$\overline{PF} + \overline{PF_2} = \overline{PF} + \overline{P_2F} = \varepsilon(\overline{Pl} + \overline{P_2l}) = 2\varepsilon \cdot x_M = \frac{2p}{1 - \varepsilon^2} = \text{const.}$$

---

[3]See Section 5.4 on page 206. As a consequence of this harmonic relation, the directrix l is the polar of the focal point F (see Exercise 7.1.3 in Section 7.1).

Hence, for $\varepsilon < 1$, the standard definition of ellipses is a consequence of the Apollonian definition. The *semimajor axis* a, the *semiminor axis* b, the parameter p, and the numerical eccentricity ε of an ellipse satisfy

$$a = \frac{p}{1 - \varepsilon^2}, \quad b = \frac{p}{\sqrt{1 - \varepsilon^2}}, \quad e = \overline{MF} = \frac{p\varepsilon}{1 - \varepsilon^2} = \sqrt{a^2 - b^2},$$

$$p = \frac{b^2}{a}, \quad \varepsilon = \frac{e}{a}, \quad \overline{Fl} = \frac{b^2}{e}, \quad \overline{Ml} = \frac{p}{\varepsilon(1 - \varepsilon^2)} = \frac{a^2}{e}. \tag{2.4}$$

The triangle inequality applied to any triangle FF_2P with $P \in c$ implies that $2e = \overline{FF_2} < \overline{PF} + \overline{PF_2} = 2a$. The circles are not among the ellipses satisfying the Apollonian definition since $e > 0$, hence $F_2 \ne F$.

Conversely, if any ellipse c is given according to the standard definition by its two different focal points F_1, F_2 and the constant $2a$, we can compute consecutively the linear eccentricity $e = \overline{F_1F_2}/2 \ne 0$, the numerical eccentricity $\varepsilon = a/e$, the semiminor axis $b = \sqrt{a^2 - e^2}$, the parameter $p = b^2/a$ and the distance a^2/e of the directrices l_1, l_2 from the midpoint M of F_1F_2. The focal point F_i lies between M and the associated directrix l_i, $i = 1, 2$. Then, the conic c' defined according to Apollonius by F_1, by the associated directrix l_1 and by the numerical eccentricity ε shares with c the focal points and the constant $2a$ which implies $c' = c$.

(3) In the hyperbolic case $\varepsilon > 1$ (see Figure 2.4, right), the x-coordinates of the two points P and P_2 have different signs. Therefore, we obtain

$$\left| \overline{PF} - \overline{PF_2} \right| = \left| \overline{PF} - \overline{P_2F} \right| = \varepsilon \left| \overline{Pl} - \overline{P_2l} \right| = 2\varepsilon \cdot |x_M| = \frac{2p}{\varepsilon^2 - 1} = \text{const}.$$

Also for hyperbolas, the standard definition follows from the Apollonian definition. The mutual dependencies between the two semiaxes a and b, the linear and the numerical eccentricity e and ε, and the parameter p are

$$a = \frac{p}{\varepsilon^2 - 1}, \quad b = \frac{p}{\sqrt{\varepsilon^2 - 1}}, \quad e = \overline{MF} = \frac{p\varepsilon}{\varepsilon^2 - 1} = \sqrt{a^2 + b^2},$$

$$p = \frac{b^2}{a}, \quad \varepsilon = \frac{e}{a}, \quad \overline{Fl} = \frac{b^2}{e}, \quad \overline{Ml} = \frac{p}{\varepsilon(\varepsilon^2 - 1)} = \frac{a^2}{e}. \tag{2.5}$$

The lines through the center M and with the slope $\pm \frac{b}{a} = \pm\sqrt{\varepsilon^2 - 1}$ are the *asymptotes* of the hyperbola (see Figures 2.1 and 2.4, right). In the following sections, we will learn that the asymptotes are the limits of tangent lines when the point of contact tends to infinity. The formulas above reveal that the directrix l passes through the pedal points of the asymptotes w.r.t. F.

With arguments analogous to the elliptic case, we can confirm that, conversely, each hyperbola c given according to the standard definition is identical with a hyperbola c' defined by the Apollonian definition.   ∎

Equations of conics

A first equation of conics was already presented in (2.1) on page 14. The corresponding coordinate frame (depicted in Figure 2.4) had its origin at the intersection point L between the directrix l and the axis s of symmetry which served as x-axis. Now we translate the origin to the vertex A (see

Figure 2.5, left) in order to eliminate the constant term in the quadratic polynomial on the left-hand side of (2.1). This means by (2.2) that we have to replace x by $x + \frac{p}{\varepsilon(1+\varepsilon)}$. Thus, we obtain

$$(1 - \varepsilon^2)\left(x + \frac{p}{\varepsilon(1 + \varepsilon)}\right)^2 - \frac{2p}{\varepsilon}\left(x + \frac{p}{\varepsilon(1 + \varepsilon)}\right) + \left(\frac{p^2}{\varepsilon^2} + y^2\right) = 0$$

which reduces to

$$y^2 = 2px - (1 - \varepsilon^2)x^2. \tag{2.6}$$

This quadratic equation is usually called a *vertex equation* since one vertex is taken as the origin. The vertex equation generalizes the standard equation

$$y^2 = 2px \tag{2.7}$$

of parabolas to the other conics. In the limiting case $\varepsilon = 0$, we get the vertex equation of a circle with radius p.

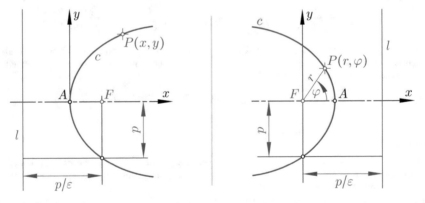

FIGURE 2.5. The coordinate frames for the vertex equation (left) and for the polar equation (right) of the conic c.

On the other hand, in the case $\varepsilon \neq 1$, we can eliminate the linear term in (2.1) by choosing the center M as the origin of our coordinate frame (Figure 2.1). This means by (2.3) that we replace x in (2.1) by $x + x_M = x + \frac{p}{\varepsilon(1-\varepsilon^2)}$. This gives

$$(1 - \varepsilon^2)^2 x^2 + (1 - \varepsilon^2)y^2 - p^2 = 0. \tag{2.8}$$

We call this the *center equation* of the conic.

Now we apply formulas from (2.4) or (2.5), respectively, in order to express the coefficients of x^2 and y^2 in terms of the semiaxes a and b. Thus, we obtain the standard equations

$$\text{Ellipse:} \quad \frac{x^2}{a^2} + \frac{y^2}{b^2} - 1 = 0, \quad \text{Hyperbola:} \quad \frac{x^2}{a^2} - \frac{y^2}{b^2} - 1 = 0. \tag{2.9}$$

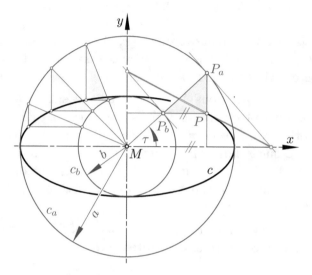

FIGURE 2.6. The construction of points P of an ellipse, named after PROCLUS as well as after DE LA HIRE.

The standard equation (2.9), left, of the ellipse c reveals that c with the semimajor axis a and the semiminor axis b can be transformed into its *principal circle* or *circumcircle* c_a, *i.e.*, the circle with center M and radius a (Figure 2.6) by the scaling (affine transformation)

$$\alpha : \ \mathbb{E}^2 \to \mathbb{E}^2, \quad (x, y) \mapsto \left(x, \frac{a}{b} y \right). \tag{2.10}$$

Conversely, one obtains the ellipse by starting with the circumcircle c_a and multiplying all y-coordinates by the factor $\frac{b}{a} < 1$. In an analogous way we can start with the *incircle* c_b with radius b and stretch the x-coordinates by the factor $\frac{a}{b} > 1$.

This results in a construction attributed to PROCLUS (Greek philosopher, 412–485), as well as to DE LA HIRE (French mathematician, 1640–1718), and which is depicted in Figure 2.6. Here the axial dilations are performed by parallel projection of the ratio $\frac{a}{b}$ formed by the aligned points $P_a \in c_a$, $P_b \in c_b$ and the center M.[4]

A *curve of degree two* in the Euclidean plane $\mathbb{E}^2$ is the set of points whose coordinates in a Cartesian coordinate system satisfy a quadratic equation

$$a_{11}x^2 + 2a_{12}xy + a_{22}y^2 + 2a_1x + 2a_2y + a = 0. \qquad (2.11)$$

The degree is invariant under changes $(x,y) \mapsto (x',y')$ of the Cartesian coordinate system, since the new coordinates x' and y' are linear functions of the initial coordinates, and *vice versa*.

When in (2.11) the polynomial on the left-hand side is reducible, *i.e.*, equals the product of two linear factors, the curve is called *reducible* or *degenerate*, otherwise *irreducible*. A reducible curve either consists of two lines, or it is a single line. The curve $x^2 + y^2 + 1 = 0$ is irreducible but contains no real points. The curve $x^2 + y^2 = 0$ contains only the origin $(0,0)$; it is irreducible over $\mathbb{R}$, but not over the field $\mathbb{C}$ of complex numbers, since $x^2 + y^2 = (x + iy)(x - iy)$.

Obviously, the conics are non-degenerate curves of degree two. The theorem below states a converse of this statement.

Theorem 2.1.2 *Each non-empty curve of degree two which is irreducible over* $\mathbb{C}$, *is a conic.*

If (2.11) is the equation of a conic c, then the sign of the discriminant $D = a_{11}a_{22} - a_{12}^2$ *determines the type of the conic c. In the case* $D > 0$ *the conic is an ellipse, for* $D = 0$ *a parabola, and* $D < 0$ *characterizes hyperbolas.*

Proof: The first part of this statement is a particular case of a standard result from linear algebra, the principal axes transformation which holds for all quadrics. This will be shown in volume 2. Here we confine ourselves to a short description of how to obtain a suitable coordinate system in the two-dimensional case which reduces the general equation (2.11) to one of the three standard equations given above.

(i) The substitution

$$x = x' \cos \varphi - y' \sin \varphi, \quad y = x' \sin \varphi + y' \cos \varphi$$

---

[4]Later we will learn that the axial dilations are perspective affine transformations which, e.g., can be used to transform tangents of the circles c_a or c_b into tangents of the ellipse c (see Figure 2.6).

represents a rotation of the coordinate frame about the origin through the angle φ. There is always an appropriate φ such that after the substitution in (2.11) the coefficient of the mixed term $x'y'$ in vanishes.

(ii) By an appropriate substitution $x' = x'' - x_0$ and $y' = y'' - y_0$ we can eliminate linear terms: if the monomials with x'^2 and y'^2 have non-vanishing coefficients we can eliminate both linear terms, otherwise only one of them which yields the standard equations (2.9) or (2.7), respectively.

The substitutions (ii) do not influence the coefficients a_{11}, a_{12} and a_{22} of the quadratic terms in (2.11). In order to figure out the effect of the substitutions (i), we rewrite the quadratic form $a_{11}x^2 + 2a_{12}xy + a_{22}y^2$ in matrix form as

$$(x\ y)\ \mathbf{A} \begin{pmatrix} x \\ y \end{pmatrix} \quad \text{with} \quad \mathbf{A} = \begin{pmatrix} a_{11} & a_{12} \\ a_{12} & a_{22} \end{pmatrix}.$$

Hence, the substitutions (i) transform the quadratic form into

$$(x'\ y')\ \mathbf{A}' \begin{pmatrix} x' \\ y' \end{pmatrix} \quad \text{with} \quad \mathbf{A}' = \mathbf{R}^T \mathbf{A} \mathbf{R}, \quad \text{where} \quad \mathbf{R} = \begin{pmatrix} \cos\varphi & -\sin\varphi \\ \sin\varphi & \cos\varphi \end{pmatrix}.$$

The discriminant $D = \det \mathbf{A}$ remains unchanged since the matrix $\mathbf{R}$ is orthogonal, and therefore, $\det \mathbf{A}' = (\det \mathbf{R})^2 \det \mathbf{A} = D$. The standard equations of the conics indicate which sign of D corresponds to which conic type.

The substitution $x = r\cos\varphi$ and $y = r\sin\varphi$ in (2.11) shows after the division by r^2 in the limit $r \to \infty$, that hyperbolas ($D < 0$) have two real points at infinity, parabolas ($D = 0$) one, while ellipses ($D > 0$) have no real ideal point. Of course, this follows also immediately by inspection of related figures. ■

We continue our discussion on equations of conics by providing some parameter representations. Let us first return to the affine transformation α in (2.10) between an ellipse c_e and its circumcircle c_a (see Figure 2.6). The inverse transformation α^{-1} maps the parametrization

$$c_a = \{(a\cos\tau,\ a\sin\tau) \mid 0 \le \tau < 2\pi\}$$

of c_a by its polar angle τ onto a parameter representation of the ellipse c_e with semiaxes a, b:

$$c_e = \{(a\cos\tau,\ b\sin\tau) \mid 0 \le \tau < 2\pi\}. \tag{2.12}$$

Herein it does not matter whether $a > b$ or not. The 'half-angle substitution' (see Figure 2.7)

$$\sin\tau = \frac{1 - t^2}{1 + t^2}, \quad \cos\tau = \frac{2t}{1 + t^2} \quad \text{with} \quad t = \tan\frac{\tau}{2}$$

gives rise to a parametrization of the ellipse,

$$c_e = \left\{ \left(a\frac{1 - t^2}{1 + t^2},\ b\frac{2t}{1 + t^2} \right) \middle| t \in \mathbb{R} \right\}, \tag{2.13}$$

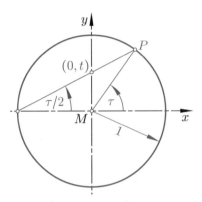

FIGURE 2.7. The half-angle substitution $t = \tan \frac{\tau}{2}$.

which is called *rational* since it contains only rational functions. However, the point $(-a, 0)$ is only obtained as the limit $t \to \infty$.

An analogous parametrization, where the trigonometric functions are replaced by hyperbolic functions

$$h_r = \{(a \cosh \tau, b \sinh \tau) \mid -\infty < \tau < \infty\} \qquad (2.14)$$

describes one branch h_r of the hyperbola. This is true since on the one hand each point of h_r satisfies the hyperbolic standard equation because of

$$\frac{(a \cosh \tau)^2}{a^2} - \frac{(b \sinh \tau)^2}{b^2} = \cosh^2 \tau - \sinh^2 \tau = 1.$$

On the other hand, for each $y \in \mathbb{R}$ there is a unique t with $y = \sinh \tau$, while $\cosh \tau > 0$. By the same token, the branch h_r of the hyperbola can also be parametrized by

$$h_r = \left\{ \left(\frac{a}{\cos \tau}, b \tan \tau \right) \mid -\frac{\pi}{2} < \tau < \frac{\pi}{2} \right\}. \qquad (2.15)$$

As an analogue to (2.13), there is also a rational parametrization of hyperbolas:

$$c_h = \left\{ \left(a \frac{1 + t^2}{1 - t^2}, b \frac{2t}{1 - t^2} \right) \mid t \in \mathbb{R} \setminus \{1, -1\} \right\}. \qquad (2.16)$$

It represents the right branch for $-1 < t < 1$. The lower part of the left branch is obtained for $t > 1$ and the upper part for $t < -1$. The point $(-a, 0)$ is again missing.

For the sake of completeness, we add the parametrization

$$c_p = \{(2p\tau^2,\, 2p\tau) \mid -\infty < \tau < \infty\} \tag{2.17}$$

for the parabola satisfying the vertex equation $y^2 = 2px$.

There is another important equation for conics:

Theorem 2.1.3 Polar equation: *The set of points with coordinates* $(x, y) = (r\cos\varphi,\, r\sin\varphi)$ *satisfying the polar equation*

$$r = \frac{p}{1 + \varepsilon \cos\varphi} \tag{2.18}$$

with constant $p, \varepsilon \in \mathbb{R} \setminus \{0\}$ *is a conic with the focus* $F = (0,0)$ *and the associated directrix* $x = p/\varepsilon$. *The conic has the parameter* $|p|$ *and the numerical eccentricity* $|\varepsilon|$. *In the case* $\varepsilon = 0$, *we obtain a circle with center* F *and radius* p.

Proof: For $p, \varepsilon > 0$, the given polar equation is equivalent to $r + \varepsilon r \cos\varphi = p$, and further, to

$$r = \varepsilon\left(\frac{p}{\varepsilon} - r\cos\varphi\right).$$

This implies that $|r| = \varepsilon\left|\frac{p}{\varepsilon} - r\cos\varphi\right|$, which means, in view of the coordinate frame depicted in Figure 2.5, right, that

$$\overline{XF} = \varepsilon \cdot \overline{Xl} \quad \text{with} \quad F = (0,0) \quad \text{and} \quad l: \; x = \frac{p}{\varepsilon}.$$

Also the converse of the implication above is valid for the following reason: When the sign of r differs from that on the right-hand side in the equation above, *i.e.*,

$$r = -\varepsilon\left(\frac{p}{\varepsilon} - r\cos\varphi\right), \quad \text{hence} \quad r = \frac{-p}{1 - \varepsilon\cos\varphi},$$

we replace simultaneously r by $-r$ and φ by $\varphi + \pi$. This does not change the defined set of points, but converts the above given polar equation into that presented in (2.18). Hence, for $p, \varepsilon > 0$, the given polar equation is equivalent to the Apollonian definition.

Changing the sign of ε in the polar equation (2.18) is equivalent to the replacement of φ by $\varphi + \pi$. Therefore, it causes a rotation of the conic about F through the angle π. The same is true for changing the sign of p, since in this case r is replaced by $-r$. Thus, we have proved that for all $p, \varepsilon \in \mathbb{R} \setminus \{0\}$ the polar equation (2.18) represents a conic. ∎

We conclude with a comment on the domain of φ in the function on the right-hand side of the polar equation (2.18):

- We obtain the entire ellipse by $-\pi < \varphi \leq \pi$ for $\varepsilon < 1$.
- For parabolas $\varepsilon = 1$, we need to exclude $\varphi = \pi$, hence $-\pi < \varphi < \pi$.

- In the hyperbolic case $\varepsilon > 1$, we have to take into account that when φ tends to $\varphi_0 := \arccos(-1/\varepsilon)$, the radius r tends to infinity. Therefore, we need to restrict the polar angle to $-\varphi_0 < \varphi < \varphi_0$ for obtaining the first branch of the hyperbola. For $\varphi_0 < \varphi < 2\pi - \varphi_0$, the radius r in (2.18) becomes negative, and we get the second branch. The two asymptotes have the slope $\pm \tan \varphi_0$.

Velocity and acceleration vectors

In order to comprehend why the trajectories of planets are conics, one needs to be familiar with velocity and acceleration vectors of curves. Therefore, we provide a brief introduction into the differential geometry of parametrized curves in the Euclidean 3-space $\mathbb{E}^3$.

With respect to any Cartesian coordinate frame (x, y, z), let

$$\mathbf{c}(\tau) = \big(x(\tau), y(\tau), z(\tau) \big), \quad \tau \in I,$$

be the trajectory of a point or a particle moving in $\mathbb{E}^3$. This curve is parametrized by time τ which passes the interval I. Then, the first derivative

$$\mathbf{v} := \dot{\mathbf{c}} =: v\mathbf{e}_1 \text{ with } \|\mathbf{e}_1\| = 1$$

is the *velocity vector*. Its norm $v = \|\dot{\mathbf{c}}\|$ equals the instantaneous *velocity*, i.e., the derivative $\mathrm{d}s/\mathrm{d}\tau$ of the curve's *arc length* s by the time τ. For $v \neq 0$, the vector $\mathbf{e}_1$ is the *unit tangent vector* to the curve, and the spanned line $\{\mathbf{c}(\tau) + \lambda\mathbf{e}_1 \,|\, \lambda \in \mathbb{R}\}$ is the *tangent line* to the curve $\mathbf{c}(I)$ at the point $\mathbf{c}(\tau)$.

The second derivative

$$\mathbf{a} := \ddot{\mathbf{c}} = \dot{v}\,\mathbf{e}_1 + v\,\dot{\mathbf{e}}_1$$

is the *acceleration vector*. According to results from elementary differential geometry, we have

$$\dot{\mathbf{e}}_1 = \frac{\mathrm{d}\mathbf{e}_1}{\mathrm{d}s} \cdot \frac{\mathrm{d}s}{\mathrm{d}\tau} = v\kappa\,\mathbf{e}_2 = \frac{v}{\varrho}\,\mathbf{e}_2, \qquad (2.19)$$

where $\kappa \geq 0$ is the *curvature* and, for $\kappa \neq 0$, $\varrho = 1/\kappa$ is the *radius of curvature*. For $\kappa \neq 0$, the unit vector $\mathbf{e}_2$ is unique and orthogonal to $\mathbf{e}_1$: A positive quarter turn in the plane spanned by $\dot{\mathbf{c}}$ and $\ddot{\mathbf{c}}$ carries $\mathbf{e}_1$ to $\mathbf{e}_2$. We call $\mathbf{e}_2$ the *principal normal vector* of the curve. Thus, we obtain

$$\mathbf{a} = \ddot{\mathbf{c}} = \dot{v}\,\mathbf{e}_1 + v^2\kappa\,\mathbf{e}_2.$$

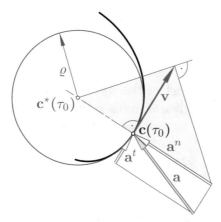

FIGURE 2.8. The geometric equivalent to the mutual dependence between the velocity vector $\mathbf{v}$, the acceleration vector $\mathbf{a} = \mathbf{a}^n + \mathbf{a}^t$, and the center of curvature $\mathbf{c}^*$ is the shaded right-angled triangle.

The first term in this vector sum defines the *tangential acceleration vector* $\mathbf{a}^t$, the second one the *normal acceleration vector* $\mathbf{a}^n$. We summarize

$$\mathbf{a} = \mathbf{a}^n + \mathbf{a}^t \quad \text{with} \quad \|\mathbf{a}^t\| = |\dot{v}|, \quad \|\mathbf{a}^n\| = \frac{v^2}{\varrho}. \tag{2.20}$$

For constant speed $(\dot{v} = 0)$ the *tangential acceleration* $\|\mathbf{a}^t\|$ vanishes. Then the second derivative $\ddot{\mathbf{c}}$ is orthogonal to $\dot{\mathbf{c}}$ everywhere in I.

In Figure 2.8, a planar curve is depicted. Beside the velocity vector $\mathbf{v}$ and the acceleration vector $\mathbf{a} = \mathbf{a}^n + \mathbf{a}^t$, also the *center of curvature* $\mathbf{c}^*(\tau_0) = \mathbf{c} + \varrho\,\mathbf{e}_2|_{\tau=\tau_0}$, *i.e.*, the center of the *osculating circle* at the point $\mathbf{c}(\tau_0)$ is shown. As a consequence of the last equation in (2.20), there is a right-angled triangle with the altitude $\mathbf{v}$ subdividing the hypotenuse in segments of lengths ϱ and $\|\mathbf{a}^n\|$.

As demonstrated in Figure 2.8, it is quite useful to visualize velocity vectors as well as acceleration vectors by arrows. However, this needs an agreement about how to represent velocities and accelerations by lengths of arrows. The simplest way is to define an appropriate scaling factor σ and to depict the scaled vector $\sigma\mathbf{v}$. For example, $\sigma = 0.005\,\text{s}$ converts a velocity of $10\,\text{m/s}$ into an arrow of length $50\,\text{mm}$.

The last equation in (2.20) reveals that, if σ is used for velocities, then σ^2 should be the scaling factor for accelerations. This guarantees that because of $(\sigma^2 \|\mathbf{a}^n\|)\varrho = (\sigma v)^2$ the right-angled shaded triangle in Figure 2.8 remains right-angled after scaling.

Often we are mainly interested in the *geometry* of curves, *i.e.*, in properties which are independent of the way how we traverse the curve. Such properties do not alter when we replace the time τ by any other parameter $\overline{\tau}$ which is related to τ by a differentiable and monotonic function

$$\tau = f(\overline{\tau}), \quad \overline{\tau} \in \overline{I}, \quad \text{with} \quad \dot{f} = \frac{\mathrm{d}f}{\mathrm{d}\overline{\tau}} \neq 0.$$

The new parameter $\overline{\tau}$ needs not be the time anymore. Nevertheless, we want to retain the notations 'velocity' and 'acceleration' since they provide the derivatives with an intuitive meaning.

The derivatives of the new parametrization $\overline{\mathbf{c}}(\overline{\tau}) := \mathbf{c}(f(\overline{\tau})), \overline{\tau} \in \overline{I}$, are

$$\dot{\overline{\mathbf{c}}} = \frac{\mathrm{d}\overline{\mathbf{c}}}{\mathrm{d}\overline{\tau}} = \frac{\mathrm{d}\mathbf{c}}{\mathrm{d}\tau} \cdot \frac{\mathrm{d}\tau}{\mathrm{d}\overline{\tau}} = \dot{f}\,\dot{\mathbf{c}} \neq \mathbf{0}, \quad \ddot{\overline{\mathbf{c}}} = \frac{\mathrm{d}^2\overline{\mathbf{c}}}{\mathrm{d}\overline{\tau}^2} = \ddot{f}\dot{\mathbf{c}} + \dot{f}^2\ddot{\mathbf{c}}.$$

We note that the velocity vectors $\mathbf{v} = \dot{\mathbf{c}}$ and $\overline{\mathbf{v}} = \dot{\overline{\mathbf{c}}}$ are linearly dependent. The spanned tangent line remains the same. The vector $\overline{\mathbf{a}} := \ddot{\overline{\mathbf{c}}}$ is a linear combination of $\dot{\mathbf{c}}$ and $\ddot{\mathbf{c}}$. The decomposition

$$\overline{\mathbf{a}} = \overline{\mathbf{a}}^t + \overline{\mathbf{a}}^n$$

into components parallel and orthogonal to the velocity vector $\overline{\mathbf{v}} := \dot{\overline{\mathbf{c}}}$ shows that $\overline{\mathbf{a}}^n = \dot{f}^2 \mathbf{a}^n$. Therefore, by virtue of (2.20), also the radius of curvature $\varrho = \|\mathbf{v}\|^2/\|\mathbf{a}^n\|$ is invariant with respect to parameter transformations.

This statement is also a consequence of the fact that among all circles passing through the point $\mathbf{c}(\tau)$, the osculating circle with center $\mathbf{c}^*(\tau_0) = \mathbf{c} + \varrho\mathbf{e}_2$, with radius $\varrho = 1/\kappa$ and located in the *osculating plane* spanned by $\dot{\mathbf{c}}(\tau)$ and $\ddot{\mathbf{c}}(\tau)$, is the best approximation of the given curve $\mathbf{c}(I)$. We prove this by showing that for any fixed parameter value $\tau_0 \in I$ the distance $d(\tau)$ between $\mathbf{c}(\tau)$ and a fixed center $\mathbf{m}$ in the osculating plane is *stationary of order* ≥ 2 at $\tau = \tau_0$ if, and only if, $\mathbf{m}$ is the center of curvature $\mathbf{c}^*(\tau_0)$.

Proof: We differentiate the equation $d^2(\tau) = (\mathbf{c}(\tau) - \mathbf{m})^2$ twice and obtain

$$d\dot{d} = \langle \mathbf{c}(\tau) - \mathbf{m}, \dot{\mathbf{c}}(\tau)\rangle, \quad \dot{d}^2 + d\ddot{d} = \|\dot{\mathbf{c}}\|^2 + \langle \mathbf{c}(\tau) - \mathbf{m}, \ddot{\mathbf{c}}(\tau)\rangle.$$

We denote the Frenet frame at $\mathbf{c}(\tau_0)$ by $(\mathbf{e}_1, \mathbf{e}_2)$, extend it by $\mathbf{e}_3$ to an orthonormal frame in $\mathbb{E}^3$, and set $\varrho_0 = \varrho(\tau_0)$ and $v_0 = v(\tau_0)$. Then, by virtue of (2.20), we get at $\tau = \tau_0$

$$d\dot{d}|_{\tau=\tau_0} = v_0\langle \mathbf{c}(\tau_0) - \mathbf{m}, \mathbf{e}_1 \rangle, \quad \dot{d}^2 + d\ddot{d}|_{\tau=\tau_0} = v_0^2 + \langle \mathbf{c}(\tau_0) - \mathbf{m}, \dot{v}(\tau_0)\,\mathbf{e}_1 + \frac{v_0^2}{\varrho_0}\,\mathbf{e}_2 \rangle.$$

Hence, $\dot{d}(\tau_0) = \ddot{d}(\tau_0) = 0$ is equivalent to $\mathbf{c}(\tau_0) - \mathbf{m} = \lambda\mathbf{e}_2 + \mu\mathbf{e}_3$ and $\lambda + \varrho_0 = 0$, i.e., $\mathbf{m} = \mathbf{c}(\tau_0) + \varrho_0\mathbf{e}_2 + \mu\mathbf{e}_3 = \mathbf{c}^*(\tau_0) + \mu, \mathbf{e}_3$ for all $\mu \in \mathbb{R}^3$. The center of curvature $\mathbf{c}^*(\tau_0)$ is the only solution with $\mu = 0$, i.e., in the osculating plane. ∎

Summing up, each reparametrization $\tau = f(\overline{\tau})$ of a given curve acts on the velocity vectors $\mathbf{v}$ like a scaling with factor $\sigma = \dot{f}$ and on the normal acceleration vectors $\mathbf{a}^n$ like a scaling with $\sigma^2 = \dot{f}^2$.

The arc length s is often called a *natural parameter* of the curve. This particular parametrization is characterized by the fact that all tangent vectors $d\mathbf{c}/ds$ are unit vectors. The second derivatives $d^2\mathbf{c}/ds^2$ are orthogonal to the curve and have length κ; they are called *curvature vectors*.

After this brief introduction to the elementary differential geometry of curves, we return to conics. We recall the polar equation (2.18) and the related coordinate frame depicted in Figures 2.5 (right) and 2.9. By choosing the polar angle φ as parameter τ, we obtain the parametrization

$$\mathbf{c}(\tau) = r(\tau)\,(\cos\tau, \sin\tau) \text{ with } r(\tau) = \frac{p}{1 + \varepsilon\cos\tau}, \text{ if } 1 + \varepsilon\cos\tau \neq 0.$$

This yields the velocity vector

$$\mathbf{v}(\tau) = \mathbf{v}_r + \mathbf{v}_r^\perp \text{ with } \mathbf{v}_r = \dot{r}\,(\cos\tau, \sin\tau), \quad \mathbf{v}_r^\perp = r\,(-\sin\tau, \cos\tau), \quad (2.21)$$

and

$$\dot{r} = \frac{p\varepsilon\sin\tau}{(1 + \varepsilon\cos\tau)^2} = \frac{\varepsilon r^2\sin\tau}{p}.$$

The first component $\mathbf{v}_r$ of the velocity vector $\mathbf{v}(\tau)$ has the direction of the position vector $\mathbf{c}(\tau)$ while the second $\mathbf{v}_r^\perp$ is orthogonal. This gives rise to a universal graphical construction of tangent lines for conics (Figure 2.9).

Theorem 2.1.4 *Let c be a conic with focus F and associated directrix l. The tangent line t_P to c at any point P different from the vertices A and A_2 intersects l at a point T such that $[F, P]$ is orthogonal to $[F, T]$.*

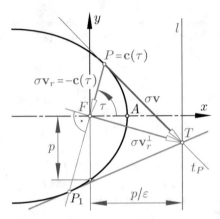

FIGURE 2.9. Construction of the tangent line t_P at the point $P = \mathbf{c}(\tau)$ to the conic c with focal point F and directrix l. We use the scaling factor $\sigma = -r/\dot{r}$ for velocity vectors.

Proof: Under the assumption $\sin \tau \neq 0$ we have $\dot{r} \neq 0$. Therefore, we can choose the scaling factor

$$\sigma = -\frac{r}{\dot{r}} = \frac{-rp}{\varepsilon r^2 \sin \tau} = \frac{-p}{\varepsilon r \sin \tau}.$$

Then,

$$\sigma \mathbf{v} = \sigma \mathbf{v}_r + \sigma \mathbf{v}_r^\perp = -r(\cos \tau, \sin \tau) - \frac{p}{\varepsilon \sin \tau}(-\sin \tau, \cos \tau) = -\mathbf{c}(\tau) + \frac{p}{\varepsilon}(1, \cot \tau).$$

In the final vector sum, the second component is orthogonal to the position vector, and its first coordinate equals that of the directrix l.   ∎

Remark 2.1.2 We have already noted that F is the pole of the directrix l. According to the properties of polarities (see Section 7.1), the point T is the pole of the connecting line $[F, P]$. This is obvious when we pay attention to the second point of intersection P_1 between the conic and the line $[F, P]$ (Figure 2.9).

Differentiating the conic's pararametrization a second time w.r.t. the polar angle τ yields the acceleration vector

$$\mathbf{a}(\tau) = (\ddot{r} - r)(\cos \tau, \sin \tau) + 2\dot{r}(-\sin \tau, \cos \tau)$$

with

$$\ddot{r} = \frac{p\varepsilon \cos \tau}{(1 + \varepsilon \cos \tau)^2} + \frac{2p\varepsilon^2 \sin^2 \tau}{(1 + \varepsilon \cos \tau)^3}.$$

In order to compute the velocity and acceleration at the vertex A, we set $\tau = 0$, to get

$$\mathbf{v}(0) = \frac{p}{1 + \varepsilon}(0, 1), \quad \mathbf{a}(0) = \frac{-p}{(1 + \varepsilon)^2}(1, 0) = \mathbf{a}^n(0).$$

We obtain $v = \|\mathbf{v}\| = p/(1 + \varepsilon)$ and, by virtue of (2.20), the curvature radius $\varrho(0) = v^2/\|\mathbf{a}^n\| = p$ which gives a new geometric meaning to the parameter p:

Theorem 2.1.5 *The parameter p of any conic c equals the radius of curvature $\varrho(0)$ of c at the vertices A on the principal axis s.*

The vector $\mathbf{a}^n(0)$ of the normal acceleration points from the vertex A towards the center of curvature A^* (Figure 2.8). Therefore, both A^* and the focal point F lie on the same side of the corresponding vertex A (Figure 2.10).

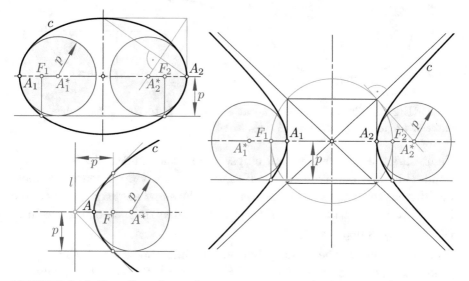

FIGURE 2.10. For all conics c, the parameter p equals the curvature radius at the vertices A_i on the principal axis.

● **Exercise 2.1.1** Osculating circle at the vertex. Confirm Theorem 2.1.5 in the following way: Based on the vertex equations (2.6) of a conic c and a circle c^* ($\varepsilon = 0$), try to specify c^* in such a way that all intersection points between c and c^* coincide with the origin.

2.2 Tangent lines of conics

The optical property

We now are going to apply the general construction of tangents for conics to ellipses and hyperbolas and pay attention to the fact that Theorem 2.1.4 holds for both focal points F_1 and F_2.

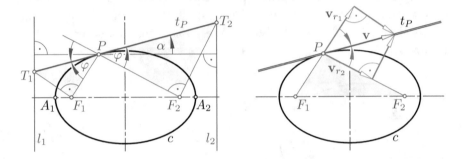

FIGURE 2.11. The tangent line t_P at any point P to the ellipse c bisects the exterior angle of the triangle F_1PF_2.

Figure 2.11, left, shows for an ellipse c how the tangent t_P to c at any point P can be constructed via F_1 as well as via F_2. Of course, P must be different from the vertices A_1 and A_2. The slope angle of t_P is denoted by α. The intersection point of t_P with the directrix l_i associated with F_i (for $i = 1, 2$) is denoted by T_i.

As a consequence, the angle φ between t_P and the segment PF_1 is congruent to the one between t_P and PF_2, since according to the Apollonian definition (page 13), we have

$$\cos\varphi = \frac{\overline{PF_1}}{\overline{PT_1}} = \frac{\varepsilon\,\overline{Pl_1}}{\overline{Pl_1}/\cos\alpha} = \varepsilon\cos\alpha = \frac{\varepsilon\,\overline{Pl_2}}{\overline{Pl_2}/\cos\alpha} = \frac{\overline{PF_2}}{\overline{PT_2}}.$$

There is another way to prove this symmetry. Let any parametrization of the ellipse be given. According to (2.21) the velocity vector $\mathbf{v}$ at P can be decomposed into a radial component $\mathbf{v}_r$ of signed length $\dot r$ and an orthogonal vector denoted by $\mathbf{v}_r^\perp$ (Figure 2.9). If $r_i(\tau)$ for $i = 1, 2$ denotes the distance of points $P \in c$ to the focal point F_i, then the standard definition $r_1 + r_2 = $ const. implies that $\dot r_1 = -\dot r_2$. Consequently, the radial components $\mathbf{v}_{r_1}$ and $\mathbf{v}_{r_2}$ of $\mathbf{v}$ in direction of F_1P and F_2P, respectively, have the same length but different signs. Therefore, independently of the scaling of velocity vectors, the triangles formed by $\mathbf{v}$ and the two decompositions are symmetric with respect to the tangent t_P (see Figure 2.11, right). We can rephrase this property by claiming that the tangent t_P bisects the exterior angle of the triangle F_1PF_2 at P.

Similar arguments hold for tangents t_P of hyperbolas. We detect again mutually similar right-angled triangles PF_1T_1 and PF_2T_2 due to the Apollonian definition (Figure 2.12, left). On the other hand, the standard definition $|r_1 - r_2| = $ const. implies that $\dot r_1 = \dot r_2$. This time the lengths of the radial components $\mathbf{v}_{r_1}$ and $\mathbf{v}_{r_2}$ of $\mathbf{v}$ have equal signs (Figure 2.12, right). Consequently, the tangent t_P at P to the hyperbola c bisects the interior angle of the triangle F_1PF_2 at P.

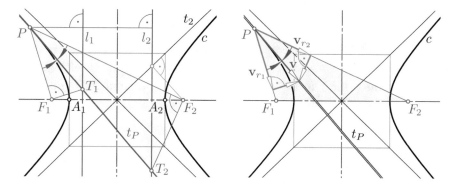

FIGURE 2.12. The tangent line t_P at any point P to the hyperbola c bisects the interior angle of the triangle F_1PF_2.

If the point $P \in c$ tends to infinity, the tangent t_P tends to one of the two asymptotes, since they intersect the directrices at the pedal points with respect to F_1 or F_2 (note the asymptote t_2 and F_2 in Figure 2.12, left).

In the case of a parabola (Figure 2.13, right), we have equal distances $\overline{PF} = \overline{Pl}$. Let G denote the pedal point (or foot) of the directrix l w.r.t. P. Then, the triangles PFT and PGT are congruent as they both are right-angled, they share the hypotenuse PT and they have equal side-lengths $\overline{PF} = \overline{Pl}$. Hence, the tangent t_P is the axis of symmetry of the kite $PFTG$ and bisects the interior angle between PF and PG.

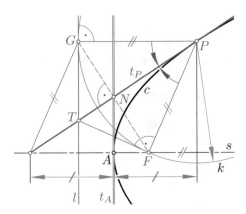

FIGURE 2.13. The tangent line t_P at any point P to the parabola c bisects an angle between $[P, F]$ and the parallel to the axis s. When reflecting the focal point F in the tangent line t_P, we obtain a point G of the directrix. The pedal point N of t_P w.r.t. F is located on the tangent line t_A to c at the vertex A.

Theorem 2.2.1 *At each point P of a conic c, the tangent t_P bisects an angle between the connecting lines $[P, F_1]$ and $[P, F_2]$ of P with the focal points, provided that, in the case of a parabola, the point F_2 is defined as the ideal point of the axis s.*

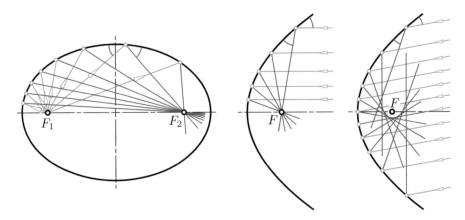

FIGURE 2.14. The optical property of ellipses and parabolas.

This is called the *optical property* of conics. In the case of an ellipse it means that rays of light going out from one focal point F_1 and being reflected in the ellipse meet all at the second focal point F_2 (Figure 2.14, left). Analogously, after reflection in a parabola all rays parallel to the axis concentrate at the focus F (Figure 2.14, middle). This is no longer true for rays being not parallel to the axis (Figure 2.14, right).

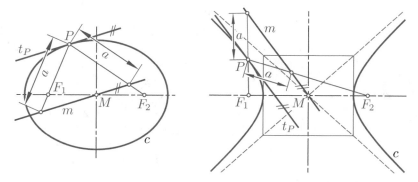

FIGURE 2.15. The line m through the center M and parallel to t_P intersects the focal lines $[P, F_1]$ and $[P, F_2]$ in points at distance a to the point P.

● Exercise 2.2.1 Construction of the semimajor axis. **Prove the following statement which holds for ellipses and hyperbolas** c **with focal points** F_1 **and** F_2 **and with the semimajor axis** a: **Let the line** m **through the center** M **of** c **be parallel to the tangent** t_P **at** P. **Then,** M **intersects the lines** $[P, F_1]$ **and** $[P, F_2]$ **at points** P_1 **and** P_2 **at distance** a **to** P **(Figure 2.15).** Hint: Project the equidistant points F_1, M, and F_2 parallel to t_P onto the line $[P, F_i]$.

Pedal curves and orthotomics

There are other consequences of the bisecting property of tangent lines, as stated in Theorem 2.2.1: In the case of a parabola, we learn immediately from Figure 2.13, that the reflection of the focal point in the tangent line t_P is a point G on the directrix l. A dilation with center F and scaling factor $1/2$ maps G onto the pedal point (or foot) N of t_P with respect to F; this pedal point is located on the tangent t_A at the vertex A.

The analogues for ellipses and hyperbolas are depicted in Figure 2.16. The reflection in the tangent t_P maps the focus F_1 onto a point G_1 on the line $[P, F_2]$ such that $\overline{PF_1} = \overline{PG_1}$. The standard definitions of ellipses and hyperbolas imply $\overline{G_1 F_2} = 2a$. Hence, the reflection G_1 lies on a circle with center F_2, the *orthotomic circle* g_1, for which we write $(F_2; 2a)$ in brief.

A dilation with center F_1 and scaling factor $1/2$ maps G_1 onto the pedal point N_1 of t_P w.r.t. F_1 and the second focus F_2 onto the center M of c.

The orthotomic circle g_1 with radius $2a$ is transformed into the principal circle $c_a = (M; a)$.

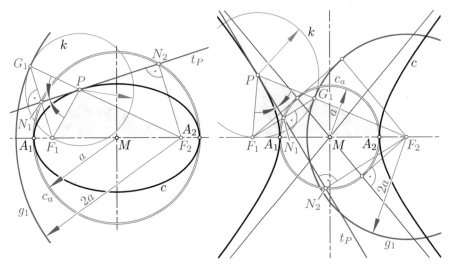

FIGURE 2.16. The reflection of the focal point F_1 in tangent lines t_P of the ellipse or hyperbola c gives points G_1 on the orthotomic circle $g_1 = (F_2; 2a)$. The pedal point N_1 of t_P with respect to F_1 is located on the principal circle $c_a = (M; a)$ which contacts c at the vertices A_1 and A_2.

Remark 2.2.1 The set of pedal points of the tangent lines of a curve d with respect to any point P is called the *pedal curve* of d with respect to P. The locus of reflections of P in the tangent lines of d is called the *orthotomic* of d with respect to P. A dilation with center P and scaling factor $1/2$ maps the orthotomic of d w.r.t. P onto the pedal curve.

Theorem 2.2.2 *The pedal curve of an ellipse or hyperbola c with respect to a focus is the principal circle $c_a = (M; a)$. The reflections G_1 of the focus F_1 in the tangent lines t_P to c are located on the orthotomic circle $g_1 = (F_2; 2a)$. The line $[G_1, F_2]$ passes through the point P of contact between c and t_P (Figure 2.16).*

The pedal curve of a parabola c with respect to the focus F is the tangent t_A at the vertex; the locus of reflections of F in the tangent lines of c is the directrix l (Figure 2.13).

This theorem as well as the following statements are also valid for circles when they are seen as limiting cases of ellipses with $F_1 = F_2 = M$.

Conversely to Theorem 2.2.2, each tangent line t_P of a conic c is the perpendicular bisector of the segment F_1G_1 where G_1 traces the circle $g_1 = (F_2; 2a)$ or the line l. The latter holds for parabolas, of course.

Corollary 2.2.1 *The envelope of the axes of symmetry between a fixed point F_1 and a point G_1 tracing a line l or a circle g_1 is a conic c.*

Note the particular case: When F_1 happens to be the center of g_1, then c is a circle.

The equation $\overline{PF_1} = \overline{PG_1}$ together with the collinearity of the points F_2, P, and G_1 implies that P is the center of a circle k which contacts g_1 at G_1 and passes through F_1 (see Figure 2.16, left and right). In the case of a parabola (see Figure 2.13), P is the center of a circle k which passes through F and is tangent to l.

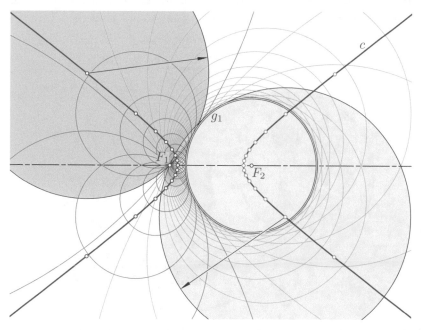

FIGURE 2.17. The displayed circles contact g_1 and pass through F_1. Their centers are located on the hyperbola c.

Corollary 2.2.2 *A conic c is the locus of centers P of circles k which are tangent to a given circle $g_1 = (F_2; 2a)$ or to a given line l and pass through a given point $F_1 \notin g_1, l$ (Figure 2.17).*

There are two kinds of contact between two different circles $(M_1; r_1)$ and $(M_2; r_2)$. Either one circle includes the other, or the point of contact lies between the two centers M_1 and M_2. In the first case we speak of *interior contact*, in the second of *exterior contact*. It makes sense to endow the radii of circles with a sign such that at interior contact the signs of the involved radii are equal and otherwise different. Then, the contact of the two circles is equivalent to

$$\overline{M_1 M_2} = |r_1 - r_2|. \tag{2.22}$$

This equation reveals that the contact is preserved when we add an arbitrary constant $r_0 \in \mathbb{R}$ to both signed radii while the centers remain fixed. Of course, when one radius changes its sign, the type of contact changes. In the case $r_1 = 0$ the center M_1 lies on the second circle.

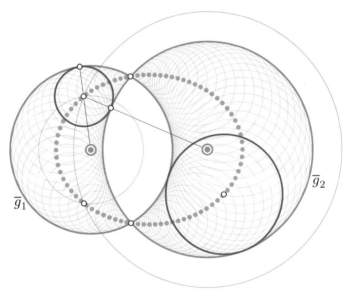

FIGURE 2.18. The displayed circles which contact $\overline{g}_1$ and $\overline{g}_2$, have their centers on an ellipse.

When, according to Corollary 2.2.2, the circles k pass through a fixed point F_1, then it depends on whether F_1 is inside or outside $g_1 = (F_2; 2a)$.

In the first (=elliptic) case, all k's are in the interior of g_1 and their signed radii are positive. Otherwise (=hyperbolic case) we have an exterior contact, and the radii of the circles k must be negative (Figure 2.17) in order to satisfy (2.22).

Now we can generalize the definition of conics c given in Corollary 2.2.2 in the following way: We add a constant $r_0 \in \mathbb{R}$ to the signed radii of all k's, replace at the same time the point F_1 by the circle $\bar{g}_2 = (F_1; r_0)$ and $g_1 = (F; 2a)$ by $\bar{g}_1 = (F; 2a + r_0)$. This yields a set of circles $\bar{k}$ being tangent to two fixed circles $\bar{g}_2, \bar{g}_1$ with respective centers F_1 and F_2. When traversing the $\bar{k}$'s, the types of contact with $\bar{g}_1$ and $\bar{g}_2$ either remain the same or change simultaneously. The transition from one type of contact to the other takes place when the signed radius of k either becomes zero (note Figure 2.18) or tends to $\pm\infty$, while the center changes from one branch of the hyperbola to the other.

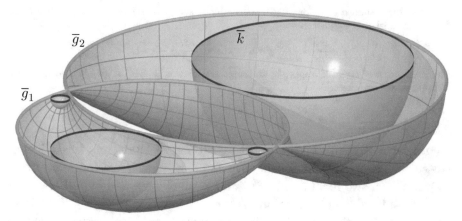

FIGURE 2.19. The lower half of a Dupin spindle cyclide.

When the circles k serve as equators of spheres, then the surface enveloped by these spheres is a *Dupin cyclide*, a *spindle cyclide* (Figure 2.19) or a *ring cyclide*. A *needle cyclide* arises when $\bar{g}_1$ contacts $\bar{g}_2$. The parabolic case with a line $\bar{g}_1$ leads to *parabolic Dypin cyclides* (Figure 2.20). We will meet the Dupin cyclides again in Section 4.2 on page 147.

• **Exercise 2.2.2** Pedal curve of an imaginary focus. In Chapter 7 we will learn that for ellipses and hyperbolas it makes sense to define beside the (real) foci with standard coordinates $(\pm e, 0)$ also the pair of complex conjugate points $(0, \pm ie)$ as focal points. Prove analytically the following counterpart to Theorem 2.2.2:

The pedal curve of an ellipse c with respect to its complex conjugate focal points is the circle

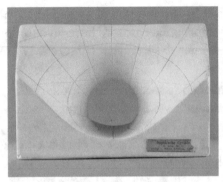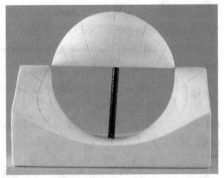

FIGURE 2.20. Two parabolic Dupin cyclides as plaster models, selected from the collection Martin Schilling, Leipzig 1880 (cf. H. WIENER , P. TREUTLEIN, *Verzeichnis mathematischer Modelle*, Teubner, Leipzig 1912).

$c_b = (M; b)$. What is the analogue for hyperbolas?

Hint: Use the standard equations (2.9). At the point $P = (x_P, y_P) \in c$ the tangent line t_P to the ellipse satisfies the equation

$$\frac{x_P}{a^2}\, x + \frac{y_P}{b^2}\, y = 1, \quad \text{where} \quad \frac{x_P^2}{a^2} + \frac{y_P^2}{b^2} = 1.$$

The pedal point of t_P with respect to an imaginary focus is $(0, \pm ie) + \lambda(x_P/a^2,\ y_P/b^2)$ with an appropriate parameter λ.

Confocal conics

Another consequence of the bisecting property of tangents to conics, as stated in Theorem 2.2.1, deals with families of *confocal* (or *homofocal*) conics: Given two focal points F_1 and F_2, all ellipses and hyperbolas sharing these focal points form an orthogonal net (Figure 2.21). This is so since through each point P off the symmetry axes there passes one ellipse and one hyperbola, and the corresponding tangent lines at P are the two bisectors of the lines $[P, F_1]$ and $[P, F_2]$, and therefore, perpendicular. As limiting curves also the two axes of symmetry can be added: the segment $F_1 F_2$ is the limit of an ellipse with vanishing semiminor axis b; the secondary axis is the limit of a hyperbola with vanishing semimajor axis a; the half-lines starting from F_1 or F_2 and pointing away from the center M are the limit of a hyperbola with vanishing b.

Similarily, there is also an orthogonal net formed by *confocal parabolas*, *i.e.*, by parabolas sharing the focal point F and the axis (Figure 2.22). On the axis, the two half-lines starting from F complete this orthogonal net.

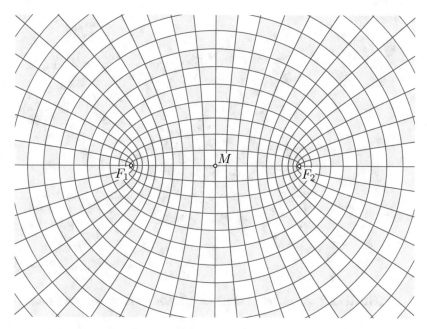

FIGURE 2.21. Confocal conics constitute an orthogonal net.

The nets of confocal conics have an additional property which is named after J. IVORY[5]. This theorem turns out to be valid also in other geometries (see, e.g., Theorem 10.1.10).

Theorem 2.2.3 **Ivory's theorem in** $\mathbb{E}^2$. *In each curvilinear quadrangle formed by two pairs of conics of the same type within a confocal net, the two diagonals have the same length.*

We leave the proof of this planar Euclidean version as an exercise.

• Exercise 2.2.3 Ivory's theorem in $\mathbb{E}^2$. Verify Ivory's theorem (Figure 2.23) for confocal ellipses and hyperbolas as well as for confocal parabolas in the following way: The parametrization

$$x = e \cos u \cosh v, \quad y = e \sin u \sinh v, \quad 0 \le u < 2\pi, \ v \in \mathbb{R}$$

defines as lines $v = \text{const.}$, $v \ne 0$, and $u = \text{const.}$, $u \ne 0$, ellipses and branches of hyperbolas, respectively which share the focal points $(\pm e, 0)$ (compare with Figure 2.21). On the other hand, the parameter lines $u = \text{const.}$ and $v = \text{const.}$ of the parametrization

$$x = u^2 - v^2, \quad y = 2uv, \quad (u, v) \in \mathbb{R}^2$$

---

[5]Sir JAMES IVORY (1765–1842), Scottish mathematician and astronomer. His famous theorem is only a by-product in his publication [39].

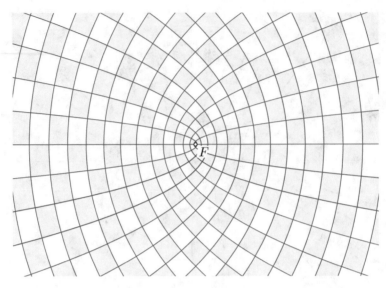

FIGURE 2.22. The net of confocal parabolas is orthogonal, too.

belong to the family of confocal parabolas (Figure 2.22). Show that for given parameter values u_1, u_2, v_1, v_2 the diagonals in the curvilinear quadrangle bounded by the curves $u = u_1$, $v = v_1$, $u = u_2$, and $v = v_2$ have the same length.

Figure 2.23 shows the mapping $c \to c'$ with $X_i \mapsto X_i'$ between conics of the same type. This mapping is induced by $v_1 \mapsto v_2$ while u remains invariant. It acts like simultaneous scalings $x \mapsto x \cosh v_2 / \cosh v_1$ and $y \mapsto y \sinh v_2 / \sinh v_1$ of both coordinates.

The statement of Ivory's theorem, $\overline{X_1 X_2'} = \overline{X_1' X_2}$, is still valid when the ellipse c' degenerates into the segment $F_1 F_2$. When the principal vertices A or B serve as points X_2 with $X_2' = F_1$ or F_2, then Ivory's theorem shows directly the standard definition of the ellipse c (Figure 2.24): $\overline{XF_1} + \overline{XF_2} = \overline{XA'} + \overline{XB'} = \overline{X'A} + \overline{X'B} = \overline{AB}$.

● **Exercise 2.2.4** Confocal conics obtained by a conformal map. Show that in the complex plane the analytic function $z \mapsto e \cos z$ transforms the rectangular coordinate grid into a net of confocal ellipses and hyperbolas. The function $z \mapsto z^2$ produces confocal parabolas.

Constructions with ruler and compass, Problem 1

In this section, we present graphic solutions for two basic geometric problems related to a conic c: which tangents of c pass through a given point Q, and in which points intersects a given line g the conic c. These constructions will provide a deeper insight into the geometry of conics.[6]

---

[6]Permanent users of geometry software might question the importance of these problems. When one can draw the conic with sufficient precision, one can immediately 'see' the points

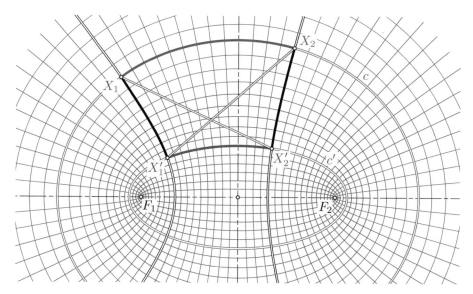

FIGURE 2.23. IVORY's theorem in the Euclidean plane $\mathbb{E}^2$: $\overline{X_1 X_2'} = \overline{X_1' X_2}$.

Problem 1: Let a conic c be given — in the case of an ellipse or hyperbola by its focal points F_1, F_2 and the length $2a$, in the case of a parabola by its focal point F and the directrix l. Which tangents can be drawn from any given point Q to the conic c? Where are the points of contact?

Let c be an ellipse. By virtue of Theorem 2.2.2, the orthotomic of c with respect to the focal point F_1 is the circle $g_1 = (F_2; 2a)$. Each tangent t to c is the bisector of F_1 and any point $G_1 \in g_1$. When t passes through Q, then Q must be equidistant to F_1 and G_1. Therefore, we can solve our problem in the following steps (Figure 2.25):

(i) Draw the auxiliary circle $h = (Q; \overline{QF_1})$ and intersect it with the orthotomic circle $g_1 = (F_2; 2a)$.

(ii) For each point G_1 of intersection, the bisector t of G_1 and F_1 is tangent to c and passes through Q.

(iii) The point T of contact between t and c is aligned with G_1 and F_2.

---

of intersection or 'draw' the tangents. Nevertheless, there is an interest in an exact graphical solution, i.e., in an algorithm using only ruler and compass for finding the solutions — even without depicting the conic.

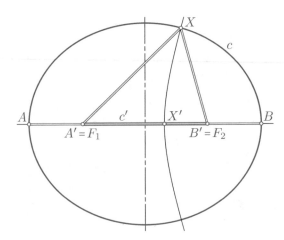

FIGURE 2.24. IVORY's theorem remains valid when the conic c' degenerates into the segment F_1F_2. Then, we obtain $\overline{XA'} = \overline{X'A}$, $\overline{XB'} = \overline{X'B}$ and, hence, $\overline{XA'} + \overline{XB'} = \overline{AB}$.

In the case displayed in Figure 2.25, the circle h intersects g_1 at two different points G_1 and $\overline{G}_1$. We obtain two different tangents t and $\overline{t}$ through Q. The point Q lies in the *exterior* of c. When Q happens to be a point of c, then the two intersection points G_1 and $\overline{G}_1$ coincide. The line $t = \overline{t}$ equals the tangent t_Q to c at Q. The circle h is an element of the set mentioned in Corollary 2.2.2. For Q in the *interior* of c there are no real points of intersection between h and g_1.

Next, we apply the center-angle-theorem to the auxiliary circle h in Figure 2.25. According to this theorem, the center angle $\sphericalangle G_1QF_1 = 2\alpha$ is twice the angle of cirmumference $\sphericalangle G_1\overline{G}_1F_1$ over the chord G_1F_1 of h. The latter angle is a normal angle of $\sphericalangle F_2Q\overline{T}$ since we have pairs of orthogonal lines involved: $[G_1, \overline{G}_1] \perp [F_2, Q]$ and $[\overline{G}_1, F_1] \perp [Q, \overline{T}]$.

These arguments cover only the case of ellipses. But a similar procedure is valid for hyperbolas (note Exercise 2.2.5). The situation for parabolas is displayed in Figure 2.26.

We summarize the results in the following generalization of Theorem 2.2.1:

Theorem 2.2.4 *The tangent lines t and $\overline{t}$ drawn from any point Q to a conic c share the angle bisectors with the lines connecting Q with the two focal points of c, provided that, in the case of a parabola, the ideal point of the axis serves as second focal point.*

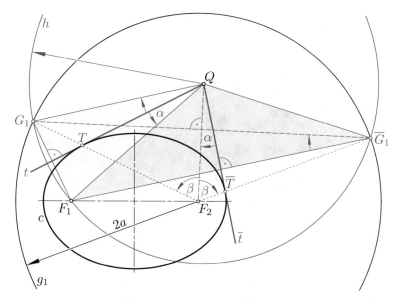

FIGURE 2.25. Construction of the tangents t and $\bar{t}$ through the given point Q to the conic c. The lines t and $\bar{t}$ share the angle bisectors with the lines connecting Q with the two focal points. $[F_2, Q]$ bisects the angle $\sphericalangle TF_2\overline{T}$.

On the other hand, when T and $\overline{T}$ are the points of contact of the two tangents, then for each focal point F of c the lines $[F, T]$ and $[F, \overline{T}]$ are symmetric with respect to $[F, Q]$. This holds also in the case of a parabola for the ideal point of the axis.

Proof: Only the second part of the statement above remains to be proved. Figure 2.25 shows immediately that for ellipses the line $[F_2, Q]$ bisects the angle $\sphericalangle G_1 F_2 \overline{G}_1$ which is equal to the angle $\sphericalangle TF_2\overline{T}$. The same must be valid for the focus F_1, since the construction via the ortho-tomic circle $g_2 = (F_1; 2a)$ must lead to the same result. The proof for hyperbolas is similar.

In the case of a parabola (Figure 2.26) we must pay attention to three symmetries: the symmetry with respect to t with $F \mapsto G$, the symmetry w.r.t. $\bar{t}$ with $F \mapsto \overline{G}$, and w.r.t. the line through Q parallel to the axis with $G \mapsto \overline{G}$: These symmetries reveal the congruence of the following angles: $\sphericalangle QFT \cong \sphericalangle QGT \cong \sphericalangle Q\overline{G}\overline{T} \cong \sphericalangle QF\overline{T}$. ∎

Remark 2.2.2 Later, in Section 7.3, we will notice that the first part of Theorem 2.2.4 is a particular case of DESARGUES's involution Theorem 7.4.1 (page 309) applied to the dual pencil of confocal conics.

We continue with some consequences of Theorem 2.2.4.

Let us apply the construction of tangents from Q to a family of confocal ellipses and hyperbolas. By virtue of Theorem 2.2.4, the points T_1 and

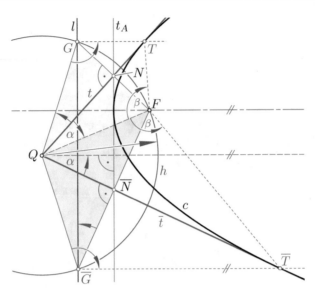

FIGURE 2.26. Construction of the tangents t and $\bar{t}$ through the given point Q to the parabola c. The lines t and $\bar{t}$ share the angle bisectors with the line $[Q, F]$ and the parallel to the axis. There are congruent angles $\sphericalangle QFT$ and $\sphericalangle QF\bar{T}$.

T_2 of contact have the property that their connections with the focal points F_1 and F_2 have an angle bisector passing through Q. The locus of points s satisfying this condition is a particular curve of degree 3 called *strophoid*. It has the point Q as a node and two orthogonal tangents there (see Figure 2.27). The normal lines at T_1 and T_2 together with the connecting lines $[T_1, T_2]$ are tangent to a parabola p which is called CHASLES's parabola of the confocal family w.r.t. Q.[7] It is the image of the line pencil Q under the quadratic transformation of conjugate normals w.r.t. any conic of the confocal family (note Example 7.5.1 on page 342).

If the point Q is particularily specified on an axis of the confocal family, then for $Q \neq M$ the cubic curve splits into the axis and a circle centered on the same axis (Figure 2.28). Only the circle serves as the locus of proper points of contact. For the proof see Exercise 2.2.6. The same circle is also the locus of contact points for point Q' diametral to Q on s. All this is also valid for confocal parabolas.

---

[7]For a proof see the references given in [43, p. 515] or [1].

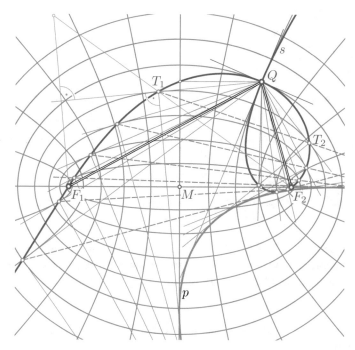

FIGURE 2.27. The points of contact (T_1, T_2) of the tangents drawn from a fixed point Q to the conics of a confocal family of ellipses and hyperbolas are located on a strophoid s. The normal lines at T_1 and T_2 as well as the connecting lines $[T_1, T_2]$ are tangent to a parabola p.

Remark 2.2.3 The strophoid s shown in Figure 2.27 and the circles s in Figure 2.28 are at the same time *locus of pedal points* of normals which can be drawn from point Q to the conics of the confocal family. This follows from the orthogonality of the confocal net: if the line $[Q, T]$ is tangent to one of the two conics passing through T, then it is orthogonal to the other.

The following generalization of the gardener's construction of ellipses is attributed to CHARLES GRAVES (1812–1899), bishop of Limerick and mathematician:

Theorem 2.2.5 *Let an ellipse e_0 be given as well as a closed piece of string strictly longer than the perimeter of e_0. Then, the locus of a pencil used to pull the string taut around e_0 is an ellipse e confocal to e_0 (Figure 2.29).*

When e_0 degenerates into the segment $F_1 F_2$, then this construction converts into the gardener's construction (Figure 1.8).

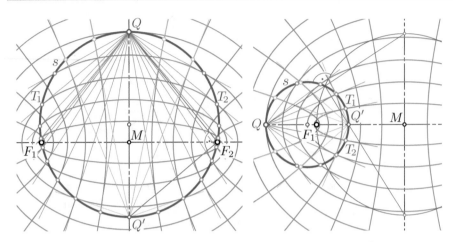

FIGURE 2.28. For points Q other than M on an axis of the confocal family, the locus s of contact points as depicted in Figure 2.27 degenerates into a circle. The same circle is also the locus of contact points for the point $Q' \in s$ opposite to Q on s.

Proof: Our arguments are similar to that displayed in Figure 2.11 for the bisecting property of conics' tangents.

We parametrize e_0 by its arc length as $\mathbf{c}(s)$, $0 \le s \le L$. Then, the velocity vector $\dot{\mathbf{c}}$ is a unit vector, and the acceleration vector $\ddot{\mathbf{c}}$ is orthogonal to $\dot{\mathbf{c}}$. Suppose, we attach the string to the ellipse e_0 any point $\mathbf{c}(0)$. Let us cut the string in our imagination at Q (Figure 2.29). First we focus on the left part with the straight segment TQ:

When λ denotes the length of the string starting at $\mathbf{c}(0)$, curved along the ellipse e_0 and aligned from point $T = \mathbf{c}(s_0)$ on, then the position vector of Q is

$$\mathbf{q}(s_0, \lambda) = \mathbf{c}(s_0) + (\lambda - s_0)\,\dot{\mathbf{c}}(s_0).$$

For variable s_0 this would parametrize the trajectory of point Q if it were fixed on the string. However, in GRAVES's construction Q varies relative to the string. Therefore, we assume λ as a function of s, too. Then, the velocity vector of Q reads

$$\mathbf{v} = \dot{\mathbf{q}}(s, \lambda) = \dot{\mathbf{c}}(s) + (\dot{\lambda} - 1)\,\dot{\mathbf{c}}(s) + (\lambda - s)\,\ddot{\mathbf{c}}(s) = \dot{\lambda}\,\dot{\mathbf{c}}(s) + (\lambda - s)\,\ddot{\mathbf{c}}(s).$$

This shows $\mathbf{v}$ as a sum $\mathbf{v}_{t_1} + \mathbf{v}_{n_1}$ of two orthogonal components. The first one of length $\dot{\lambda}$ is in direction of $\dot{\mathbf{c}}$ and expresses the instant velocity of point Q with respect to the left portion of the string. The second component is orthogonal to $\dot{\mathbf{c}}$.

The same holds for the right section of the string with the straight segment $\overline{T}Q$. We obtain a second decomposition $\mathbf{v} = \mathbf{v}_{t_2} + \mathbf{v}_{n_2}$ where $\mathbf{v}_{t_2}$ shares the length $\dot{\lambda}$ with $\mathbf{v}_{t_1}$ but points towards $\overline{T}$, as it represents the instant velocity of Q against the right portion of the string. The symmetry of the two decompositions of $\mathbf{v}$ reveals (see Figure 2.29) that $\mathbf{v}$ bisects the exterior of the angle $\sphericalangle TQ\overline{T}$ and by Theorem 2.2.4 also that of $\sphericalangle F_1QF_2$.

Now we reverse the arguments displayed in Figure 2.11, right: The orthogonal projection of $\mathbf{v}$ onto the lines $[F_1, Q]$ and $[F_2, Q]$ gives two vectors of equal length. Therefore, the sum of distances $\overline{F_1Q} + \overline{F_2Q}$ must remain constant in GRAVES's construction. The point Q has to trace an ellipse with focal points F_1 and F_2.   ∎

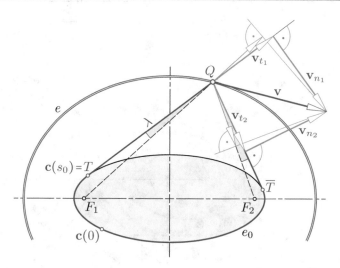

FIGURE 2.29. GRAVES's string construction of an ellipse e confocal to e_0.

Figures 2.25 and 2.26 are also the basis for the following statement. It deals with *orthoptic curves* which by definition are the loci of points from which two orthogonal tangents to a given curve can be drawn.

Theorem 2.2.6 *The orthoptic curve of an ellipse or hyperbola, i.e., the locus of intersection points between orthogonal tangents, is the director circle with radius $\sqrt{a^2 + b^2}$ or $\sqrt{a^2 - b^2}$, respectively. In the case of a parabola, the orthoptic curve coincides with the directrix l.*

Proof: Firstly, we study ellipses and hyperbolas: Once more we analyze Figure 2.25 with the construction of the tangents t and $\bar{t}$ passing through the given point Q. Since there are two equilateral triangles G_1QF_1 and $\overline{G}_1QF_1$ involved, the angle $\angle G_1Q\overline{G}_1$ is twice the angle between the two tangents t and $\bar{t}$ passing through point Q. Therefore, these tangents are orthogonal if, and only if, the points G_1, Q and $\overline{G}_1$ are aligned, *i.e.*, the chord of the orthotomic circle $g_1 = (F_2; 2a)$ and the auxiliary circle $h = (Q; \overline{QF_1})$ is a diameter of h.

The common chord is orthogonal to the line $[Q, F_2]$ connecting the centers of g_1 and h. Since $\overline{QG_1} = \overline{QF_1}$, the orthogonality of t and $\bar{t}$ is equivalent to the relation

$$\overline{QF_1}^2 + \overline{QF_2}^2 = 4a^2.$$

Now we us use the Cartesian standard coordinate frame of the given conic c: The origin is specified at the center M, and we obtain coordinates $F_1 = (-e, 0)$, $F_1 = (e, 0)$, and $Q = (x, y)$. The sum of the two equations

$$\overline{QF_1}^2 = (x + e)^2 + y^2 \quad \text{and} \quad \overline{QF_2}^2 = (x - e)^2 + y^2$$

gives after division by 2

$$2a^2 = x^2 + e^2 + y^2.$$

By (2.4), we conclude for ellipses $\overline{QM}^2 = x^2 + y^2 = a^2 + b^2$ (note Figure 9.36 on page 424).

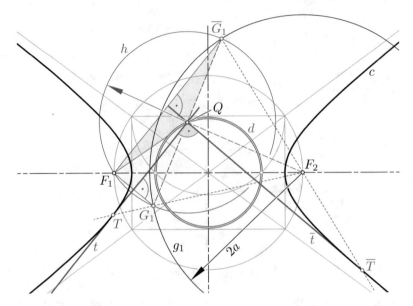

FIGURE 2.30. The orthoptic curve of a hyperbola with $a > b$ is a circle, the director circle d.

For hyperbolas we obtain from (2.5) $\overline{QM}^2 = x^2 + y^2 = a^2 - b^2$ (see Figure 2.30). For an equilateral hyperbola ($a = b$), the orthoptic curves shrinks to the point M, in the case $a < b$ the orthoptic curve is empty.

When, secondly, in the case of a parabola (see Figure 2.26) the points $G_1, \overline{G}_1$ on the directrix l are diametral on the auxiliary circle h, then also the center Q of h must be located on l. ∎

● Exercise 2.2.5 Tangents to confocal conics. Complete the proof of Theorem 2.2.4 for hyperbolas and parabolas. In the first case adapt Figure 2.25 for a hyperbola c. In the parabolic case extract the proof from Figure 2.26.

● Exercise 2.2.6 Prove for points T of the circles s in Figure 2.28, left and right, that the lines TQ and TQ' are angle bisectors of the lines connecting T with the two focal points. When Q is aligned with the focal points (Figure 2.28, right), then s is the *circle of* Apollonius which is defined by a constant ratio of distances $\overline{TF_1} : \overline{TF_2}$ = const. for all $T \in s$, including Q and Q'.

● Exercise 2.2.7 Isoptic curve of a parabola. Prove that the *isoptic curve* of a parabola, *i.e.*, the set of points from which a parabola with focal point F and directrix l can be seen under a given angle φ, under the conditions $0 < \varphi < \frac{\pi}{2}$ or $\frac{\pi}{2} < \varphi < \pi$, is a branch of a hyperbola with the focal point F, the associated directrix l and the numerical eccentricity $1/\cos\varphi$.

Hint: We refer to Figure 2.26. The depicted parabola c is seen from point Q under the angle $\varphi = \sphericalangle TQ\overline{T}$, and we have here $\varphi < \frac{\pi}{2}$. A rotation of t and $\overline{t}$ about Q through the marked angle α together with $\overline{QF} = \overline{QG}$ yields $\overline{Ql} = \overline{QF}\cos\varphi$. The same is true in the case $\varphi > \frac{\pi}{2}$. The rest follows from Definition 2.1.1. The orthoptic curve of a parabola which is mentioned in Theorem 2.2.6, is obviously a limiting case: The hyperbola degenerates into a line.

Constructions with ruler and compass, Problem 2

Before we start with the second problem, we recall a term from elementary geometry: The *power* $p_k(Q)$ *of any point* Q *with respect to the circle* $k = (M; r)$ is defined as

$$p_k(Q) = \overline{QM}^2 - r^2.$$

Of course, $p_k(Q) = 0$ is equivalent to $Q \in k$, and $p_k(Q) > 0$ characterizes points Q exterior to k.

Let a line g of distance d to M intersect the circle k in two points S_1 and S_2 (Figure 2.31). Then, by PYTHAGORAS's theorem we obtain

$$\overline{QS_i} = \left| \sqrt{\overline{QM}^2 - d^2} \pm \sqrt{r^2 - d^2} \right|, \quad i = 1, 2,$$

which immediately implies that

$$\overline{QS_1} \cdot \overline{QS_2} = \left| \overline{QM}^2 - r^2 \right| = |p_k(Q)|.$$

For signed distances on g this equation is also valid without using the absolute values since for Q in the interior of k the signs of the distances $\overline{QS_1}$ and $\overline{QS_2}$ are different. In the case $d = r$, the line g contacts the circle k at a point T, and $p_k(Q) = \overline{QT}^2$.

In Cartesian coordinates, with $M = (m, n)$ and $Q = (p, q)$, we obtain

$$p_k(Q) = (p - m)^2 + (q - n)^2 - r^2.$$

So, to find the power of Q w.r.t. the circle k, we only need to plug the coordinates of Q into the standard equation of k.

Given a second circle $k' = (M'; r')$ with $M \neq M'$, a point $X = (x, y)$ with equal powers with respect to k and k' satisfies

$$(x - m)^2 + (y - n)^2 - r^2 = (x - m')^2 + (y - n')^2 - r'^2,$$

which is equivalent to the linear equation

$$2(m - m')x + 2(n - n')y = m^2 - m'^2 + n^2 - n'^2 - r^2 + r'^2.$$

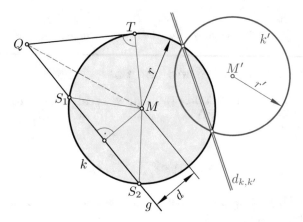

FIGURE 2.31. The power of the point Q with respect to the circle k is defined as $p_k(Q) = \overline{QM}^2 - r^2$. For Q in the exterior of k, the power $p_k(Q)$ satisfies $p_k(Q) = \overline{QT}^2$. On lines g intersecting k, we obtain $p_k(Q) = \overline{QS_1} \cdot \overline{QS_2}$, provided these distances are signed.

This defines a line, the *radical axis* $d_{k,k'}$ of k and k'. It is orthogonal to the common diameter $[M, M']$. When there are two intersection points $P_1, P_2 \in k \cap k'$ (see Figure 2.31), the radical axis $d_{k,k'}$ equals the common chord $[P_1, P_2]$ of the two circles. If k and k' touch each other at T, then $d_{k,k'}$ is the common tangent at T.

Given a third circle k'', when the radical axes $d_{k,k'}$ and $d_{k,k''}$ intersect at a point C, then C is called the *radical center* of the three circles k, k', and k''. Also the third radical axis $d_{k',k''}$ passes through C. If C is exterior to k, then the circle $o = (C; \sqrt{p_k(C)})$ intersects all three given circles orthogonally.

Problem 2: Let a conic c be given — in the case of an ellipse or hyperbola by its focal points F_1, F_2 and the length $2a$, in the case of a parabola by its focal point F and the directrix l. Where does a given line q intersect the conic c?

According to Corollary 2.2.2, the conic c is the locus of centers P of circles k which pass through the focal point F_1 and contact the orthotomic g_1 of F_1. The point G_1 of contact with g_1 is the reflection of F_1 in the tangent t_P. When P lies on q, then the corresponding circle k passes also through the reflection of F_1' in q.

Hence, our problem is reduced to the following problem: Find circles k which contact g_1 and pass through the points F_1 and F_1'. We first discuss the elliptic or hyperbolic case where the orthotomic g_1 of F_1 w.r.t. c is the circle $(F_2; 2a)$ (Figure 2.32).

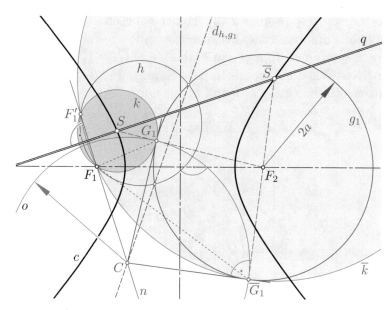

FIGURE 2.32. Points of intersection between a conic c and a line q.

Each point on the line $n = [F_1, F_1']$ has the same power w.r.t. all circles through F_1 and F_1'. Let h be one of these circles. The intersection point C between the radical axis d_{h,g_1} and n satisfies $p_{g_1}(C) = p_k(C)$ for all circles k through F_1 and F_1'. Hence, all chords $d_{g_1,k}$ pass through C. When k contacts the circle g_1, then the common tangent at the point of contact passes through the radical center C too. Therefore, we can solve our problem in the following steps (Figure 2.32):

(i) Draw the orthotomic circle $g_1 = (F_2; 2a)$.

(ii) Draw through F_1 the normal line n to the given line q and determine on n the reflection F_1' of F_1 w.r.t. q.

(iii) Draw any circle h through F_1 and F_1' (preferably one which intersects g_1) and let the radical axis d_{h,g_1} meet n at C.

(iv) The tangents through C to the orthotomic circle g_1 contact g_1 at points $G_1, \overline{G}_1$ which are respectively aligned with F_2 and the requested points of intersection $S, \overline{S} \in q \cap c$.

If C is exterior to g_1, then C is the center of a circle o which intersects g_1 and all circles through F_1 and F_1' orthogonally. The circle o passes through G_1 and $\overline{G}_1$ (Figure 2.32). For C between F_1 and F_1' there are no real points of intersection between the conic c and the given line q. In the case $C = F_1'$, i.e., $F_1' \in g_1$, we obtain $G_1 = \overline{G}_1 = F_1'$ and the line q is tangent to c. When q happens to pass through F_1, then the normal line n is tangent to the circles h, k and $\overline{k}$.

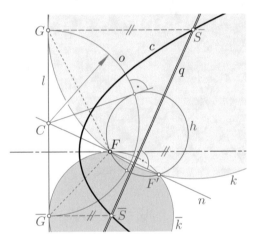

FIGURE 2.33. Points of intersection between a parabola c and a line q.

Figure 2.33 shows the construction of the intersection points between a parabola c (given by its focus F and the directrix l) and a line q. Again, the problem can be reduced to the determination of circles which are centered on q which pass through F and contact the directrix l.

● Exercise 2.2.8 Intersection between a parabola and a line. Extract from Figure 2.33 the ruler-and-compass construction of intersection points between a parabola and a line.

● Exercise 2.2.9 Relation between two points of a conic and the foci. Prove the following statement: If a conic with focal points F_1 and F_2 intersects a line in two different points S and $\overline{S}$ (see Figure 2.32), then S and $\overline{S}$ are focal points of another conic which passes through F_1 and F_2. In this case the sides of the quadrilateral $F_1 S F_2 \overline{S}$ are tangent to a circle. Find the analogue statement for parabolas.

2.3 Mechanisms tracing conics

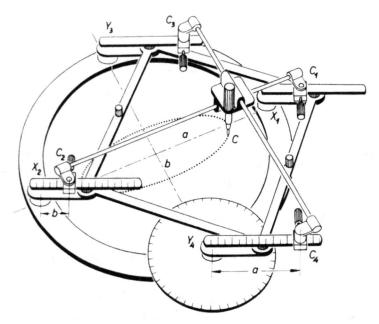

FIGURE 2.34. Elliptic compass after HOECKEN from [66].

In times of computer graphics, mechanisms for the accurate drawing of conic sections seem only to be curiosities from former centuries. In fact, no one will nowadays draw ellipses or parabolas with "conic compasses". In order to cut an elliptic mirror, however, LEONARDO DA VINCI's invention is still of importance. What is more: The theory behind the mechanisms leads to deeper insights and a better understanding of conics.

A special elliptic motion

If a rod of fixed length is directed such that its endpoints X and Y move on the perpendicular straight lines x and y, then it is undergoing a very special constrained motion, see Figure 2.35.

We are concerned with the path curve of an arbitrary point $P \in XY$ (but $P \neq X, Y$) with $\overline{YP} = a$ and $\overline{XP} = b$, cf. Figure 2.35. Let $O = x \cap y$ be the origin of a Cartesian coordinate system and φ be the rod's angle of inclination. Thus, C's path is given by $(x, y) = (a \cos \varphi, \, y = b \sin \varphi)$ which is a parametric representation of an ellipse with axis lengths $2a$ and $2b$.

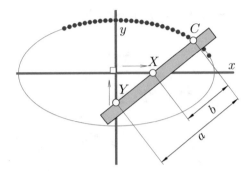

FIGURE 2.35. The simplest case of an elliptic motion and the trammel construction: A rod XY of constant length is moving along perpendicular straight lines x and y. Each point C on the rod has an ellipse for its path.

Remarkably, the midpoint M of the segment XY $(a = b)$ describes a circle.

This constrained motion is henceforth called *elliptic motion*. The construction of points of the ellipse using a strip of paper or a rod guided in this way is sometimes referred to as the *trammel construction*.

As we shall show soon, it can be generalized in such a way that the straight guidances need not be perpendicular and the point C that is connected with the rod need not lie on the rod itself.

Hoecken's mechanism

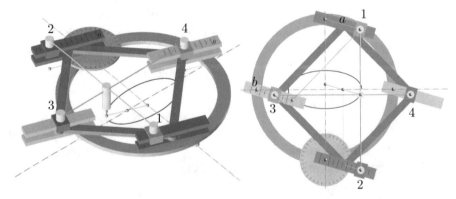

FIGURE 2.36. A computer rendering of HOECKEN's mechanism.

HOECKEN's mechanism – depicted in the introduction of this section (cf. Figure 2.34) – can be explained in a similar way. With the naming and colors of Figure 2.36, we have the following: Two yellow rods 12 and 34 are directed such that their endpoints move along circles with the adjustable radii a and b. After rotation about the angle φ, the point of intersection has the coordinates $(a\cos\varphi, b\sin\varphi)$. Thus, it lies on an ellipse.

Remark 2.3.1 In this case, the mathematical explanation is easier than a purely geometric one, because the intersection point of the rods is, from a kinematic point of view, not so easy to capture. However, one can consider the technically not realized points C_1 and C_2, see Figure 2.36 on the right, and see a "virtual elliptic motion" as described before.

A generalization in space

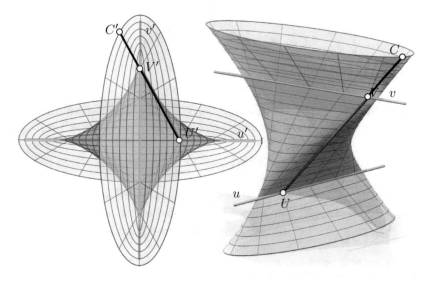

FIGURE 2.37. A generalization in space: The straight guidances u and v are skew. Each point $C \in UV$ traces an ellipse in a plane perpendicular to the common normal of u and v.

If U and V have constant distance, and the given straight guidances $u \ni U$ and $v \ni V$ are skew lines in space, we can still speak of an elliptic motion as is illustrated in Figure 2.37: We chose the common normal of u and v vertical. Then, u and v lie in horizontal planes with height difference $\Delta z \neq 0$. The top view (Figure 2.37, left) shows that the rod UV has constant length $\overline{U'V'} = \sqrt{\overline{UV}^2 - (\Delta z)^2}$ during the motion. Since orthogonal projections preserve affine ratios, any fixed point $C \in UV$ moves in constant height, *i.e.*, in a horizontal plane as well. Since its top view C' traces an ellipse, C traces a congruent horizontal ellipse.

Remark 2.3.2 The surface generated by the straight line $[U, V]$ is a quartic ruled surface.

The general elliptic motion

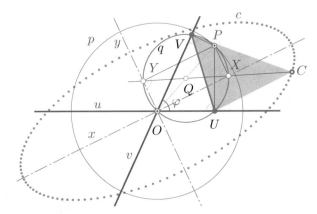

FIGURE 2.38. The general elliptic motion: The straight guidances u and v need not be perpendicular, the generating point C need not lie on the line $[U, V]$.

We now generalize the constrained motion of the last subsection as follows.

Theorem 2.3.1 *Let UV be a segment of constant length. When U is guided along a straight line u and V is guided along another straight line $v \nparallel u$, then any point C on the line $[U, V]$ traces an ellipse.*

Proof: The synthetic proof, *i.e.*, a proof based on purely geometric arguments without calculation or with a minimum of calculation, will not only verify the fact that any path curve is an ellipse. Additionally, it will provide the axis directions and the lengths of the axes of the ellipse.

- The straight lines u and v intersect at an angle $\varphi \neq 0$. From the point $O = u \cap v$, the segment UV appears under the angle φ. According to the *theorem of the angle of circumference*, the locus of all points with this property is the circle q with center Q that is given by the three points O, U, and V. This circle has constant size, independent of the position of the segment UV, and always touches another fixed circle p around O with constant radius that is twice the other radius.

- Both q and p can easily be found via the point of contact P on the diameter OQ which, due to THALES's theorem,[8] is found as the intersection of the normals to u and v in U

---

[8] THALES's theorem states that the endpoints of the diameter of a circle are seen under a right angle from all the points of the circle (except the endpoints of the diameter).

For THALES's theorem see Example 6.4.7 (Corollary 6.3.3, page 232), and in connection with centers of involutions (special case of FRÉGIER's theorem, cf. Theorem 6.4.6 on page 255).

and V. This point P can be interpreted as the so-called *instantaneous pole of the motion*: At the current position, the segment UV moves as if it was rotated around P. Thus, the distances $\overline{PU}$ and $\overline{PV}$ indicate how fast the points move: The closer to P the slower. The path tangents of U and V are (trivially) in direction of u and v.

- Now we consider an arbitrary point C, rigidly connected to the line $[U,V]$ (which is indicated by the green triangle UVC in Figure 2.38). If we intersect $[C,Q]$ with the circle q, we get a pair X and Y of opposite points on q ($\overline{XY}$ has constant length). The lines $x = [O,X]$ and $y = [O,Y]$ are perpendicular and parallel to $[P,Y]$ and $[P,X]$ according to THALES's theorem. Therefore, $[P,X]$ and $[P,Y]$ are the path normals of X and Y through P. Thus, X and Y move exactly towards O. Since this is true for any position of U and V, X and Y run on fixed perpendicular straight lines x and y through O.

We have now fulfilled the conditions of the already proven special case (shown in Section 2.3) of an elliptic motion and can be sure that $C \in XY$ runs on an ellipse with the axes x and y. The lengths of the semi-axes of the path are $a = \overline{CY}$ and $b = \overline{CX}$. ■

Some examples of practical relevance

Elliptic motions occur comparatively frequently in practice. The straight line guidance is not always easy to achieve, however, and can be "hidden" as in the following practical examples shown in Figure 2.39.

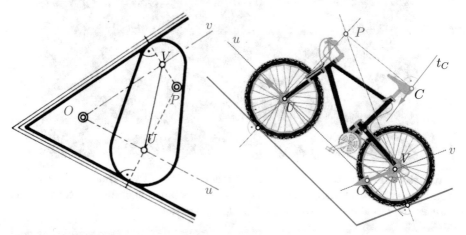

FIGURE 2.39. Ellipse motion, not immediately apparent: Two circles (or wheels) are moved along two straight lines. The lengths of the (green) velocity-vectors are proportional to the distance from the instantaneous pole P.

■ **Example 2.3.1** Gliding of rigidly connected circles along straight lines. In Figure 2.39 on the left, two circles with constant distance glide along straight lines $\overline{u}$ and $\overline{v}$. Thus, their centers U and V move on straight lines u and v, and we have an elliptic motion: Every point that is rigidly connected with the moving system has an ellipse as path curve.

■ Example 2.3.2 Rolling wheels. It does not make a difference whether the circles glide or roll: When a cyclist moves from one plane to another like in Figure 2.39 on the right, every point on the bicycle's frame is undergoing an elliptic motion. The guiding lines u and v are the parallels to the given planes through the centers of the rolling wheels (or gliding circles), the center $O = u \cap v$ of the ellipse is their intersection.

■ Example 2.3.3 Another trammler. Figure 2.40 shows another mechanism to draw ellipses: A big circular disk is moved in parallel guide rails within a rectangular boundary. Its center is, thus, moving on the perpendicular bisector of the shorter side of the boundary. A smaller circular disk is moved analogously so that its center remains on the perpendicular bisector of the longer side of the boundary. The distance of the centers of the circle is adjustable by slot guides, just as the distance of the drawing pen may be varied.

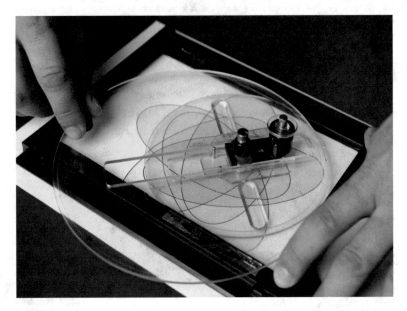

FIGURE 2.40. Another mechanism to trace ellipses: Two circular disks with a rigid connection of their centers are moved within a rectangular frame. This causes the circle centers to be guided along straight lines.

The synthetic proof of Theorem 2.3.1 implies a number of consequences:

Theorem 2.3.2 *If a circle q rolls inside a fixed circle p with double radius, each point C rigidly connected to q has an ellipse as orbit. Points on the circumference of q run on straight lines through the center of p.*

Proof: We only have to prove that q rolls inside p without gliding. This is due to the fact that p and q are the *fixed polhode* and the *moving polhode* of the constrained motion. For any motion in the plane, the moving polhode rolls on the fixed polhode without gliding [66]. ■

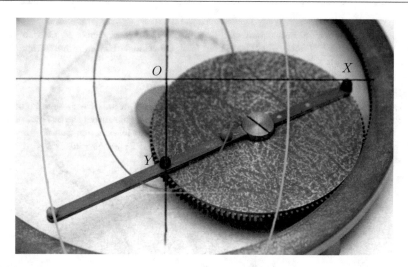

FIGURE 2.41. Ellipse motion, forced by the rolling of the CARDAN circles: A circle inside a fixed circle of double size. Photograph of a kinematic model. The black points X and Y on the rod lie on the diameter of the small circle, and thus, have straight paths.

Theorem 2.3.2 is illustrated in Figure 2.41. The circles p and q are often called *Cardan circles,* named after GEROLAMO CARDANO (1501–1576), although they were known earlier to Arabian mathematicians.

■ **Example 2.3.4** Revolving a Reuleaux triangle inside a square.

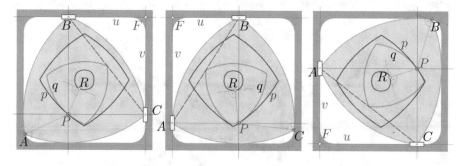

FIGURE 2.42. A Reuleaux triangle ABC is revolved inside a square: The edges are rounded off "almost circularly" by elliptic arcs.

Consider a regular triangle ABC (side length r). It can be "rounded off" by means of three circular arcs (radius r = side length), and then has "constant thickness" r. Thus, it is possible to revolve it inside a square with side length r. There are always two vertices that glide along two perpendicular sides of the square. Figure 2.42 shows three subsequent positions of ABC

when revolving clockwise. Subsequent elliptic motions (axes u, v, center F, Cardan circles p and q) "shake hands" periodically. The center R of the triangle has an almost circular orbit, consisting of four elliptic arcs.

■ **Example 2.3.5** A mechanism that draws an ellipse.

The focal points F_1 and F_2 and the length $\overline{F_1G} = 2a$ of the major axis are chosen (Figure 2.43). The rhomb F_1UGV with (in principle arbitrary) side length s guarantees the bisector direction. The path of G is the circle g, U, and V run on the circle u. The intersection point of the bisector with $[F_1, G]$ is a point on the ellipse since $\overline{PF_2} = \overline{PG}$ and $\overline{PF_1} + \overline{PF_2} = 2a$.

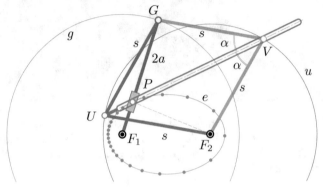

FIGURE 2.43. The mechanism from Example 2.3.5 that draws an ellipse e.

■ **Example 2.3.6** A mechanism that draws a parabola.

The focal point F and the straight guidance g are chosen (Figure 2.44). The rhomb $FUGV$ with (in principle arbitrary) side length s guarantees the bisector direction. The point of intersection of the bisector with the normal n to g in G is a point on the parabola since $\overline{PF} = \overline{Pg}$.

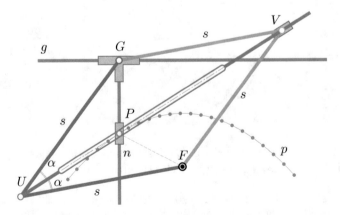

FIGURE 2.44. The mechanism from Example 2.3.6 that draws a parabola p.

3 Differential Geometry

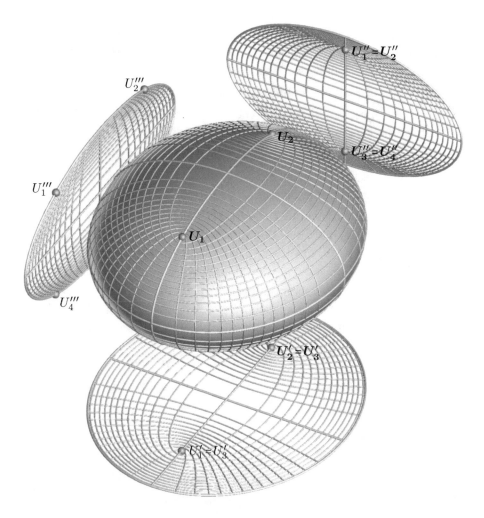

The picture shows a triaxial ellipsoid with its curvature lines. The top and front view of the curvature lines of the ellipsoid are affine images of confocal conics. The four umbilical points $U_1, \ldots, U_4$ (singularities of the curvature line parametrization) map to the common focal points in the top and front view.

3.1 Conics as orbits of planets

Historical background

The problem of understanding the motion of the planets is possibly the greatest historical problem of mathematical sciences. In 1609, JOHANNES KEPLER[1] stated his first two laws concerning the orbits of planets:

- The orbits of planets are ellipses with the Sun at one focus.
- The line segment joining a planet and the Sun sweeps equal areas during equal intervals of time (note Figure 3.2 on page 66).

Ten years later KEPLER published his third rule:

- The square of the orbital period of a planet is directly proportional to the cube of the semimajor axis of its orbit.

KEPLER had no explanation for these facts which he had deduced by analyzing the extremely accurate measurements taken by TYCHO BRAHE[2]. It took almost 80 years, until ISAAC NEWTON[3] finally derived these laws as consequences of his famous Law of Gravitation.

In what follows, we present an approach to KEPLER's laws via calculus, and recall W.R. HAMILTON's[4] short geometric derivation. We refer to the differential geometric properties of parametrized space curves

$$\mathbf{c}(t) = \big(x(t), y(t), z(t)\big)^T, \quad t \in I \subset \mathbb{R},$$

which have been presented in Section 2.1 and visualized in Figure 2.8 on page 25.

Two fundamental laws from mechanics

Newton's Law of Gravitation: *Between any two particles P_1, P_2 with masses m_1, m_2 at distance r there acts a mutually attractive force $\overrightarrow{F}$ with*

$$\|\overrightarrow{F}\| = G\,\frac{m_1 m_2}{r^2},$$

---

[1] Astronomer and mathematician, 1571–1630, lived in Graz, Prague, Linz, and Zagan.
[2] Born as TYGE OTTESEN BRAHE, astronomer, 1546–1601, lived in Copenhagen, Basel, Hamburg, Leipzig, and Prague.
[3] Mathematician, physicist, astronomer, alchemist, theologian, philosopher, 1642–1727, lived in London.
[4] Sir WILLIAM ROWAN HAMILTON, 1805–1865, was an Irish physicist, astronomer, and mathematician.

where G is a constant, called gravitational constant.[5]

NEWTON's Law of Gravitation is also valid for any two disjoint spherical masses with homogeneous mass density; the attractive force $\overrightarrow{F}$ acts on the sphere centers. In this way NEWTON's Law describes the falling speed on Earth as well as the orbits of planets around the Sun or the orbit of the Moon around the Earth.

Additionally, we need

Newton's Second Law of Motion: *For any particle with mass m and acceleration vector* **a** *the scalar multiple* $m\,\mathbf{a}$ *equals the sum of all affecting forces, i.e.,* $\overrightarrow{F} = m\,\mathbf{a}$.

Now, we focus on the Sun and on one planet only, and we assume that there are no other forces acting on this pair of particles than the mutual gravitational attraction. Our assumption means that we examine the *two-body problem* instead of the *n-body problem*.

Moreover, we simplify by the assumption that the center of the Sun is fixed and the origin of a fixed coordinate frame. Thus, we ignore that even when the mass of the planet under consideration is small compared to that of the Sun[6], the center of gravity of the union of the two masses remains fixed. This would be the true origin of an appropriate coordinate system.

Two-body problem

How does a particle P_2 with mass m_2 move around a central mass point P_1 with mass m_1 fixed at the origin of our coordinate system?

Let $\mathbf{c}(t)$ be the position vector of P_2 as a function of time t with $\mathbf{c} = r\mathbf{e}$, $r = \|\mathbf{c}\| \neq 0$. Then, according to NEWTON's laws, the acceleration vector $\mathbf{a} = \ddot{\mathbf{c}}$ obeys the vectorial differential equation

$$m_2\,\ddot{\mathbf{c}} = -G\,\frac{m_1 m_2}{r^2}\,\mathbf{e}\,. \tag{3.1}$$

---

[5] In the metrical system we have $G = 6.67260 \cdot 10^{-11}\,\frac{\text{N m}^2}{\text{kg}^2}$.

[6] The mass of the Sun is about 333000 times greater than the mass of the Earth and 750 times greater than the sum of the masses of all planets in our solar system.

We know from the theory of ordinary differential equations of second order that for given initial conditions $\mathbf{c}(t_0)$ and $\dot{\mathbf{c}}(t_0)$ there is a unique solution.

Theorem 3.1.1 (Kepler's First Law)

When the motion of a particle P_2 is determined by the attractive force of a single mass with center P_1, then the orbit of P_2 is either on the line connecting P_1 and P_2, or it is a conic having P_1 as a focus.

Proof: From (3.1), we conclude that for $\mathbf{c} = r\mathbf{e}$ with $\|\mathbf{e}\| = 1$

$$\frac{d}{dt}(\mathbf{c} \times \dot{\mathbf{c}}) = \mathbf{c} \times \ddot{\mathbf{c}} = -G\frac{m_1}{r}(\mathbf{e} \times \mathbf{e}) = \mathbf{0},$$

hence

$$\mathbf{c} \times \dot{\mathbf{c}} = \mathbf{n} = \text{const.} \tag{3.2}$$

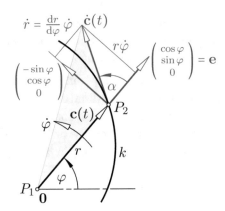

FIGURE 3.1. The set up for the proof of KEPLER's First Law.

(i) We first focus on the case $\mathbf{n} \neq \mathbf{0}$:

The scalar product $\langle \mathbf{n}, \mathbf{c} \rangle = 0$ shows that the orbit is planar. We specify our coordinate frame such that $\mathbf{n} = (0, 0, C)^T$ with constant $C > 0$ and set

$$\mathbf{c} = r\mathbf{e} = r \begin{pmatrix} \cos\varphi \\ \sin\varphi \\ 0 \end{pmatrix};$$

hence (Figure 3.1)

$$\dot{\mathbf{c}} = \dot{r} \begin{pmatrix} \cos\varphi \\ \sin\varphi \\ 0 \end{pmatrix} + r\dot{\varphi} \begin{pmatrix} -\sin\varphi \\ \cos\varphi \\ 0 \end{pmatrix}, \quad \mathbf{c} \times \dot{\mathbf{c}} = r^2\dot{\varphi} \begin{pmatrix} 0 \\ 0 \\ 1 \end{pmatrix}, \tag{3.3}$$

which implies that

$$r^2\dot{\varphi} = C = \text{const.} \tag{3.4}$$

Differentiating a second time, we obtain

$$\ddot{\mathbf{c}} = (\ddot{r} - r\dot{\varphi}^2)\begin{pmatrix} \cos\varphi \\ \sin\varphi \\ 0 \end{pmatrix} + (2\dot{r}\dot{\varphi} + r\ddot{\varphi})\begin{pmatrix} -\sin\varphi \\ \cos\varphi \\ 0 \end{pmatrix}. \tag{3.5}$$

Note that the second summand vanishes, since by (3.4)

$$\frac{\mathrm{d}}{\mathrm{d}t}(r^2\dot{\varphi}) = r(2\dot{r}\dot{\varphi} + r\ddot{\varphi}) = 0$$

and $r \neq 0$. Due to (3.1) we have reduced our problem to the differential equation

$$\ddot{r} - r\dot{\varphi}^2 = -G\frac{m_1}{r^2}. \tag{3.6}$$

Equation (3.4) defines already the time-dependence of our orbit. Therefore, we only need to solve our differential equation for the function $r(\varphi)$. Let primes denote the differentiations w.r.t. φ. Then, the chain rule implies

$$\dot{r} = r'\dot{\varphi} = C\frac{r'}{r^2}, \quad \ddot{r} = r''\dot{\varphi}^2 + r'\ddot{\varphi} = C^2\left(\frac{r''}{r^4} - \frac{2r'^2}{r^5}\right). \tag{3.7}$$

Thus, we obtain from (3.6)

$$C^2\left(\frac{r''}{r^4} - \frac{2r'^2}{r^5}\right) - \frac{C^2}{r^3} = -G\frac{m_1}{r^2}$$

and, after multiplication by r^2,

$$\frac{r''}{r^2} - \frac{2r'^2}{r^3} - \frac{1}{r} = -G\frac{m_1}{C^2}. \tag{3.8}$$

Now we substitute

$$r = u^{-1}, \quad r' = -u^{-2}u', \quad r'' = 2u^{-3}u'^2 - u^{-2}u''$$

and end up with the harmonic differential equation

$$u'' + u = G\frac{m_1}{C^2} \quad \text{for} \quad u = \frac{1}{r}. \tag{3.9}$$

Each solution can be written as

$$u = A\cos(\varphi + B) + G\frac{m_1}{C^2}$$

with constants A and B. After an appropriate rotation of our coordinate frame, we may specify $B = 0$, so that we come up with the solution

$$r = \frac{1}{G\frac{m_1}{C^2} + A\cos\varphi} = \frac{p}{1 + \varepsilon\cos\varphi} \quad \text{for} \quad p = \frac{C^2}{Gm_1} \quad \text{and} \quad \varepsilon = \frac{AC^2}{Gm_1}. \tag{3.10}$$

This is exactly the polar equation (2.18) of a conic with parameter p and numerical eccentricity ε.

(ii) In the case $\mathbf{c} \times \dot{\mathbf{c}} = \mathbf{n} = \mathbf{0}$ the vectors $\mathbf{c}$ and $\dot{\mathbf{c}}$ are always linearly dependent, i.e., $\dot{\mathbf{c}} = \lambda\mathbf{c}$. This implies that

$$\mathbf{c} = e^{\int \lambda \mathrm{d}t}\,\mathbf{a}$$

with a constant vector $\mathbf{a}$. So point P_2 moves along a line through P_1. By virtue of (3.4), this particular case with $C = 0$ shows up if, and only if, the initial vectors $\{\mathbf{c}(t_0), \dot{\mathbf{c}}(t_0)\}$ are linearly dependent. ∎

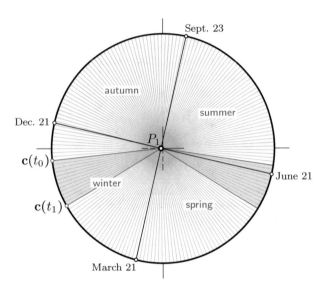

FIGURE 3.2. The approximate orbit of the Earth (scale $1 : 5.0 \cdot 10^{12}$) with positions marked every second day. The two red-shaded sectors of equal area should illustrate KEPLER's Second Law.

Theorem 3.1.2 (Kepler's Second Law)

When a particle P_2 traverses its orbit around P_1 according to KEPLER's First Law, then it moves with constant areal velocity $\frac{1}{2} r^2 \dot{\varphi} = \frac{C}{2}$. This means that in time intervals of equal duration, the line segment $P_1 P_2$ sweeps sectors of equal areas.

Proof: We specify two points $\mathbf{c}(t_0)$ and $\mathbf{c}(t_1)$ of the planar orbit and denote the polar coordinates of $\mathbf{c}(t)$ by $(r(t), \varphi(t))$. We know from calculus that the area $A(t_0, t_1)$ swept by the line segment connecting the origin $\mathbf{0}$ with the point $\mathbf{c}(t)$, for t between t_0 and t_1 (Figure 3.2), can be computed by

$$A(t_0, t_1) = \frac{1}{2} \int_{t_0}^{t_1} r^2 \dot{\varphi} \, dt.$$

This yields for the *areal velocity* at t_0, by virtue of (3.4),

$$\lim_{t_1 \to t_0} \frac{A(t_0, t_1)}{t_1 - t_0} = \frac{1}{2} r^2(t_0) \dot{\varphi}(t_0) = \frac{1}{2} C.$$

The areal velocity equals the area of the triangle with vertices $\mathbf{0}$, $\mathbf{c}$, and $\mathbf{c} + \dot{\mathbf{c}}$ (note the shaded area in Figure 3.1) with base length r and height $r\dot{\varphi}$ — provided the scaling factor for velocity vectors is specified as $\sigma = 1$. ∎

Elliptical orbits

All types of conics are possible as orbits solving the two-body problem. Ellipses occur for the motion of the planets and recurring comets around

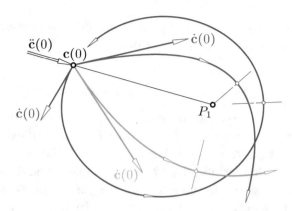

FIGURE 3.3. Different orbits starting from the same point $\mathbf{c}(0)$.

the Sun. The orbit of HALLEY's comet around the Sun with an orbital period of about 76 years is almost parabolic ($\varepsilon = 0.966$). Figure 3.3 shows three orbits with the same initial point $\mathbf{c}(0)$ but different initial velocity vectors $\dot{\mathbf{c}}(0)$: an ellipse (in blue), a parabola (in green) and a hyperbola (in red).

For $\varepsilon < 1$ the orbit is an ellipse with semiaxes

$$a = \frac{p}{1 - \varepsilon^2}, \quad b = \frac{p}{\sqrt{1 - \varepsilon^2}}. \tag{3.11}$$

This follows, by virtue of (2.4), from $p = b^2/a$ and $\varepsilon = e/a$ since

$$p = \frac{b^2}{a} = \frac{a^2 - e^2}{a} = a(1 - \varepsilon^2) \quad \text{and} \quad b = \sqrt{a^2 - e^2} = a\sqrt{1 - \varepsilon^2}.$$

The area of this ellipse is $ab\pi$, and the complete ellipse is swept by the position vector $\mathbf{c}(t)$ during one orbital period T. The constant C in (3.4) is twice the areal velocity. This implies for the orbital period

$$T = \frac{2ab\pi}{C} = \frac{2\pi}{C} \frac{p^2}{\sqrt{(1 - \varepsilon^2)^3}} = \frac{2\pi}{C} p^2 \left(\frac{a}{p}\right)^{\frac{3}{2}} = \frac{2\pi}{C} \sqrt{p}\, a^{\frac{3}{2}} = \frac{2\pi}{\sqrt{Gm_1}} a^{\frac{3}{2}}.$$

For the last equality we substituted from (3.10). This proves

Theorem 3.1.3 (Kepler's Third Law)

The orbital period of elliptic orbits solving the two-body problem is

$$T = \frac{2\pi}{\sqrt{Gm_1}} a^{\frac{3}{2}}.$$

Hence, the squares of the orbital periods are directly proportional to the cubes of the semimajor axes of the ellipses.

The velocity diagram of planetary orbits

In 1847 W.R. HAMILTON introduced in [34] the *velocity diagram* or *hodograph* of any parametrized curve $\mathbf{c}(t)$, $t \in I$, as the curve with the first derivative $\dot{\mathbf{c}}(t)$ as position vector. We verify that the hodograph of the planetary orbit is a circular arc. This yields a simple graphical construction of velocity vectors and a formula for the radii of curvature in the case of conics — valid simultaneously for all types of conics.

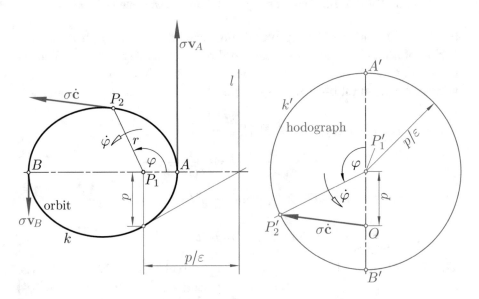

FIGURE 3.4. While $P_2 = \mathbf{c}(t)$ is tracing the planetary orbit k (left) with respect to P_1, the point $P_2' = \dot{\mathbf{c}}(t)$ is running with the same angular velocity $\dot{\varphi}$ along the circular hodograph k' (right). Note the orthogonality between $P_1 P_2$ and $P_1' P_2'$. We use the scaling factor $\sigma = p^2/C\varepsilon$ for velocities.

The function $\dot{\varphi}(t)$ is the basis for computing the velocity at different positions along an orbit. From (3.4) and (3.10) we obtain

$$\dot{\varphi} = \frac{C}{r^2} = \frac{C}{p^2}(1 + \varepsilon \cos \varphi)^2.$$

In the case $\varepsilon \neq 0$, we obtain from (3.3) and (3.7) the coordinates of the velocity vector $\dot{\mathbf{c}}$:

$$
\begin{aligned}
\dot{\mathbf{c}} &= r'\dot{\varphi}\begin{pmatrix} \cos\varphi \\ \sin\varphi \end{pmatrix} + r\dot{\varphi}\begin{pmatrix} -\sin\varphi \\ \cos\varphi \end{pmatrix} = \frac{Cr'}{r^2}\begin{pmatrix} \cos\varphi \\ \sin\varphi \end{pmatrix} + \frac{Cr}{r^2}\begin{pmatrix} -\sin\varphi \\ \cos\varphi \end{pmatrix} \\
&= \frac{Cp\varepsilon\sin\varphi}{(1+\varepsilon\cos\varphi)^2}\frac{(1+\varepsilon\cos\varphi)^2}{p^2}\begin{pmatrix} \cos\varphi \\ \sin\varphi \end{pmatrix} + \frac{C(1+\varepsilon\cos\varphi)}{p}\begin{pmatrix} -\sin\varphi \\ \cos\varphi \end{pmatrix} \\
&= \frac{C}{p}\begin{pmatrix} \varepsilon\sin\varphi\cos\varphi - \sin\varphi - \varepsilon\cos\varphi\sin\varphi \\ \varepsilon\sin^2\varphi + \cos\varphi + \varepsilon\cos^2\varphi \end{pmatrix} \\
&= \frac{C}{p}\begin{pmatrix} -\sin\varphi \\ \varepsilon + \cos\varphi \end{pmatrix} = \frac{C\varepsilon}{p^2}\begin{pmatrix} -\frac{p}{\varepsilon}\sin\varphi \\ p + \frac{p}{\varepsilon}\cos\varphi \end{pmatrix}.
\end{aligned}
$$

This formula leads to a simple graphical construction of velocity vectors $\dot{\mathbf{c}}$ (Figure 3.4, right): If we multiply each vector $\dot{\mathbf{c}}$ by the scaling factor $\sigma := p^2/C\varepsilon$, then it can be represented by the arrow pointing from the fixed point O to a point P_2' which runs along the circle k' with radius p/ε with the same angular velocity $\dot{\varphi}$ as does the point $\mathbf{c}(t)$ along the conic k. The point O is at distance p from the center P_1' of k'. The auxiliary line $P_1'P_2'$ is perpendicular to the line P_1P_2 connecting the fixed particle with the moving one.

The hodograph reveals in particular the maximum velocity at the vertex A ($\varphi = 0$) and the minimum at B ($\varphi = \pi$).[7] The scaled velocity vector $\sigma\mathbf{v}_A$ at A equals $\overrightarrow{OA'}$; vector $\sigma\mathbf{v}_B$ at B equals $\overrightarrow{OB'}$ (Figure 3.4, right).

In the case of a parabola k, (the case $\varepsilon = 1$), the point O is located on the circle k'. This means that the velocity of the moving point P_2 tends to zero when P_2 tends to infinity. In the hyperbolic case ($\varepsilon > 1$), the fixed point O lies outside k'.

When the velocity vector $\dot{\mathbf{c}}$ is seen as a position vector tracing the circle k' with angular velocity $\dot{\varphi}$, then $\ddot{\mathbf{c}}$ is a tangent vector to k'. Hence, there is also a graphical way to determine $\ddot{\mathbf{c}}$, as shown in Figure 3.5 using the scaling factor σ^2. Consequently, we can also find the normal acceleration $\mathbf{a}^n$, and by Figure 2.8 the center of curvature P_2^* of k at the point P_2.

---

[7]The velocities of the Earth around the Sun range between 29.3 km/s in summertime and 30.3 km/s in wintertime (w.r.t. the Northern hemisphere). The data for the orbit of the Earth (Figure 3.2) are as follows (in metrical units): $a = 149\,600\,000$ km = 1 astronomic unit, $\varepsilon = 0.01672$, $b = 149\,580\,000$ km, $p = 149\,560\,000$ km, $T = 365.25964$ days.

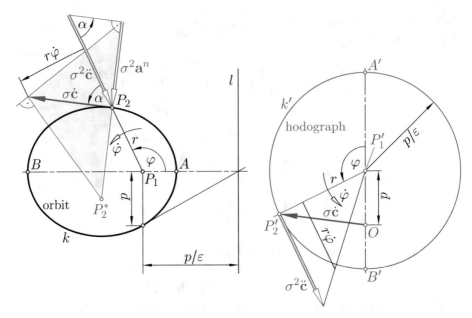

FIGURE 3.5. The acceleration vector $\ddot{\mathbf{c}}$ is the velocity vector of the hodograph. This gives rise to a graphical construction of the orbit's center of curvature P_2^*. We use the scaling factor $\sigma = p^2/C\varepsilon$ for velocities and σ^2 for accelerations.

This is equivalent to the following analytic approach: Let α denote the angle between $\mathbf{c}$ and $\dot{\mathbf{c}}$ (Figure 3.5, left). The two components of $\dot{\mathbf{c}}$ in (3.3) can be expressed in terms of $v := \|\dot{\mathbf{c}}\|$ as

$$r'\dot{\varphi} = v\cos\alpha \quad \text{and} \quad r\dot{\varphi} = v\sin\alpha.$$

The second equation implies

$$C = r^2\dot{\varphi} = rv\sin\alpha, \quad \text{hence} \quad v = \frac{C}{r\sin\alpha}.$$

On the other hand, by (2.20) we obtain the radius of curvature ϱ as $\varrho = v^2/a^n$ with $a^n = \|\mathbf{a}^n\|$ as the normal acceleration. This yields, together with (3.1),

$$\frac{v^2}{\varrho} = \|\mathbf{a}^n\| = \|\ddot{\mathbf{c}}\|\sin\alpha = \frac{Gm_1}{r^2}\sin\alpha,$$

and therefore, by (3.10)

$$\varrho = \frac{r^2}{Gm_1}\cdot\frac{v^2}{\sin\alpha} = \frac{r^2}{Gm_1}\cdot\frac{C^2}{r^2\sin^2\alpha}\cdot\frac{1}{\sin\alpha} = \frac{C^2}{Gm_1}\cdot\frac{1}{\sin^3\alpha} = \frac{p}{\sin^3\alpha}.$$

Theorem 3.1.4 *Let k be any conic with focal point F and parameter p. For any given point $X \in k$ let α denote the angle between the line FX and the tangent line at X (see Figure 3.1 and Figure 3.5, left). Then, the radius of curvature of k at X is*

$$\varrho = \frac{p}{\sin^3 \alpha} .$$

Note that this formula holds for all types of conics. It is even independent of the numerical eccentricity ε. The formula includes in particular the statement that the parameter p of the conic equals the radius of curvature at the principal vertices with $\alpha = \frac{\pi}{2}$ (Theorem 2.1.5).

Kepler's equation

For each Kepler orbit there is a simple geometric relation between the position $\mathbf{c}(t)$ and the corresponding velocity vector $\dot{\mathbf{c}}(t)$, as Figure 3.5 demonstrates. However, the computation of $\mathbf{c}(t)$ as a function of time t can only be carried out numerically since the function $\varphi(t)$ cannot be expressed in closed form.

We present a method valid for ellipses k, *i.e.*, in the case $0 < \varepsilon < 1$. Let our moving point P_2 pass the vertex A at minimal distance to P_1 at time t_0. Due to the constant areal velocity, the ratio between $(t - t_0)$ and the orbital period T equals the ratio between the areas of the elliptical sector from $\mathbf{c}(t_0)$ to $\mathbf{c}(t)$ (shaded in Figure 3.6) and the area $ab\pi$ of the ellipse.

We compute the area of the elliptical sector by inspecting its image under the affine transformation from the ellipse k to its principal circle $\tilde{k}$ with radius a. Under this transformation, areas are multiplied by the factor a/b.

Let β denote the center angle of the circular arc from $\mathbf{c}(t_0)$ to the affine image $\tilde{\mathbf{c}}(t)$. For $0 < \beta < \pi$ we can compute the area of the affine image by subtracting the area of a triangle from the area of the circular sector. Thus, we obtain the expression

$$\frac{a^2 \beta}{2} - \frac{ea \sin \beta}{2} ,$$

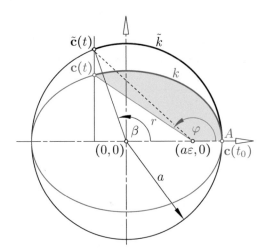

FIGURE 3.6. The Kepler equation allows the computation of the polar angle φ for a given time t.

which turns out to be valid also for $\pi < \beta < 2\pi$. This leads to the proportion

$$\frac{\left(a^2\beta - ea\sin\beta\right)\frac{b}{2a}}{ab\pi} = \frac{t - t_0}{T},$$

and hence, to the *Kepler equation*

$$\beta - \varepsilon\sin\beta = 2\pi\,\frac{t - t_0}{T}\,. \tag{3.12}$$

The procedure for computing the polar angle φ for a given t runs as follows. First we determine β from (3.12), where the function of β on the left-hand side is monotonic. Now the comparison of the polar coordinates of $\mathbf{c}(t)$ with that of the affine pre-image of $\tilde{\mathbf{c}}(t)$, *i.e.*,

$$(a\cos\beta,\; b\sin\beta) = (a\varepsilon + r\cos\varphi, r\sin\varphi)\,,$$

yields $r\cos\beta = a(\cos\beta - \varepsilon)$. We substitute r from the polar equation of the conic and obtain finally

$$\cos\varphi = \frac{\cos\beta - \varepsilon}{1 - \varepsilon\cos\beta}, \quad \text{while}\;\; \sin\beta\sin\varphi \geq 0.$$

Historical geometrical proofs revisited

It is worthwhile to check how NEWTON in his *"Philosophiae Naturalis Principia Mathematica"* in 1686 could prove KEPLER's First Law without any support from calculus. The Nobel Prize winner RICHARD P. FEYNMAN demonstrated this in his famous 1964 lecture at the California Institute of Technology [32].

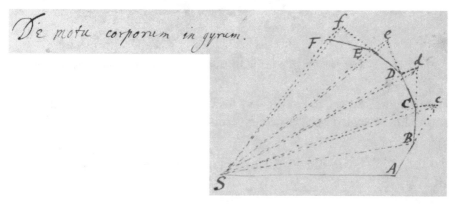

FIGURE 3.7. From NEWTON's *"De motu corporum in gyrum"*, early draft — reproduced with the kind permission of the Syndics of Cambridge University Library.

At the time of NEWTON and still decades later 'differential geometry' was rather 'difference geometry'. One imagined an orbit as a polygon where the segments represented the velocity vectors of the moving point. We demonstrate this way of reasoning in a proof for the constancy of the areal velocity.

We assume the fixed mass point P_1 as origin of our coordinate frame and start with a uniform subdivision $t_0, t_1, \ldots, t_n$ of the given time interval, i.e., with $t_i - t_{i-1} = h$ for $i = 1, \ldots, n$. The polygon $\mathbf{c}(t_0)\mathbf{c}(t_1)\ldots\mathbf{c}(t_n)$ approximates the orbit (Figures 3.7 and 3.8). The difference vectors $\mathbf{v}_i :=$ $\mathbf{c}(t_i) - \mathbf{c}(t_{i-1})$ approximate scaled velocity vectors (with scaling factor h) since

$$\dot{\mathbf{c}}(t_i) = \lim_{h \to 0} \frac{1}{h}\left(\mathbf{c}(t_i) - \mathbf{c}(t_i - h)\right) = \lim_{h \to 0} \frac{1}{h}\left(\mathbf{c}(t_i) - \mathbf{c}(t_{i-1})\right) = \lim_{h \to 0} \frac{1}{h}\mathbf{v}_i \approx \frac{1}{h}\mathbf{v}_i.$$

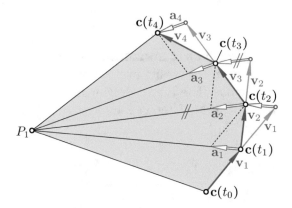

FIGURE 3.8. NEWTON's proof of the constancy of the areal velocity

Furthermore, we set $\mathbf{a}_i := \mathbf{v}_{i+1} - \mathbf{v}_i$ for $i = 1, \dots, n-1$, and note that they approximate scaled acceleration vectors (with scaling factor h^2) since

$$\ddot{\mathbf{c}}(t_i) = \lim_{h \to 0} \frac{1}{h}(\dot{\mathbf{c}}(t_{i+1}) - \dot{\mathbf{c}}(t_i)) = \lim_{h \to 0} \frac{1}{h^2}(\mathbf{v}_{i+1} - \mathbf{v}_i)$$
$$= \lim_{h \to 0} \frac{1}{h^2}\mathbf{a}_i \approx \frac{1}{h^2}\mathbf{a}_i.$$

Since the attractive force on the point $\mathbf{c}(t_i)$ acts along the connecting line with P_1, we can assume that $\mathbf{a}_i$ is parallel to the position vector $\mathbf{c}(t_i)$. Hence, the vector $\mathbf{c}(t_i)$ spans with $\mathbf{v}_i$ as well as with $\mathbf{v}_{i+1}$ parallelograms of equal area (note Figure 3.8)[8]. The halves of these parallelograms, the triangles $P_1\mathbf{c}(t_{i-1})\mathbf{c}(t_i)$ and $P_1\mathbf{c}(t_i)\mathbf{c}(t_{i+1})$ have the same area too. Hence, all sectors $P_1\mathbf{c}(t_{i-1})\mathbf{c}(t_i)$ of the orbit are equi-areal. The limit $h \to 0$ shows that the areal velocity must be constant.

Now we recall HAMILTON's elegant geometric explanation why the hodograph is circular and the orbit is a conic [34]. However, we will rephrase his way of reasoning in the language of differential geometry and modify some parts.

Theorem 3.1.5 (HAMILTON, 1847) *Let P_1 be a fixed particle in space and let a particle P_2 move under the influence of a single force $\overrightarrow{F}$.*

---

[8]By the "parallelogram spanned by vectors $\mathbf{a}$ and $\mathbf{b}$" we understand the parallelogram with the vertices $\mathbf{0}$, $\mathbf{a}$, $\mathbf{a} + \mathbf{b}$, and $\mathbf{b}$.

1. *The force $\overrightarrow{F}$ acts along the connecting line $[P_1, P_2]$ if, and only if, (i) the orbit of P_2 is located in a plane through P_1, and (ii) P_2 moves with constant areal velocity about P_1.*

2. *Let $\overrightarrow{F}$ always act along $[P_1, P_2]$. Then, in the case of non-vanishing areal velocity, the following three statements are equivalent:*

   a) *The force $\|\overrightarrow{F}\|$ is proportional to the inverse of the squared distance r between P_1 and P_2.*

   b) *The hodograph of the orbit of P_2 is a circular arc.*

   c) *The orbit of P_2 is a conic with P_1 as a focal point.*

Proof: According to our proof of Theorem 3.1.1, the linear dependence of $\{\mathbf{c}, \ddot{\mathbf{c}}\}$ implies that $\mathbf{c} \times \dot{\mathbf{c}} = \mathbf{n}$ is constant. In the case $\mathbf{n} \neq \mathbf{0}$, the orbit is planar and the areal velocity is constant. Otherwise the point P_2 moves along a line through P_1.

Conversely, any planar curve can be represented in polar coordinates. The condition $r^2 \dot{\varphi} = C = $ const. shows by (3.5) the linear dependence of $\mathbf{c}$ and $\ddot{\mathbf{c}}$.

a) $\Leftrightarrow$ b): In the case of a central attractive force there is a kind of symmetry between the orbit and the hodograph (see Figure 3.5): The position vector of one curve is a tangent vector to the other.

For any planar curve, the Frenet equation (2.19) $\dot{\mathbf{e}}_1 = v\kappa\mathbf{e}_2$ implies that $\omega = v\kappa$ is the signed angular velocity of the tangent vector. Hence

$$\varrho = \frac{1}{|\kappa|} = \frac{v}{|\omega|}$$

is the radius of curvature.

The angular velocity of $\mathbf{c}$ and at the same time that of the tangent vector $\ddot{\mathbf{c}}$ to the hodograph is $\dot{\varphi}$, and we confine our attention to the case $C = r^2\dot{\varphi} \neq 0$. Therefore, we obtain for the hodograph in the case of an arbitrary attraction law $\|\ddot{\mathbf{c}}\| = a(r)$ the radius of curvature

$$\varrho_h = \frac{\|\ddot{\mathbf{c}}\|}{|\dot{\varphi}|} = \frac{a(r)}{|\dot{\varphi}|} = \frac{r^2 a(r)}{|C|}.$$

This equation reveals immediately that ϱ_h is constant if, and only if, $r^2 a(r)$ is constant, and this holds only when the magnitude of the attractive force, namely $a(r)$, is proportional to r^{-2}.

b) $\Rightarrow$ c): Let the circle k' with center P_1' and radius ϱ_h be the given hodograph. The origin O of the velocity diagram is assumed to be at a distance d to P_1', and we start with the case $d > 0$. We use polar coordinates centered at P_1' such that O has the polar coordinates (π, d) (Figure 3.9).

Let φ be the polar angle of any point $P_2' \in k'$. Then, the arrow pointing from O to P_2' represents the velocity vector $\dot{\mathbf{c}}$ of the corresponding point P_2 of the orbit k. Since the acceleration vector $\ddot{\mathbf{c}}$ of P_2 acts along $[P_2, P_1]$, the tangent vector to k' is parallel to $-\mathbf{c}$, the negative position vector of the point P_2. For the orbit k, we now introduce polar coordinates with origin P_1 such that the point $A \in k$ corresponding to $A' = (0, \varrho_h)$ of the hodograph k' gets the polar angle 0. Then, the polar angle φ of $P_2' \in k'$ occurs also at the polar coordinates (r, φ) of the corresponding point P_2 of the orbit k.

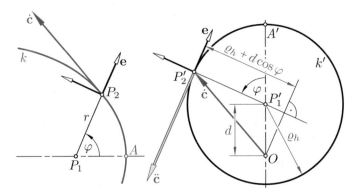

FIGURE 3.9. For a circular hodograph k' one component of $\dot{\mathbf{c}}$ immediately yields the polar equation of the orbit k.

We learned from Figure 3.1 that $\dot{\mathbf{c}}$ can be decomposed into a radial component of signed length $r'\dot{\varphi}$ and an orthogonal component of signed length $r\dot{\varphi}$. The same decomposition can be performed on the hodograph (see Figure 3.9) and we obtain

$$\varrho_h + d\cos\varphi = r\dot{\varphi} = r\frac{C}{r^2} = \frac{C}{r},$$

where $C = r^2\dot{\varphi}$ still denotes twice the constant areal velocity. This yields immediately the polar equation of k

$$r = \frac{C}{\varrho_h + d\cos\varphi},$$

which confirms that the orbit k is a conic with focal point P_1, with $p = C/\varrho_h$ and $\varepsilon = d/\varrho_h$.[9]

In the excluded case $d = 0$, the vector $\dot{\mathbf{c}}$ is permanently orthogonal to $\ddot{\mathbf{c}}$, and therefore, also to $\mathbf{c}$, and consequently k is a circle.

c) $\Rightarrow$ b): Whenever a conic with polar equation (3.10) is traced with constant areal velocity, then, according to the computation on page 69,

$$
\dot{\mathbf{c}} = r'\dot{\varphi}\begin{pmatrix}\cos\varphi \\ \sin\varphi\end{pmatrix} + r\dot{\varphi}\begin{pmatrix}-\sin\varphi \\ \cos\varphi\end{pmatrix} = \frac{Cr'}{r^2}\begin{pmatrix}\cos\varphi \\ \sin\varphi\end{pmatrix} + \frac{Cr}{r^2}\begin{pmatrix}-\sin\varphi \\ \cos\varphi\end{pmatrix} =
$$

$$
= \frac{C}{p}\begin{pmatrix}-\sin\varphi \\ \varepsilon + \cos\varphi\end{pmatrix} = \frac{C}{p}\left[\begin{pmatrix}0 \\ \varepsilon\end{pmatrix} + \begin{pmatrix}-\sin\varphi \\ \cos\varphi\end{pmatrix}\right].
$$

This shows that the hodograph is a circle with radius $\varrho_h = \frac{|C|}{p}$. ∎

---

[9]HAMILTON proved this step by revealing a direct relation between the power of O with respect to the circle k' and the Apollonian property of k.

3.2 Conics in planar differential geometry

Arc length of an ellipse

The circumference l of a circle c of radius $r > 0$ can be computed with the simple formula

$$l = 2\pi r.$$

The latter equation can be used as a definition of π. On the other hand, we can start with a parametrization of c like the standard trigonometric parametrization

$$\mathbf{c}(t) = r(\cos t, \sin t) \quad \text{with} \quad t \in J = [0, 2\pi[.$$

The arc length $l(\mathbf{c}, J)$ of the parametrized curve $\mathbf{c}(t)$ in the interval J is given by

$$l(\mathbf{c}, J) := \int_J \|\dot{\mathbf{c}}\| \mathrm{d}t \tag{3.13}$$

where $\dot{\mathbf{c}}$ denotes the velocity vector, *i.e.*, the derivative of the parametrization $\mathbf{c}(t)$ with respect to t. Thus, we find

$$l(\mathbf{c}, J) = \int_0^{2\pi} r \ \mathrm{d}t = 2r\pi.$$

● **Exercise 3.2.1** Verify the above result by using the rational parametrization of the circle like the one given in (2.13) in the case $a = b = r$.

Now we are interested in the circumference l of an ellipse with semimajor axis a and semiminor axis b, and thus, $a > b$. We use the trigonometric parametrization (2.12) of

$$\mathbf{c}(t) = (a \cos t, b \sin t) \quad \text{with} \quad t \in J = [0, 2\pi[. \tag{3.14}$$

With (3.13), we find the circumference l

$$p = 4 \int_0^{\frac{\pi}{2}} \sqrt{a^2 \sin^2 t + b^2 \cos^2 t} \ \mathrm{d}t = 2a \int_0^{\frac{\pi}{2}} \sqrt{1 - \varepsilon^2 \sin^2 t} \ \mathrm{d}t, \tag{3.15}$$

where

$$\varepsilon = \sqrt{1 - \frac{b^2}{a^2}}$$

is the *numerical eccentricity*. Note that $0 < \varepsilon < 1$ for $a > b$ and $a > 0$. In the last step, we have used the linear substitution $t \to \frac{\pi}{2} - t$. The integral in (3.15) is an elliptic integral of the second kind. Therefore, it is not possible to give a formula for the circumference of an ellipse (different from a circle) in terms of elementary functions.

However, one can derive approximations from this elliptic integral.

Approximations of the circumference of an ellipse

A rough approximation of the circumference of an ellipse with semimajor axis a and semiminor axis b is given by

$$l \approx \pi(a + b)$$

which somehow imitates the analogous formula for the circle's circumference. A much better approximation is given by

$$l \approx 2\pi \sqrt{\frac{a^2 + b^2}{2}}$$

which is not more than five per cent away from the actual circumference as long as $a < 3b$. Still better approximations are given by

$$l \approx \pi \left(3(a + b) - \sqrt{(3a + b)(a + 3b)} \right)$$

or by

$$l \approx \pi(a + b) \left(1 + \frac{3h}{10 + \sqrt{4 - 3h}} \right)$$

with $h = (a - b)^2/(a + b)^2$. The latter two formulas are due to the Indian mathematician SRINIVASA RAMANUJAN IYENGAR (1887–1920).

Using the numerical eccentricity ε one can evaluate the exact formula

$$l = 2b\pi \left(1 - \sum_{k=1}^{\infty} \frac{(2k)!^2}{(2k \cdot k!)^4} \cdot \frac{\varepsilon^{2i}}{2i - 1} \right).$$

The series starts with

$$2b\pi \left(1 - \left(\frac{1}{2} \right)^2 \varepsilon^2 - \left(\frac{1 \cdot 3 \cdot 5}{2 \cdot 4 \cdot 6} \right)^2 \varepsilon^3 - \ldots \right).$$

Another exact formula for the circumference l of an ellipse reads

$$l = \pi(a + b) \sum_{k=0}^{\infty} \binom{\frac{1}{2}}{k} h^k$$

with $h = (a - b)^2/(a + b)^2$. The initial terms read

$$\pi(a + b)\left(1 + \frac{1}{4}h + \frac{1}{64}h^2 + \frac{1}{256}h^3 + \ldots\right).$$

Curvature of conics and the construction of osculating circles

In Example 6.4.4, we learn that at any point P of a conic c there exists a circle o being the image of c under a perspective collineation with center P and axis through P. The circle o shares the tangent (and of course the normal) with c and we called it the osculating circle. In this section, we shall see that the osculating circle obtained from the differential geometric approach yields the same circle.

Vertices of an ellipse

According to (2.19) and (2.20), the curvature of a planar curve with the parametrization $\mathbf{c}(t) = (x(t), y(t))$ and $t \in J$ can be computed as $\kappa = \frac{a^n}{v^2}$, where $v = \|\dot{\mathbf{c}}\|$ is the speed and a^n is the length of the component of the acceleration vector $\ddot{\mathbf{c}}$ with respect to the normal vector $\mathbf{e}_2 = \frac{1}{v}(-\dot{y}, \dot{x})$. This yields a formula for the signed curvature

$$\kappa(t) = \frac{\det(\dot{\mathbf{c}}(t), \ddot{\mathbf{c}}(t))}{\|\dot{\mathbf{c}}\|^3}, \tag{3.16}$$

cf. [23, 58]. The sign of κ depends on the orientation of $\mathbf{c}(t)$: A positive curvature means that the curve turns left, when traversing the curve with increasing parameter t.

We apply the formula (3.16) to the parametrization of the ellipse c given in (2.12) and find for $a > b$

$$\kappa(t) = \frac{ab}{\sqrt{(a^2 \sin^2 t + b^2 \cos^2 t)^3}}. \tag{3.17}$$

In the case $a = b = r$ we get $\kappa(t) = \frac{1}{r}$, i.e., the circle's curvature is constant. Therefore, the center of curvature with position vector $\mathbf{c}^\star = \mathbf{c} + \frac{1}{\kappa}\mathbf{e}_2$ is the center of a circle o which has the same curvature as c at $P = c(t)$. Thus, o is called the *osculating circle* and we shall learn that o is a good approximation of c at P.

The first derivative of $\kappa(t)$ with respect to its parameter t yields

$$\dot{\kappa}(t) = \frac{-3ab(a^2 - b^2)\sin t \cos t}{\sqrt{(a^2 \sin^2 t + b^2 \cos^2 t)^5}}$$

with zeros
$$t_1 = 0, \quad t_2 = \frac{\pi}{2}, \quad t_3 = \pi, \quad t_4 = \frac{3\pi}{2}.$$

These correspond to the points
$$V_1 = (a,0), \quad V_2 = (0,b), \quad V_3 = (-a,0), \quad V_4 = (0,-b)$$

where the second derivatives of the curvature of c satisfy
$$\ddot{\kappa}(t_1), \ddot{\kappa}(t_3) < 0 \quad \text{and} \quad \ddot{\kappa}(t_2), \ddot{\kappa}(t_4) > 0.$$

This tells us that the points V_1 and V_3 are points of *maximal curvature* on the ellipse c, whereas V_2 and V_4 are points of *minimal curvature* on c. Note that the points V_1 and V_3 are the *principal vertices*, and V_2 and V_4 are the *auxiliary vertices* of the ellipse c (cf. Figure 3.10).

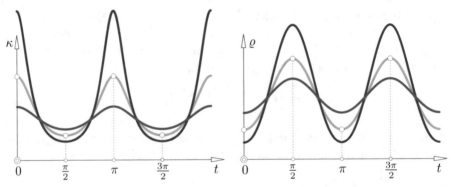

FIGURE 3.10. Left: Curvature plot of ellipses for different ratios of the semimajor and semiminor axes: $3 : 2$ (blue), $2 : 1$ (violet), and $6 : 5$ (magenta). Right: the respective plots of the curvature radii.

The respective curvatures κ and curvature radii $\varrho = \frac{1}{\kappa}$ are
$$\kappa_1 = \kappa_3 = \frac{a}{b^2} \iff \varrho_1 = \varrho_2 = \frac{b^2}{q}, \quad \kappa_2 = \kappa_4 = \frac{b}{a^2} \iff \varrho_2 = \varrho_4 = \frac{a^2}{b}, \quad (3.18)$$

compare with Theorem 2.1.5 on page 29.

We can give a simple construction for the centers of curvature at the vertices of an ellipse c, as can be seen in Figure 3.11: The axes of the ellipse c and the tangents at the vertices V_1 and V_2 bound a rectangle MV_1UV_2. We add the diagonal joining V_1 and V_2 and draw the line perpendicular to $[V_1, V_2]$ through the fourth corner U (not the center M of c).

Corollary 3.2.1 *The line through U and perpendicular to $[V_1, V_2]$ meets the principal and auxiliary axes in the centers $V_1^\star$ and $V_2^\star$ of curvature at the vertices V_1 and V_2.*

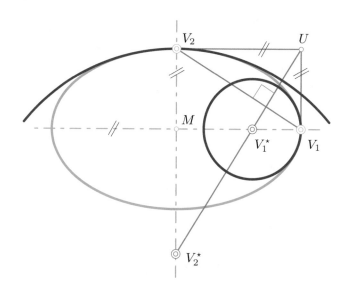

FIGURE 3.11. The construction of the centers of curvature $V_1^\star$ and $V_2^\star$ at the vertices V_1 and V_2.

Proof: It is easy to see that this construction yields the desired centers of curvature and the proper radii: Let M denote the center of the ellipse c. From Figure 3.11, we can read that

$$a : b = \overline{MV_1} : \overline{MV_2} = b : \varrho_1 \quad \text{and} \quad a : b = a : \varrho_2$$

since the right-angled triangles MV_1V_2 and $V_1UV_1^\star$ are similar. The same holds true for the two right-angled triangles MV_1V_2 and $V_2V_2^\star U$. It is easy to verify that the circles around $V_i^\star$ through V_i hyperosculate the ellipse (cf. Definition 6.4.1). A simple computation shows that the osculating circles at the vertices share only the vertices with c (cf. Exercise 2.1.1). ∎

We shall meet this construction again in Section 7.5 when we deal with the quadratic transformation of *conjugate normals*.

Remark 3.2.1 We defined the vertices of conics in Section 2.1 as the points where a conic meets its lines of symmetry. This yields four points on the ellipse, two points on the hyperbola, and one point on the parabola. The only proper way of generalizing the notion of a vertex for other curves is the following: Look for regular points with stationary curvature, *i.e.*, regular points on a curve with $\kappa \neq 0$, $\dot{\kappa} = 0$, and $\ddot{\kappa} \neq 0$ provided that the curve is of class C^4.

Metric relations between curvature radii and tangential distances

Curves can also be defined as envelopes of one-parameter families of lines. Each line is uniquely defined by a unit normal vector $\mathbf{n} = (\cos\varphi, \sin\varphi)$ and the oriented distance $d(\varphi)$ to the origin of the coordinate system such that $d\mathbf{n}$ is the position vector of the line's pedal point with respect to the origin. The function $d(\varphi)$ is called the *support function* of the curve.

The curvature and the support function of an ellipse fulfill a simple relation:

Theorem 3.2.1 *The curvature function κ and the support function d of an ellipse with semimajor axis a and semiminor axis b are related via*

$$\kappa = \frac{d^3}{a^2 b^2}. \tag{3.19}$$

Proof: We shall have in mind that $x = a\cos t$ and $y = b\sin t$ (cf. (2.12)) holds for any point on the ellipse

$$\frac{x^2}{a^2} + \frac{y^2}{b^2} = 1. \tag{3.20}$$

After eliminating $\cos t$ and $\sin t$ from the curvature function given in (3.17), we arrive at an algebraic expression for the curvature:

$$\kappa(t) = \frac{a^4 b^4}{\sqrt{(b^4 x^2 + a^4 y^2)^3}}. \tag{3.21}$$

Via the velocity vector of the ellipse at some point $P = (\xi, \eta)$, we obtain the Hesse normal form of the tangent at P as

$$t_P : \frac{-b^2\xi\, x - a^2\eta\, y + a^2 b^2}{\sqrt{b^4\xi^2 + a^4\eta^2}} = 0. \tag{3.22}$$

If we insert $\xi = \eta = 0$, we obtain the signed distance of t_P to the center of the ellipse,

$$d(x, y) = \frac{a^2 b^2}{\sqrt{b^4 x^2 + a^4 y^2}} \tag{3.23}$$

Note that x and y in (3.23) have to fulfill (3.20). Combining (3.21) and (3.23) we can eliminate the nasty square root and obtain (3.19). ∎

● Exercise 3.2.2 Analyze the curvature function of the ellipse given in (3.19) and show that the vertices of the ellipse appear where the support function is stationary.

● Exercise 3.2.3 Derive the analogous formula to (3.19) for hyperbolas.

A deeper result relating the curvature radii with metric values is due to the French mathematician JOSEPH LIOUVILLE (1809–1882) and given in

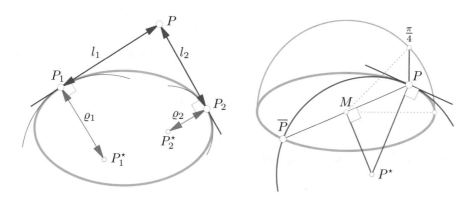

FIGURE 3.12. Left: The cube of the ratio $l_1 : l_2$ equals the ratio $\varrho_1 : \varrho_2$ of the curvatures at the two contact points P_1 and P_2. Right: For the particular point P, the osculating circle o_P meets the ellipse c at the antipode $\overline{P}$ of P.

Theorem 3.2.2 *Let P be an exterior point of a conic c. The two tangents from P to c touch at P_1 and P_2. Let $l_1 = \overline{P_1 P}$ and $l_2 = \overline{P_2 P}$ and let further $\varrho_1 = \kappa_1^{-1}$ and $\varrho_2 = \kappa_2^{-1}$ be the curvature radii of the conic at P_1 and P_2. Then,*

$$\frac{\varrho_1}{\varrho_2} = \left(\frac{l_1}{l_2}\right)^3$$

holds independently of the choice of P and the conic.

● **Exercise 3.2.4** Give a proof of Theorem 3.2.2 by assuming P_1 and P_2 are arbitrary points on an ellipse / hyperbola / parabola with the parametrizations

$$\text{ellipse } \left(a\frac{1-t^2}{1+t^2}, b\frac{2t}{1+t^2}\right), \quad \text{hyperbola } \left(a\frac{1+t^2}{1-t^2}, b\frac{2t}{1-t^2}\right), \quad \text{parabola } \left(2qt^2, 2qt\right)$$

with $a, b, p \in \mathbb{R}$. Compute the intersection P of the tangents at P_1 and P_2, the distances $\overline{P_1 P}$ and $\overline{P_2 P}$, and the curvatures at the respective points on the conic.

In Figure 3.12 (left), the metric values mentioned in Theorem 3.2.2 are marked. Though Figure 3.12 shows only the case of an ellipse, Theorem 3.2.2 holds for hyperbolas and parabolas as well.

Figure 3.12 (right) shows another curious result. There are four points P on the ellipse (corresponding to the parameter values $t = \pm\frac{\pi}{4}$ and $t = \pm\frac{3\pi}{4}$ in the trigonometric parametrization (2.12)) whose osculating circles meet the ellipse in a pair $(P, \overline{P})$ of opposite points. In this special case, the center $P^\star$ of the osculating circle lies on the diameter line which is orthogonal to the diameter $[P, \overline{P}]$.

- Exercise 3.2.5 Find a proof for the facts illustrated in Figure 3.12 (right).

- Exercise 3.2.6 Vertices of a hyperbola. Show that there are two vertices $V_1 = (a, 0)$ and $V_2 = (-a, 0)$ on the hyperbola

$$h : \frac{x^2}{a^2} - \frac{y^2}{b^2} = 1 \quad \text{with} \quad ab \neq 0$$

by using the two parametrizations

$$\mathbf{h}(t) = (\pm a \cosh t, b \sinh t), \quad t \in \mathbb{R}$$

and compute the curvature (cf. the curvature plot in Figure 3.13)

$$\kappa = \mp \frac{ab}{(a^2 \sinh^2 t + b^2 \cosh^2 t)^{\frac{3}{2}}}.$$

The radius of the osculating circle at V_1 and V_2 equals $\varrho_1 = \varrho_2 = \frac{b^2}{a}$. As shown in Figure

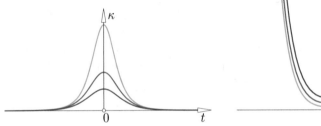
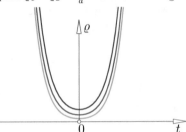

FIGURE 3.13. Curvature plot and the curvature radius function of the hyperbola for three different ratios of the semimajor and semiminor axes: $3 : 2$ (blue), $1 : 1$ (magenta), and $3 : 4$ (violet).

3.14 (left), the centers V_1^* and V_2^* of the osculating (hyperosculating) circles at the hyperbola's vertices can be constructed in a simple way, provided that the vertex and at least one asymptote is known.

- Exercise 3.2.7 Vertex of a parabola. A parabola $p : x^2 = 2qy$ (with $q \in \mathbb{R} \setminus \{0\}$) can be given in parametric form as

$$\mathbf{p}(t) = (2qt, 2qt^2), \quad t \in \mathbb{R}.$$

Thus, the curvature radius function $\varrho = \frac{1}{\kappa}$ reads

$$\varrho = q(1 + 4t^2)^{\frac{3}{2}}.$$

Show that p has only one vertex $V = (0, 0)$ which corresponds to $t = 0$. The radius of the osculating circle at V equals $\varrho_0 = q$.

In Section 4.2 (Theorem 4.2.3, page 145) we learn that the *subnormals* (the orthogonal projections of the normal's segment between the curve and its principal axis onto the latter) of conics are linear functions in the abscissa. The parabola is more special: The subnormal is constant, and thus, independent of the point on the parabola as shown in Figure 3.14.

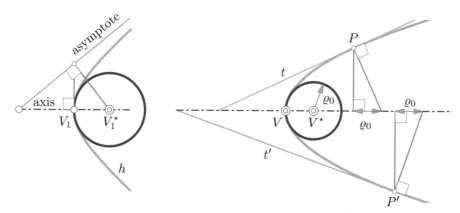

FIGURE 3.14. Left: The construction of the center $V_1^\star$ of the hyperbola's osculating circle at the vertex V_1. Right: The radius ϱ_0 of the osculating circle at a parabola's vertex V can be found if a line element, *i.e.*, a point $P \neq V$ plus incident tangent, is known.

Evolute of an ellipse

What about the osculating circle o_P for a generic point $P = \mathbf{c}(t_0)$ of an ellipse c? Assume that o is such a circle with center $P^\star$ (with coordinate vector $\mathbf{c}^\star$ and radius R). Locally, that means, in a sufficiently small neighborhood of a generic point $P = \mathbf{c}(t_0)$ on the ellipse c, the circle and the ellipse share the point, the tangent, and the curvature.

We can express these three conditions algebraically: The first condition yields

$$\langle \mathbf{c}(t) - \mathbf{c}^\star, \mathbf{c}(t) - \mathbf{c}^\star \rangle|_{t=t_0} = R^2,$$

i.e., a circle through $\mathbf{c}(t)$ (at any t) with the yet unknown center $\mathbf{c}^\star$ and radius R. Note that both, center and radius are constant at a certain point $P = \mathbf{c}(t_0)$. We differentiate once with respect to t and obtain

$$\langle \dot{\mathbf{c}}(t), \mathbf{c}(t) - \mathbf{c}^\star \rangle|_{t=t_0} = 0 \quad \Longrightarrow \quad \langle \mathbf{e}_1, \mathbf{c}(t) - \mathbf{c}^\star \rangle = 0 \qquad (3.24)$$

where $(\mathbf{e}_1, \mathbf{e}_2)$ is the Frenet frame at P. The last equation tells us that the desired circle o is tangent to c at $P = \mathbf{c}(t_0)$, for its radius vector $\mathbf{c}(t) - \mathbf{c}^\star$ is perpendicular to the tangent t_P of c at P that aims in the direction $\dot{\mathbf{c}}(t)$ or $\mathbf{e}_1$. Thus, the center $P^\star$ of o is a point on the curve's normal at P.

We differentiate once again. With $\dot{\mathbf{c}} = v\kappa\mathbf{e}_2$ and $\langle \mathbf{e}_1, \mathbf{e}_1 \rangle = 1$, we find

$$v\langle \kappa\mathbf{e}_2, \mathbf{c}(t) - \mathbf{c}^\star \rangle|_{t=t_0} + v = 0 \qquad (3.25)$$

which yields

$$\mathbf{c}^\star = \mathbf{c} + \frac{1}{\kappa}\mathbf{e}_2. \tag{3.26}$$

At this point we recall that, by virtue of (3.16), κ can be expressed in terms of the derivatives of the parametrization $\mathbf{c}(t)$ (with an arbitrary parameter t which does not necessarily agree with the arc length parameter).

We learned that $\mathbf{c}^\star$ is the center of the osculating circle or the *center of curvature*. The term *osculation* is explained for pairs of conics in a completely different way in Definition 6.4.1 (see Section 6.3) although the two separate approaches lead to the same circle.[10]

If t runs through the entire interval $J = [0, 2\pi[$, then $\mathbf{c}(t)$ parametrizes the ellipse and $\mathbf{c}^\star(t)$ parametrizes a curve called the *evolute of the ellipse*. The tangents of the evolute are the normals of the ellipse which follows from $\frac{\mathrm{d}}{\mathrm{d}t}\mathbf{c}^\star = v\mathbf{e}_1 - \frac{\dot\kappa}{\kappa^2}\mathbf{e}_2 - \frac{v\kappa}{\kappa}\mathbf{e}_1 = -\frac{\dot\kappa}{\kappa^2}\mathbf{e}_2$. Figure 3.15 shows some ellipses together with their evolutes.

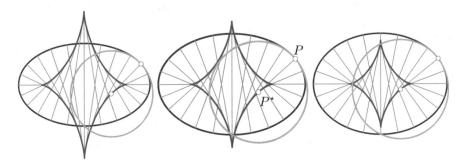

FIGURE 3.15. Evolutes of different ellipses being the envelopes of the ellipses' normals. The cusps of the evolute correspond to the vertices of the ellipse.

---

[10]Two C^r curves c and d are said to be in n-th order contact (with $n \le r$) if there exist parametrizations of both curves such that the derivatives of c and d agree up to order n. Sometimes this is called a C^n-contact. Two C^r curves are said to be in G^n-contact (geometric continuity of order $n \le r$) if the span of the first n derivative vectors of both curves agree. C^n-contact always implies G^n-contact, the converse is not true.

The projective differential geometric version deals with the first $n + 1$ derivative points starting with $c(t_0)$ and $d(t_0)$ as the first derivatives.

We insert the parametrization (2.12) of the ellipse c into (3.26) and obtain a parametrization of the evolute as

$$\mathbf{c}^\star(t) = \frac{a^2 - b^2}{ab}\left(b\cos^3 t, -a\sin^3 t\right), \quad t \in J. \tag{3.27}$$

The evolute of an ellipse is an algebraic curve of degree six and class[11] four. The parametric representation (3.27) fulfills the algebraic equation

$$\left(\frac{x}{\alpha}\right)^{\frac{2}{3}} + \left(\frac{y}{\beta}\right)^{\frac{2}{3}} = 1 \tag{3.28}$$

with $\alpha = \frac{a^2-b^2}{a}$ and $\beta = \frac{a^2-b^2}{b}$ which is equivalent to the polynomial equation

$$\left((a^2 - b^2)^2 - (a^2x^2 + b^2y^2)\right)^3 - 27a^2b^2x^2y^2(a^2 - b^2)^2 = 0.$$

The evolute of an ellipse has four cusps which are the centers of curvature of the four vertices, see Figure 3.15.

The scaling $(x, y) \mapsto (x', y') = \left(\frac{x}{\alpha}, \frac{y}{\beta}\right)$ transforms (3.28) into the *astroid* with the equation

$$x'^{\frac{2}{3}} + y'^{\frac{2}{3}} = 1.$$

For a generic point P on the ellipse $\mathbf{c}(t)$, there exists also an elementary construction of the center of curvature $P^\star = \mathbf{c}^\star(t)$, cf. Figure 3.16 (left):

Lemma 3.2.2 *Draw the tangent t_P and the normal n_P at $P \in c$. The normal intersects the principal axis in the point 1. The line parallel to t_P through 1 meets the ellipse's diameter through P in the point 2. The line $[2, P^\star]$ is parallel to the auxiliary axis of c, and $P^\star$ is the center of curvature which has to lie on the curve's normal n_P.*

Proof: We can verify the construction shown in Figure 3.16 (left) by a simple computation. We start with the trigonometric parametrization of c given in (2.12) and set $P = (a\cos t, b\sin t)$. For fixed t_0 we find a point P on c. Then, the tangent t_P and the normal n_P of e at P are given by the Cartesian equations

$$t_P: \frac{x\cos t}{a} + \frac{y\sin t}{b} = 1 \quad \text{and} \quad n_P: -\frac{x\sin t}{b} + \frac{y\cos t}{a} = \frac{b^2 - a^2}{ab}\cos t\sin t.$$

---

[11]The class of a planar algebraic curve is, roughly speaking, the number of tangents to the curve that can be drawn from a generic point to the curve. In Section 7.5, we shall obtain the evolute of a conic as the image of the set of the conic's tangents under a quadratic Cremona transformation. Then, it will be immediately clear why the class of the evolute is four.

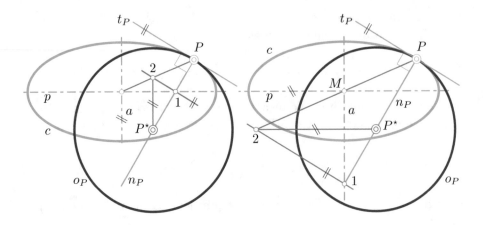

FIGURE 3.16. Two constructions of the center of curvature of an ellipse c. The construction also works if we interchange the roles of the axes.

The normal n_P and the principal axis with equation $y = 0$ meet at 1 with coordinates

$$\left(\frac{a^2 - b^2}{a} \cos t, 0 \right),$$

and thus, the line l parallel to the tangent t_P has the equation

$$l : \frac{x \cos t}{a} + \frac{y \sin t}{b} = \frac{a^2 - b^2}{a^2} \cos^2 t.$$

The diameter through P is given by $xb \sin t - ya \cos t = 0$. Therefore, its intersection 2 with l has the same x-coordinate as $P^\star$

$$x_{P^\star} = \frac{a^2 - b^2}{a} \cos^3 t$$

which, inserted into the normal's equation, yields $y_{P^\star}$, and thus, $\mathbf{p}^\star(t)$ agrees with $\mathbf{c}^\star$ from (3.27). ∎

● Exercise 3.2.8 Alternative curvature construction. In Figure 3.16 (right), we can see that there is a second construction that leads to the center $P^\star$ of curvature for a generic point P on the ellipse c. Show that this construction leads to the same $P^\star$, i.e., it is equivalent to the first construction given above.

Theorem 3.2.3 *When in the case of an ellipse or a hyperbola c the tangent t_P and the normal n_P at a generic point $P \in c$ intersect the principal axis at points 1, 1′ and the auxiliary axis at 2, 2′, then the signed distances to P and the center $P^\star$ of curvature satisfy*

$$\overline{P1} : \overline{P2} = \overline{P^\star 1'} : \overline{P^\star 2'}. \tag{3.29}$$

• **Exercise 3.2.9** Proof of Theorem 3.2.3. Show that the construction of $P^\star$ from Theorem 3.2.3 as shown in Figure 3.17 is equivalent to that given in Theorem 3.2.2.

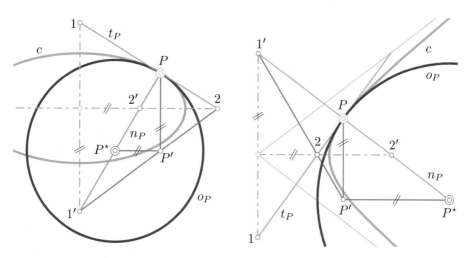

FIGURE 3.17. An alternative constructive approach to the centers of curvature of ellipses and hyperbolas. P divides the segment 12 on t_P at the same ratio as $P^\star$ does with $1'2'$ on n_P.

• **Exercise 3.2.10** Evolute of an ellipse as the envelope of the ellipse's normals. Show that the evolute of an ellipse can be found as the envelope of its normals. Start with the equations of lines depending on $t \in [0, 2\pi[$

$$n_P: \quad -\frac{x \sin t}{b} + \frac{y \cos t}{a} = \frac{b^2 - a^2}{ab} \cos t \sin t$$

and intersect the lines n_P with their derivatives $\dot{n}_P$ (with respect to t). The points $n_P \cap \dot{n}_P$ are precisely the points of the ellipse's evolute.

• **Exercise 3.2.11** Evolutes of hyperbola and parabola as envelopes. Find the evolute of a hyperbola and a parabola as the envelopes of the normals.

The evolute of a parabola

Assume that we are given a parabola $p: x^2 = 2qy$ (with $q \in \mathbb{R} \setminus \{0\}$) which can be parametized as

$$\mathbf{p}(t) = (2qt, 2qt^2), \quad t \in \mathbb{R}. \tag{3.30}$$

We insert (3.30) into (3.26) and find a parametrization of its evolute

$$\mathbf{p}^\star(t) = q(-8t^3, 6t^2 + 1) \tag{3.31}$$

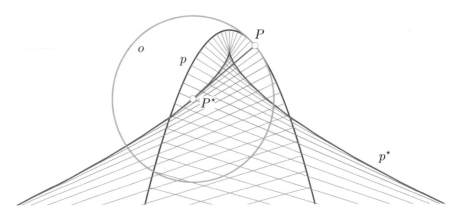

FIGURE 3.18. The evolute of a parabola as the envelope of the parabola's normals.

which is even polynomial, and thus, it can be written as a Bézier curve (see Section 8.2). This curve is sometimes called a *semicubical parabola*. An example of a parabola's evolute can be seen in Figure 3.18.

The curve given in (3.31) is algebraic and of degree three and of class three, although it is the evolute of a parabola. Its only cusp is the point $(q,0)^\mathrm{T}$ which is the center of the osculating circle at the vertex. An implicit equation of the parabola's evolute from (3.31) reads

$$8(y-q)^3 - 27qx^2 = 0$$

which can be found by eliminating t from $x = -8qt^3$ and $y = 6qt^2 + q$.

● Exercise 3.2.12 Center of curvature for a generic point on a parabola Assume $P \in p$ is a generic point (not the vertex). We can find the center $P^\star$ of p's osculating circle at P in two different ways shown in Figure 3.19:

The construction shown on the left-hand side of Figure 3.19 is the analogue to the construction of the center of curvature of the ellipse (cf. Figure 3.16). The parabola's normal n_P at P meets the axis of the parabola at a point D. There we draw the parallel to the tangent t_P. In the case of the ellipse, the point P has to be joined with the center of the curve. In the present case of the parabola, there is no center. However, there is a diameter through P, *i.e.*, the line parallel to the parabola's axis which meets the aforementioned parallel to the tangent at the point 2, such that $[2, C]$ is orthogonal to all diameters of p, especially to its axis.

Prove this by direct computation.

Theorem 3.2.4 *The normal n_P at any point P of a parabola meets the parabola's directrix at a point D. If $P^\star$ is the parabola's center of curvature*

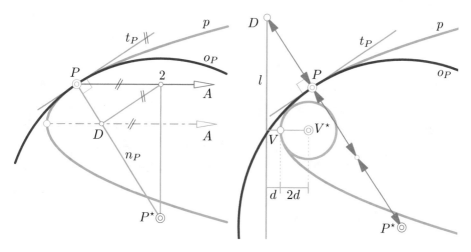

FIGURE 3.19. Two constructions of the center of curvature of parabolas.

at P, then the lengths of the segments DP and DP* fulfill

$$\overline{DP} : \overline{DP^\star} = 1 : 3.$$

• **Exercise 3.2.13** Alternative approach to the center of curvature. The construction in Figure 3.19, right, for the center $P^\star$ of curvature of a parabola is also valid at the vertex (cf. Theorem 3.2.4). Give a proof of Theorem 3.2.4.

Evolute of a hyperbola

With the preparations from the previous sections, it is easily verified that the evolute of the hyperbola parametrized by

$$(\pm a \cosh t, b \sinh t), \quad t \in \mathbb{R}$$

admits the parametrization

$$\frac{a^2 + b^2}{ab} \left(\pm b \cosh^3 t, \mp a \sinh^3 t \right), \quad t \in \mathbb{R}.$$

An implicit equation of the evolute of the hyperbola reads either

$$\left(\frac{ax}{a^2 + b^2} \right)^{\frac{2}{3}} - \left(\frac{by}{a^2 + b^2} \right)^{\frac{2}{3}} = 1 \quad \text{or} \quad \left(1 - \frac{a^2 x^2 - b^2 y^2}{(a^2 + b^2)^2} \right)^3 = 27 \frac{(abxy)^2}{(a^2 + b^2)^4}.$$

Like the evolute of the ellipse, the evolute of a hyperbola is a sextic curve of class four. It has two real cusps being the centers of the osculating

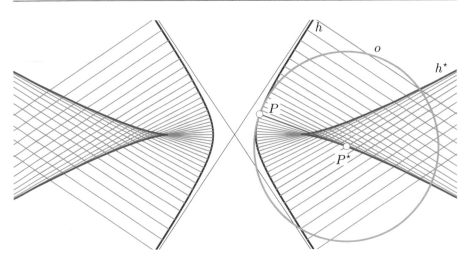

FIGURE 3.20. A hyperbola h with its evolute $h^\star$.

circles at the hyperbola's vertices. Figure 3.20 shows a hyperbola with its evolute.

The construction of the center $P^\star$ of curvature for a generic point on the hyperbola is very similar to the case of the ellipse. In Figure 3.21, we have illustrated the two equivalent constructions of the center of curvature for a generic point of a given hyperbola. The construction mirrors that of the ellipse as shown in Figure 3.16.[12]

● Exercise 3.2.14 A circle on three infinitesimally close points. Another approach to the osculating circle of a C^2-curve at a regular point P (that is not a point of inflection) could be the following: Assume $P = \mathbf{c}(t_0)$ is a point on $\mathbf{c}$ and let $\varepsilon > 0$ be sufficiently small such that the Taylor series of $\mathbf{c}(t)$ is arbitrarily close to $\mathbf{c}(t)$ at least in the interval $J := [t_0 - \varepsilon, t_0 + \varepsilon]$. Then, there exists a circle through the three points $\mathbf{c}(t_0 - \varepsilon)$, $\mathbf{c}(t_0)$, and $\mathbf{c}(t_0 + \varepsilon)$, provided that there is no point of inflection on $\mathbf{c}(J)$ (see Figure 3.22).

If the circle degenerates into a line (the radius becomes infinitely large and the center disappears), then P_0 is a *point of inflection*. This will not occur on conics. Show that the limit of the circle for $\varepsilon \to 0$ is the osculating circle o_P.

● Exercise 3.2.15 Circle with intersection of multiplicity three or four. When dealing with algebraic curves c, the osculating circle at a regular non-inflection point P can be defined as a circle whose intersection at P with c is at least of multiplicity three.

A vertex V of an algebraic curve is a regular point where the curve and the osculating circle intersect at least with multiplicity four. The definition of a vertex of a generic curve as given on page 81 still holds. Conics are algebraic curves of degree two, and thus, this definition of

---

[12]The constructions of the center of curvature shown in Figures 3.16 and 3.21 are sometimes ascribed to K.H. SCHELLBACH, see [22, p. 72].

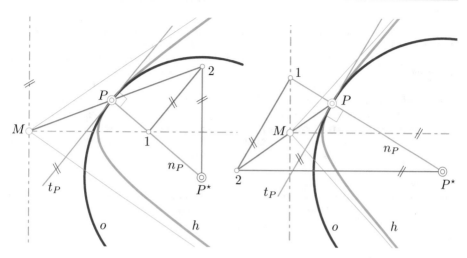

FIGURE 3.21. The construction of the center of curvature of an ellipse shown in Figure 3.16 also applies to the hyperbola. Again, there are several equivalent ways to the center of curvature.

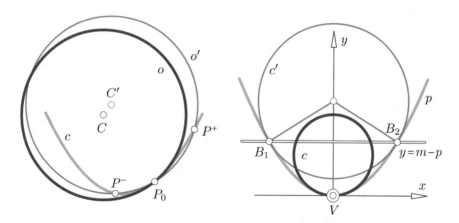

FIGURE 3.22. Left: The circumcircle o' (with center C') of three sufficiently close points $P^-, P_0, P^+ \in c$ converges to the osculating circle o (with center C) at P_0 if $P^- \to P_0$ and $P^+ \to P_0$ provided that the discretization P^-, P_0, P^+ comes from a C^2-curve and no point of inflection occurs in this interval. Right: A circle c' with double contact (at the points B_1 and B_2) converges to the osculating circle c at the parabola's vertex V. The parabola p and c are in third-order contact at V, i.e., they intersect with multiplicity four at P.

a vertex also applies to conics. At the vertex of a conic (and any other algebraic curve) the osculating circle is hyperosculating.

The case of a parabola shall demonstrate a way how to find a vertex with the help of the hyperosculating circle. We assume that $p : x^2 = 2py$ is the parabola in question. Assume $c : x^2 + (y - m)^2 = r^2$ is a circle which is in double contact with p, see Figure 3.22 (right). Consequently, m, q, and r are related via $q^2 + r^2 = 2qm$, and thus, the y-coordinate of the two contact points equals $y = m - q$. The x-coordinates are the solutions of the bi-quadratic equation $\frac{x^4}{4q^2} + x^2 \left(1 - \frac{m}{q}\right) + m^2 - r^2 = 0$. If the two points of contact fall in one, then we have one point of intersection with multiplicity $2 \cdot 2 = 4$. This is the case if, and only if, $m = q$, i.e., the circle is centered at the point $(0, q)$ and has radius $r = q$. The vertex is the point $(0, 0)$ of contact. The parabola p and c are in third-order contact at V, i.e., they intersect with multiplicity four at P. This verifies the results of Exercise 3.2.6.

● **Exercise 3.2.16** Osculating parabola. An alternative to the approximation of a curve by an osculating circle could be the following: Let $P = \mathbf{c}(t_0)$ be a regular point on a curve $\mathbf{c}$. Now there is a uniquely defined parabola p with P for its vertex and o for its osculating circle at the vertex P. How does the curvature $\kappa_0 = \kappa(t_0)$ of $\mathbf{c}$ at P enter the equation of the parabola p?

● **Exercise 3.2.17** Centers of curvature for parabolas and hyperbolas. Figure 3.23 shows a construction of the centers of curvature $P^\star$ of an ellipse e, a parabola p, and a hyperbola h at a generic point P. In the case of a conic with center, i.e., a conic with two (real) focal points, the construction is as follows: Draw the normals to the focal rays at the focal points F_1 and F_2. These normals meet the conic's normal n at P in two points 1 and 2. Find the fourth harmonic point $P^\star$ of P with respect to 1 and 2. The case of the parabola is simpler (see Figure 3.23, middle). Give a proof for these three constructions.

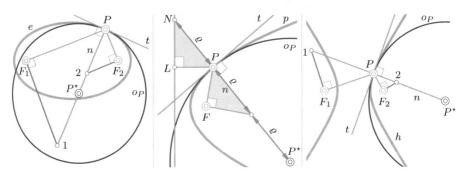

FIGURE 3.23. Left and right: The center $P^\star$ of curvature at the point P on a conic is the harmonic conjugate of P with respect to the points 1 and 2. The points 1 and 2 lie on the conic's normal n such that the triangle $PF_1 1$ and $PF_2 2$ have right angles at the focal points F_1 and F_2. Middle: The parabola differs somehow. Nevertheless, there is a similar but simpler construction.

The fourth common point

Assume that we are given the parabola $p : x^2 = 2qy$ with $q \in \mathbb{R} \setminus \{0\}$. Let $P = (\xi, \eta)$ be a point on the parabola. According to Theorem 3.2.4 (see

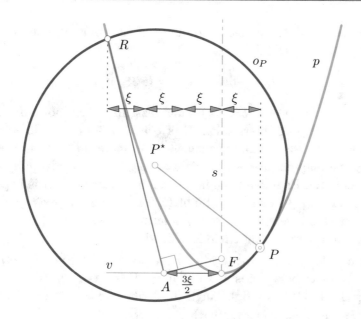

FIGURE 3.24. The linear construction of the remaining common point R of a parabola p and one of its osculating circles o_P.

Section 3.2), we construct the center $P^\star$ of the osculating circle o_P. We know that o_P and p intersect at P with multiplicity three since P is a generic point. However, there must be one further common point R which is a transversal intersection of o_P and p. How do we find it?

It is clear that R can be found by means of a linear construction: The osculating circle o_P has the equation

$$o_P: \ (x + 8qt^3)^2 + (y - 6qt^2 - q)^2 = q^2(1 + 4t^2)^3$$

for its center is given in (3.31) provided that p is parametrized as in (3.30). The resultants[13] of o_P and p with respect to y and x are

$$\mathrm{res}(o_P, p, y) = (6qt + x)(x - 2qtx)^3,$$
$$\mathrm{res}(o_P, p, x) = (18qt^2 - y)(2qt^2 - y)^3$$

---

[13]The resultant of two polynomials with respect to a common variable is a polynomial defined by the coefficients of the given polynomials. Resultants can be used to eliminate variables, as is the case here. All common zeros of the two polynomials are zeros of the resultant. For a precise definition and properties of resultants we refer to [20].

whose zeros are the x- and y-coordinates of the common points P (with multiplicity three) and R:

$$R = (-6qt, \ 18qt^2) = (-3\xi, \ 9\eta).$$

The mapping $(\xi, \eta) \mapsto (-3\xi, \ 9\eta)$ which assigns to the point P of osculation the remaining intersection R of o and p is linear. Moreover, if t is considered to be an inhomogeneous coordinate on p, then $P \mapsto R$ is given by the mapping $t \mapsto -3t$. This mapping is a projective mapping on the parabola (cf. Example 5.4.3 in Section 5.4, page 248). It is hyperbolic with the vertex $\mathbf{p}(0)$ as a fixed point $(1,0)$. The second fixed point does not matter when we speak about osculating circles: It is the ideal point $\mathbf{p}(\infty)$ where no osculating circle is defined.

Now we can deduce a construction for the point R (cf. Figure 3.24): Draw a parallel to parabola's axis s at distance $3\overline{Ps}$ on the opposite side of P. The triangle FAR has a right angle at A where $A = (-\frac{3\xi}{2}, 0)$ is the point on the tangent v at p's vertex with $\overline{As} = \frac{1}{2}\overline{Rs}$.

In the case of an ellipse e with equation

$$e: \quad \frac{x^2}{a^2} + \frac{y^2}{b^2} = 1$$

we can use the trigonometric parametrization $P = (a\cos t, b\sin t)$ as given in (2.12). Let $c_t := \cos t$ and $s_t := \sin t$. Then, (3.27) yields the equation of e's osculating circle at the points P

$$o_P: \quad \left(x - \frac{a^2 - b^2}{a}c_t^3\right)^2 + \left(y + \frac{a^2 - b^2}{b}s_t^3\right)^2 = \frac{1}{a^2 b^2}\left(a^2 s_t^2 + b^2 c_t^2\right)^3.$$

We compute the point $R = e \cap o_P \setminus \{P\}$ in the same way as we did for the parabola. Therefore, we compute the resultants of e and o_P, with respect to y and x, cancel out the constants and the obvious cubic factors corresponding to P. Thus, we find

$$R = (ac_t(4c_t^2 - 3), -bs_t(4c_t^2 - 1)).$$

Now we rewrite $c_t(4c_t^2 - 3)$ and $s_t(4c_t^2 - 1))$ as follows:

$$4c_t^3 - 3c_t = c_t^3 - c_t(1 - s_t^2) - 2c_t s_t^2 = c_{2t}c_t - s_{2t}s_t = \cos 3t,$$

$$s_t(4c_t^2 - 1) = s_t(c_t^2 - 1 + c_t^2 + 2c_t^2) = s_t c_t^2 - s_t^3 + 2s_t c_t^2 = s_t c_{2t} + s_{2t}c_t = \sin 3t.$$

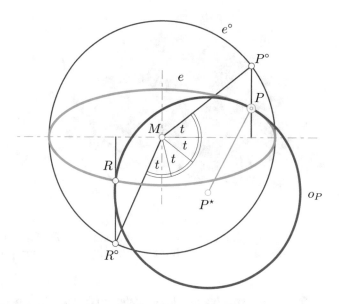

FIGURE 3.25. The central angle t corresponding to the point P of osculation is to be reoriented and multiplied by three in order to find the central angle of R. Note that the central angles are measured for the affine images of P and R on the principal circle $e°$ of the ellipse.

Thus, the point R can be written as

$$R = (a \cos 3t, -b \sin 3t).$$

Figure 3.25 shows how to find the point R. The parameter t corresponding to the point P of osculation can be viewed as a central angle on the principal circle $e°$ which is e's image under a scaling of the y-coordinates (cf. PROCLUS's construction in Figure 2.6. Section 2.1). We reverse the orientation of t and multiply it by three which gives the central angle of the point $R° \in e°$. The point $R°$ is the affine image of the desired point R.

The construction given for the point R works even if $t = 0$, $i.e.$, the point P is a vertex of e with $P = R$. Note that conversely there are three points P_1, P_2, and P_3 with osculating circles through the same point R since t can be replaced by $t \pm \frac{2}{3}\pi$ without changing R (see Figure 3.26). In Section 9.1, we shall learn that e is STEINER's circumellipse of the triangle $P_1P_2P_3$.

Consequently, we have:

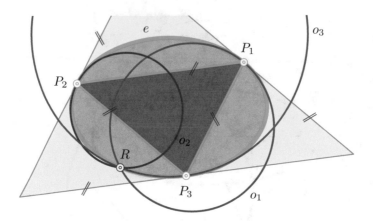

FIGURE 3.26. Three osculating circles of e passing through R. The ellipse e is STEINER's circumellipse of the triangle built by the three osculation points P_1, P_2, and P_3.

Theorem 3.2.5 *There are always three different points of an ellipse e whose osculating circles share the remaining point of intersection R. Conversely, each point $R \in e$ can serve as the remaining point.*

The case of hyperbolas is similar:

● **Exercise 3.2.18** Intersection of a hyperbola and an osculating circle. Use either the rational parametrization

$$\left(a\frac{1+t^2}{1-t^2}, b\frac{2t}{1-t^2} \right) \quad \text{with } t \in \mathbb{R}$$

or the parametrization in terms of hyperbolic functions ($\pm a \cosh u, b \sinh u$) with $u \in \mathbb{R}$ of the hyperbola

$$h : \frac{x^2}{a^2} - \frac{y^2}{b^2} = 1$$

and show that the osculating circle o_P at P intersects h in the point R (Figure 3.27) with either

$$R = \left(2bt\frac{(1+3t^2)(3+t^2)}{(t^2-1)^3}, a(1+t^2)\frac{t^4+14t^2+1}{(t^2-1)^3} \right)$$

or

$$R = (a \cosh^3 u, -b \sinh^3 u).$$

We conclude with a geometric interpretation of this result. Assume A_P is the area of the triangle MVP with M being the hyperbola's center, V being its principal vertex, and P the point on the hyperbola. The edge from V to P is an arc of the hyperbola. We compute the area A_P as

$$A_P = \frac{1}{2}a \cosh t \cdot b \sinh t - \int_a^{a \cosh t} b\sqrt{\frac{x^2}{a^2} - 1} \, dx = \frac{1}{2}abt.$$

Note that the function $A_P(t)$ is linear in t. Therefore, the area A_R of the larger triangle MVR with R as the remaining point $h \cap o_P \setminus \{P\}$ equals

$$A_R = \frac{1}{2}ab \cdot 3t = \frac{3}{2}abt.$$

Thus, we have

$$A_P : A_R = 3 : 1.$$

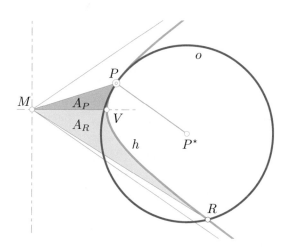

FIGURE 3.27. The area A_R of the curved triangle MVR equals $3A_P$ with A_P being the area of the curved triangle MVP.

Taylor parabolas and Ghys's theorem

There is a relatively new result dealing with the approximants of order $n - 1$ of a polynomial function of degree n. We shall consider here just the case $n = 3$, because the Taylor approximation in this case is of degree $n - 1 = 2$, *i.e.*, a parabola. A theorem which is due to the French mathematician ÉTIENNE GHYS (born 1954) says:

Theorem 3.2.6 *(cf. [31]) The Taylor approximants of degree 2 to a real polynomial function of degree 3 are disjoint parabolas with the same ideal point.*

An illustration of Theorem 3.2.6 is given in Figure 3.28. The Taylor approximants of degree two are parabolas including the tangent at the point

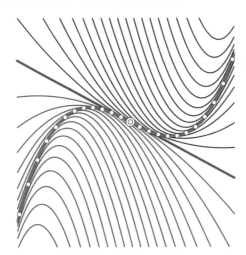

FIGURE 3.28. The degree two Taylor approximants to a cubic polynomial function are disjoint parabolas together with the tangent at the inflection point.

of inflection. In this particular point, the Taylor parabola degenerates into a straight line, namely the tangent to the graph of the polynomial function. Note that these parabolas share the ideal point, and thus, their axes are parallel. If we drop this condition, we find much more osculating parabolas. However, those mentioned in Theorem 3.2.6 form a family of disjoint parabolas.

The hyperosculating parabola

The osculating circle o of a curve c at a point P is a good approximation of the curve, at least locally around P. At P, the osculating circle and the curve share the point P, the tangent t_P, and the curvature κ. However, we can find better approximating conics.

Assume $\mathbf{c}(s)$ is the arc length parametrization of a C^4-curve c. Let $P = \mathbf{c}(t_0)$ be a regular point that is not a point of inflection. We choose the Frenet frame at P as coordinate (see Figure 3.29). In this frame, a parabola p through P and tangent to t_P can be parametrized by

$$\mathbf{p}(t) = \left(t + \frac{1}{2}a_1 t^2, \frac{1}{2}a_2 t^2\right). \tag{3.32}$$

Obviously, the axis of p is parallel to (a_1, a_2). Now we determine a_1 and a_2 such that p and c do not only agree in P, t_P, and the curvature κ at P. Moreover, the derivatives of the respective curvatures with respect

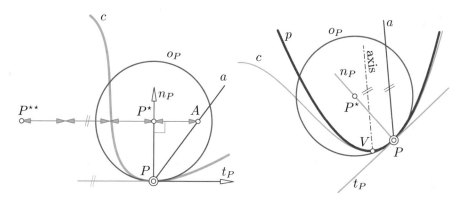

FIGURE 3.29. Left: With respect to the Frenet frame of the curve c, the center $P^\star$ of curvature has coordinates $(0, \varrho)$ and the evolute's center of curvature $P^{\star\star}$ has coordinates $(-\dot\varrho\varrho, \varrho)$. The point A on the affine normal a has the coordinates $(\frac{1}{3}\dot\varrho\varrho, \varrho)$ (see (3.33)). Right: The hyperosculating parabola p at P shares the osculating circle o_P with c. It is uniquely determined by the center $P^\star$ of curvature, the line element (P, t_P), and the diameter a.

to their arc lengths shall be equal. We compute $\dot{\mathbf{p}}(t) = (1 + a_1 t, a_2 t)$ and $\ddot{\mathbf{p}}(t) = (a_1, a_2)$. With (3.16) the signed curvature κ_p of the parabola at $t = 0$ and find

$$\kappa_p = \frac{a_2}{\dot\sigma^{3/2}}$$

with $\sigma = \|\dot{\mathbf{p}}\|$. At $t = 0$, we obtain $\dot\sigma(0) = 1$, $\ddot\sigma(0) = -a_1$, and $\kappa_P(0) = a_2$. The derivative $\kappa' = \frac{\mathrm{d}\kappa}{\mathrm{d}\sigma} = \dot\kappa/\dot\sigma$ of the parabola's curvature at $t = 0$ equals

$$\kappa' = -3a_1 a_2.$$

By assumption, the curvature κ_c and its derivative κ'_c of the given curve c at P satisfy

$$\kappa_c = \kappa_p = a_2 \quad \text{and} \quad \kappa'_c = \kappa'_p = -3a_1 a_2$$

with $\varrho = \kappa_p^{-1}$ and $\varrho' = (\kappa_p^{-1})' = -\frac{\kappa'_p}{\kappa_p^2}$ being the radius of curvature and its derivative. This yields

$$a_1 = -\frac{\dot\kappa_p}{3\kappa_p} = \frac{1}{3}\dot\varrho\varrho \quad \text{and} \quad a_2 = \kappa_p = \frac{1}{\varrho}.$$

Therefore, the axis of the parabola is parallel to

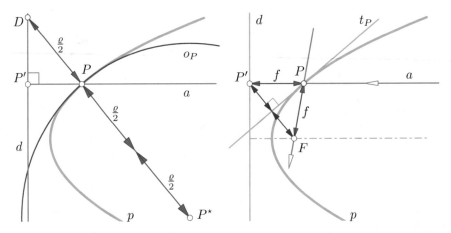

FIGURE 3.30. Left: The center $P^\star$ of curvature and the point P define a point on the directrix d (according to Theorem 3.2.4). The directrix d is perpendicular to any diameter (especially to a). Right: The focus F of the parabola is the reflection of P' in the tangent t_P where P' is the orthogonal projection of P onto the directrix d.

$$\mathbf{a} = \left(\frac{1}{3}\dot{\varrho}\varrho, \varrho\right). \tag{3.33}$$

The line $\mathbf{l}(t) = \mathbf{p} + \lambda \cdot \mathbf{a}$ (with $\lambda \in \mathbb{R}$) is a diameter of the hyperosculating parabola. This line is called the *affine normal* of c at P, since it is invariant under affine transformations, *i.e.*, an affine mapping α applied to c sends this line to the affine normal of the image curve $\alpha(c)$.

The construction of the hyperosculating parabola if the point P, the osculating circle o_P, and the affine normal a (diameter through P) are given, can be seen in Figure 3.30 (left): From Theorem 3.2.4, we know that P divides the segment on the parabola's normal between the directrix d and the center $P^\star$ of curvature with ratio $1 : 2$. Thus, we find a point D on the directrix d by adding a segment of length $\frac{\varrho}{2} = \frac{1}{2}\overline{PP^\star}$ on the normal from P outside the parabola. The directrix d is orthogonal to all diameters of p and so it is orthogonal to a (note the left-hand side of Figure 3.30). The axis and the vertex of the (osculating) parabola p are found by first determining the focus F. For that purpose, we use the focal property of the parabola. Lines parallel to the axis of p (such as the affine normal a) are reflected in p. The reflections pass through the focus F. This gives a line through the yet unknown point F. Since $\overline{Pd} = \overline{PP'} = \overline{PF}$, we find the focus F. The point P' is the orthogonal projection of P onto the directrix d. Moreover, it is the reflection of F in the tangent t_P (Figure 3.30, right).

According to Definition 6.4.1 in Section 6.4, the osculating circle o_P and the hyperosculating parabola p are related via a perspective collineation, with the affine normal as an axis and the center on it. Note that the osculating circle o_P osculates the curve c as well as the hyperosculating parabola at P. The hyperosculating parabola hyperosculates c at P.

The affine normals of conics are their diameters, *i.e.*, they are either parallel to each other (in the case of a parabola), or they are concurrent in the conic's center. For curves different from conics, the totality of affine normals envelopes the *affine evolute* which can be seen as an affine counterpart of the evolute. The affine evolute of curve is invariant with respect to affine transformations.

The conic with local five-fold intersection

The hyperosculating parabola p of a curve c at some (regular, non-inflection) point P is only one hyperosculating conic at P. In Definition 6.4.1 (cf. Section 6.4), we shall learn that there is a one-parameter family of conics hyperosculating p at P. Among these conics there is one conic f that approximates the curve c locally at P better than any other conic. In terms of algebraic geometry, f intersects c at P with multiplicity five. This cannot be the case for p since it is only intersecting with multiplicity four. Note that four is the maximal multiplicity of intersection of two conics.

The conic f with local five-fold intersection can be seen as a limit of a conic on five points on c which are sufficiently close together. If f is a conic with center, then its center is a point on the affine normal of c at P. In the following, we shall derive the center and an equation of f.

Assume that the C^4-curve c is parametrized by its arc length which is possible for any C^1-curve, at least locally. Let further $(\mathbf{e}_1, \mathbf{e}_2)$ be the Frenet frame of c at P. Then, we use the Frenet equations

$$\dot{\mathbf{e}}_1 = \kappa \mathbf{e}_2 \quad \text{and} \quad \dot{\mathbf{e}}_2 = -\kappa \mathbf{e}_1$$

where the dot indicates differentiation with respect to the arc length s. We can expand c at P in a Taylor series:

$$\mathbf{c}(s) = \mathbf{c}(0) + s\dot{\mathbf{c}}(0) + \frac{s^2}{2}\ddot{\mathbf{c}}(0) + \frac{s^3}{6}\dddot{\mathbf{c}}(0) + \dots .$$

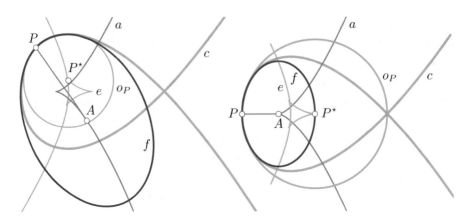

FIGURE 3.31. Comparison of the conic f with local five-fold intersection at P and the osculating circle o_P. The evolute e and the affine evolute a are also shown. The point $A \in a$ is the center of f. $P^\star$ is the center of curvature, and thus, the center of the osculating circle o_P. Left: P is not a vertex of c, *i.e.*, a point with $\dot{\kappa} = 0$. Right: P is a vertex of c. In this case, the conic f and the curve intersect with multiplicity six at P.

We write down the expansion in the Frenet frame, *i.e.*, the point P coincides with the origin of the coordinate system and the tangent t_P and normal n_P aim in the direction $\mathbf{e}_1$ and $\mathbf{e}_2$. Thus, we have

$$
\mathbf{c}(s) = \begin{pmatrix} s & -\dfrac{\kappa^2}{6}s^3 & -\dfrac{\kappa\dot{\kappa}}{8}s^4 + \dfrac{\kappa^4 - 3\dot{\kappa}^2 - 4\kappa\ddot{\kappa}}{120}s^5 + \dots \\[2ex] -\dfrac{\kappa}{2}s^2 & +\dfrac{\dot{\kappa}}{6}s^3 + \dfrac{\ddot{\kappa} - \kappa^3}{24}s^4 & +\dfrac{\dddot{\kappa} - 6\dot{\kappa}\kappa^2}{120}s^5 + \dots \end{pmatrix}. \tag{3.34}
$$

The equation of a conic that touches c at P reads

$$
k: \ Ax^2 + 2Bxy + Cy^2 + 2Ey = 0, \quad \text{with } E \neq 0.
$$

We substitute the first and second coordinates of $\mathbf{c}(s)$ from (3.34) for x and y. This yields a polynomial condition on the coefficients A, B, C, and E of k in order to annihilate the equation of k. We extract the coefficients of powers of s. The first three, start at s^2, and read

$$
A + \kappa E = 0, \quad 3\kappa B + \dot{\kappa}E = 0, \quad 4\kappa^2 A - 4\dot{\kappa}B - 3\kappa^2 C + (\kappa^3 - \ddot{\kappa})E = 0.
$$

Therefore, the conic k has fourth-order contact (or equivalently a local five-fold intersection) if, and only if,

$$
k: \ 9x^2 - 6\dot{\varrho}xy + (9 + 2\dot{\varrho}^2 - 3\ddot{\varrho})y^2 + 18\varrho y = 0.
$$

Figure 3.31 shows a conic with local five-fold intersection centered at

$$A = \left(\frac{-3\dot{\varrho}\varrho}{\ddot{\varrho} - \dot{\varrho}^2 - 3}, \frac{9\varrho}{\ddot{\varrho} - \dot{\varrho}^2 - 3} \right)$$

which is a point on the affine normal m at P. It is the point of contact between m and the affine evolute a.

Remark 3.2.2 If we take just the linear and quadratic terms of the local expansion of $\mathbf{c}(s)$ from (3.34), we obtain as second order Taylor approximant the parametrization

$$\mathbf{p}(s) = \left(s, \frac{\kappa}{2} s^2 \right) \tag{3.35}$$

which gives a parabola that osculates the given curve at $\mathbf{c}(0)$. The vertex of the parabola equals the point $\mathbf{c}(0)$, the parabola's axis coincides with the curve's normal. Later, in Section 3.3, we shall see how this can be generalized to the case of two-dimensional surfaces.

Mannheim curve, curvature diagram

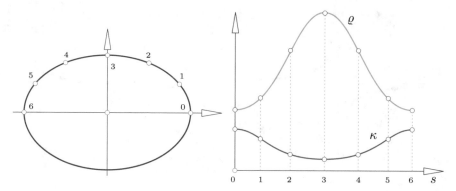

FIGURE 3.32. The curvature diagram (blue) and the Mannheim diagram (red) of the ellipse on the left-hand side. The two diagrams are drawn with different scales.

From conics, we can deduce further curves as we shall see in later sections. On the other hand, conics can define some curves in the plane: According to the fundamental theorem of curves in the Euclidean plane, the curvature function $\kappa(s)$ depending on the arc length s defines a planar curve uniquely up to Euclidean motions. This means that we can define planar curves by prescribing a curvature function $\kappa(s): I \subset \mathbb{R} \to \mathbb{R}$.

The graph of the function $\kappa(s)$ is usually referred to as the *curvature diagram*. Furthermore, with $\kappa(s)$ we also know the curvature radius function $\varrho = \frac{1}{\kappa}$. This defines another diagram showing how the curvature radius depends on the arc length s. Figure 3.32 shows both. The second diagram showing the graph $(s, \varrho(s))$ is called the *Mannheim diagram* named after the French engineer and mathematician AMÉDÉE MANNHEIM (1831–1906). Sometimes, the diagram $(s, \kappa(s))$, or equivalently $(s, \varrho(s))$ is called the CESÀRO *curve* of a planar curve, named after the Italian mathematician ERNESTO CESÀRO (1859–1906).

Let now the conics enter the scene. We shall interpret conics as the Mannheim diagram or the curvature diagram of curves in the Euclidean plane and ask for the curves defined in this way. We give just one example in detail in order to show how to find the curve c. For various other choices of diagrams showing a conic, we just present the results. The list provided in Table 3.1 is far from being complete.

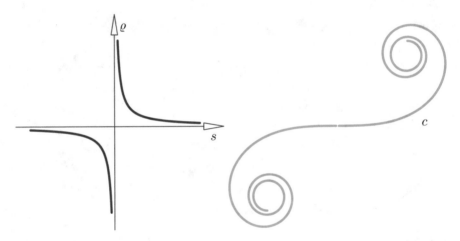

FIGURE 3.33. An equilateral hyperbola (left) as a Mannheim diagram of a Cornu spiral (right).

What does the curve look like if the Mannheim diagram is an equilateral hyperbola given by

$$\varrho = \frac{a}{s}$$

with $a \in \mathbb{R} \setminus \{0\}$?

It is well known that the curvature κ of a planar curve c and the slope angle φ of the tangent are related via

$$\kappa = \frac{d\varphi}{ds}.$$

Consequently, we have in this particular case

$$\varphi = \int_0^s \kappa(\sigma)\mathrm{d}\sigma = \frac{s^2}{2a}.$$

Without loss of generality, we can assume the tangent at $\mathbf{c}(0)$ equals the x-axis of a Cartesian frame. Then, we can write the tangent vector depending on s as

$$\dot{\mathbf{c}}(s) = \big(\cos\varphi(s), \sin\varphi(s)\big) = \left(\cos\frac{s^2}{2a}, \sin\frac{s^2}{2a}\right).$$

Integrating once yields $\int_0^s \dot{\mathbf{c}}(\sigma)\mathrm{d}\sigma = c(s)$ which in the present case gives

$$\mathbf{c}(s) = \left(\int_0^s \cos\frac{\sigma^2}{2a}\mathrm{d}\sigma, \int_0^s \sin\frac{\sigma^2}{2a}\mathrm{d}\sigma\right).$$

The latter integrals are called *Fresnel integrals* (named after the French physicist AUGUSTIN-JEAN FRESNEL, 1788–1827). The parametrized curve we obtain is called a *Clothoid* or *Cornu spiral* after the French physicist MARIE ALFRED CORNU (1841–1902). It is used in curve design, especially for streets and railroad tracks in order to make smooth transitions (at most C^2) between differently curved profiles.

So, we have seen that the Cornu spiral has an equilateral hyperbola for its Mannheim diagram. Figure 3.33 shows an example of a Cornu spiral and the corresponding equilateral hyperbola.

Figure 3.34 shows some of the curves mentioned in Table 3.1. The *epicycloid* and the *hypocycloid* are trajectories of points on a circle rolling outside or inside a fixed circle. The fixed circle is shown in Figure 3.34

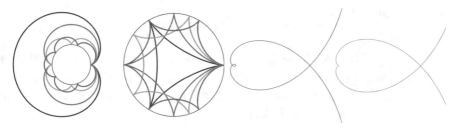

FIGURE 3.34. From left to right: epicycloids, hypocycloids, a paracycloid, and a hypercycloid.

(thin violet circle). Choosing various radii for the rolling circle (not displayed), we obtain different cycloids. The *paracycloid* and the *hypercycloid* can also be seen as traces of points undergoing superposed rotations. However, in these cases the circles are not real, *i.e.*, at least the radii of the involved circles are complex numbers (different from real numbers) and still some points move on real curves.

Finally, we collect pairs of conics being either Mannheim diagrams or curvature diagrams and the corresponding curves in Table 3.1.

conic	equation	corresponding curve
equilateral hyperbola	$\kappa = \frac{a}{s}$	logarithmic spiral
equilateral hyperbola	$\varrho = \frac{a}{s}$	Cornu spiral
circle	$s^2 + \varrho^2 = 1$	cycloid
ellipse	$\frac{s^2}{a^2} + \frac{\varrho^2}{b^2} = 1$	hypocycloid if $a > b$ epicycloid if $a > b$
hyperbola	$-\frac{s^2}{a^2} + \frac{\varrho^2}{b^2} = 1$	hypercycloid
hyperbola	$\frac{s^2}{a^2} - \frac{\varrho^2}{b^2} = 1$	paracycloid
parabola	$\varrho^2 = 2rs$	involute of a circle with radius r

TABLE 3.1. Conics being Mannheim curves or curvature diagrams and the thus defined curves.

Multifocal curves

Ellipses can be defined by means of a constant sum of the two distances to fixed focal points as we have seen in Section 2.1. Replacing the sum by a difference leads to hyperbolas. There are many ways to generalize these distance properties:

Sum of distances

Let $\mathbf{x} = (x, y)$ be the Cartesian coordinates of a point X in the Euclidean plane. Assume further that F_1, F_2, ... F_n are n mutually distinct points in the Euclidean plane. Now, we can define an *n-ellipse* or a *multifocal ellipse* e with *focal points* F_i by

$$e := \{X \mid \sum_{i=1}^{n} \overline{XF_i} = \text{const.}\}. \tag{3.36}$$

This definition includes the ellipses as we have seen in Section 2.1. If now $n > 2$, we obtain a new class of curves.

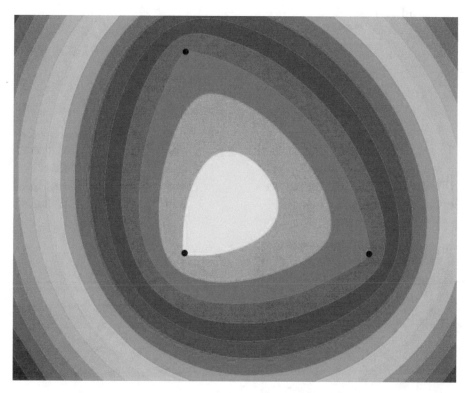

FIGURE 3.35. Multifocal ellipses: The black dots indicate the three focal points. The boundaries of the differently colored areas are multifocal ellipses.

In Section 2.1, we learned that the tangent and normal of an ellipse or a hyperbola at a point X are the bisectors of the focal rays. It is easy to see that the well-known construction of tangents (or normals) to ellipses and hyperbolas is a special case of the construction for the multifocal ellipses: The defining equation (3.36) for points $X = (x, y) \in c$ can be written as

$$E(x, y) := \sum_{i=1}^{n} \overline{XF_i} - d = 0.$$

Now, we assume that $\mathbf{f}_i = (f_{x_i}, f_{y_i})$ are the coordinates of the focal point F_i and compute the gradient of the function E. This yields

$$\mathrm{grad}\,E = \left(\frac{x - f_{x1}}{XF_1} + \ldots + \frac{x - f_{xn}}{XF_n}, \frac{y - f_{y1}}{XF_1} + \ldots + \frac{y - f_{yn}}{XF_n} \right) = \sum_{i=1}^{n} \frac{\mathbf{x} - \mathbf{f}_i}{XF_i}.$$

Since the gradient points in the direction of the normal n at X of the implicitly given curve (3.36), we have

$$\mathbf{n} = \sum_{i=1}^{n} \frac{1}{XF_i} (\mathbf{x} - \mathbf{f}_i). \qquad (3.37)$$

Thus, the normal vector $\mathbf{n}$ is the *sum of properly oriented unit vectors* pointing from the n focal points to the point X on the multifocal ellipse (Figure 3.36). So we can say that even the construction of tangents to ellipses is generalized. Note that the case of the ordinary ellipse is included.

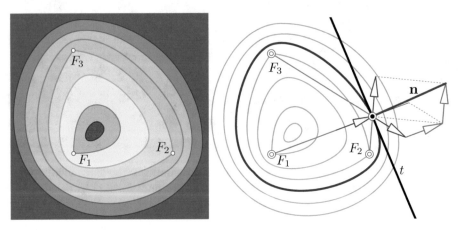

FIGURE 3.36. Left: Among the various confocal trifocal ellipses we find curves passing through the focal points. Right: The tangent construction for multifocal ellipses generalizes that of ordinary ellipses: Equally long vectors pointing from the focal points towards the point on the curve sum up to a normal $\mathbf{n}$ vector of the curve.

An algebraic treatment of the multifocal ellipses is challenging. An implicit equation of a trifocal ellipse reads

$$\left((d^2 - w_1^2 - w_2^2 - w_3^2)^2 - 4(w_1^2 w_2^2 + w_2^2 w_3^2 + w_3^2 w_1^2) \right)^2 - 64 d^2 w_1^2 w_2^2 w_3^2 = 0 \quad (3.38)$$

where $w_i = \sqrt{(x - f_{xi})^2 + (y - f_{yi})^2}$ (for $i \in \{1, 2, 3\}$) and is, therefore, of degree 8. However, this curve is of genus 9 and has the absolute points $(0 : 1 : \pm i)$ of Euclidean geometry for its singularities. These are four-fold points on e with δ-invariant[14] 6. The boundaries of the differently shaded areas shown in Figure 3.35 are only parts of the level sets of the algebraic variety described by (3.38).

There are some exceptions: Choose d such that the sum $w_1 + w_2 + w_3$ is minimized. Then, the real branch of e consists only of the *Fermat point* or *Torricelli point* X_{13} of the triangle $F_1 F_2 F_3$[15]. If we choose d such that it equals the sum of the distances of one focal point F_i to the other two focal points F_j and F_k, then the multifocal ellipse passes through F_i since these points satisfy the distance relation.

Product of distances

Instead of the constancy of the sum or difference of distances to focal points we can define a family of curves by the product

$$\prod_{i=1}^{n} \overline{XF_i} = \text{const.} \tag{3.39}$$

For $n = 2$ and a slight modification of (3.39), these curves were probably first studied by the Greek geometer PERSEUS who lived around 150 BC. He named these curves *spiric curves*, for he found them as the planar intersections of a torus which was called a *spire* at that time (see Figure 3.37).

The curves described by (3.39) (without modification) are called CASSINI's *curves*, named after the French-Italian astrologer, astronomer, engineer, and mathematician GIOVANNI DOMENICO CASSINI (1625–1712).

The curves of CASSINI contain some isoptics and pedal curves (cf. Sections 9.2 and 9.4) of an ellipse, especially BOOTH's lemniscate and BERNOULLI's lemniscate.

The spiric curves are quartic curves. For arbitrary n, the curves defined by (3.39) are of degree $2n$. The left-hand side of Figure 3.38 shows the graph surface with contour lines at constant product of three distances.

---

[14] Definitions and properties of the δ-invariant among others can be found in: J. MILNOR: *Singular points of complex hypersurfaces*. Annals Math. Stud. 61. Princeton University Press & University of Tokyo Press, Princeton - Tokio (1968), ISBN 0-691-08065-8.

[15] The Kimberling notation of triangle centers is explained in the footnote on page 377.

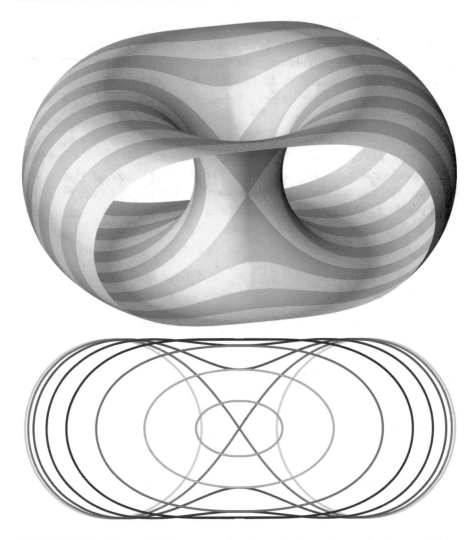

FIGURE 3.37. Above: Spiric curves as planar intersections of a torus. Below: Some spiric curves.

The right-hand side of Figure 3.38 shows a top view of the surface. There, the contour lines appear as the curves where the product of distances to three focal points is constant.

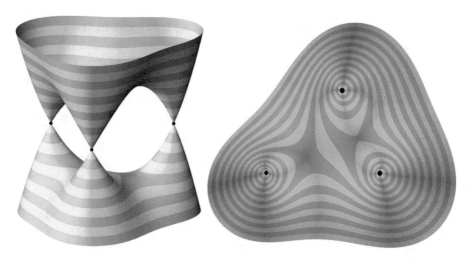

FIGURE 3.38. Left: Graph surface with contour lines being curves of constant product of distances. Right: The top view shows the curves of constant product of distances. The black spots indicate the focal points.

Weighted sum or product of distances

Other generalizations are possible: such as the sets of all points with constant weighted sum of distances

$$\sum_{i=1}^{b} w_i \cdot \overline{XF_i} = \text{const.}$$

or constant weighted sum of products

$$\Pi_{i=1}^{b} \cdot \overline{XF_i}^{e_i} = \text{const.}$$

which can also be seen as generalizations of conics.

Superellipses, superhyperbolas, superparabolas

A further generalization of conics can be found by manipulating the equation of an ellipse or a hyperbola. In the standard equation of the ellipse, we replace the power 2 of the monomials by an arbitrary rational number $n \neq 0$. This gives the equations

$$\left(\frac{x}{a}\right)^n + \left(\frac{y}{b}\right)^n = 1, \quad n \in \mathbb{Q} \setminus \{0\}, \tag{3.40}$$

of a class of curves called *Lamé curves*, named after the French mathematician and physicist GABRIEL LAMÉ (1795–1870).

The Lamé curves are obviously symmetric with respect to the x- and y-axes, and thus, also with respect to the origin of the Cartesian coordinate system if n is even. For $n = 2$, we find ellipses among the Lamé curves. For $n = \frac{2}{3}$, we obtain the star-shaped evolutes of ellipses (3.28) in Section 3.2. So, we can say that even the evolute of an ellipse is a generalization of an ellipse since these two curves belong to the same class of curves, namely the Lamé curves.

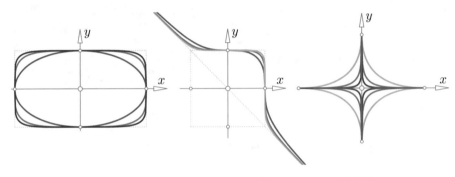

FIGURE 3.39. Lamé curves for various n: Left: $n = 2, 4, 6$. Middle: $n = 3$, $n = 5$, $n = 7$. Right: $n = \frac{1}{2}$, $n = \frac{1}{3}$, $n = \frac{1}{4}$.

Some examples of Lamé curves are displayed in Figure 3.39. We observe that for $n > 1$ and n even the curves are getting more and more "rectangular", *i.e.*, the bounding rectangle with vertices $(\pm a, \pm b)^{\mathrm{T}}$ is the limit of these curves if $n \to \infty$. The Danish mathematician and inventor PIET HEIN (1905–1996) suggested to use Lamé curves with $n = \frac{5}{2}$ (Figure 3.40) to make tables with round corners. PIET HEIN called his special type of Lamé curve a *superellipse*.

Lamé curves with $n \in \mathbb{Z} \setminus \{0, \pm 1, 2\}$ cannot be parametrized by rational functions.

The Lamé curves with $0 < n < 1$ are star-shaped, as can be seen on the right-hand side of Figure 3.39. The limit of the Lamé curves for $n \to 0$ is a pair of line segments.

Sometimes the equations of Lamé curves are given in the form

$$\left| \frac{x}{a} \right|^n + \left| \frac{y}{b} \right|^n = 1.$$

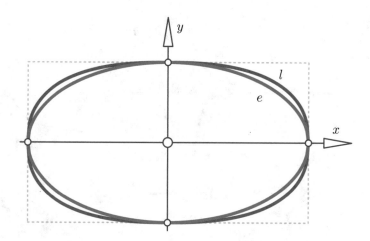

FIGURE 3.40. PIET HEIN's Lamé curve l (red) compared with the ellipse e (blue) having the same lengths of their axes. HEIN's superellipse comes closer to the corners of the common tangent rectangle of both curves.

This does not change the shape of the curves with even n, but forces the curves with odd n into the bounding rectangle. Of course, these curves are symmetric with respect to the x- and y-axes.

Naturally, Lamé-like curves can also be obtained by replacing the + with a − in (3.40). We shall not discuss these curves here since this is far beyond the scope of our book.[16]

The curves with equation

$$y = qx^n, \quad n \in \mathbb{Q} \setminus \{0\}, \quad q \in \mathbb{R} \setminus \{0\} \tag{3.41}$$

are generalizations of parabolas in some sense. Like the usual parabola ($n = 2$), they touch the ideal line at the ideal point of the y-axis with homogeneous standard coordinates $(0 : 0 : 1)$ (see Section 5.3). Furthermore, all these curves intersect the ideal line at this point with multiplicity $n-1$ provided that $n \in \mathbb{N}$ and $n \geq 2$. FELIX KLEIN (German mathematician, 1849–1925) called these curves W-curves, because of their remarkable property involving a cross ratio[17] described in Theorem 3.2.7. Apparently,

---

[16] For further details see M. MATSUURA: *Asymptotic Behaviour of the Maximum Curvature of Lamé Curves.* J. Geom. Graphics **18** (2014), 45–59.

[17] FELIX KLEIN used the German word *Wurf* (throw) instead of cross ratio which is the reason for the notation 'W'.

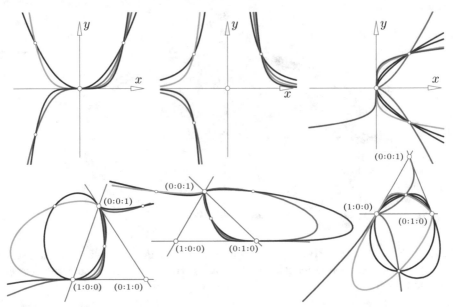

FIGURE 3.41. Some W-curves: affine images (above), projective images (below).

some W-curves with different exponents n are collinear images of each other, $e.g.$, $y = x^3$ and $y = x^{\frac{1}{3}} = \sqrt[3]{x}$ can be mapped via a reflection $x \to y$, $y \to x$ onto each other. Figure 3.41 shows some examples of W-curves. The affine images together with the respective collinear images are displayed in order to illustrate the characteristic behavior of the W-curves at infinity.

The following theorem uses terms which will be explained in Section 5.3.

Theorem 3.2.7 *Let c be a W-curve (3.41). The tangent t_P at $P \in c$, $P \neq (0,0)$, intersects the base lines $x_0 = 0$, $x_1 = 0$, and $x_2 = 0$ of the projective standard frame in points B_0, B_1, and B_2, respectively. Then, the cross ratio $\mathrm{cr}(B_2, B_1, B_0, P)$ is the same for all curves with fixed n:*

$$\mathrm{cr}(B_2, B_1, B_0, P) = n. \tag{3.42}$$

Proof: Use the parametrization $P = (1 : t : \alpha^{-1}t^n)$ of c, compute the tangent t_P, intersect it with the lines, and compute the cross ratio. ∎

The W-curves are the orbits of points under the action of smooth one-parameter subgroups of the group of projective transformations in $\mathbb{P}^2(\mathbb{F})$.

Finally, we shall point out that the evolute of a parabola is also a W-curve. It is, up to a homothety and a translation, given by the equation $y = x^{\frac{3}{2}}$, and thus, it is one of the curves given in (3.41).

Offset curves of conics

This section is devoted to the offsets of conics. The *offset curve* c_d of a curve c in the (Euclidean) plane at distance d is defined as the set of all points (in the plane of c) which are at a fixed distance $d \in \mathbb{R}$ to the points of c. We can find the offset to a general C^1-curve by simply attaching the normal vector $d\mathbf{e}_2(t)$ (of length d) at $\mathbf{c}$ which gives a parametrization of the offset (as shown in Figure 3.42)

$$\mathbf{c}_d(t) = \mathbf{c}(t) + d\mathbf{e}_2(t). \tag{3.43}$$

The tangent t_d to the offset at P_d is parallel to the tangent of c at P independent of the choice of the value $d \in \mathbb{R}$. Therefore, the one-parameter family of curves c_d (with $d \in \mathbb{R}$) is also called the family of *parallel curves*. Obviously, the offsets of a circle c are circles concentric with c. Circles are

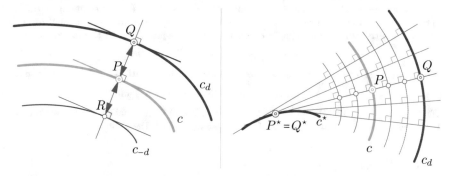

FIGURE 3.42. Left: The curve c and its offset c_d at distance d share normals and have parallel tangents. The offset d is measured along the common normals. Right: The family of offset curves of c (including c) share the evolute $c^\star$ as the envelope of the common normals.

the only conics whose non-trivial offset curves are conics.

Since the offset d can be chosen with different signs, we obtain two branches of an offset curve. One corresponds to $+d$, the other one to $-d$. Points P and Q that lie on the same normal of c share the center $P^\star = Q^\star$ of curvature (Figure 3.42, right).

The offset c_d of a planar curve c can also be found as the envelope of all circles of radius d centered at c. This approach to the offset is shown in Figure 3.43, left.

■ **Example 3.2.1** Contour of a torus under an orthogonal projection. Assume that the ellipse

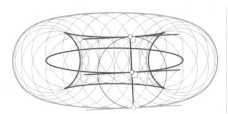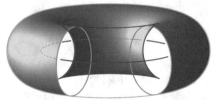

FIGURE 3.43. Left: The two branches of the offset of an ellipse e form the envelope of a one-parameter family of circles. Right: The contour of a ring-shaped torus consist of an "inner" (red) and "outer" (cyan) branch.

$e(t) = (a\cos t, b\sin t)$ is the orthogonal projection of a circle with radius a. We compute the offset at distance d which can be considered the contour of a torus with major radius a and minor radius d.

The torus can be generated in many different ways. For the moment, we consider the torus as the envelope of the one-parameter family of spheres with radius d centered at the spine curve, *i.e.*, the circle with radius a. The contours of the spheres under orthogonal projection are circles with radius d and centers on $e(t)$. Thus, the equation of all theses circles reads

$$c: \quad (x - a\cos t)^2 + (y - b\sin t)^2 = d^2 \quad \text{with } t \in [0, 2\pi[. \tag{3.44}$$

We compute the envelope of these circles by differentiating c with respect to t which gives

$$\dot{c}: \quad a\sin t(x - a\cos t) - b\cos t(y - b\sin t) = 0. \tag{3.45}$$

Eliminating t from (3.44) and (3.45), we arrive at an implicit equation of the contour of the torus, and thus, of the offset curves of the ellipse $e(t)$. With the abbreviations

$$A = a^2 + b^2, \quad E = a^2 y^2 + b^2 x^2, \quad \Omega = x^2 + y^2$$

this equation reads

$$e_d: \ E^2\Omega^2 + 4E^3 - 2(d^2 + A)E^2\Omega + 2(a^4 - A(a^2 + d^2))E\Omega^2 + 4a^2 d^2(A - a^2)\Omega^3$$
$$(12a^2(a^2 - A) - 10Ad^2 + A^2 + d^4)E^2 + (22a^2 d^2(A - a^2) + 4A(d^2 + a^2)(d^2 + b^2))E\Omega +$$
$$+\Omega^2(a^4(a^2 + 12d^4) - 2Aa^2(a^2 - 2d^2)(a^2 - 3d^2) + A^2(a^4 - 10a^2 d^2 + 2d^4)) +$$
$$+2(a^2 - b^2)((a^2 + d^2)A^2 - (5a^4 - a^2 d^2 + 4d^4)A + 2a^6 + 3a^4 d^2 + 2a^2 d^4 + d^6)E +$$
$$+((a^4 - 4a^2 d^2 + d^4)A^2 + (6a^4 d^2 + a^2 d^4 + d^6)A + a^8 - 8a^6 d^2 + 3a^4 d^4 - 4a^2 d^6)\Omega$$
$$-2a^2 x^2(a^2 - d^2)^3 + (b^2 - d^2)^2(a^2 - d^2)^2(a^2 - b^2)^2 = 0.$$

The parallel curves of ellipses are algebraic curves of degree eight. Only in the case of a circle, *i.e.*, $a = b$, the parallel curves degenerate into pairs of circles.

Consequently, we are able to construct points, tangents, and even osculating circles of a torus's contour under an orthogonal projection.

The offset curves of hyperbolas are also of degree eight. Only in case of the parabola, we can observe a reduction of the degree:

● **Exercise 3.2.19** Offset curves of a parabola. Show that the offset curves p_d of the parabola $p: x^2 - 2qy = 0$ at distance d are algebraic curves of degree six and have the equation

$$p_d: 2P^2(2\Omega + P) + 8d^2P\Omega + 4d^2(d^2 - 3q^2 - 4x^2)\Omega - 4d^2(2d^2 + 2q^2 - 3x^2)P$$

$$+(d^2 + q^2 - 2x^2)P^2 - d^2(4(d^2 + q^2)^2 - 16x^2 - 7x^4) = 0$$

where $P = x^2 - 2qy$ and $\Omega = x^2 + y^2$. Some offset curves of a parabola can be seen in Figure 3.44 (right).

Figure 3.44 shows some offset curves to an ellipse and a parabola. Offset curves tend to have cusps which happens if at some point on the initial curve the curvature radius equals the offset d. Usually, these cusps are cusps of the first kind, *i.e.*, like on the curve (t^2, t^3) at $(0,0)$. Especially, if the offset d is chosen equal to the curvature radius at a vertex, then the cusps become cusps of the second kind, *i.e.*, like on the curve (t^3, t^4) at $(0,0)$.

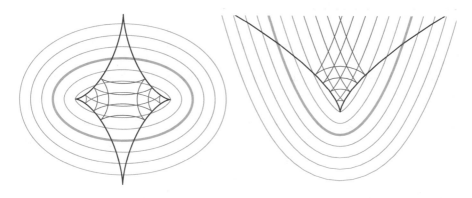

FIGURE 3.44. The offset curves of an ellipse (left) and a parabola (right) have cusps. These cusps gather on the evolute of the original curve.

3.3 Conics in differential geometry of surfaces

Dupin indicatrix

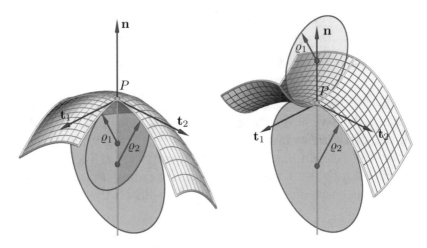

FIGURE 3.45. The principal curvature radii ϱ_1, ϱ_2 are the radii of the osculating circles of the normal sections in the principal directions t_1, t_2. Left: At an elliptic point P, both osculating circles are on the same side of the tangent plane. Right: At a hyperbolic point P, the osculating circles are on different sides of the tangent plane.

Conics also play an important role in the study of the local normal curvatures of a surface at a regular point P. For that, we assume that $\mathcal{S}$ is a surface in Euclidean three-space $\mathbb{E}^3$ given by a parametrization $\mathbf{S} : U \to \mathbb{E}^3$ over some domain $U \subset \mathbb{R}^2$. Let $P = \mathbf{S}(u_0, v_0)$ be some point on the surface with $(u_0, v_0) \in U$. We use a common notation for partial derivatives:

$$\mathbf{S}_u = \frac{\partial \mathbf{S}}{\partial u}, \quad S_v = \frac{\partial \mathbf{S}}{\partial v}, \quad \mathbf{S}_{uu} = \frac{\partial^2 \mathbf{S}}{\partial u^2}, \quad \mathbf{S}_{uv} = \frac{\partial^2 \mathbf{S}}{\partial u \partial v}, \quad \mathbf{S}_{vv} = \frac{\partial^2 \mathbf{S}}{\partial v^2}.$$

Points with $\mathbf{S}_u \times \mathbf{S}_v \neq \mathbf{0}$ are called *regular*. In the following we consider only regular points on $\mathcal{S}$ where the *first* and *second fundamental form* are

$$\mathrm{I} \;\; = \;\; E(u,v)\dot{u}\dot{u} + 2F(u,v)\dot{u}\dot{v} + G(u,v)\dot{v}\dot{v},$$

$$\mathrm{II} \;\; = \;\; L(u,v)\dot{u}\dot{u} + 2M(u,v)\dot{u}\dot{v} + N(u,v)\dot{v}\dot{v}$$

where $E(u,v) := \langle \mathbf{S}_u, \mathbf{S}_u \rangle$, $F(u,v) := \langle \mathbf{S}_u, \mathbf{S}_v \rangle$, $G(u,v) := \langle \mathbf{S}_v, \mathbf{S}_v \rangle$ are the coordinate functions of the first fundamental form, $L(u,v) := \langle \mathbf{S}_{uu}, \mathbf{n} \rangle$,

$M(u, v) := \langle \mathbf{S}_{uv}, \mathbf{n} \rangle$, $N(u, v) := \langle \mathbf{S}_{vv}, \mathbf{n} \rangle$ are the coordinate functions of the second fundamental form, and $\mathbf{n} : U \to \mathbb{S}^2$ is the unit normal vector field on S parallel to $\mathbf{S}_u \times \mathbf{S}_v$.

For any vector $\mathbf{t} = \dot{u}\mathbf{S}_u + \dot{v}\mathbf{S}_v$ in the tangent plane of S at the regular point P, the *normal curvature*, *i.e.*, the curvature of all planar sections of S formed by planes through P spanned by the normal vector $\mathbf{n}$ and the tangent vector $\mathbf{t}$, can be found via

$$\kappa_n = \frac{\mathrm{II}}{\mathrm{I}}. \tag{3.46}$$

Obviously, at any point P, the normal curvature κ_n depends on $\dot{u} : \dot{v}$ only. In other words: All curves on S with the same tangent line at P have the same normal curvature there. The normal curvature can be positive, zero, or negative.

Among the directions $(\dot{u}, \dot{v})$, we would like to find those which correspond to the maxima and minima of κ_n. For the sake of simplicity, we define

$$\mathbf{F}_1 := \begin{pmatrix} E & F \\ F & G \end{pmatrix} \quad \text{and} \quad \mathbf{F}_2 := \begin{pmatrix} L & M \\ M & N \end{pmatrix}.$$

The extremal value problem $(\dot{u} \ \dot{v})\mathbf{F}_2 \begin{pmatrix} \dot{u} \\ \dot{v} \end{pmatrix} \longrightarrow$ extr. under the constraint $(\dot{u} \ \dot{v})\mathbf{F}_1 \begin{pmatrix} \dot{u} \\ \dot{v} \end{pmatrix} = 1$ is solved by introducing the Lagrange function $(\dot{u} \ \dot{v})(\mathbf{F}_2 - \lambda\mathbf{F}_1) \begin{pmatrix} \dot{u} \\ \dot{v} \end{pmatrix}$ with the Lagrange multiplier λ. Differentiating the Lagrange function with respect to the vector $(\dot{u}, \dot{v})$ shows that the maxima or minima correspond to values of λ such that the 2×2-matrix

$$\mathbf{F}_2 - \lambda\mathbf{F}_1 \quad \text{or equivalently} \quad (\mathbf{F}_2 \cdot \mathbf{F}_1^{-1} - \lambda\mathbf{I}_2)\mathbf{F}_1$$

becomes singular. Note that $\det \mathbf{F}_1 = \|\mathbf{S}_u \times \mathbf{S}_v\|^2 > 0$.

The matrix $\mathbf{F}_1$ is positive definite since P is assumed to be regular, and so, the λs are eigenvalues of $\mathbf{W} := \mathbf{F}_2 \cdot \mathbf{F}_1^{-1}$. Since both matrices $\mathbf{F}_1$ and $\mathbf{F}_2$ are symmetric, $\mathbf{F}_1^{-1}$ and $\mathbf{W}$ are symmetric, too. The linear mapping of the tangent plane of S at P onto itself described by $\mathbf{W}$ is usually referred to as the *Weingarten map* (after the German mathematician JULIUS

WEINGARTEN, 1836–1910). Since $\mathbf{W}$ is symmetric, it can be diagonalized. Using the basis $(\mathbf{t}_1, \mathbf{t}_2)$ of $\mathbf{W}$'s eigenvectors, this diagonal form is

$$\begin{pmatrix} \kappa_1 & 0 \\ 0 & \kappa_2 \end{pmatrix}$$

with κ_1 and κ_2 are called the *principal curvatures* at P. Then,

$$H := \operatorname{tr} \mathbf{W}/2 = (\kappa_1 + \kappa_2)/2 \quad \text{and} \quad K := \det \mathbf{W} = \kappa_1 \kappa_2$$

are the *mean curvature* and the *Gaussian curvature* of S at P. The eigenvectors of $\mathbf{W}$ define the *principal tangents* t_1 and t_2 at P (Figure 3.45).

For the unit tangent vector we have $\mathbf{t} = \dot{u}\mathbf{t}_1 + \dot{v}\mathbf{t}_2$ with $\dot{u}^2 + \dot{v}^2 = 1$. In case $\kappa_1 = \kappa_2 \neq 0$ and $\kappa_1 = \kappa_2 = 0$ the point P is called an *umbilic* or a *flat point*, respectively. In these cases κ_n is indpendent of the tangent vector at P. Written within the eigenvector basis of $\mathbf{W}$ (which is not uniquely defined if $\kappa_1 = \kappa_2$), the expression for the normal curvature from (3.46) simplifies to $\kappa_n = \kappa_1 \dot{u}\dot{u} + \kappa_2 \dot{v}\dot{v}$. Without loss of generality, we may set $(\dot{u}, \dot{v}) = (\cos\alpha, \sin\alpha)$, *i.e.*, the tangent vector $\mathbf{t}$ encloses the angle α with the first principal tangent. Then, we have

$$\kappa_n(\alpha) = \kappa_1 \cos^2\alpha + \kappa_2 \sin^2\alpha \tag{3.47}$$

which is called EULER's *formula*, due to LEONHARD EULER (1707–1783).

Now we study the following diagram (Figure 3.46): We look for the points Q on all (non asymptotic) surface tangents of S at P with $\overline{PQ} = \sqrt{|c \cdot \varrho_n|}$ where $\varrho_n = \kappa_n^{-1}$ is the radius of curvature defined by the tangent $[P, Q]$. Here, $c \in \mathbb{R} \setminus \{0\}$ is an arbitrarily chosen fixed constant. The set $i(c)$ of all points Q in the tangent plane to S at P is called the *Dupin indicatrix* (CHARLES DUPIN, French mathematician, 1784–1873). The diagram $i(c)$ shall be drawn in the Cartesian frame centered at P with the first and second principal tangent for its x- and y-axes. Comparing with (3.47), we can see that the x- and y-coordinates of Q, α, and ϱ_n are related via

$$\cos\alpha = \frac{x}{\sqrt{|c \cdot \varrho_n|}} \quad \text{and} \quad \sin\alpha = \frac{y}{\sqrt{|c \cdot \varrho_n|}},$$

and in the case $\kappa_1 \kappa_2 \neq 0$, we thus have

$$i(c) : \quad \frac{x^2}{\varrho_1} + \frac{y^2}{\varrho_2} = \pm c \tag{3.48}$$

with an appropriate choice of the sign on the right-hand side. This is the equation of the Dupin indicatrix which defines up to two conics (including one singular type).

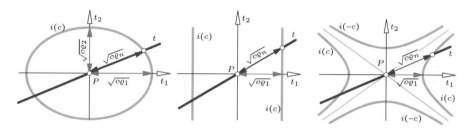

FIGURE 3.46. Dupin indicatrices at an elliptic (left), a parabolic (middle), and a hyperbolic point P (right).

The type of conic showing up as the Dupin indicatrix depends on the surface point P (which is still assumed to be regular). The point P is called *elliptic*, if $\kappa_1\kappa_2 > 0$ and then $i(c)$ is an *ellipse* provided that $c \cdot \varrho_i > 0$ for $i \in \{1, 2\}$. The point P is called *hyperbolic*, if $\kappa_1\kappa_2 < 0$ and $i(c)$ is a pair of conjugate *hyperbolas*. If one principal curvature is zero while the other one is not, then P is called a *parabolic* point. The case of a parabolic point P leads to a singular conic: Without loss of generality, we can assume that $\kappa_2 = 0$. Then, the equation of the indicatrix changes to

$$i(c): \quad x^2 = c \cdot \varrho_1$$

and $i(c)$ is the *pair of lines* $x = \pm\sqrt{c \cdot \varrho_1}$ parallel to the second principal tangent. Note that c has to be chosen such that $c \cdot \varrho_1 > 0$. Figure 3.46 shows the indicatrices of all regular types of surface points.

In the case of a hyperbolic point P, there are two hyperbolas shown in Figure 3.46, depending on whether c is positive or negative. It makes sense to draw both curves: As can be seen in Figure 3.45, the surface changes the sides of the tangent plane at P. That means the surface intersects the tangent plane at P along curves that touch the asymptotic tangents which correspond to tangent vectors with $\kappa_n = 0$. These tangents are the asymptotes of the indicatrix $i(c)$ at a hyperbolic point P and define the *asymptotic directions* at P. The integral curves of the two fields of asymptotic tangent vectors are the *asymptotic lines*. Any normal section of S in the asymptotic direction has a point of inflection at P.

Assume that the surface tangent t at P is rotating about the surface normal n. There are two instances where t becomes an asymptote of $i(c)$. In between the two asymptotes, the center of the osculating circle defined by t is on one side of the tangent plane. It changes to the other side of the tangent plane if we rotate t to the next sector. So, the two parts $i(c)$ and $i(-c)$ of the indicatrix at P allow us to decide on which side of the tangent plane we have to fix the center of the osculating circle.

The two parts $i(c)$ and $i(-c)$ of the Dupin indicatrix at a hyperbolic point are said to form a pair of *conjugate hyperbolas*, cf. Section 8.1.

When treating indicatrices constructively, we choose $c = \varrho_1$ (or $c = \varrho_2$). Thus, one semiaxis of the indicatrix equals one principal curvature radius and we only have to construct $\sqrt{\varrho_1 \varrho_2}$, which is elementary.

Two surface tangents are called *conjugate* if they are conjugate diameters of the indicatrix (cf. page 266), no matter if P is elliptic, parabolic, or hyperbolic.

At an *umbilical point*, the indicatrix $i(c)$ is a circle. There, any tangent vector is an eigenvector of $\mathbf{W}$ and the only eigenvalue $\kappa_1 = \kappa_2$ has algebraic multiplicity two. If the Weingarten map degenerates completely, *i.e.*, $\mathbf{W} = \mathbf{0}$ and then we call P a *flatpoint*. The Dupin indicatrix of a flatpoint is either empty or the ideal line of the tangent plane, depending on whether we have performed the projective closure of the tangent plane or not.

Osculating paraboloid

In a sufficiently small neighborhood of a regular surface point P, any C^2-surface $\mathcal{S}$ can be approximated by a quadratic function $f(x, y)$ over the tangent plane. We choose a Cartesian coordinate system such that P equals the origin and such that the axes aim in the direction of the principal tangents. Then, $f(x, y) = h_{11}(0, 0)x^2 + 2h_{12}(0, 0)xy + h_{22}(0, 0)y^2$, and obviously, the Dupin indicatrices to variable constants are the level sets of a paraboloid $\mathcal{P}$ with the equation $f(x, y) = z$. $\mathcal{P}$ shares P and the tangent plane at P with $\mathcal{S}$. Moreover, the first and second fundamental forms of $\mathcal{P}$ and $\mathcal{S}$ agree at P. The paraboloid $\mathcal{P}$ is called the *osculating (vertex) paraboloid*, for P is $\mathcal{P}$'s vertex. The osculating paraboloid is an elliptic or hyperbolic paraboloid if the point P is elliptic or hyperbolic. In the case of a parabolic surface point P, the osculating paraboloid $\mathcal{P}$ is a parabolic cylinder touching the tangent plane at P along the asymptotic tangent. Figure 3.47 shows the three types of (regular) surface points

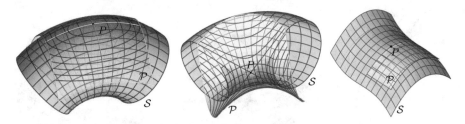

FIGURE 3.47. Osculating paraboloids with their level sets (from left to right): at an elliptic point, at a hyperbolic point, and at a parabolic point.

with their respective osculating paraboloids. In Remark 3.2.2, we have derived the two-dimensional analogon (3.35) to the osculating paraboloid by simply cutting off the local expansion (3.34) of a curve. Details on the construction of hyperosculating parabolas of curves as well as the osculating paraboliods of surfaces can be found in [45].

Meusnier's theorem and the osculating circle of an ellipse

The following Theorem 3.3.1 is due to the French mathematician JEAN BAPTISTE MARIE CHARLES MEUSNIER DE LA PLACE (1754–1793):

Theorem 3.3.1 *Let S be a C^2-surface in Euclidean three-space, let further P be a point on S and let t be a non asymptotic surface tangent of S at P. Then, all curves c on S that touch t at P have the same normal curvature κ_n. The osculating circles of all such curves c trace a sphere Σ if the osculating planes rotate about t.*

Proofs of this important theorem can be found in almost all textbooks on differential geometry, see *e.g.* [23, 58].

Figure 3.48 shows the consequences of Theorem 3.3.1, especially for the orthogonal projection of a circle c. Here, the surface S is the cylinder formed by the projecting lines through points of c. In Figure 3.48 (left), we can see that the orthogonal projection (in the direction p) of a circle c with radius a is an ellipse c'. The osculating circle o of c' at the auxiliary vertex P' has the center O' which is the intersection of c's axis x with the principal plane π through p. The ellipse c' (orthogonal projection of the circle c) has the semimajor axis length a and the semiminor axis length $a\cos\varphi$. According to (3.18), the radius of o equals $a/\cos\varphi$ which equals the length $\overline{P'O'}$ as can be read off from Figure 3.48 (left). Note that c' and o are hyperosculating. That will be of importance in Section 4.4.

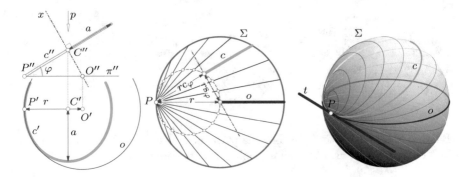

FIGURE 3.48. Left: The hyperosculating circle o of the orthogonal projection of a circle c. Middle and right: The osculating circles of all regular surface curves through the line element (P, t) cover the sphere Σ.

In the middle of Figure 3.48, we can see that the centers of the osculating circles of all curves with the same normal curvature (at P in the direction t) form a circle. Thus, all these osculating circles fill a sphere Σ which is frequently called the *Meusnier sphere*. The right-hand side of Figure 3.48 shows these particular circles on Σ sharing the line element (P, t).

● **Exercise 3.3.1** Hyperosculating circles of a hyperbola. **Figure 3.49 shows a hyperbola as a**

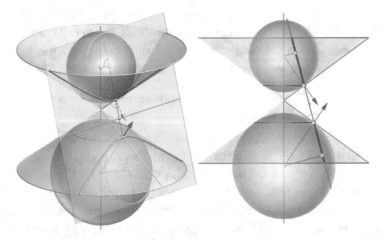

FIGURE 3.49. MEUSNIER's theorem can be used to construct the centers of curvature at the vertices of a planar section of a cone.

planar section of a right cone together with the Meusnier spheres at the vertices. Figure out the construction of the centers of curvature at the vertices. By the same token, the two radii of curvature are equal.

4 Euclidean 3-space

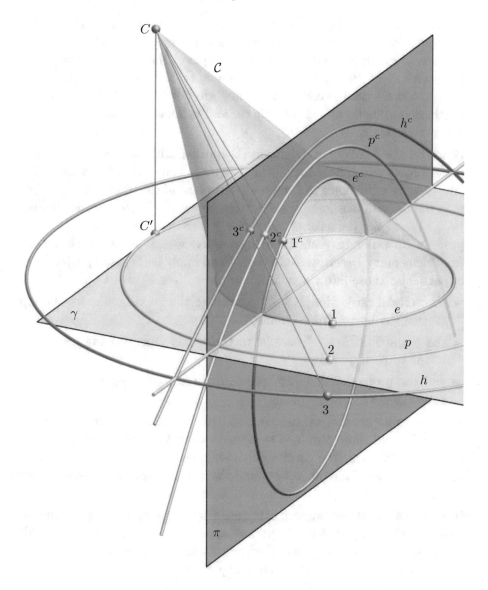

The central projection of circles with different radii may result in conics of any type. Depending on whether the projection cone $\mathcal{C}$, *i.e.*, the connection of the circles and the center C of the projection, avoids, touches, or intersects the vanishing plane, the image of the circle is an ellipse, a parabola, or a hyperbola.

4.1 Planar intersections of cones of revolution

Conics are also called *conic sections*, and this name expresses directly that they are closely related to sections of a cone, more precisely, to planar sections of quadratic cones. This is the topic of the present chapter. We start with *right cones* $\mathcal{C}$, *i.e.*, with *cones of revolution*. Cones are always understood as 'double cones', *i.e.*, they are not bounded by their apex. Furthermore, cones as well as spheres and cylinders are seen as surfaces and not as solids.

Based on the Apollonian definition (Definition 2.1.1 on page 13), we still prefer proofs which hold for all three types of conics simultaneously. But now there is a difference to Chapter 2: the plane of the conic lies somewhere in the three-dimensional space $\mathbb{E}^3$. Hence, we use Cartesian coordinates (x, y, z), and for the sake of simplicity we assume the z-axis in vertical position. In order to visualize 3D-geometry, we sometimes take benefit from elementary Descriptive Geometry and apply orthogonal projections, in particular those into the coordinate planes (see Figure 4.1):

- the vertical projection $P \mapsto P'$ parallel to the z-axis generates the *top view* in the horizontal xy-plane,
- the projection $P \mapsto P''$ parallel to the x-axis generates the *front view* in the vertical yz-plane,
- and the projection $P \mapsto P'''$ parallel to the y-axis generates the *side view* in the vertical xz-plane.

In order to be precise, we strictly distinguish between pre-images in the 3-space $\mathbb{E}^3$ and those obtained by a projection: The images are indicated by primes. We omit the primes only, when the pre-image already lies in the image plane of the projection, and therefore, image and pre-image are coinciding.

In many cases, we use appropriate coordinate frames and corresponding views, so that, *e.g.*, the cutting plane appears as a line. Then, we speak of an 'edge view' of the plane. Such particular views have the advantage that they show immediately what is significant.

Cones with a vertical axis

We begin with the standard case: A cone of revolution ($=$ right cone) with a vertical axis has to be intersected with a plane.

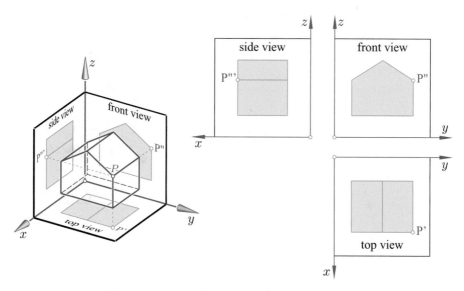

FIGURE 4.1. Top view, front view and side view.

Theorem 4.1.1 *The curve c of intersection between a right cone $\mathcal{C}$ and a plane π is a conic, provided, the plane π does not pass through the cone's apex S and is not orthogonal to the cone's axis s. When, under the assumption of a vertical axis s of $\mathcal{C}$, α and β are the angles of inclination of π and of the generators of $\mathcal{C}$, then the numerical eccentricity of c is $\varepsilon = \sin\alpha/\sin\beta$.*

Suppose, a sphere $\mathcal{S}$ contacts the cone along a circle k and touches the plane π at a point F: Then, F is a focal point of c, and the line of intersection between π and the plane of k is the associated directrix l. The orthogonal projection onto π maps the cone's axis s onto the principal axis of c.

Proof: The axis s of the given right cone $\mathcal{C}$ is supposed to be vertical. We can ignore horizontal cutting planes as they intersect along circles. Therefore, we may assume that the cutting plane π is inclined at the angle α with $0 < \alpha \leq \frac{\pi}{2}$. The generators of the cone $\mathcal{C}$ are inclined at β, $0 < \beta < \frac{\pi}{2}$. Figure 4.2 shows the front view with the image plane passing through the cone's axis s. The cutting plane π is in an edge-view.

Let M be the center of a sphere $\mathcal{S}$ which is inscribed in the cone $\mathcal{C}$ and contacts π in the point F. We call this sphere a *Dandelin sphere*, since the following proof dates back to G.P. DANDELIN[1]. $\mathcal{S}$ contacts the cone along a circle k in a plane which intersects the given cutting

---

[1]GERMINAL PIERRE DANDELIN, 1794–1847, professor of mathematics in Liège and lieutenant of the Belgian army.

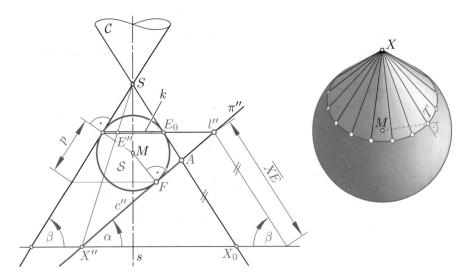

FIGURE 4.2. The planar sections of right cones are conics.

plane π along a line l. In Figure 4.2 the line appears as a point l'', and the contour of $\mathcal{S}$ is the incircle of the triangle formed by π'' and the contours of the cone $\mathcal{C}$.

In order to prove that the section $c = \mathcal{C} \cap \pi$ is a conic with focal point F and associated directrix l, we utilize what is depicted in Figure 4.2 on the right-hand side: Given a sphere $\mathcal{S} = (M; r)$ and an exterior point X, for all tangents drawn from X to $\mathcal{S}$, the points T of contact are at the same distance to X. The square of this distance $\overline{XT}^2 = \overline{XM}^2 - r^2$ is the *power of X w.r.t.* $\mathcal{S}$ (compare with Figure 2.31 on page 50).

Now we choose a point $X \in \pi \cap \mathcal{C}$. There are two particular tangents to the Dandelin sphere $\mathcal{S}$ available through X: One touches at F, the other along the cone's generator through X contacts at a point $E \in k$. In our front view we see the "true" distance $\overline{XE}$ after rotating the cone about its axis until X comes to X_0 and E to E_0 on the right contour line of $\mathcal{C}$. On the other hand, the front view shows already the "true" distance between X and the line l. Now we apply the sine law to the triangle which is shaded in Figure 4.2 and obtain that for all $X \in \pi \cap \mathcal{C}$

$$\overline{XF} : \overline{Xl} = \overline{XE} : \overline{Xl} = \overline{X_0 E_0} : \overline{X'' l''} = \sin\alpha : \sin\beta = \varepsilon = \text{const.}$$

The quotient $\varepsilon = \sin\alpha / \sin\beta$ is the numerical eccentricity of the conic $c = \pi \cap \mathcal{C}$. Hence, for $\alpha < \beta$ we obtain an ellipse, for $\alpha = \beta$ a parabola, and for $\alpha > \beta$ a hyperbola (Figure 4.3). ■

Figure 4.2 shows on the left contour line of $\mathcal{C}$ how the parameter $p = \varepsilon \cdot \overline{Fl}$ of the conic c can be found. Another construction of p, based on MEUSNIER's Theorem 3.3.1, is presented in Exercise 4.1.2 (Figure 4.9, left).

The proof given above can easily be adapted to the limiting case $\beta = \pi/2$. The limit of $\mathcal{C}$ is a right cylinder $\mathcal{L}$ (Figure 4.3, right). We leave the details of the proof to the reader and claim as follows:

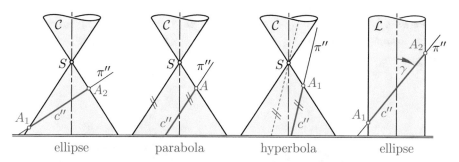

| ellipse | parabola | hyperbola | ellipse |

FIGURE 4.3. The type of the conic section $c = \pi \cap C$ depends on how the inclination of the cutting plane π relates to that of the cone's generators.

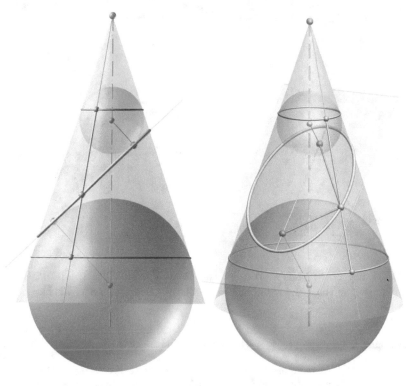

FIGURE 4.4. DANDELIN's proof in the case of an ellipse: The intersecting plane is less inclined than the tangent planes of the cone (Exercise 4.1.1).

Theorem 4.1.2 *Let $\mathcal{L}$ be a right cylinder and π a plane which is neither parallel nor orthogonal to the axis of the cylinder. Then, the curve $c = \mathcal{L} \cap \pi$ is an ellipse with its minor axis orthogonal to the cylinder's axis.*

In terms of the radius r of $\mathcal{L}$ and the angle γ between the cylinder's axis and π, the semimajor and semiminor axes, and the eccentricities of c are, respectively,

$$a = r/\sin\gamma, \quad b = r, \quad \varepsilon = \cos\gamma, \quad e = \varepsilon a = r\cot\gamma.$$

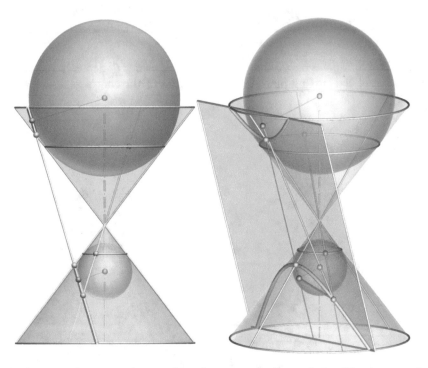

FIGURE 4.5. DANDELIN's proof in the case of a hyperbola: The intersecting plane is steeper than the tangent planes of the cone (Exercise 4.1.1).

● Exercise 4.1.1 Separate Dandelin proofs for ellipses and hyperbolas. In the proof above, we presented a 'Dandelin proof' of Theorem 4.1.1 which works simultaneously for all types of conics. Since ellipses and hyperbolas have two focal points, we can apply the fundamental equation $\overline{XF} = \overline{XE}$ (note Figure 4.2) two-times. In this way we directly obtain that the related curves of intersection satisfy the standard definition of ellipses or hyperbolas. Formulate these proofs on the basis of Figures 4.4 and 4.5.

Projection of a conic section parallel to the cone's axis

Theorem 4.1.3 *Let $\mathcal{C}$ be a right cone with a vertical axis s and with generators with the inclination angle β. If the conic $c = \mathcal{C} \cap \pi$ lies in a*

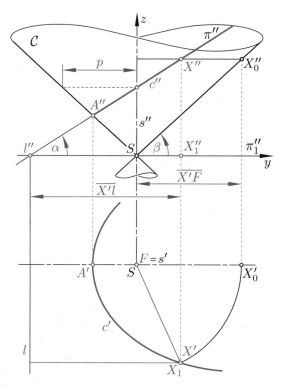

FIGURE 4.6. The top view c' of the conic section c is a conic with the focus S.

plane π which is inclined under the angle α, $\alpha < \frac{\pi}{2}$, then the top view c' of c is a conic with $F = s'$ as a focus and with the numerical eccentricity $\varepsilon = \tan\alpha / \tan\beta$. The horizontal plane π_1 through the apex S intersects π in a line l which appears in the top view as the directrix l associated with the focus F.

Proof: As shown in Figure 4.6, we choose the apex S of $\mathcal{C}$ as the origin of our coordinate frame and the cone's axis as z-axis. In the front view, the cutting plane π appears as a line π''.

For each point $X \in \mathcal{C} \cap \pi$ with the pedal point X_1 in the xy-plane, we obtain by virtue of Figure 4.6

$$\tan\beta = \frac{\overline{XX_1}}{\overline{X_1S}} = \frac{\overline{X''X_1''}}{\overline{X_1S}} = \frac{\overline{X_1l}\tan\alpha}{\overline{X_1S}}, \quad \text{hence} \quad \frac{\overline{X'S}}{\overline{X'l}} = \frac{\overline{X_1S}}{\overline{X_1l}} = \frac{\tan\alpha}{\tan\beta} = \text{const.}$$

Contrary to the eccentricity $\varepsilon = \sin\alpha / \sin\beta$ of the conic c itself, the top view c' has the numerical eccentricity $\tan\alpha / \tan\beta$. This shows that c and c' are of the same type. Here we excluded hyperbolas in vertical planes ($\alpha = \frac{\pi}{2}$); their top view is a line, of course. ∎

Figure 4.7 shows a right cone with a vertical axis and different planar sections (compare also with the figure on page 11). All intersecting planes

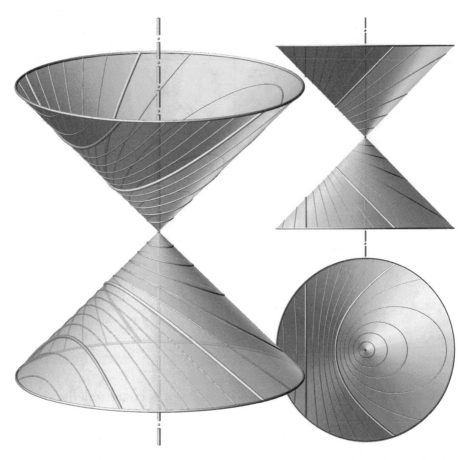

FIGURE 4.7. A cone of revolution, intersected by a pencil of planes. In the green area, all intersections are ellipses, in the yellow area, only hyperbolas occur. When the intersecting plane is parallel to a tangent plane of the cone, the intersection curve is a parabola (blue color).

pass through the same line in the horizontal plane through the apex. By virtue of Theorem 4.1.3, the top views of the conic sections share a focus and the associated directrix. Hence, the top views of these conics form exactly the same family which has been depicted in Figure 2.3 on page 15.

Remark 4.1.1 Theorem 4.1.3 reveals that, once a cone $\mathcal{C}$ with a vertical axis s is fixed, all conics c' with focus s' (including circles with center s') show up as top views of appropriate planar sections $\mathcal{C} \cap \pi$. Hence, $\mathcal{C}$ induces a one-to-one correspondence between *monofocal conics*, *i.e.*, conics sharing one focus, and planes π in $\mathbb{E}^3$ not passing through S.

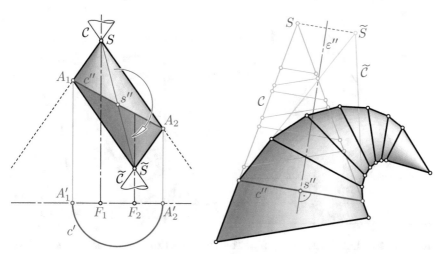

FIGURE 4.8. The planar section c is symmetric w.r.t. its secondary axis s. Therefore, the cone $\mathcal{C}$ still passes through c after a 180°-rotation about s or after reflection in the plane ε which contains s and is orthogonal to the plane spanned by c.

Ellipses and hyperbolas c' have also a second focus. Where does it come from?

The conic c on the cone $\mathcal{C}$ has a second axis s of symmetry which appears in the front view (Figure 4.8, left) as the midpoint between the vertices A_1, A_2 on the contour lines of $\mathcal{C}$.[2] A rotation about this axis s through 180° maps c onto itself while the cone $\mathcal{C}$ is transformed into a second cone $\widetilde{\mathcal{C}}$ with vertical axis which again passes through c. Therefore, the apex $\widetilde{S}$ of $\widetilde{\mathcal{C}}$ has a top view which coincides with the second focus.

Figure 4.8, right, shows how the symmetry of the conic c with respect to the plane ε through the minor axis s can be used to find another right cone $\widetilde{\mathcal{C}}$ passing through the same conic c. This procedure can be iterated.

● Exercise 4.1.2 Parameter p of planar sections of cones. Confirm for the conic section $c = \mathcal{C} \cap \pi$ the construction of the parameter p, as demonstrated in Figure 4.9, left (compare with Figure 4.2). By virtue of Theorem 2.1.5, this construction yields as a by-product also the

---

[2]It is not so obvious that planar sections of a right cone have a second axis of symmetry. One might conjecture that at the principal vertex which is closer to the cone's apex, the radius of curvature of the planar section is smaller than at the other vertex. It seems that even DÜRER had this expectation (see Figure 1.10, left). Figure 3.49 shows for a hyperbola the centers of curvature at the two vertices. Of course, they deliver the same radius of curvature, though the respective constructions look quite different.

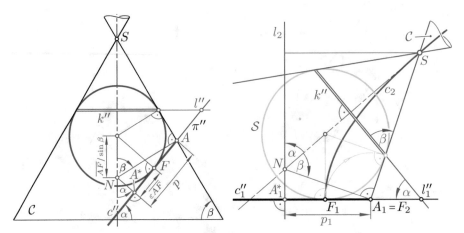

FIGURE 4.9. Left: parameter p of the conic section $c = \mathcal{C} \cap \pi$. Right: universal proof of Theorem 4.2.1 (Exercise 4.1.3).

center of curvature A^* for the vertex A of c (Figure 3.49). The latter can also be concluded by virtue of MEUSNIER's Theorem 3.3.1.

Hint: Note that $p = (1 + \varepsilon)\overline{AF}$ and $\varepsilon = \sin\alpha/\sin\beta$.

● Exercise 4.1.3 Apollonian property of planar sections of cones. Let a conic c_1 be given with vertex A_1, focus F_1, and associated directrix l_1 (Figure 4.9, right). Prove Theorem 4.2.1 simultaneously for all types of conics c_1 by verifying that the apices S of right cones passing through c_1 satisfy the Apollonian definition $\overline{SA_1} = \varepsilon_2 \cdot \overline{Sl_2}$, where l_2 is the axis of the circle which osculates c_1 at A_1. Hence, all points S lie on the focal conic c_2 of c_1. The numerical eccentricities ε_1 and ε_2 of c_1 and c_2 are reciprocal. As parameter p_2 of c_2 we obtain $p_2 = p_1/\varepsilon_1 = \overline{F_1 l_1}$.

Hint: We choose any sphere S which contacts the plane of c_1 at the given focus F_1. Then, we determine a cone $\mathcal{C}$ which contacts S along a circle k in a plane passing through the given directrix l_1, such that one contour line of $\mathcal{C}$ contains the vertex A_1. By virtue of Theorem 4.1.1, in terms of the inclination angles α and β of the cone's generators and the cutting plane (relative to the plane of k), the excentricity of c_1 is $\varepsilon_1 = \sin\alpha/\sin\beta$. We apply the construction presented in Exercise 4.1.2 (Figure 4.9, left) and conclude, using the notation of Figure 4.9, right, that[3]

$$\overline{SN} = \frac{\overline{SA_1}}{\sin\beta}, \quad \overline{Sl_2} = \overline{SN}\sin\alpha, \text{ hence } \overline{SA_1} = \frac{1}{\varepsilon_1}\overline{Sl_2}.$$

---

[3]By virtue of Theorem 3.3.1, the point N is the center of the Meusnier sphere corresponding to the horizontal tangent at A_1.

4.2 Pairs of focal conics, Dupin cyclides

We are going to report about a symmetric pairing of conics in $\mathbb{E}^3$ which has some remarkable geometric properties. We will meet these pairs again in volume 2 in connection with confocal quadrics in $\mathbb{E}^3$.

Definition 4.2.1 Given a conic c_1 in $\mathbb{E}^3$, the conic c_2 is called *focal conic* of c_1 if the two conics lie in orthogonal planes sharing the principal axis, and on this common axis each vertex of one conic is a focus of the other.

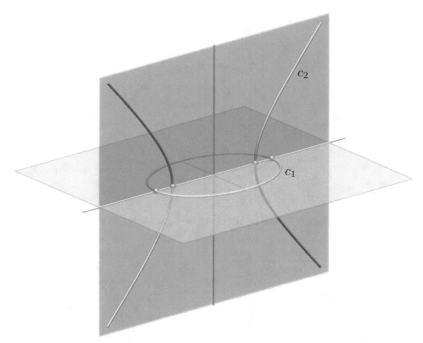

FIGURE 4.10. A pair of focal conics.

The focal conic of an ellipse with semimajor axis a and semiminor axis b is a hyperbola with the semimajor axis $\sqrt{a^2 - b^2}$ and the semiminor axis $\sqrt{a^2 - (a^2 - b^2)} = b$ (see Figure 4.10), and *vice versa*. The focal conic of a parabola is a parabola with the same parameter; however, the two parabolas open to different sides. We do not speak of a pair of focal conics in the limiting case consisting of a circle c_1 and its axis c_2, though some of the following theorems are still valid in this case.

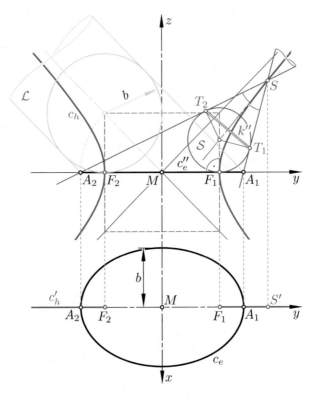

FIGURE 4.11. Which cones of revolution pass through a given ellipse c_e ?

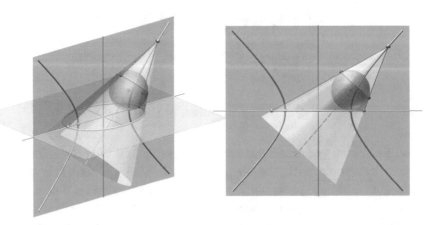

FIGURE 4.12. The cones of revolution which pass through a given ellipse, have their apices on the focal hyperbola.

Focal conics and cones of revolution

Theorem 4.2.1 *For any given conic c_1 in $\mathbb{E}^3$, the locus of apices S of right cones, which pass through c_1, is the focal conic c_2 (up to the focal points of) c_1. The axis of the cone through c_1 with apex $S \in c_2$ is tangent to c_2 at S.*

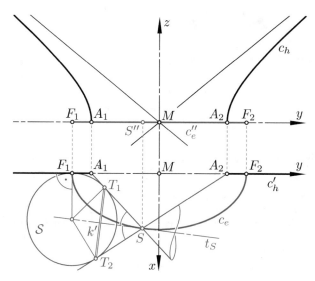

FIGURE 4.13. Which cones of revolution pass through a given hyperbola c_h?

Proof: At this time we discuss the three types of conics separately, since the universal proof of Theorem 4.2.1 is a little bit tricky and not so instructive (see Exercise 4.1.3 and the related Figure 4.9, right).

Case 1: c_1 is an ellipse c_e in the horizontal xy-plane with axes a, b and the y-axis as principal axis (see top view and front view in Figure 4.11). By virtue of Theorem 4.1.1, it is necessary that any right cone passing through c_1 has its axis in the vertical plane through the principal axis, i.e., in the yz-plane. Furthermore, there must be a sphere S inscribed in this cone and contacting the xy-plane at one focus F_1 of c_e.

So, let us choose such a sphere S with radius r and inspect the front view in Figure 4.11. The second tangents drawn from the vertices A_1 and A_2 of c_e to the contour of S intersect at a point S — except the case where these tangents are parallel, *i.e.*, the sphere S has the radius $r = b$. The lines $[S, A_1]$ and $[S, A_2]$ are the contour of a right cone which contacts S along a circle k. In fact, this cone already passes through the ellipse c_e, since by virtue of Theorem 4.1.1 the curve of intersection with the xy-plane has the principal vertices A_1, A_2 and the focus F_1. Thus, it is identical with c_e.

Let T_i for $i = 1, 2$ denote the point of contact between $[S, A_i]$ and S. Then, we learn from the front view a series of equal distances:

$$\overline{A_1 T_1} = \overline{A_1 F_1}, \quad \overline{A_2 T_2} = \overline{A_2 F_1}, \quad \overline{ST_1} = \overline{ST_2}.$$

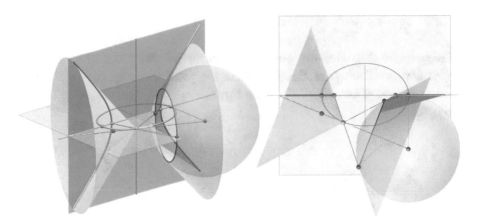

FIGURE 4.14. All cones of revolution passing through a given hyperbola have their apices on the focal ellipse.

In the depicted case $r < b$ we conclude that

$$\overline{SA_2} - \overline{SA_1} = \overline{ST_2} + \overline{T_2A_2} - \overline{ST_1} - \overline{T_1A_1} = \overline{A_2F_1} - \overline{A_1F_1} = \overline{F_1F_2} = \text{const.}$$

In the case $r > b$ the point S lies closer to the second focus F_2. Therefore, we use the second sphere: It is inscribed in the same cone but touches the xy-plane at the focus F_2, and it has a radius smaller than b. Hence, we obtain a similar result, but under $\overline{SA_2} < \overline{SA_1}$. So, in all cases the requested locus of apices satisfies the standard definition of a hyperbola c_h in the yz-plane with the foci A_1 and A_2 (see also Figure 4.12).

The asymptotes of c_h are axes of right cylinders $\mathcal{L}$ passing through c_e (Figure 4.11).

Case 2: Let the given conic c_1 be a hyperbola c_h in the yz-plane with the vertices A_1 and A_2. Then, the apices of the right cones passing through c_h lie in the xy-plane. Figure 4.13 shows from both the top and the front view only one half. The rest can be obtained by reflection in the y-axis.

We follow the same strategy like in case 1: We specify a sphere $\mathcal{S}$ which is in contact with the yz-plane at the focus F_1 of c_h and inspect the top view. The tangents drawn from the vertices A_1 and A_2 to the contour of $\mathcal{S}$ (contact points T_1 and T_2) intersect at a possible apex S which satisfies obviously the following conditions:

$$\overline{A_1S} + \overline{A_2S} = \overline{A_1T_1} + \overline{ST_1} + \overline{A_2T_2} - \overline{ST_2} = \overline{A_1F_1} + \overline{A_2F_1} = \overline{F_1F_2} = \text{const.}$$

This characterizes the focal ellipse c_e of the given hyperbola c_h (see also Figure 4.14).

Case 3: Figure 4.15 shows that all right cones $\mathcal{C}$ passing through a given parabola c_1 have their apex S on the focal parabola c_2, since $\overline{SA} = \overline{ST_1} + \overline{T_1A} = \overline{ST_2} + \overline{T_2l} = \overline{Sl}$.   ■

We conclude this subsection with some additional theorems concerning focal conics.

Theorem 4.2.2 *Given a pair of focal conics (c_1, c_2), each point $P_i \in c_i$ for $i = 1, 2$ defines a point Q_i on the common principal axis such that*

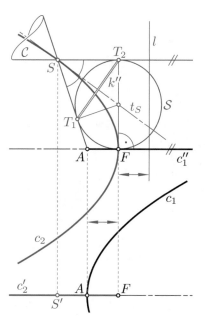

FIGURE 4.15. Which cones $\mathcal{C}$ of revolution pass through a given parabola c_1?

$\overline{P_1 P_2} = \overline{Q_1 Q_2}$ *holds. The point* Q_i *is a vertex of the second conic* d_i *which is confocal with* c_i *and passes through* P_i *(Figures 4.16 and 4.17).*

Remark 4.2.1 In volume 2, we shall learn that this theorem is a special case of IVORY's theorem in three dimensions. It will be useful for proving thread constructions of quadrics.

Proof: We start with two parabolas, one in the xy-plane and the other in the yz-plane. We adjust the standard parametrization of parabolas from (2.17) and set (Figure 4.16)

$$P_1 = (2pu, \ 2pu^2, \ 0), \quad P_2 = \left(0, \ \frac{p}{2} - 2pv^2, \ 2pv\right), \quad (u, v) \in \mathbb{R}^2.$$

This yields

$$\begin{aligned}\overline{P_1 P_2}^2 &= 4p^2u^2 + \left(2pu^2 - \tfrac{p}{2} + 2pv^2\right)^2 + 4p^2v^2 \\ &= 2p^2u^2 + 2p^2v^2 + 4p^2u^4 + 4p^2v^4 + 8p^2u^2v^2 + \tfrac{p^2}{4} = p^2\left[2(u^2 + v^2) + \tfrac{1}{2}\right]^2\end{aligned}$$

Now, we verify that

$$\overline{P_1 P_2} = |p|\left[2(u^2 + v^2) + \frac{1}{2}\right] = |y_1 - y_2 + p| = \overline{Q_1 Q_2}, \tag{4.1}$$

where $y_1 = 2pu^2$ and $y_2 = \frac{p}{2} - 2pv^2$ are the respective y-coordinates of P_1 and P_2, $Q_1 = (0, \ y_1 + \frac{p}{2}, \ 0)$, and $Q_2 = (0, \ y_2 - \frac{p}{2}, \ 0)$. The second parabola through P_1 which is confocal with c_1 has the vertex Q_1. This follows for $v = 0$ from (4.1) since the directrix of this parabola has the y-coordinate $y_1 + 2pu^2 + \frac{p}{2}$. Similarily we can confirm that Q_2 is the vertex of the confocal parabola through C_2.

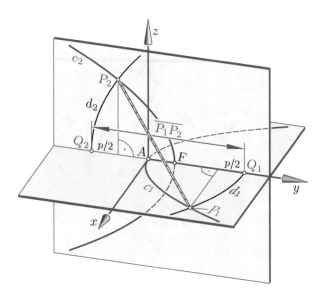

FIGURE 4.16. $\overline{P_1P_2} = \overline{Q_1Q_2}$ for points P_1 and P_2 on a pair of focal parabolas.

When c_1 is assumed as an ellipse and c_2 as its focal hyperbola, then we refer to the coordinate frame of Figure 4.11 with the y–axis as the common principal axis and the origin at the common center M. We recall the parametrization of ellipses given in (2.12) and of hyperbolas given in (2.14). Thus we can set up the points $P_1 \in c_1$ and $P_2 \in c_2$ as

$$P_1 = (b\sin u,\ a\cos u,\ 0), \quad P_2 = (0,\ \pm\sqrt{a^2 - b^2}\cosh v,\ b\sinh v), \quad u \in [0, 2\pi), \ v \in \mathbb{R},$$

where the upper sign at the y-coordinate of P_2 defines the right branch and the lower sign the left branch of the hyperbola c_2. Hence, we compute

$$
\begin{aligned}
\overline{P_1P_2}^2 &= b^2\sin^2 u + (a^2 - b^2)\cosh^2 v \mp 2a\sqrt{a^2 - b^2}\cos u\cosh v + a^2\cos^2 u + b^2\sinh^2 v \\
&= (a^2 - b^2)\cos^2 u \mp 2a\sqrt{a^2 - b^2}\cos u\cosh v + a^2\cosh^2 v.
\end{aligned}
$$

With $e = \sqrt{a^2 - b^2}$ we obtain

$$\overline{P_1P_2} = |e\cos u \mp a\cosh v| = \left| \frac{e}{a}y_1 - \frac{a}{e}y_2 \right|, \tag{4.2}$$

where y_i denotes the y-coordinate of the point P_i for $i = 1, 2$. The point $Q_1 = (0, \frac{e}{a}y_1, 0)$ is a vertex of the hyperbola which is confocal with c_e and passes through P_1. This can be verified by direct computation. But we can also recall the particular case of IVORY's theorem which is depicted in Figure 2.24: The scaling $(x, y, 0) \mapsto (0, \frac{e}{a}y, 0)$ maps the ellipse c_e onto the segment bounded by its focal points while each point remains on the same hyperbola of the confocal family. In the same way we can prove that Q_2 is a vertex of the ellipse in the yz-plane which passes through P_2 and is confocal with the hyperbola c_2.   ∎

Figure 4.16 shows the equal distances in the case of parabolas. The case of an ellipse and its focal hyperbola is displayed in Figure 4.17, however with a slightly modified notation. Here, Theorem 4.2.2 gives rise to a generalization of the standard definitions of ellipses and hyperbolas, which might explain the choice of the name 'focal' hyperbola or ellipse.

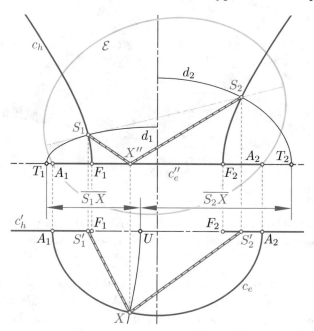

FIGURE 4.17. Generalized gardener's construction: For points S_1, S_2 fixed on different branches of the focal hyperbola c_h we have $\overline{S_1X} + \overline{S_2X} = \text{const.}$ for all $X \in c_e$.

Corollary 4.2.1 *Given an ellipse c_e, let S_1 and S_2 be points chosen on different branches of the focal hyperbola c_h of c_e. Then, for all points $X \in c_e$, the sum $\overline{S_1X} + \overline{S_2X}$ is constant. Conversely, given any two points $R_1, R_2 \in c_e$, for all points $Y \in c_h$ the difference $|\overline{R_1Y} - \overline{R_2Y}|$ is constant.*

Proof: With Theorem 4.2.2, we recognize in Figure 4.17 equal distances $\overline{S_iX} = \overline{T_iU}$ for $i = 1, 2$. For S_1 on the right branch and S_2 on the left branch of the hyperbola c_h, we obtain by (4.2)

$$\overline{S_1X} = a\cosh v_1 - e\cos u, \quad \overline{S_2X} = a\cosh v_2 + e\cos u.$$

The sum $\overline{S_1X} + \overline{S_2X} = \overline{T_1T_2}$ is independent of the choice of $X \in c_e$.

On the other hand, for the two points $R_1, R_2 \in c_e$ with corresponding parameter values u_1 and u_2 we get for all $Y \in c_h$ $|\overline{R_1Y} - \overline{R_2Y}| = |e(\cos u_1 - \cos u_2)|$. By the same token, such a constant difference can also be identified at a pair of focal parabolas. ∎

The constant sum $\overline{S_1X} + \overline{S_2X} = a(\cosh v_1 + \cosh v_2)$ gives rise to a generalized gardener's construction (Figure 4.17). If we fix the endpoints of a

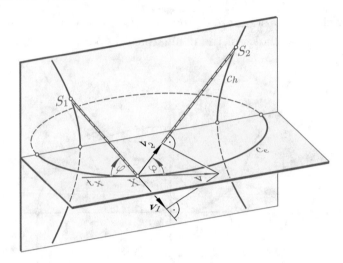

FIGURE 4.18. The constant length of the string from S_1 via X to S_2 is also a consequence of congruent angles between S_1X or S_2X and the tangent t_X.

string of length $a(\cosh v_1 + \cosh v_2)$ at S_1 and S_2 on c_h and pull the string taut by a point X in the xy-plane, then X will trace the ellipse. Or, in other words: Let us imagine the two curves as wires and choose a rubber band instead of the string. Now we fix one endpoint at $S_1 \in c_h$, pull the band around the ellipse c_e and span it at the point S_2 on the other branch of c_h. Surprisingly, the point $X \in c_e$ will not tend to a particular point on c_e where the total length attends its minimum, since for all X we get the same length. This means, the point X can move on the ellipse without changing the tension along the band (Figure 4.18). This phenomenon is a consequence of the congruent angles between the tangent t_X to c_e at R and the segments XS_1 and XS_2. Both are lying on a right cone with apex X and axis t_X. Similar to the proof of GRAVES's theorem (Theorem 2.2.5, see Figure 2.29) we can conclude that the sum of distances $\overline{S_1X} + \overline{S_2R}$ must be constant, since the velocity vector $\mathbf{v}$ of X has equal components $\mathbf{v}_1$ and $\mathbf{v}_2$ in direction of XS_1 and XS_2, but with different sign.

What happens, when the point X which pulls the string between S_1 and S_2 taut is not kept within the xy-plane? Then, in each plane through the line $[S_1, S_2]$ the point X traces an ellipse with the foci S_1 and S_2. Rotation about S_1S_2 generates an ellipsoid $\mathcal{E}$ of revolution which is depicted in the front view of Figure 4.17. This ellipsoid $\mathcal{E}$ intersects the xy-plane exactly along the ellipse c_e.

Corollary 4.2.2 *Let (c_1, c_2) be a pair of focal conics. Then, for all common secants $[X_1, X_2]$ with $X_i \in c_i$, $i = 1, 2$, the connecting planes with the tangents t_i at X_i to c_i are orthogonal.*

Proof: This is trivial when X_2 is a focus of c_1 since t_2 is orthogonal to the plane of c_1. Otherwise, by virtue of Theorem 4.2.1, there is a right cone $\mathcal{C}$ passing through c_1, with apex X_2 and axis t_2. The plane connecting $[X_1, X_2]$ with t_2 is a diameter plane of $\mathcal{C}$ while the connection with t_1 is tangent to $\mathcal{C}$ along the generator $[X_1, X_2]$. ∎

Remark 4.2.2 It can be proved (see vol. 2) that orthogonal projections map a pair (c_1, c_2) of focal conics onto *confocal* conics (see, e.g., Figures 4.16 and 4.18), when edge views of one of the conics' planes are excluded. By virtue of Corollary 4.2.2, we can already confirm that the views of c_1 and c_2 under an orthogonal projection must intersect orthogonally. This follows, since each point X of intersection is the image of any secant X_1X_2, and the tangents at X are edge views of the orthogonal planes mentioned in the corollary above.

Theorem 4.2.3 *Let the x-axis of a Cartesian coordinate frame be an axis of symmetry of a conic c. If the normal line to c at the point $P = (x, y)$ intersects the x-axis at $P_n = (x_n, 0)$, then x_n is a linear function of x. For parabolas the difference $x_n - x$ is constant.*
The point P_n is the center of a circle k which touches c twice: at P and at its reflection in the x-axis. The squared radius r^2 of k is a quadratic function of x.

Proof: By virtue of Theorem 2.1.2, conics are irreducible curves of degree two. When the x-axis is an axis of symmetry of the conic c, then in the general equation (2.11) the unknown y shows up only with an even exponent. Hence, we may set up

$$c: \quad F(x, y) = a_{11}x^2 + a_{22}y^2 + 2a_1 x + a = 0,$$

where $a_{22} \neq 0$, since otherwise c would be reducible. The gradient

$$\operatorname{grad}(F) = \left(\frac{\partial F}{\partial x}, \frac{\partial F}{\partial y} \right) = (F_x, F_y) = (2a_{11}x + a_1, \ 2a_{22}y)$$

at the point $P = (x, y) \in c$ is orthogonal to c, since for each parametrization $(x(t), y(t))$ of c the substitution $F(x(t), y(t))$ is the zero-function in t, and therefore, $F_x \dot{x} + F_y \dot{y} = 0$. We set up the point P_n as

$$\begin{pmatrix} x \\ y \end{pmatrix} + \lambda \begin{pmatrix} 2a_{11}x + a_1 \\ 2a_{22}y \end{pmatrix} = \begin{pmatrix} x_n \\ 0 \end{pmatrix}.$$

This yields for points P with $y \neq 0$

$$\lambda = \frac{-1}{2a_{22}} \quad \text{and} \quad x_n = x - \frac{a_{11}}{a_{22}}x - \frac{a_1}{2a_{22}},$$

which is the stated linear function. The squared distance $\overline{PP_n}^2 = (x_n - x)^2 + y^2$ is a quadratic function of x.

The signed distance $x_n - x$ is called the *subnormal* of point P. It is constant if, and only if, $a_{11} = 0$, hence, by virtue of Theorem 2.1.2, the conic c is a parabola.

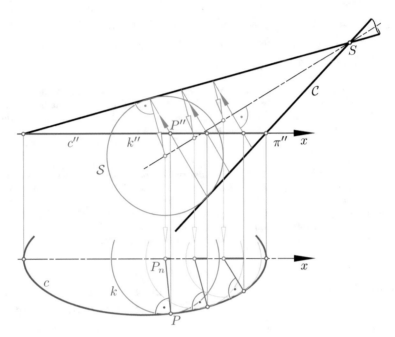

FIGURE 4.19. For the conic $c = \mathcal{C} \cap \pi$ the subnormal $x_n - x$ of $P = (x, y)$ is a linear function of x.

If the given axis of symmetry is the principal axis, then Theorem 4.2.3 can also be concluded, when c is given as a planar section of a right cone $\mathcal{C}$ (see Figure 4.19).

We choose a sphere $\mathcal{S}$ which is inscribed in $\mathcal{C}$ and intersects the plane π along a circle k. This circle with center P_n has a double contact with c. The points P of contact belong to the circle of contact between $\mathcal{S}$ and $\mathcal{C}$. Obviously, the point P_n is the point, where the normal line at P intersects the axis of symmetry. In Figure 4.19, the front view shows the x-coordinates of P and of P_n. Suppose, the inscribed sphere $\mathcal{S}$ varies. Then, the point P'' is mapped onto P_n by a sequence of three parallel projections from lines to lines. This causes the stated linearity of the function $x_n(x)$. ∎

● **Exercise 4.2.1** A flexible bipartite framework. For any pair (c_1, c_2) of focal conics, let points $X_1, \ldots, X_n \in c_1$ and $Y_1, \ldots, Y_m$ be the knots of a bipartite bar-and-joint framework with $m, n \geq 2$ such that no knot lies on the common principal axis of the conics. Why can all knots move along their conics such that all mutual distances $\overline{X_i Y_j}$ remain unchanged?
Hint: By virtue of (4.1), in the case of two parabolas the distance $\overline{X_i Y_j}$ depends only on the y-coordinates of the points X_i and Y_j. Choose a sufficiently small $k \in \mathbb{R}$ and add k to the y-coordinates of all X_i while k is subtracted from the y-coordinates of all knots Y_j. Due to (4.2), a similar method works for ellipses and hyperbolas.[4]

---

[4]See W. WUNDERLICH: Fokalkurvenpaare in orthogonalen Ebenen und bewegliche Stabwerke. Sitzungsber., Abt. II, österr. Akad. Wiss., Math.-Naturw. Kl. 185 (1976), 275–290.

Dupin cyclides

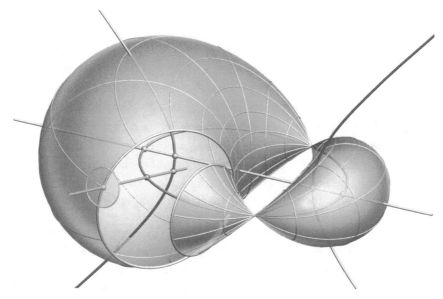

FIGURE 4.20. Dupin cyclides $\mathcal{D}$ are canal surfaces (cf. Remark 4.2.3) in two ways. The spine curves are focal conics. The circles of contact with the enveloped spheres form an orthogonal net.

We introduced non-parabolic Dupin cyclides on page 37 in the following way. Given two circles g_1 and g_2 with different centers, the envelope of a one-parameter family of spheres, whose equators h contact g_1 and g_2, is called a Dupin cyclide, provided, the type of contacts between the equators and the given circles is either everywhere the same or changes simultaneously.

We can reformulate the restrictions on the kinds of contact by using signed radii for all involved circles. In accordance with (2.22), the equation $\overline{M_1 M_2} = |r_1 - r_2|$ characterizes the contact between the circles $(M_1; r_1)$ and $(M_2; r_2)$. In case of an interior contact the signs of r_1 and r_2 are equal, while in the case of an exterior contact the signs differ. This convention which will also be used for spheres in 3-space allows us to define Dupin cyclides more precisely.

Definition 4.2.2 Let two circles $g_i = (F_i; r_i)$, $i = 1, 2$, be given with signed radii r_1 and r_2, where $F_1 \neq F_2$ and $r_1 \neq r_2$. Then, the envelope of the family of spheres, whose equators $h = (X; \varrho)$ satisfy $\overline{XF_i} = |r_i - \varrho|$ for $i = 1, 2$, is a (non-parabolic) *Dupin cyclide* $\mathcal{D}$, provided the set of equators is not empty.

All equators h are in contact with g_1 and g_2. For variable $\varrho \in \mathbb{R}$, the kind of contact can only change at $\varrho = 0$, and then this happens for g_1 and g_2 simultaneously. This is exactly what we wanted to achieve. A simultaneous change of signs for r_1 and r_2 has no effect on the cyclide $\mathcal{D}$.

In order to obtain also *parabolic cyclides*, we have to replace in Definition 4.2.2 the circle g_2 by a line l such that $\overline{Xl}$ is the signed distance from point X to l, i.e., positive in one half-plane of l and negative in the other. Then, the circles $h = (X; \varrho)$ satisfying $\overline{XF_1} = |r_1 - \varrho|$ and $\varrho = \overline{Xl}$ are the equators of a parabolic Dupin cyclide (see Figure 2.20).

Theorem 4.2.4 *Each Dupin cyclide $\mathcal{D}$ is the envelope of two one-parameter families of spheres. The centers of the spheres form a pair of focal conics (c_1, c_2). All normal lines of $\mathcal{D}$ are common secants of (c_1, c_2) (Figure 4.20). Each sphere of one family contacts all spheres of the other family.*

For each family, the carrier planes of the circles of contact have a line in common. Along each of these circles there is a right cone tangent to the cyclide, and the apices of these cones are located on another line. When one family is replaced by the other, the two lines change their roles. The circles of contact constitute an orthogonal net on $\mathcal{D}$.

Remark 4.2.3 The envelope of a one-parametric family of spheres is called a *canal surface*. The centers of the spheres form the associated *spine curve*. Dupin cyclides together with the torus are the only surfaces which are canal surfaces in two different ways. Another approach to Dupin cyclides via pentaspheric coordinates can be found in [42, p. 53ff].

Proof: We proceed in six steps and treat only the non-parabolic case. A similar proof for parabolic Dupin cyclides is left to the readers as Exercise 4.2.2.

1) Let two circles $g_1 = (F_1; r_1)$ and $g_2 = (F_2; r_2)$ with $F_1 \neq F_2$ and $r_1 \neq r_2$ be given in a horizontal plane. The two radii are signed, and these signs remain fixed. We show first that the circles $h = (X; \varrho)$ satisfying $\overline{XF_i} = |r_i - \varrho|$ have their centers on a conic c_1 with focal points F_1 and F_2 (Figure 4.21).

To this end, we use the following transformation: We subtract from all involved circles the signed radius r_1 of g_1, while the centers remain fixed. This preserves each mutual contact, but transforms the circles h into circles which pass through F_1 and contact the circle $(F_2; r_2 - r_1)$. The rest follows from Corollary 2.2.2.

2) The sphere S with equator $h = (X; \varrho)$ contacts the enveloped surface along a circle k. This circle is defined as the limit of the circle of intersection between S and a neighboring sphere,

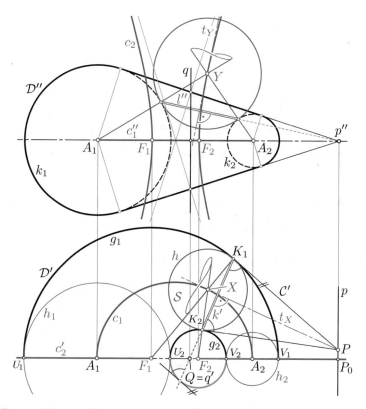

FIGURE 4.21. Top and front view of the Dupin cyclide $\mathcal{D}$ which envelopes the spheres with equators h being tangent to two given circles g_1 and g_2.

when the latter tends towards $\mathcal{S}$ within the family of spheres. The plane of k is vertical. Furthermore, the axis t_X of k is tangent to the conic c_1 at X. We know already two points $K_1, K_2 \in k$, the respective points of contact of the equator h and the given circles g_1 and g_2.

The lines connecting the center X of the sphere $\mathcal{S}$ with points of the circle $k \subset \mathcal{S}$ are orthogonal to $\mathcal{S}$ and to the envelope $\mathcal{D}$. These lines form a right cone with apex X and axis t_X. Since the centers F_1 and F_2 of the given circles are focal points of the conic c_1, the comparison with Figures 4.11, 4.13 and 4.15 reveals that this cone passes through the focal conic c_2 of c_1. Hence, all normals of the Dupin cyclide $\mathcal{D}$ are common secants of the pair of focal conics (c_1, c_2).

3) There is a right cone $\mathcal{C}$ which contacts $\mathcal{S}$ and $\mathcal{D}$ along k. Let P be its apex. The power $\overline{PK_1}^2$ of P w.r.t. the equator $h = (X; \varrho)$ is the same as with respect to g_1 and g_2. Hence, for all $X \in c_1$, the apex P is a point of the horizontal radical axis p of the given circles g_1 and g_2.

4) The common diameter $[F_1, F_2]$ of the given circles is the principal axis of c_1. At the principal vertices A_1, A_2 of c_1, the radii of the corresponding equators h_1, h_2 attain extremal values ϱ_1, ϱ_2, respectively. $[F_1, F_2]$ is also an axis of symmetry of the Dupin cyclide $\mathcal{D}$.

The tangents to h at K_1 and K_2 are symmetric with respect to the axis t_X of the cone $\mathcal{C}$ (Figure 4.21). The line $[K_1, K_2]$ intersects g_2 at a second point, and there the tangent to

g_2 is parallel to the tangent to g_1 at K_1. Therefore, the point Q of intersection between the common diameter $[F_1, F_2]$ and the plane of k is the center of a similarity which maps g_1 onto g_2. Thus, the point Q is independent of the choice of $X \in c_1$. For all circles k, the spanned planes share the vertical line q through Q.

Let U_i and V_i for $i = 1, 2$ denote the points of intersection between $[F_1, F_2]$ and g_i. Then, the point Q satisfies the equation $\overline{QU_1} : \overline{QV_1} = \overline{QV_2} : \overline{QU_2}$. This implies that $\overline{QU_1} \cdot \overline{QU_2} = \overline{QV_1} \cdot \overline{QV_2}$. Hence, the power of Q with respect to h_1, i.e., $\overline{QA_1}^2 - \varrho_1^2 = \overline{QU_1} \cdot \overline{QV_2}$, is the same as that with respect to h_2. The vertical line q through Q is the radical axis of the two circles k_1, k_2 of contact with the centers A_1 and A_2 at the principal vertices of c_1. In the same way it follows that the point P_0 of intersection between the common diameter and the line p is the center of a similarity with $h_1 \mapsto h_2$ and $k_1 \mapsto k_2$.

5) We choose the principal axis $[F_1, F_2]$ of c_1 as the y-axis and the secondary axis as x-axis of a Cartesian coordinate frame (like in Figure 4.11). By virtue of Theorem 4.2.2, the distance between any two points $X \in c_1$ and $Y \in c_2$ depends only on the respective y-coordinates y_1 of X and y_2 of Y. If c_1 is an ellipse with semimajor axis a and numerical eccentricity e, then according to (4.2) $\overline{XY} = \left| \frac{e}{a} y_1 - \frac{a}{e} y_2 \right|$.

Let all points $X \in c_1$ and $Y \in c_2$ be centers of spheres with respective signed radii

$$r_X = \frac{e}{a} y_1 + r_0, \quad r_Y = \frac{a}{e} y_2 + r_0 \tag{4.3}$$

with any real constant r_0. This defines two families of spheres, one centered on c_1 and the other centered on c_2. Because of $\overline{XY} = |\varrho_1 - \varrho_2|$ any two spheres $(X; r_X)$ and $(Y; r_Y)$ taken from different families are in contact. The point of contact lies on the connecting line $[X, Y]$.

In particular, all spheres centered on c_1 contact the spheres with centers F_1 and F_2, i.e., where $y_2 = \pm e$. Their equators in the xy-plane have the radii $r_Y = r_0 \pm a$.

Conversely, if these two equators with centers F_1, F_2 and respective signed radii $r_1 = r_0 + a$ and $r_2 = r_0 - a$ are given, we obtain $r_0 = (r_1 + r_2)/2$ and $a = |r_1 - r_2|/2$. According to Definition 4.2.2, the envelope of both families of spheres is the same Dupin cyclide $\mathcal{D}$ which has been studied above. By the same token, when the constant r_0 changes, $\mathcal{D}$ is replaced by an offset surface.

6) The normal lines passing through a fixed point $Y \in c_2$ form a cone of revolution. Its curve of intersection with the sphere $(Y; r_Y)$ is a circle l which is at the same time the circle of contact between this sphere and $\mathcal{D}$. For $Y \to \infty$ the sphere becomes a plane. There are two such planes passing through p. The top view in Figure 4.21 shows them in an edge view. All spheres $(X; r_X)$ of the first family contact these two planes which reveals again that the radius r_X is a linear function of the y-coordinate y_1 of X.

Let $T \in \mathcal{D}$ on the normal line $[X, Y]$ be the point of contact between the two spheres $(X; r_X)$ and $(Y; r_Y)$. There are two circles of $\mathcal{D}$ passing through T: The circle k of contact between $(X; r_X)$ and $\mathcal{D}$ lies in a plane through the vertical line q; the circle l of contact between $(Y; r_Y)$ and $\mathcal{D}$ lies in a plane through the line p (Figure 4.22).

The tangents to k and l at T span the tangent plane of $\mathcal{D}$ which is orthogonal to $[X, Y]$. The tangent to l must intersect p and passes, therefore, through the apex P of the right cone $\mathcal{C}$, which contacts $\mathcal{D}$ along k (compare with Figure 4.21). The tangent to l at T is, therefore, orthogonal to k. The circles of contact of both families of spheres form an orthogonal net on $\mathcal{D}$.[5]   ∎

---

[5] In fact, the circles of contact between the spheres of both families and the Dupin cyclide $\mathcal{D}$ are the curvature lines of $\mathcal{D}$.

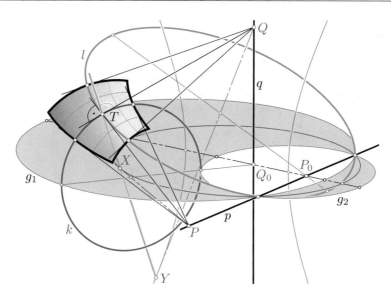

FIGURE 4.22. The circles of contact between the Dupin cyclide and its two families of spheres constitute an orthogonal net.

Remark 4.2.4 According to the proof given above, the point Q has the same power with respect to the two circles h_1 and h_2 (Figure 4.21). Therefore, Q is the center of an inversion (see Section 7.5) which maps h_1 onto itself as well as h_2 . This inversion exchanges g_1 with g_2, while all circles h remain fixed. Consequently, also in space there is an inversion with center Q mapping $\mathcal{D}$ onto itself since all spheres of the first family remain fixed. Similarly, the point P_0 is the center of an inversion which maps each sphere of the second family onto itself.

By the way, Dupin cyclides can also be obtained by applying a 3D inversion onto a torus, a right cone, or a right cylinder.

● Exercise 4.2.2 Parabolic Dupin cyclides. Modify the proof of Theorem 4.2.4 as provided above, for the case of a parabolic Dupin cyclide.

● Exercise 4.2.3 Equation of Dupin cyclides. Prove that a Dupin cyclide $\mathcal{D}$ with an ellipse c_1 and its focal hyperbola c_2 as spine curves satisfies in the coordinate frame of Figure 4.11 the equation
$$\left(x^2 + y^2 + z^2 + b^2 - r_0^2\right)^2 - 4b^2x^2 - 4(ay + er_0)^2 = 0$$
of degree four with $e = \sqrt{a^2 - b^2}$ and any constant $r_0 \in \mathbb{R}$, when, by virtue of (4.3), the enveloping spheres have the radii $r_1 = \frac{e}{a}\,y_1 + r_0$ and $r_2 = \frac{a}{e}\,y_2 - r_0$.

Hint: The spheres centered on c_1 satisfy the equation
$$(x - b\sin u)^2 + (y - a\cos u)^2 + z^2 = (e\cos u + r_0)^2,$$
with $u \in [0, 2\pi[$ as parameter. Because of $(a^2 - e^2)\cos^2 u + b^2 \sin^2 u = b^2$, we can rewrite it as
$$F(x, y, z, u) := x^2 + y^2 + z^2 - 2bx\sin u - 2\cos u(ay + er_0) + b^2 - r_0^2 = 0.$$
The circles of contact with the enveloped cyclide $\mathcal{D}$ satisfy, beside $F(x, y, z, u) = 0$, also $\frac{\partial}{\partial u} F(x, y, z, u) = 0$ which leads to $2bx\cos u - 2\sin u(ay + er_0) = 0$. The stated equation of $\mathcal{D}$ is the result after the elimination of u from $F(x, y, z, u) = 0$ and its partial derivative.

We obtain the same equation when starting with the spheres centered on the focal hyperbola c_2. This gives

$$x^2 + y^2 + z^2 \mp 2\cosh v(ey + ar_0) - 2bz \sinh v - b^2 - r_0^2 = 0\,,$$

and for the planes through the circles of contact $2\sinh v(ey + ar_0) \mp 2bz \cosh v = 0$.

Hence, Dupin cyclides with an ellipse and hyperbola as spine curves are of degree four, while parabolic cylides are cubic surfaces. The latter can be proved in a similar way.

4.3 Perspective images of conics

The main goal of this section is to prove that not only circles, but all conics have the property that their perspective images are again conics, provided the carrier plane of the conic does not pass through the center of the projection.

Quadratic cones

In analogy to the definition of curves of degree two on page 20, we call in the Euclidean 3-space $\mathbb{E}^3$ the set $\mathcal{Q}$ of points whose Cartesian coordinates (x, y, z) satisfy a given quadratic equation

$$b_{11}x^2 + 2b_{12}xy + 2b_{13}xz + \cdots + b_{33}z^2 + 2b_1x + 2b_2y + 2b_3z + b = 0 \quad (4.4)$$

a *surface of degree two*. Again, the degree is invariant under changes $(x, y, z) \mapsto (x', y', z')$ of the Cartesian coordinate system, since the new coordinates are linear functions of the initial coordinates, and *vice versa*.

The surface with the equation (4.4) is called *reducible* if the polynomial on the left-hand side splits into two linear polynomials. Then, the surface $\mathcal{Q}$ consists of two planes or only of one. Similarily to the planar case, a polynomial can be irreducible over $\mathbb{R}$, but reducible over $\mathbb{C}$, and the point set $\mathcal{Q}$ can be empty, *e.g.*, in the case $x^2 + y^2 + z^2 + 1 = 0$.

A surface $\mathcal{Q}$ of degree two is called a *quadratic cone* $\mathcal{C}$ if there exists exactly one point $S \in \mathcal{C}$ such that with each point $P \in \mathcal{C} \setminus \{S\}$ all points of the connecting line $[S, P]$ belong to $\mathcal{C}$. The point S is called the *apex* of $\mathcal{C}$. A quadratic cone must be irreducible since otherwise the apex would not be unique.

Let (4.4) be the equation of a quadratic cone $\mathcal{C}$ with its apex S at the origin. Then, the polynomial $F(x, y, z)$ on the left-hand side satisfies $F(0, 0, 0) = b = 0$. Furthermore, $P \in \mathcal{C} \setminus \{0\}$, *i.e.*, $F(x_P, y_P, z_P) = 0$, must imply

$$F(\lambda x_P, \lambda y_P, \lambda z_P) = \lambda^2(b_{11}x_P^2 + \cdots + b_{33}z_P^2) + 2\lambda(b_1x_P + b_2y_P + b_3z_P) = 0$$

for all $\lambda \in \mathbb{R}$. This is only possible, when for all $P \in \mathcal{C}$ in the equation above the quadratic term and the linear term vanish. Being a quadratic polynomial, $F(x, y, z)$ must be homogeneous of degree two, and therefore, $b_1 = b_2 = b_3 = 0$.

The lemma below focuses on two important properties of quadratic cones.

Lemma 4.3.1 *The lines in $\mathbb{E}^3$ connecting a given point S with the points of a conic which is not coplanar with S, form a quadratic cone. Conversely, a quadratic cone intersects all planes not passing through the apex along conics.*

Proof: We choose a Cartesian coordinate frame with the origin at the point S such that the given conic lies in the plane $z = 1$. If the conic satisfies the equation

$$a_{11}x^2 + 2a_{12}xy + a_{22}y^2 + a_1 x + a_2 y + a = 0,$$

then

$$a_{11}x^2 + 2a_{12}xy + a_{22}y^2 + a_1 xz + a_2 yz + az^2 = 0$$

is the equation of a quadratic cone. In the reverse direction, this confirms the second statement of Lemma 4.3.1, since without loss of generality the intersecting plane can be specified with the equation $z = 1$. ∎

Remark 4.3.1 As a consequence of Theorem 2.1.2, each irreducible homogeneous quadratic polynomial with any zero other than $(0,0,0)$ defines as its set of zeros a quadratic cone.

By virtue of the principal axes transformation which will be provided in volume 2, there is always an appropriate coordinate system such that the corresponding homogeneous equation of the quadratic cone with apex $S = (0,0,0)$ contains only squares; all terms with mixed products xy, yz or yz vanish. Since this equation must admit other zeros than the origin, quadratic cones have the standard equation

$$\mathcal{C}: \frac{x^2}{a^2} + \frac{y^2}{b^2} - z^2 = 0 \quad \text{with} \quad a \geq b > 0. \tag{4.5}$$

It turns out that among the planar sections of quadratic cones there are always circles. We can prove this by checking the intersection between the cone $\mathcal{C}$ in (4.5) and the sphere

$$\mathcal{S}: x^2 + y^2 + (z - 1 - a^2)^2 = a^2(1 + a)^2$$

(Figure 4.23). We eliminate x^2 from the two equations and obtain

$$(a^2 - b^2)y^2 - b^2(1 + a^2)(z - 1)^2 = 0.$$

This is the equation of a pair of planes through the point $(0,0,1)$ which lies in the interior of $\mathcal{S}$. Hence, the two planes which are displayed in red color in Figure 4.23, intersect $\mathcal{S}$ along two circles which also belong to $\mathcal{C}$. All planes which are parallel to one of these two planes, intersect the

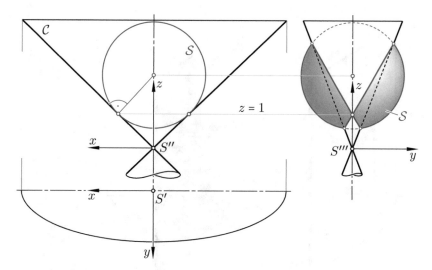

FIGURE 4.23. Each quadratic cone has circular sections and (at least) three planes of symmetry.

quadratic cone C along circles. We call them *circular sections* of C. This means also that each quadratic cone can be defined as the connection of a circle with a non-coplanar apex S. Therefore, quadratic cones other than right cones can also be called *oblique circular cones*.

Central projection of conics

The basic term in this subsection is the *central projection*. It is defined by an *image plane* π and a *center* $C \notin \pi$ (Figure 4.24). The pedal point of C with respect to π is called *principal point H*; the distance $d = \overline{CH}$ is the *principal distance*. The image X^c of any point X other than C is the point of intersection between the 'line of sight' $[C, X]$ and the image plane π. The image of any scene is called *(central or linear) perspective*.

It can happen that $[C, X]$ is parallel to π. Then, we define the image X^c as a point at infinity. In this case, X lies in the *vanishing plane π^v* which passes through C and is parallel to π. On the other hand, points at infinity can have a finite image: Figure 4.25 shows that the images of lines parallel to the z-axis have a common point Z_u^c whose pre-image is at infinity (for the extension of the Euclidean plane by points at infinity see Section 5.1).

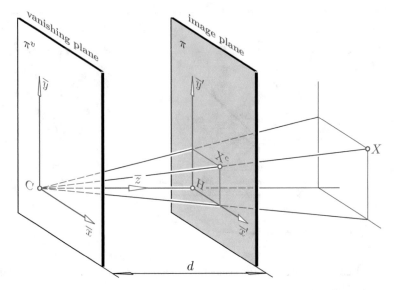

FIGURE 4.24. Central projection $X \mapsto X^c$ with center C, principal point H and camera frame $(\overline{x}, \overline{y}, \overline{z})$.

Theorem 4.3.1 *Central projections map conics c onto conics c^c, provided the plane of c does not pass through the projection center C. Depending on the number 0, 1, or 2 of intersection points between c and the vanishing plane, the image c^c is an ellipse, a parabola, or a hyperbola.*

Proof: This follows directly from Lemma 4.3.1: The cone connecting the projection center with the conic c is quadratic, and this cone intersects the image plane π along a conic c^c. Exactly the points of intersection between c and the vanishing plane are mapped onto the points at infinity of the conic c^c. ∎

Of course, Theorem 4.3.1 is also valid for circles. Figure 4.26 shows the perspective of a packing of congruent circles over a square grid. The circles depicted in Figure 4.27 form a parabolic pencil. They are located in the xy-plane and contact the y-axis at the origin O. All types of conics can be found among the images. Figure 4.25 shows a perspective of circle packings in all three coordinate planes. The points X_u^c, Y_u^c, and Z_u^c are the images of the ideal points of the coordinate axes.

The central projection is the geometric idealization of the *photographic mapping* with C as the *focal center* of the lenses, with π as the plane carrying the film or the CCD sensor, and with d as *focal length*. However,

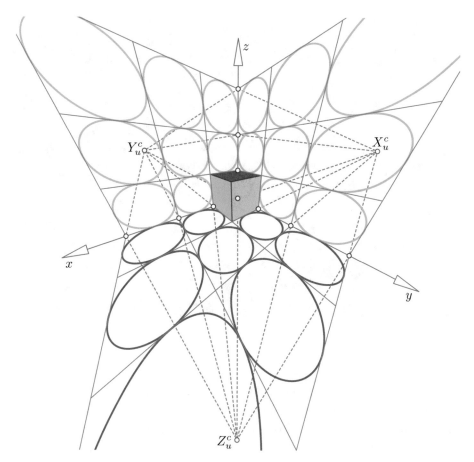

FIGURE 4.25. Perspective of unit circle packings in the coordinate planes.

one has to note that the central projection is defined for all points $X \neq C$ in space, while the photographic mapping depicts only points inside a four-sided pyramid with apex C in one half-space of the vanishing plane. In Figure 4.27, some of the circles are mapped onto hyperbolas; their second branch would not be visible under a photographic mapping.

It does not matter that — contrary to Figure 4.24 — at the photographic mapping the image plane does not belong to this half-space. As demonstrated in Figure 4.28, the plane π shows the same image as its reflection $\overline{\pi}$ in the center C. There is just one difference: The *negative plane* $\overline{\pi}$ shows the image upside down.

FIGURE 4.26. Perspective of a circle packing over a square grid.

Each photographic mapping defines an associated coordinate system in space, the *camera frame* $(\overline{x}, \overline{y}, \overline{z})$ (Figure 4.24). Its origin is placed at the center C. The *principal ray of sight* pointing from C into the visible halfspace is the $\overline{z}$-axis. The horizontal and vertical directions of the image sensor define the $\overline{x}$ and $\overline{y}$-axis spanning the vanishing plane π^v.

When the principal point H is specified as the origin of a coordinate frame $(\overline{x}', \overline{y}')$ in the image plane π, then the photographic mapping and the associated central projection obey the matrix equation

$$
\begin{pmatrix} \overline{x} \\ \overline{y} \\ \overline{z} \end{pmatrix} \mapsto \begin{pmatrix} \overline{x}' \\ \overline{y}' \end{pmatrix} = \frac{d}{\overline{z}} \begin{pmatrix} \overline{x} \\ \overline{y} \\ \overline{z} \end{pmatrix} = \frac{d}{\overline{z}} \begin{pmatrix} 1 & 0 & 0 \\ 0 & 1 & 0 \end{pmatrix} \begin{pmatrix} \overline{x} \\ \overline{y} \\ \overline{z} \end{pmatrix}.
$$

Now, we bring this in a more general form: We replace the camera frame by arbitrary *world coordinates* (x, y, z), namely

$$
\begin{pmatrix} \overline{x} \\ \overline{y} \\ \overline{z} \end{pmatrix} = \begin{pmatrix} \overline{a} \\ \overline{b} \\ \overline{c} \end{pmatrix} + \mathbf{R} \begin{pmatrix} x \\ y \\ z \end{pmatrix},
$$

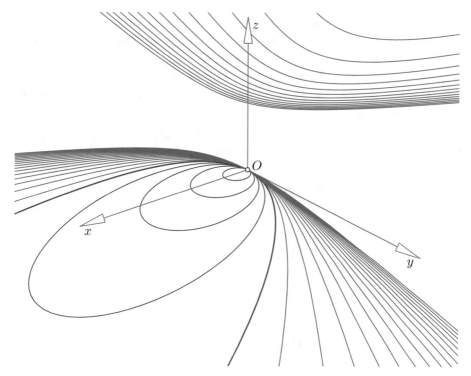

FIGURE 4.27. Perspective of a parabolic pencil of circles.

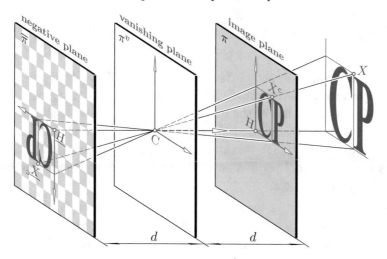

FIGURE 4.28. Central projection $X \mapsto \overline{X}^c$ onto the negative plane $\overline{\pi}$.

where $\mathbf{R}$ is an orthogonal 3×3-matrix. A similar coordinate transformation $(\overline{x}',\overline{y}') \rightarrow (x',y')$ can be performed in the image plane π. This reveals that finally the coordinates x' and y' of X^c can be expressed as rational functions of the world coordinates (x,y,z) of the pre-image, such that the numerator and the denominator are linear in x, y, and z, and the denominators in x' and y' are equal.

These formulas can be used to transform any rational parametrization of a conic c onto a rational parametrization of the image c^c.

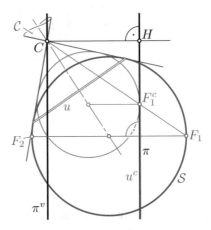

FIGURE 4.29. The perspective contour u^c of the sphere $\mathcal{S}$ is a conic u^c. The points $F_1, F_2 \in \mathcal{S}$ are projected onto the focal points of u^c.

In order to obtain the perspective of any sphere $\mathcal{S}$, we draw the tangents from the center C to the sphere $\mathcal{S}$. They form a right cone $\mathcal{C}$, provided the projection center C lies in the exterior of $\mathcal{S}$. The cone $\mathcal{C}$ contacts $\mathcal{S}$ along a circle u and intersects the image plane π in a conic u^c, the *contour* of $\mathcal{S}$, whose principal axis passes through the principal point H. Figure 4.29 shows the orthogonal projection of the scene onto a plane through C, H and the sphere's center.

When $\mathcal{S}$ intersects the vanishing plane π^v along a circle, then u^c is a hyperbola. In the case of contact between $\mathcal{S}$ and π^v the contour u^c is a parabola. We obtain an ellipse u^c under $\mathcal{S} \cap \pi^v = \varnothing$ (note Figure 4.30).

By virtue of Theorem 4.1.1, the focal points of the contour $u^c = \mathcal{C} \cap \pi$ are the points of contact of the image plane π and the Dandelin spheres. Each Dandelin sphere can be transformed onto the given sphere $\mathcal{S}$ by a similarity with center C. This similarity sends the corresponding focal point onto a point on $\mathcal{S}$, whose tangent plane is parallel to π. Therefore, the projections of the points $F_1, F_2 \in \mathcal{S}$ with extreme distance to the image plane π are the focal points of the contour u^c (Figure 4.29).

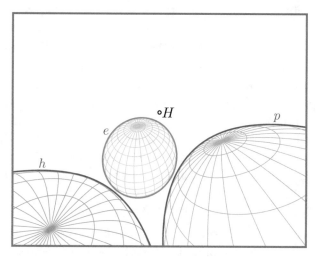

FIGURE 4.30. Ellipse e, parabola p and hyperbola h as contours of spheres.

Similar results are valid for *parallel projections* which are the limiting cases of central projections, when the center C tends to infinity: conics are projected onto conics of the same type, and the perspective of a sphere is an ellipse. We will get a deeper insight in chapter 8 which treats affine geometry.

We speak of a *linear image*, whenever a scene in $\mathbb{E}^3$ is mapped into a plane by a central or parallel projection.

4.4 Spatial interpretation of conic constructions

In this section, we shall learn that some constructions of conics and related to conics can be solved by simply interpreting the given planar figures as some projection of a geometric object in a three-dimensional space and solving the transferred problem there. The projection of the solution to the spatial problem yields the solution to the initial problem.

Unfortunately, there is no general concept for this method. Some interpretations are obvious, some seem to be tricky. The method is restricted to special problems: Any given configuration of points, lines, and conics that can be interpreted as a linear image of points and conics on an auxiliary quadric (surface of degree two, including cones and cylinders) or lines tangent to the auxiliary surface can be treated that way. The major advantage of this method is that it allows us to find the solutions without any knowledge in Projective Geometry and is, therefore, accessible to everybody who is familiar with three-dimensional geometry.

In the following we shall solve some problems in order to show to which problems the technique of spatial interpretation applies to.

Ellipses on three points and two parallel tangents

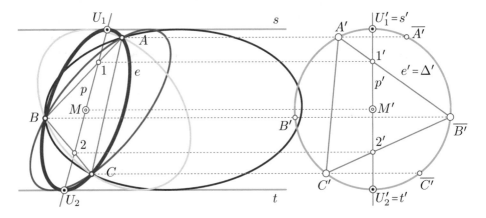

FIGURE 4.31. The ellipses on three points A, B, C and two parallel tangents s, t are orthogonal projections of four planar intersections of a cylinder of revolution.

How to find ellipses on three points A, B, C tangent to two tangents s, t in admissible position? Figure 4.31 shows the initial situation: We are given three points A, B, C and two parallel tangents s, t. We assume that none of the given points is incident with a given line. Further, A, B, and C shall not be collinear and lie entirely in the strip bounded by s and t.

The key idea is the following: The two parallel tangents s and t can be viewed as the contour of a cylinder Δ of revolution under an orthogonal projection to some plane parallel to the cylinder's axis. If s and t are displayed in a side-view which is the result of an orthogonal projection in the direction of s and t, then any generator of the cylinder Δ appears as a point. Consequently, the cylinder Δ shows up as a circle Δ' whose radius is half the distance between s and t. The points A, B, C are now considered to be images of points on Δ. Therefore, they can be seen in the side-view as points A', B', C' on the circle $e' = \Delta'$, see Figure 4.31.

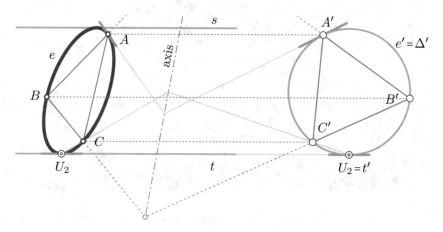

FIGURE 4.32. There is a perspective affine mapping between the circle e' and the requested ellipse e. Pairs (A, A') of corresponding points are joined by mutually parallel fixed lines of the affine mapping. The axis of the affine mapping is the locus of points that have coinciding images in both orthogonal projections.

Note that there are two choices for $A' \in \Delta'$: on the left or on the right-hand side (Figure 4.31), $i.e.$, on the back or on the front side of the cylinder Δ. This is also the case for the pre-images of B and C. We choose the particular configuration of A', B', and C' shown in Figure 4.31. The pair (A, A') determines a point P in Euclidean three-space which is located on the cylinder Δ. The same holds true for (B, B') and (C, C'). We call these points Q and R.

The points P, Q, and R span a plane that meets Δ along an ellipse e. The side-view e' of e agrees with the side-view Δ' of Δ. In order to complete the conic e, $i.e.$, the solution to the initial problem, one has to construct at least the points U_1 and U_2 of contact of e with s and t. The latter points are the contour points on e, and therefore, they are located on the *principal line*[6] with the side view p' (cf. Figure 4.31).

Axes and vertices can be found in many ways. A constructive way is displayed and explained in Figure 9.7. We skip the details of the construction here, since many CAD-systems have tools to construct ellipses from various pieces. On the other hand, there is a *perspective affine* mapping that maps e' onto e (Figure 4.32). It can be used to complete the drawing.

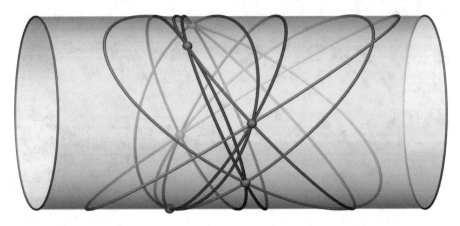

FIGURE 4.33. Eight ellipses on the cylinder Δ: Any two equally colored ellipses map to the same ellipse under a certain orthogonal projection.

More important is the fact that there are, in general, four solutions. This is clearly seen when we consider the following: As mentioned above, there are two choices for the side-view of a given point. In Figure 4.31, the different choices are distinguished by bars on top of the symbol. Thus, there are eight planes that can be defined on these points and we have eight intersections with Δ, see Figure 4.33. However, there are four pairs of planes that result in coinciding images: For example the planes $\varepsilon :=$ $[P, Q, R]$ and $\overline{\varepsilon} := [\overline{P}, \overline{Q}, \overline{R}]$ are symmetric with respect to the principal

---

[6]In Descriptive Geometry, a principal line is a line that is parallel to the image plane of a parallel projection. Similarily, a principal plane is parallel to the image plane of any projection.

plane through Δ's axis. Thus, the two curves $\varepsilon \cap \Delta$ and $\overline{\varepsilon} \cap \Delta$ are mapped to the same curve under the orthogonal projection that yields the initial figure (Figure 4.31, left).

There exists no ellipse on three points touching s and t if at least one of the given points lies outside the strip bounded by s and t.

The choice of the auxiliary surface for obtaining a spatial pre-image should be made such that constructions become as simple as possible. In the present example, we could also take an elliptic cylinder (different from a cylinder of revolution), but this would not simplify the construction. Only theoretically, it would be the same way of solving the problem, and it would yield the same four ellipses.

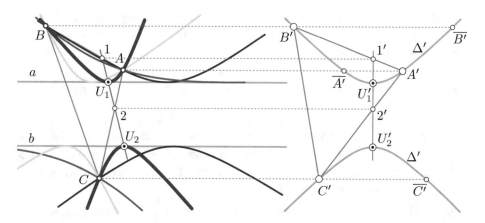

FIGURE 4.34. Hyperbolas on three points A, B, C and two parallel tangents s, t may be found as an orthogonal projection of planar intersections of a hyperbolic cylinder (cf. Exercise 4.4.1). The fat hyperbola on the left-hand side corresponds to the particular choice of A', B', C' in the side view on the right-hand side.

• Exercise 4.4.1 Hyperbolas on three points and two parallel tangents. Discuss the type and number of solutions to the example displayed in Figure 4.31 depending on the position of A, B, C relative to s and t.

• Exercise 4.4.2 Modify the configuration of given points and lines from Example 4.4.1: Assume that the points A, B, C lie outside of the strip bounded by the parallel lines s and t.

Hint: In this case, the auxiliary cylinder is a hyperbolic cylinder, i.e., its base is a hyperbola. A short version of the construction may look as shown in Figure 4.34.

• Exercise 4.4.3 Find all parabolas on three points and a given tangent. Use a parabolic cylinder as the auxiliary surface. An example is shown in Figure 4.38.

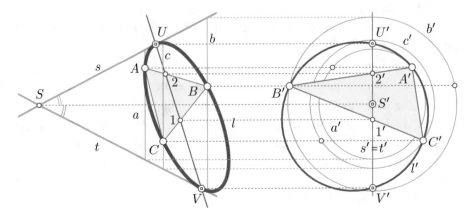

FIGURE 4.35. The ellipse l is one of four conics passing through A, B, C touching the lines s and t. It is the orthogonal projection of a planar intersection of a cone or revolution.

Conics on three points and a pair of intersecting tangents

What happens if s and t are intersecting in a proper point? Then, we cannot expect to find conics of a certain affine type interpolating the given points and lines. From the constructive point of view, we have to replace the auxiliary cylinder by a quadratic cone Γ. For the sake of simplicity, Γ should be chosen as a cone of revolution.

Again the two given lines s and t are interpreted as the contour lines (see Figure 4.35) and the conics (ellipses, hyperbolas, and, in some cases, even parabolas) are found as the orthogonal projections of planar intersections of the cone Γ with its axis parallel to the image plane.

Figure 4.35 shows the given lines s, t together with the points A, B, C. To the given configuration of points and lines (interpreted as an orthogonal projection of a cone of revolution), we attach another orthogonal view in the direction of the cone's axis. In the attached view, we can see the parallel circles a, b, c on the cone Γ carrying the points A', B', C' that map to A, B, C. Like in the previous cases, there are two possible choices for A', B', and C'. For the moment, we treat one certain choice.

Our goal is to find two further points of a solution l. Then, the conic l is well-defined (or even over-determined). Here, we pay attention to the contour points U and V which are located on a principal line. Once the contour points (points of contact of a solution l with s and t) are found, the solution l is over-determined by two line elements and three further points. As mentioned before, there are several different ways to find axes and vertices of the solution l: the approach via Constructive Geometry, or via Projective Geometry, or even with purely planar and elementary techniques like the perspective affinity joining the two orthogonal projections. We skip the details here.

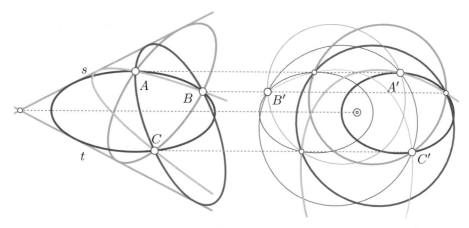

FIGURE 4.36. All four solutions to the problem given in Figure 4.35 are orthogonal projections of eight planar sections of the cone Γ. Curves with the same color on the right-hand side map to the same curve on the left-hand side.

Figure 4.36 shows all four conics that touch the two intersecting lines s and t and pass through the three points A, B, C. Any solution is the image of two conics which are displayed in the same color in Figure 4.36.

In a similar way, *i.e.*, using cylinders and cones as auxiliary surfaces, we can solve the following problems: For the construction of conics on two points and three tangents, we interpret two out of the three tangents as the contour lines of a cone or cylinder (if two of these lines happen to be parallel). For construction purposes it is much simpler to assume that the contour is that of a surface of revolution. The third line is then considered as a tangent of the auxiliary surface, and the two given points shall lie on the surface.

• Exercise 4.4.4 Conics on line elements, points, and tangents. Apply the method of spatial interpretation to the following configuration of given elements. Discuss the number of solutions depending on the position of lines and points relative to each other: Figure 4.37 shall give an idea how possible configurations and solutions may look like.

1. Find all conics on one line element (A, a) plus two tangents b, c, and one point B.
2. Find all conics on one line element (A, a) plus two points B, C, and one tangent b.

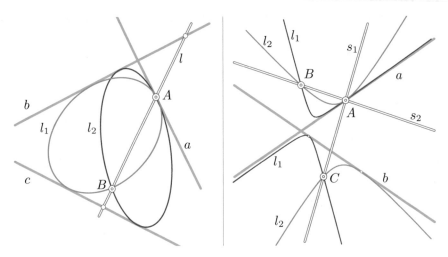

FIGURE 4.37. Conics determined by given line elements, points or lines as mentioned in Exercise 4.4.4: Left: Given one line element (A, a), two tangents b, c, and one point B. Right: Given one line element (A, a), two points B, C, and one tangent b. In both cases, some solutions are degenerate: a repeated l (left-hand side) and a pair (s_1, s_2) of lines (right-hand side). These lines meet the requirements of solutions algebraically.

The example of ellipses tangent to two parallel lines given at the very beginning of this chapter (see Figure 4.31) is somehow special. *A priori*, the affine type of the conics showing up as solutions is known. Another special case in the sense of affine geometry would be the following variation: Find all parabolas on one tangent (proper line) and three (proper and non-collinear) points. In this case, the given line is interpreted as the contour line of a parabolic cylinder. The solutions to the initial problem are orthogonal projections of planar intersections of the cylinder. The image plane is parallel to the cylinder's generators. An example is displayed in Figure 4.38. The construction is indicated and reduced to the construction of the contact point T_1 of one particular solution l_1.

Conics on three points and touching a given conic twice

There is a second group of problems that can be solved by means of spatial interpretation. However, there may be configurations of points that may

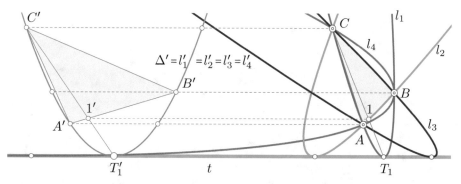

FIGURE 4.38. A parabolic cylinder Δ serves as an auxiliary surface when we have to find all parabolas on three points A, B, C, and a given tangent t.

not yield a single solution. We only deal with those cases that show (real) solutions and start with a circle c and three points A, B, C in the interior of c (Figure 4.39, right). We are looking for all conics that pass through the given three points and touch c twice. Consequently, we interpret c as the contour of a sphere Σ (indeed any regular quadric of revolution with the same contour would also be a possible choice provided that its points are mapped to the interior of c). The points A, B, and C are the orthogonal projections of points P, Q, R on Σ. (Again, there are two possible choices of P as pre-images of A, and similar for Q and R.) The plane ε spanned by P, Q, and R meets Σ in a circle k_1 whose orthogonal projection gives one solution l_1 (see Figure 4.39, left). Note that the solution l_1 touches c twice while the pre-image k_1 meets c transversely.

Again, the situation in three-dimensional space shows eight conics that map to only four solutions. This is due to the symmetry of the auxiliary surface Σ with respect to the carrier plane of c. The red solution l_4 shown in Figure 4.39 (on the right) is not a *wrong* solution. The double contact happens in a pair of complex conjugate points joined by a real line.

Figure 4.40 shows that a one-sheeted hyperboloid of revolution can serve as an auxiliary surface if the three given points are exterior points of c. Again, one solution seems to be wrong. However, like in Figure 4.39, the seemingly wrong solution is in double contact with c at a pair of complex conjugate points.

● Exercise 4.4.5 What to do if c is not a circle? The technique of spatial interpretation is not restricted to the case of a circle. If c happens to be an ellipse, the auxiliary surface may no longer be a surface of revolution. However, in such cases it is useful to interpret c as the

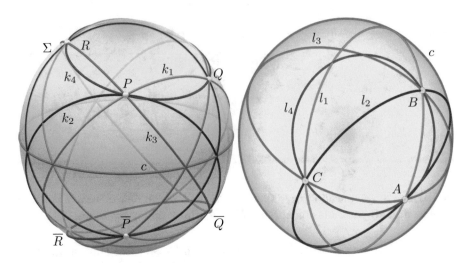

FIGURE 4.39. Left: Eight circles on the sphere that map to four ellipses touching the contour circle twice. Right: Four ellipses l_1, l_2, l_3, l_4 are in double contact with the circle c and pass through three points A, B, C. These ellipses are obtained from the eight ellipses on the left by an orthogonal projection.

contour of an ellipsoid (which may still be one of revolution) or as the contour of a one-sheeted hyperboloid (which cannot be of revolution in this case). An ellipsoid has to be chosen if the given points A, B, and C are interior points of c. In case of exterior points, we have to use the one-sheeted hyperboloid.

● **Exercise 4.4.6** What to do if c is a parabola? If there is a parabola c to be touched twice by conics on three points A, B, and C, there are again two cases to be distinguished: If the given points A, B, and C are interior points, we interpret c as the contour of an elliptic paraboloid (which could be, for the sake of simplicity, a paraboloid of revolution). In the case of three exterior points, we view c as the contour of a hyperbolic paraboloid (which could be an orthogonal one). The last case is illustrated in Figure 4.41. Figure 4.42 shows the four conics l_1, l_2, l_3, and l_4 on A, B, C that touch the parabola c. None of the four conics can be an ellipse. Each solution is the orthogonal projection of a planar intersetion of a hyperbolic paraboloid which carries only parabolas and hyperbolas.

Note that all examples given in this section have one property in common: There are no real solutions if the given conic c separates one of the given points from the others. A mixture of interior and exterior points of c cannot be interpolated by a conic that touches a given conic twice.

In the case of lines and line elements, the given points have to lie in the same wedge enclosed by two given tangents to allow real solutions.

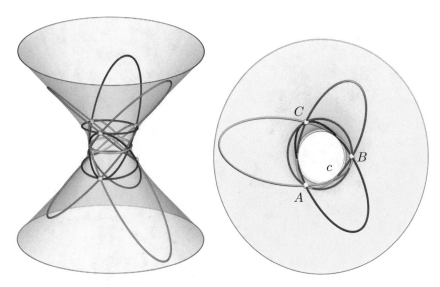

FIGURE 4.40. Left: Eight conics (ellipses) on the one-sheeted hyperboloid that map to four ellipses touching the contour circle c twice (cf. Exercise 4.4.5). Right: The four ellipses on A, B, C in double contact with c are obtained from the eight ellipses on the left by an orthogonal projection to the carrier plane of c.

Hyperosculating conics on two points touching a given conic

How to find conics which hyperosculate a given conic c and interpolate two given points A and B. Hyperosculation can be viewed as a limiting case of double contact: Let the two contact points converge towards each other, and the double contact becomes one contact but with contact order three, *i.e.*, algebraically speaking, the two curves intersect at one point with multiplicity four. Another important fact follows from Theorem 3.3.1: Assume (P, t) is a line element on a sphere Σ. As already shown in Figure 3.48, the planes through t intersect Σ along circles which share the line element (P, t). The orthogonal projection onto the plane of the great circle u through (P, t) yields ellipses with the common hyperosculating circle u. In Definition 6.4.1 (see Section 6.3), we shall give a different, but nevertheless equivalent definition of hyperosculation.

Figure 4.43 shows the two conics that hyperosculate a circle c and pass through two given points A and B. As in the aforementioned cases, we cannot expect real solutions if one of the given points is an interior point and the other one is an outer point of c.

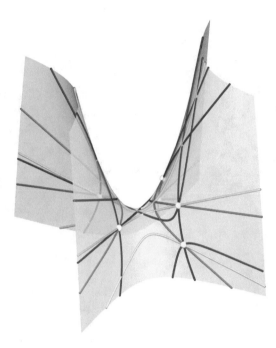

FIGURE 4.41. The spatial interpretation (cf. Exercise 4.4.6) shows eight conics (parabolas and hyperbolas) on a hyperbolic paraboloid meeting the contour (cyan) transversely. Any pair of conics with the same color map onto one conic under the orthogonal projection onto the contour's plane.

Again the (affine) type of the auxiliary surface depends on the given conic c and on whether the given points are interior or outer points. For a circle c the simplest auxiliary surface is a surface of revolution: a sphere (or ellipsoid) if A and B are interior points of c, a one-sheeted hyperboloid otherwise. If c is a hyperbola, we use a hyperboloid, and if c is a parabola, we choose a paraboloid as auxiliary surface. Figure 4.44 shows the situation in three-space, $i.e.$, the result of the spatial interpretation. Here, the simple case of conics that osculate a circle c is treated. The auxiliary surface is a sphere.

The key idea is to interpret the two given points A and B as points on the auxiliary quadric Φ which again leaves two possibilities for each point (Figure 4.44). The line $s = [P, Q]$ connecting the pre-images of A and B meets the carrier (principal) plane ε of c in a point T. The two tangents from T to c touch c at the contact points H_1 and H_2 of the conics l_1 and l_2

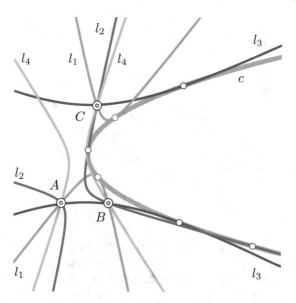

FIGURE 4.42. Conics in double contact with a parabola (cf. Exercise 4.4.6). The orange solution l_4 is in double contact with c at a pair of complex conjugate points.

we are looking for. Note that the position of T relative to c is responsible for the number (or existence) of real solutions.

The choice of $\overline{s} = [P, \overline{Q}]$, *i.e.*, one point on either side of ε, results in a point $S = \overline{s} \cap \varepsilon$ in the interior of c. Consequently, there are no real tangents from S to c. Therefore, the contact points as well as the conics interpolating A and B and hyperosculating c are not real.

● Exercise 4.4.7 Parabolas or hyperbolas on two points which hyperosculate a given conic. Replace the circle from above with an ellipse, a parabola, or a hyperbola and look for all conics on two points that hyperosculate the given conic c. What could possible auxiliary surfaces look like depending on the affine type of the conic c?

Conics on three points with a given focus

Conics on three points A, B, C and a given focus F can also be found via spatial interpretation. We consider the given three points as the top view of three points on a cone Γ of revolution with vertical axis. The common focal point F is the top view of Γ's apex (cf. Theorem 4.1.3).

Figure 4.45 shows the construction of the conic on three points A', B', C', and prescribed focus F'. (The initial points got primes, because they are

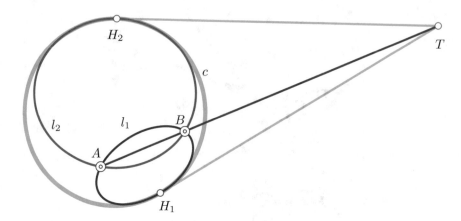

FIGURE 4.43. Spatial interpretation also helps to find hyperosculating conics interpolating two additional points. Note that the line $[A, B]$ is concurrent with the two tangents at the hyperosculation points H_1, H_2.

interpreted as the top view of three points in $\mathbb{E}^3$.) The front view (points labeled with a double prime) shows the straight contour of the cone and the three points A'', B'', C''. The right-side view (triple primes) attached to the top view is chosen such that the plane $\varepsilon = [A, B, C]$ is in edge-view and can be seen as a straight line ε'''. The right-side view $l''' = \varepsilon'''$ of the particular solution l coincides with the image of l. Vertices and the center (if at all present) can be found in the side view.

Figure 4.45 shows the constructive approach to one particular solution l. Once a choice is made whether the corresponding points A, B, C in three-space are on the "upper" or "lower" half of the cone Γ, the plane ε is determined. The intersection $l = \varepsilon \cap \Gamma$ maps to a conic that passes through A', B', C' and has F' for a focal point. Figure 4.46 shows all solutions to the initial problem. As in some examples before, there are eight planes determined by these two times three points in Euclidean three-space. These planes can be arranged in four pairs of planes $(\varepsilon, \bar{\varepsilon})$ such that ε and $\bar{\varepsilon}$ are symmetric with respect to the horizontal plane through the apex of Γ. Therefore, the curves $\varepsilon \cap \Gamma$ and $\bar{\varepsilon} \cap \Gamma$ have coinciding top views and we find four different solutions. Example 7.2.2 in Section 7.2 shows a different way to find conics on three points with a common focal point by using the polarity with regard to a circle centered at F.

Once the pre-images of A, B, C on Γ are specified on different half-cones, the corresponding solutions are hyperbolas with A, B, C on different

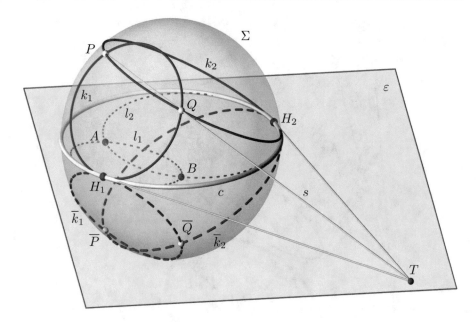

FIGURE 4.44. Spatial interpretation also helps to find hyperosculating conics interpolating two additional points: The dotted lines are the solutions to the initial problem. The dashed lines are the symmetric copies of the two circles on the upper hemisphere.

branches. Hence, there is only one solution out of the four that gives a continuous transition between the given points along an ellipse, or a parabola, or one branch of a hyperbola. This is the solution which was mentioned in the preface on page vi concerning the dwarf planet Ceres.

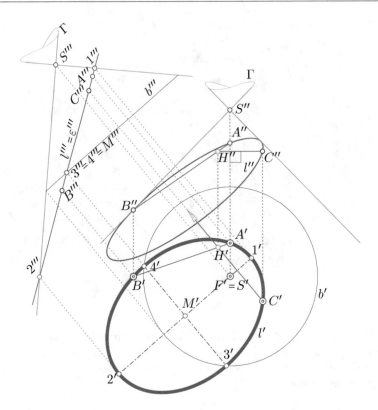

FIGURE 4.45. Any conic on A', B', C' with focus F' is a top view of a planar section of a cone Γ of revolution. The focus F' is the top view of Γ's apex S. The constructive solution for one of the in total four solutions is shown.

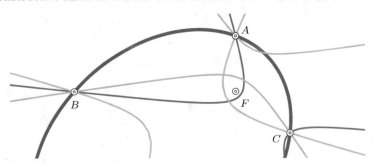

FIGURE 4.46. The top view of a point on Γ may correspond to two different points in Euclidean three-space. Thus, there are eight different planar intersections which lead to four different conics with the focus F and passing through A, B, C.

5 Projective Geometry

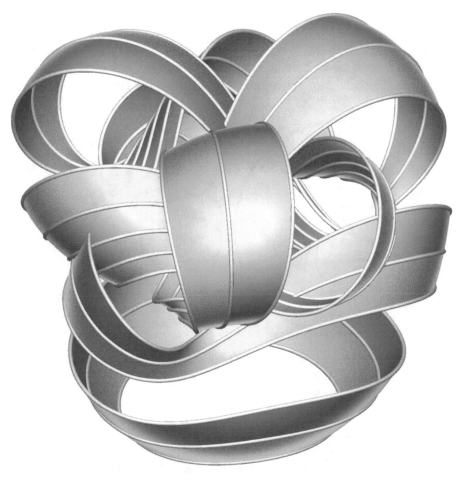

Projective Geometry is the proper frame work for understanding the geometry of conics. It differs from Euclidean Geometry and allows us to treat points and lines in a unifying way: There is no difference between points at infinity and proper points. The lines in a projective plane are closed like any conic, and the line at infinity is a line like any other. The above surface was discovered in 1901 by WERNER BOY and is an immersion of the real projective plane into three-space. Thus, BOY answered a question raised by DAVID HILBERT whether the projective plane can be embedded into a three-dimensional space or not.

5.1 Projective planes

Models of projective planes

In this section, we shall briefly describe projective planes. Therefore, we give a short overview of the axiomatic approach to projective planes. We also describe the vector space model of a projective plane.

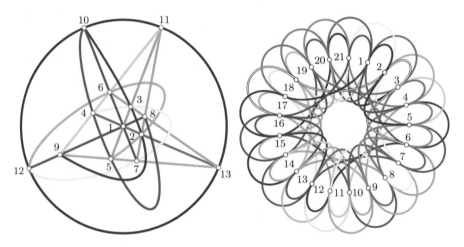

FIGURE 5.1. Models of finite projective planes. Left: The plane of order three has thirteen points $\{1, 2, \ldots, 13\}$ and as many lines (drawn in different colors). Right: The plane of order four has twenty-one points and lines.

From the incidence geometric point of view, a projective plane consists of a set $\mathcal{P}$ of *points* and a set $\mathcal{L}$ of *lines*. Lines are always considered as sets of points. Sometimes we emphasize this fact by calling them *range of points*. Closely related to ranges of points are *pencils of lines* which are defined as the set of all lines through a certain point, say V. The point V is frequently called the *vertex* or *carrier* of the pencil. Briefly, we say *pencil V*. Figure 5.1 shows two examples of finite projectve planes. Each range of points has a certain color in order to emphasize that certain points belong to this set.

If for some $P \in \mathcal{P}$ and some $l \in \mathcal{L}$ the relation $P \in l$ holds, we say that the point P *is contained in* the line l, or P *is incident with* l, or *lies on* l, or l *goes through* P, or l *passes through* P. Any set of points on the same line is called *collinear*. Any set of lines through the same point is called *concurrent*.

What makes the pair $(\mathcal{P}, \mathcal{L})$ a projective plane $\mathbb{P}^2$? It is a list of axioms which is as follows:

(A1) Any two different points $P \neq Q$ can be joined by a unique line $l = [P, Q]$, the line connecting P and Q.

(A2) Any two different lines $l \neq m$ intersect in one point $S = l \cap m$, the point of intersection of l and m.

(A3) There exists a quadrangle, *i.e.*, four points no three of which are collinear.

Two projective planes $\mathbb{P}^2 = (\mathcal{P}, \mathcal{L})$ and $\mathbb{P}^{2'} = (\mathcal{P}', \mathcal{L}')$ are called *isomorphic* if there exists a bijective mapping $\kappa : \mathcal{P} \to \mathcal{P}'$ and $\mathcal{L} \to \mathcal{L}'$ such that $P \in l$ holds if, and only if, $\kappa(P) \in \kappa(l)$ for all $P \in \mathcal{P}$ and all $l \in \mathcal{L}$. Sometimes, we say that $(\mathcal{P}, \mathcal{L})$ and $(\mathcal{P}', \mathcal{L}')$ are *models* of the same abstract projective plane.

From the axioms (A1) and (A2), we can deduce: Any two different lines meet in *precisely* one point, otherwise, if they met in two different points, these lines would be equal by (A1). Further, we can infer that any line carries at least three points, or through any point, there are at least three lines. A range of points carries as many points as there are lines in a pencil. Moreover, the sets $\mathcal{P}$ and $\mathcal{L}$ are *equipotent*, *i.e.*, there exists a bijective mapping between $\mathcal{P}$ and $\mathcal{L}$. A set of four lines with the property that no three of them are concurrent, *i.e.*, no three have a point in common, is also called a quadrangle.

■ **Example 5.1.1** Some projective planes.

Projective minimal plane. The fact that there exists a quadrangle together with the unique point of intersection of any two different lines allows us to infer that a projective plane has at least seven points: the points A, B, C, D of the quadrangle together with the three *diagonal points* $[A, B] \cap [C, D]$, $[A, C][\mathcal{B}, \mathcal{D}]$, $[A, D] \cap [B, C]$. Actually, there exists a projective plane with exactly seven points. It is called the *Fano plane*, named after the Italian mathematician GINO FANO (1871–1952). An example of the "minimal projective plane" can be seen in Figure 5.6. For more details, see Example 5.1.2.

The projectively closed real plane. All the drawings we produce take place in the *real plane* $\mathbb{E}^2$. There, we observe that some lines are not intersecting, see Figure 5.2. Usually, we call such lines *parallel*. Parallelity of lines in a plane is an *equivalence relation*.[1] Therefore, we can

---

[1] A *relation* on a set S is a subset R of the Cartesian product $S \times S$, and therefore, R is the union of ordered pairs of elements of S. With $x \sim y$ we indicate that $(x, y) \in R$ or x is related to y. An *equivalence relation* R satisfies the following conditions for all $a, b, c \in S$: (1) $a \sim a$

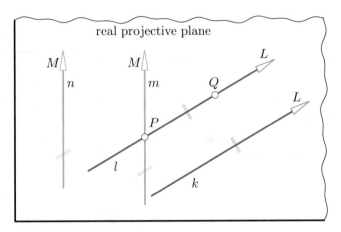

FIGURE 5.2. A scene from the real projective plane: The unique line l connecting P and Q intersects the parallel line k in the ideal point L. The parallels m and n share the ideal point M. (In drawings, we indicate ideal points by arrows. Note that arrows in opposite directions define the same ideal point.)

speak of *classes of parallel lines*. In order to perform the *projective closure of the real plane*, we add a unique *ideal point* or *point at infinity* to each class of parallel lines. The set of all ideal points is called *ideal line* or *line at infinity*. The thus extended plane is called the *real projective plane* and we denote it by $\mathbb{P}^2(\mathbb{R})$.

The real projective plane is, indeed, a projective plane: The axiom (A1) is fulfilled since any point P in the real projective plane can be joined with any other point. In practice: To join P with an ideal point means to draw a parallel to some line through P (cf. figure 5.2). Further, the axiom (A2) is fulfilled since any two lines are either intersecting or parallel. If two lines are parallel, then they intersect in the common ideal point. Finally, there is no doubt about the existence of a quadrangle.

The bundle model of a projective plane. From the real plane $\mathbb{P}^2(\mathbb{R})$, we can easily create a further model of a projective plane. Let O be a point which is not contained in $\mathbb{P}^2(\mathbb{R})$. Then, we project the points and lines in the projective plane from O. This means that we connect the points and lines of $\mathbb{P}^2(\mathbb{R})$ with the point O. Of course, points at infinity yield line parallel to $\mathbb{P}^2(\mathbb{R})$. Any point $P \in \mathbb{P}^2(\mathbb{R})$ is now "blown up" to a line $[O, P]$ through O; and any line $l \in \mathbb{P}^2(\mathbb{R})$ becomes a plane $[O, l]$ through O. The points and lines of the real projective plane are in a one-to-one correspondence with the lines and planes through O. We call this the *bundle model* of the real projective plane.[2] The point O is sometimes called the *carrier of the bundle*. Figure 5.3 shows points, lines, and a quadrangle in the bundle model of a projective plane.

(reflexive), (2) $a \sim b$ implies $b \sim a$ (symmetric), and (3) $a \sim b$ and $b \sim c$ together imply $a \sim c$ (transitive). These three conditions are completely independent.

[2] The name *bundle* is a translation of the German word *Bündel*. The word *Bündel* is used for two or more-parametric linear families of geometric objects. The manifold of lines through a common point in a three-dimensional space is called a *star*. Hence, one could also speak of the *star model of a projective plane*.

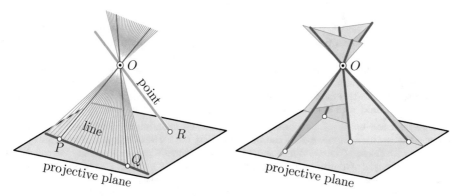

FIGURE 5.3. The bundle in $\mathbb{E}^3$ is a projective plane isomorphic to the real projective plane. Left: The "lines" and "points" are represented by planes and lines through the vertex O of the bundle. Right: A "quadrangle" of the projective plane is a four-sided pyramid with apex O.

It is not hard to verify that the bundle also serves as a model of a projective plane: For any two points p and q in the bundle (lines through O), there is a unique line in the bundle (plane through O) joining these two, and thus, (A1) is fulfilled. Any two lines in the bundle (planes through O) intersect in a point (line through O) which fits to axiom (A2). Four planes through O constitute a quadrangle in the bundle model provided that no three of these planes are coaxial (have a common line). So, the existence of quadrangles (as requested in (A3)) is guaranteed, too.

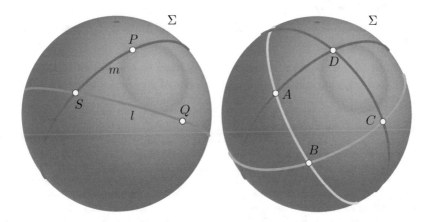

FIGURE 5.4. Left: Two lines l and m in the spherical model intersect at a point S. Right: The lines of a quadrangle in the spherical model are great circles of the sphere Σ.

The sphere model of a projective plane. From the bundle model, we can create a further model of a projective plane. We take a sphere Σ centered at the carrier O of the bundle. It means no restriction to assume that the radius of Σ equals one. Any line p in the bundle intersects Σ in a *pair of opposite* or *antipodal points*. Thus, the points in the sphere model are the pairs of opposite or antipodal points on Σ. The lines in the bundle are planes through the carrier O. Since O is the center of Σ, any plane of the bundle intersects Σ along a *great circle* which represents a line in this model. The quadrangles in this model appear as the intersection of a four-sided pyramid with apex O with the sphere Σ, see Figure 5.4.

The vector space model of a projective plane. Let $\mathbb{F}$ be an arbitrary field[3] and $\mathbb{F}^3$ a three-dimensional vector space over $\mathbb{F}$. It can be shown that the set $\mathcal{P}$ of one-dimensional subspaces together with the set $\mathcal{L}$ of two-dimensional subspaces defines a projective plane. We call it the *vector space model* of a projective plane and denote it by $\mathbb{P}^2(\mathbb{F})$. Points P spanned by non-zero vectors $\mathbf{p} \in \mathbb{F}^3$. Instead of $P = \{\lambda \mathbf{p} | \lambda \in \mathbb{F}\}$ we write $P = \mathbf{p}\mathbb{F}$ or briefly even $P = \mathbf{p}$.

Later, we shall use a vector space in order to find an algebraic model of a projective plane. The bundle model of a projective plane is the geometric realization of a vector space model, cf. page 194.

A very important concept of Projective Geometry is the

Principle of Duality. *Any valid theorem that is formulated in terms of points, lines, connection, intersection, and incidence remains true if we interchange the words points and lines, intersection and connection, but keep incidence between points and lines.*

This "new" theorem is called the *dual theorem*. The fact that dualizing valid theorems produces new valid theorems in Projective Geometry was possibly first observed by the French mathematician JOSEPH DIAZ GERGONNE (1771–1859).

In order to make the geometry of a projective plane more meaty, one has to add further axioms:

The axiom of FANO, as given below, is added to the list of axioms if the projective plane should display a property of quadrangles that is known from $\mathbb{P}^2(\mathbb{R})$:

---

[3] A (non-empty) set $\mathbb{F}$ with an operation + is called a *(commutative) group* if for all $x, y, z \in \mathbb{F}$: (1) $x + y \in \mathbb{F}$ and $x + y = y + x$ for commutativity, (2) there is a unique element 0 (zero element), such that $x + 0 = 0 + x = x$, (3) there is a unique element $-x$ (additive inverse) such that $x + (-x) = 0$, (4) and $x + (y + z) = (x + y) + z$ (associativity), and additionally, for commutativity $x + y = y + x$. A (non-empty) set $\mathbb{F}$ is called a *(commutative) field* if there are two different operations, say + and $\cdot$, acting on $\mathbb{F}$ such that $\mathbb{F}$ together with + is a commutative group and $\mathbb{F} \setminus \{0\}$ together with $\cdot$ fulfills for all $x, y, z \in \mathbb{F}$: (1) $x \cdot y \in \mathbb{F}$. (2) There is a unique element $1 \neq 0$ (neutral element) such that $x \cdot 1 = 1 \cdot x = x$, (3) there is a unique element x^{-1} (left multiplicative inverse element) such that $x^{-1} \cdot x = 1$, (4) $x \cdot (y + z) = x \cdot y + x \cdot z$ and $(x+y) \cdot z = x \cdot z + y \cdot z$ (distributive laws), and additionally, for commutative fields $x \cdot y = y \cdot x \in \mathbb{F}$.

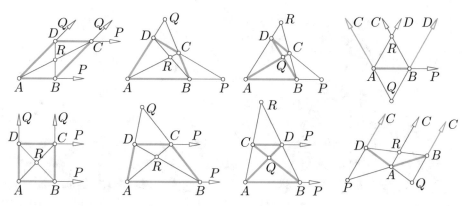

FIGURE 5.5. Quadrangles in the real projective plane: The diagonal points P, Q, and R, are never collinear.

(FA) The three diagonal points of a quadrangle, *i.e.*, the points of intersection of opposite sides of the quadrangle are never collinear.

In Figure 5.5, we can see different appearances of quadrangles in the real projective plane. It does not matter if vertices or diagonal points are ideal points (lie at infinity) or not. The three diagonal points are never collinear. So, in the real projective plane the axiom (FA) is fulfilled.

The dual of a quadrangle is called a *quadrilateral*. It is a set of four lines such that no three of them are concurrent. A quadrilateral defines six vertices and three *diagonals*. The axiom (FA) implies that the diagonals of a quadrilateral are never concurrent.

■ **Example 5.1.2** The Fano plane. We shall continue the first item of Example 5.1.1.

Synthetic approach. In this example, we use the notation given in Figure 5.6. According to the axiom (A3), there exists a quadrangle $P_1 P_2 P_3 P_7$ in the Fano plane together with the diagonal points $P_4 = [P_2, P_3] \cap [P_1, P_7]$, $P_5 = [P_3, P_1] \cap [P_2, P_7]$, and $P_6 = [P_1, P_2] \cap [P_3, P_7]$. The axiom (A1) says that there is a unique line l_7 joining P_4 and P_5. Further, l_7 has to have a unique point P_6 of intersection with $[P_1, P_2]$.

Analytic approach. We identify the points of the Fano plane with the following vectors, cf. Figure 5.6 (right):

$$P_1 = (1,0,0), \quad P_2 = (0,1,0), \quad P_3 = (0,0,1), \quad P_4 = (0,1,1),$$

$$P_5 = (1,0,1), \quad P_6 = (1,1,0), \quad P_7 = (1,1,1).$$

We use 0 and 1 as the elements of the field $\mathbb{Z}_2$, *i.e.*, they satisfy the following rules: $0 + 0 = 1 + 1 = 0$, $0 + 1 = 1 + 0 = 1$, $1 \cdot 1 = 1$, $0 \cdot 0 = 0 \cdot 1 = 1 \cdot 0 = 0$. Now, we apply the usual addition of vectors. By the latter rules, we find that three points are collinear if, and only if, any two of the corresponding vectors sum up to the third. This allows us to prove all collinearities shown in Figure 5.6. The fact that the diagonal points of a quadrangle are collinear is equivalent to

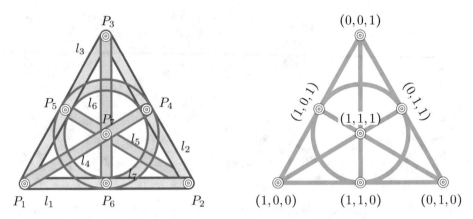

FIGURE 5.6. Left: The Fano plane with its seven points $P_1, \ldots, P_7$ and its seven lines $l_1, \ldots, l_7$. In the Fano plane, the axiom of Fano is not valid. Right: The analytic model of the Fano plane.

the fact that the *characteristic*[4] of the field, upon which the algebraic model is built, equals two. The coordinatization of a projective plane will be described on page 194.

The axiom (FA) is not valid. The two equivalent models of the Fano plane given above show the characteristic property of this projective plane: The three diagonal points P_4, P_5, and P_6 of the quadrangle $P_1 P_2 P_3 P_7$ are collinear, and thus, the Fano axiom (FA) is not fulfilled in the minimal projective plane. It can be shown that $\mathbb{P}^2(\mathbb{F})$ is a Fano plane if, and only if, the characteristic of the field $\mathbb{F}$ equals 2.

In this book, we mainly concentrate on Fano planes without mention. Only in exceptional cases, we will assume that (FA) does not hold.

For the axiomatic approach to Projective Geometry, the axiom of DE-SARGUES (named after the French architect and mathematician GÉRARD DESARGUES (1591–1661)) and the axiom of PAPPUS are very important. Both axioms are usually stated as theorems which are, indeed, theorems in the real projective plane. A proof of the latter will be given on page 191.

DESARGUES's axiom reads:

(DE) If two triangles are perspective to a point, then their sides are perspective to a line (note Figure 5.7).

---

[4]The characteristic of a field $\mathbb{F}$ is denoted by char$\mathbb{F}$. The symbol char$\mathbb{F} = n$ means that the n-fold sum of the unit element 1 equals 0, *i.e.*, $\underbrace{1 + 1 + \ldots + 1}_{n \text{ times}} = 0$ and $n \in \mathbb{N} \setminus \{0\}$ is the smallest number with this property.

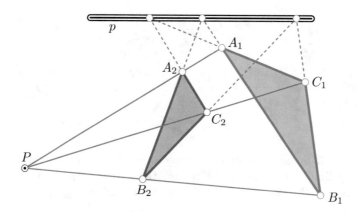

FIGURE 5.7. A Desargues configuration: The triangles $A_1B_1C_1$ and $A_2B_2C_2$ are perspective. The point P is the perspector. The sides are perspective with the perspectrix p.

The axiom (DE) together with the axioms (A1) – (A3) implies its dual which is exactly its converse. The configuration described in the axiom/theorem of Desargues is frequently called a *Desargues configuration* or *Desargues figure*. The point to which the triangles are perspective is called the *perspector*. The line where assigned sides of the triangles intersect is called the *perspectrix*. Figure 5.7 shows such a configuration including the perspector and the perspectrix. It is *self-dual*, since each line (in the configuration) carries three points of the configuration, and through each point (of the configuration) there pass exactly three lines. We call a projective plane a *Desarguesian plane* if the axiom of DESARGUES holds. In fact, (DE) holds in any projective plane $\mathbb{P}^2(\mathbb{F})$.

The axiom of DESARGUES guarantees the existence of perspective collineations (cf. page 233). Furthermore, it can be proved that in any Desarguesian plane coordinates from a field $\mathbb{F}$ can be introduced such that there is an isomorphism to a projective plane $\mathbb{P}^2(\mathbb{F})$ over a field $\mathbb{F}$ which needs not be commutative (see [11]).

Figure 5.8 shows special forms of Desargues figures in the real projective plane and their relations to elementary transformations. A *translation* is defined by a Desargues figure with an ideal point P as the perspector and the ideal line as perspectrix.

The two triangles in the middle of Figure 5.8 are *similar*, i.e., corresponding sides are parallel and the lengths of corresponding sides have a

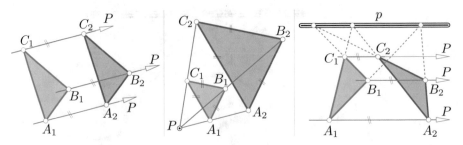

FIGURE 5.8. Some versions of Desargues configurations in the (projectively extended) Euclidean plane and their relations to elementary transformations: translation (left), central similarity (middle), shearing (right).

constant ratio. Since this is a Desargues figure, corresponding points are collinear with the perspector P, and thus, P is the center of a *central similarity*. The perspectrix p coincides with the ideal line.

On the right-hand side of Figure 5.8, we see a *shearing* or *shear transformation* relating the two triangles of the Desargues figure. The perspectrix p is a proper line and the perspector P is the ideal point of the perspectrix.

Finally, we formulate the axiom of PAPPUS. PAPPUS OF ALEXANDRIA (290–350) was a Greek mathematician who left us an eight-volumes book on geometry and mathematics. The following axiom was originally formulated as a theorem. It is closely related to projectivities and to conics. It states:

(PP) Assume that 1, 3, 5 are three points on a line l, and let further 2, 4, 6 be three points on another line m. Then, the points

$$A := [1,2] \cap [4,5], \quad B := [2,3] \cap [5,6], \quad C := [3,4] \cap [6,1]$$

are collinear.

Later, we shall see that (PP) also holds for six points on a conic. So, the union of the two lines mentioned in (PP) can be seen as a conic.

Figure 5.9 illustrates the axiom (PP). It further displays the dual version of PAPPUS's theorem which is usually ascribed to CHARLES JULIEN BRIANCHON (1783–1864) and is related to PAPPUS's theorem in a natural way. We call a projective plane a *Pappian plane* if the axiom of PAPPUS is valid.

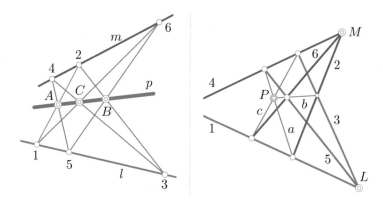

FIGURE 5.9. PAPPUS's theorem (left) and its dual form (right).

The axiom (PP) is stronger than (DE). A classical result which is due to GERHARD HESSENBERG (1874–1925) states that (A1) – (A3) and (PP) imply (DE). Hence, each Pappian projective plane is also isomorphic to a projective plane $\mathbb{P}^2(\mathbb{F})$. Furthermore, it can be proved that the projective plane $\mathbb{P}^2(\mathbb{F})$ is Pappian if, and only if, $\mathbb{F}$ is commutative. This means that the axiomatic approach to Pappian projective planes is equivalent to the analytic approach via coordinates which come from a commutative field. In the sequel, we will use both, because each of them, the analytic approach as well as the so-called *synthetic* approach, has its advantages. If not stated otherwise, the projective planes we are dealing with are always Pappian.

5.2 Perspectivities, projectivities

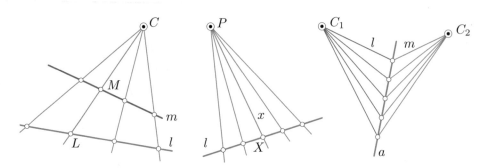

FIGURE 5.10. The basic mappings in Projective Geometry (as described in Definition 5.2.1): Left: the perspectivity $l \overset{C}{\barwedge} m$ with perspector C. Middle: the perspectivity $l \barwedge P$ between a range of points and a pencil of lines. Right: the perspectivity $C_1 \overset{a}{\barwedge} C_2$ between two pencils of lines with perspectrix a.

The basic mappings in Projective Geometry are *perspectivities*. These mappings are one-to-one and onto, and they relate ranges of points with pencils of lines, pencils with pencils, and ranges with ranges. The simplest among these mappings (as shown in Figure 5.10, in the middle) relates a line l and a pencil of lines with carrier $P \notin l$ such that corresponding elements $X \in l$ and $x \ni P$ are incident. Here are the fundamental notions:

Definition 5.2.1 *A perspectivity* $\alpha : l \to m$ *with the center* $C \notin l, m$ *sends* $L \in l$ *to* $M \in m$ *whenever* L, M, *and* C *are collinear. We use the symbol*

$$l \overset{C}{\barwedge} m.$$

The dual version is also called a perspectivity and the symbol

$$S \overset{a}{\barwedge} T$$

means that corresponding elements are lines $s \ni S$ *and* $t \ni T$ *sharing a point on an* axis a. *We also call the pencil of lines through* P *perspective to the range* l $(P \notin l)$ *of points if for each line* x *through* P *the corresponding point* $X \in l$ *is incident with* x. *In this case, we use the symbol*

$$l \barwedge P \quad or \quad P \barwedge l.$$

Figure 5.10 illustrates the contents of Definition 5.2.1. In the middle, we see the basic mapping $l \barwedge P$. The inverse mapping is symbolized by $P \barwedge l$. The left-hand side shows the action of the perspectivity $\alpha : l \to m$ which can be viewed as the composition of two perspectivities of the basic type. The right-hand side shows a perspectivity between two pencils of lines. The vertices of the pencils are the points C_1 and C_2, while the axis of the perspectivity is the line a.

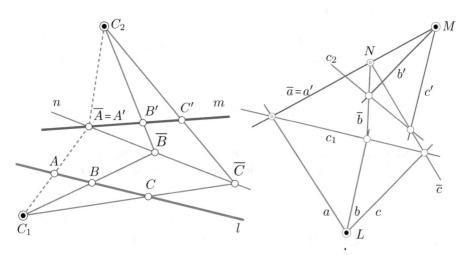

FIGURE 5.11. The reduction of projectivities. Left: A projectivity between two lines l and m given by $A \mapsto A'$, $B \mapsto B'$, $C \mapsto C'$ can be realized as the composition of only two perspectivities: $l \overset{C_1}{\barwedge} n \overset{C_2}{\barwedge} m$. Right: The dual version, the projectivity between two pencils of lines with vertices L and M given by $a \mapsto a'$, $b \mapsto b'$, $c \mapsto c'$, can be realized as a composition of two perspectivities $L \overset{c_1}{\barwedge} N \overset{c_2}{\barwedge} M$.

Perspectivities between ranges/pencils are uniquely defined by prescribing two pairs of corresponding elements. In general, the composition of two perspectivities is no longer a perspectivity:

Definition 5.2.2 *The composition of finitely many perspectivities between ranges of points and pencils of lines is called a projectivity.*

In order to indicate that two ranges of points l and m are joined by the projectivity α (different from a perspectivity) we write $l \overset{\alpha}{\barwedge} m$. The symbol $(A, B, C) \barwedge (A', B', C')$ means that A, B, and C are mapped to A', B', and C', respectively, via a projectivity.

The projectivities act transitively on ordered triplets of collinear points (or concurrent lines), *i.e.*, there is a projectivity for prescribed three mutually distinct collinear points (or concurrent lines) and their images. In Pappian planes, *i.e.*, in projective planes where PAPPUS's theorem (cf. page 191) is valid, this projectivity is even uniquely determined, *i.e.*, projectivities act sharply transitive on ordered triples of collinear points and pencils of concurrent lines. Hence, the following "fundamental theorem" ((FT) in brief) is valid:

Theorem 5.2.1 **(FT)** *A projectivity* $\alpha : l \to m$ *in a Pappian projective plane is uniquely defined by prescribing three pairs* (A_i, A_i') *of points with* $\alpha(A_i) = A_i'$ *for all* $i \in \{1, 2, 3, \}$.

In any projective plane, (PP) is equivalent to (FT), for details, see [11, 36]. The chains of subsequent perspectivities can be arbitrarily long, but only finitely many perspectivities are allowed. However, in any Pappian plane, there is a possibility to shorten chains of perspectivities with the help of the following *reduction theorem*:

Theorem 5.2.2 *In projective planes with (FT), any projectivity between two different ranges* l *and* m *of points can be realized as the product of at most two perspectivities between ranges.*

A projectivity $\alpha : l \to m$ *whith* $l \neq m$ *is a perspectivity if, and only if, the point* $l \cap m$ *is mapped onto itself.*

Proof: Assume A, B, C are three different points on a line l, and A', B', C' are three different points on a line m. Without loss of generality, we can suppose $A \notin m$ and $A' \notin l$. By (FT), there exists exactly one projectivity $\alpha : l \to m$ with $\alpha(A) = A'$, $\alpha(B) = B'$, and $\alpha(C) = C'$.

In order to show that α can be decomposed into two perspectivities between ranges of points, we choose a line $n \neq m$, $n \neq [A, A']$ through A' and further a point C_1 on the line $[A, A']$ with $C_1 \neq A, A'$ as can be seen in Figure 5.11 (left). Now, we have a first perspectivity $\beta_1 := l \overset{C_1}{\overline{\wedge}} n$ that maps A, B, C to $\overline{A} = A'$, $\overline{B} := [C_1, B] \cap n$, $\overline{C} := [C_1, C] \cap n$. Because of the choice of C_1 and n, the points $\overline{B}, \overline{B}, B', C'$ form a quadrangle, and the point $C_2 := [\overline{B}, B'] \cap [\overline{C}, C']$ is different from all other points and can, therefore, serve as the center of a second perspectivity $\beta_2 : n \overset{C_2}{\overline{\wedge}} m$ that maps $\overline{A} = A'$, $\overline{B}, \overline{C}$ to the points A', B', C', respectively. Thus, $\alpha = \beta_2 \circ \beta_1$. (FT) guarantees that for any suitable choice of n and C_1 the product of the two corresponding perspectivities remains the same.

The case of a projective mapping between two pencils of lines can be treated in the same way by simply dualizing the construction as illustrated in Figure 5.11 (right).

If $\alpha(l \cap m) = l \cap m$, then we may set $C = C' = l \cap m$. Now, A, A', B, B' is a quadrangle and for $C_1 = [A, A'] \cap [B, B']$ we get α as the perspectivity $l \overset{C_1}{\overline{\wedge}} m$. ■

The statements of Theorem 5.2.2 are also valid in their dual formulation (illustrated in Figure 5.11). In the case $l = m$ the minimal number of perspectivities is two only if there is a point which is mapped onto itself. Otherwise, we need at least three perspectivities.

Later, we shall use the following result:

Theorem 5.2.3 *For any four collinear points A, B, C, D there exists a projectivity with the following property*

$$A \mapsto B, \quad B \mapsto A, \quad C \mapsto D, \quad D \mapsto C.$$

Proof: Let $m \neq l$ be a line through D and $Z \notin \{l \cup m\}$. Now, we define the points $\overline{A} = [Z, A] \cap m$, $\overline{B} = [Z, B] \cap m$, $\overline{C} = [C, Z] \cap m$, and $B' = [A, \overline{B}] \cap [C, Z]$, as displayed in Figure 5.12. The existence of m and Z and all further points is guaranteed by the axiom (A3).

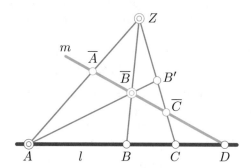

FIGURE 5.12. The proof of Theorem 5.2.3.

Then, we follow the chain of perspectivities

$$l \overset{Z}{\barwedge} m \overset{A}{\barwedge} [Z, C] \overset{\overline{B}}{\barwedge} l$$

with $A \mapsto \overline{A} \mapsto Z \mapsto B$, $B \mapsto \overline{B} \mapsto B' \mapsto A$, $C \mapsto \overline{C} \mapsto \overline{C} \mapsto D$, and $D \mapsto D \mapsto C \mapsto C$ and the proof is finished. ∎

Pappus's theorem and the axis of a projectivity

A useful result on projectivities between lines depends on whether the theorem of PAPPUS is valid in the underlying projective plane or not. In the real and complex projective plane, PAPPUS's theorem holds because the underlying fields $\mathbb{R}$ and $\mathbb{C}$ are commutative. However, there are some projective planes which are not Pappian, *e.g.*, the projective plane $\mathbb{P}^2(\mathbb{H})$

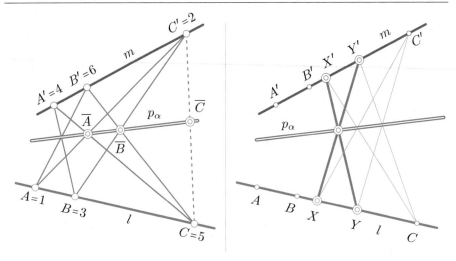

FIGURE 5.13. The proof of Theorem 5.2.4: the construction of the axis of a projectivity (left), the important property of the axis (right).

where $\mathbb{H}$ denotes the Hamiltonian quaternions. When dealing with conics, we always assume that the axioms of PAPPUS and FANO are valid.

The original version of the axiom (PP) (cf. page 186) is in close relation to projectivities between two different ranges of points. We show

Theorem 5.2.4 *Let* $\alpha : l \to m$ *with* $l \neq m$ *be a projectivity between two ranges of points in a Pappian projective plane. For any two pairs* (X, X') *and* (Y, Y') *of corresponding points with* $X, Y \in l$, $X \neq Y$, *and* $X' = \alpha(X)$, $Y' = \alpha(Y)$, *the lines*

$$[X, Y'] \quad \text{and} \quad [X', Y]$$

meet on a fixed line p_α *called the* axis *of the projectivity.*

In the case of a projectivity α *between two different pencils of lines, we can find a* center P_α *of the projectivity.*

Proof: Let $\alpha : l \to m$ be given by $A \mapsto A'$, $B \mapsto B'$, $C \mapsto C'$ with $A, B, C \in l$ and $A', B', C' \in m$ where $C \notin m$ and $C' \notin l$, cf. Figure 5.13. According to Definition 5.2.2, the projectivity α should decompose into a finite sequence of perspectivities.

In fact, when we set $\overline{A} = [1, 2] \cap [4, 5]$, $\overline{B} = [2, 3] \cap [5, 6]$, and $p_\alpha = [\overline{A}, \overline{B}]$ we have $\alpha : l \overset{C'}{\barwedge} p_\alpha \overset{C}{\barwedge} m$ because $A \mapsto \overline{A} \mapsto A'$, $B \mapsto \overline{B} \mapsto B'$, and $C \mapsto \overline{C} = [C, C'] \cap p_\alpha \mapsto C'$.

For any two points $X, Y \in l \setminus \{C\}$ we find $X' = \alpha(X)$ by running through the chain of perspectivities (fig 5.13, right). Thereby, we find that $[X, C'] \cap [X', C]$ and $[Y, C'] \cap [Y', C]$ are points of p_α. Now, we label the points in the way $Y' = 1$, $C = 2$, $X' = 3$, $Y = 4$, $C' = 5$, $X = 6$ and conclude by (PP) also $[X, Y'] \cap [X', Y] \in p_\alpha$. ■

A projectivity may not always map points from one line to another line. When we deal with conics, we are sometimes concerned with the case of *projectivities on a line* or in a pencil. This will be of importance in Section 7.3, especially in Theorem 7.4.1. For more clarity, we shall analyze some examples on page 210.

As a consequence of the Fundamental Theorem 5.2.1, we can say: If a projectivity $\alpha : l \to l$ has three mutually distinct *fixed points*, *i.e.*, points F with the property $\alpha(F) = F$, then α is the *identity mapping* id_l on l. The mapping id_l has only fixed points.

5.3 Coordinatization

In Projective Geometry, there is a need for a suitable coordinatization that allows us to describe the objects and transformations in a simple way. Homogeneous coordinates turned out to be the right choice. This kind of coordinates goes back to AUGUST FERDINAND MÖBIUS (1790–1868), a German mathematician and astronomer. In his famous book *Der barycentrische Calcul* (cf. [48]), he introduced *barycentric coordinates* as a method to assign coordinates to points in a plane with respect to a base of three points.

In the following, we show how to assign homogeneous coordinates to points which are given in Cartesian coordinates. Thus, we perform the projective extension of the Euclidean plane $\mathbb{E}^2$ analytically and reveal in a constructive way its isomorphy to $\mathbb{P}^2(\mathbb{R})$. Before that, we give a very short summary on affine and Cartesian coordinates.

Inhomogeneous and homogeneous coordinates on a line

Let us first explain the term *affine coordinates* of a point X on a line l. We fix two different points O and E on l, as can be seen in Figure 5.14. The point O is called the *origin* and E is called the *unit point*. (It means no restriction to assume that O is on the left-hand side of E as shown in Figure 5.14.) We assign the coordinates 0 and 1 to the points O and E. If X is a point on the line $l = [O, E]$, then its inhomogeneous coordinate x is defined as the *affine ratio*

$$x := \overline{OX} : \overline{OE} =: \operatorname{ar}(O, E, X) \tag{5.1}$$

provided the distances are signed: The coordinate x is negative if we find X left from O; and it is positive, if it is right from O. If the point X lies between O and E, then $0 < x < 1$.

Now, we turn to the case where X is a point in a plane and use two *coordinate axes*, *i.e.*, a pair of non-parallel lines. One of them shall be the x-axis, the other one shall be the y-axis. Though it does not really matter, we may assume that the x-axis is "horizontal" and the y-axis is "vertical" as illustrated in Figure 5.14. Their point of intersection O is called *origin* of the coordinate system.

On the x-axis, we pick a point $E_x \neq O$; similarly, we choose $E_y \neq O$ on the y-axis. These two points are usually referred to as the *unit points* on the

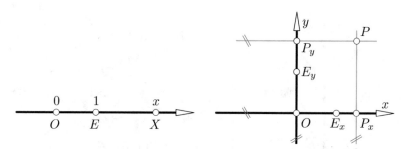

FIGURE 5.14. Affine coordinates on a line and in the plane.

coordinate axes and the lengths $\overline{OE_x}$ and $\overline{OE_y}$ are called the *unit lengths* (on the axes and on all lines parallel to the axes).

The coordinates (x_P, y_P) of a point P in the plane are now found in the following way: Draw two lines through P, one parallel to each axis. The respective intersections with the coordinate axes shall be labeled with P_x and P_y. Then, the coordinates x_P and y_P are the coordinates of P_x and P_y on the lines $[O, E_x]$ and $[O, E_y]$, *i.e.*,

$$x_P = \mathrm{ar}(O, E_x, P_x) \quad \text{and} \quad y_P = \mathrm{ar}(O, E_y, P_y). \tag{5.2}$$

The generalization to affine coordinates in n-dimensional $(n > 3)$ space is straightforward.

Note that the x-axis and the y-axis are not necessarily orthogonal. The coordinates x_P and y_P are defined as ratios in (5.2). The pair (x_P, y_P) is called *affine coordinates of the point P*, in general. If the two axes enclose a right angle and the two unit lengths are equal, then we speak of *Cartesian coordinates* named after the French mathematician, philosopher, and writer RENÉ DESCARTES (1596–1650).

Homogeneous coordinates

Affine coordinates are no longer useful in the projective extension $\mathbb{P}^2(\mathbb{R})$ of $\mathbb{E}^2$ since points at infinity would get coordinates (∞, ∞) and could not be distinguished. Therefore, we introduce homogeneous coordinates. The previously mentioned coordinates are called *inhomogeneous* just in order to emphasize the difference.

We start with homogeneous coordinates of points on a line l. We assume that a projective line, *i.e.*, a line in a projective plane is embedded as the line with the equation $x_0 = 1$ in the x_1x_0-plane, as can be seen in

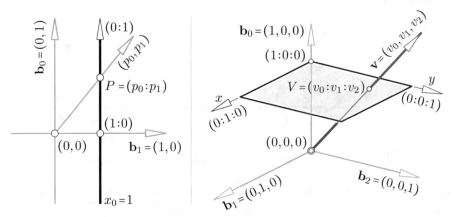

FIGURE 5.15. Homogeneous coordinates on a line and in a plane.

Figure 5.15. Any vector $(p_0, p_1) \in \mathbb{R}^2 \setminus \{(0,0)\}$ starting from $(0,0)$ aims at a point on the line l. For $p_0 \neq 0$ this point on l is finite and has the coordinates $(1, p_1 p_0^{-1})$ considered as a point in the $x_1 x_0$-plane. However, the vector $\mathbf{p} = (p_0, p_1)$ points in the same direction as any of its scalar (non-zero) multiple does. We say that $\mathbf{p}$ represents the point P on the line, but also any scalar multiple of $\mathbf{p}$ can serve as a *representative* of P. So, only the ratio $p_0 : p_1$ matters and we say that $(p_0 : p_1)$ are the homogeneous coordinates of P on l and write briefly $P = (p_0 : p_1)$. We call $\mathbf{p}$ coordinate vector of P and use the notation $P = \mathbf{p}\mathbb{R}$ in order to express that we have chosen some representative of P. More precisely,

$$\mathbf{p}\mathbb{R} = \{\mathbf{p} \cdot \lambda \mid \lambda \in \mathbb{R}\}$$

is a one-dimensional subspace of $\mathbb{R}^2$. Note that $(\lambda p_0 : \lambda p_1)$ are the *homogeneous coordinates* of the same point for any $\lambda \in \mathbb{R} \setminus \{0\}$. On the other hand, any vector $(p_0, p_1) \neq (0,0)$ defines a unique point on l. The vector $(0,0)$ must be excluded since it does not define a line through $(0,0)$ and is not directed to a point on l.

Because of the special choice of the coordinate system in the $x_1 x_0$-plane, the line l is parallel to the x_1-axis. Thus, the vector $\mathbf{b}_1 = (0,1)$ points to the direction of l. Therefore, $\mathbf{b}_1 = (0,1)$ is a representative of the *ideal point* or *point at infinity* of l. The homogeneous coordinates of l's point at infinity are thus $(0 : 1)$.

The key idea of the coordinatization with homogeneous coordinates of points on a line can now easily be carried over to a homogeneous coordinatization of the points in the real projective plane. All the vectors $\mathbf{v}$ in the three-dimensional vector space $\mathbb{R}^3$ are triplets (v_0, v_1, v_2) of numbers taken from $\mathbb{R}$. For practical reasons, we assume that the projective plane is embedded as the plane $x_0 = 1$ into the space $\mathbb{E}^3$. In order to support our imagination, the x_0-axis shall point upwards as shown in Figure 5.15, on the right-hand side.

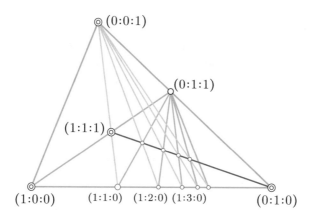

FIGURE 5.16. The fundamental triangle and a projective scale on one side.

Any point P in the plane $x_0 = 1$ has coordinates $(1, p_x, p_y)$. However, it is not only the vector $\mathbf{v} = (1, p_x, p_y)$ emanating from $(0, 0, 0)$ that points to P. Any scalar multiple $(\lambda, \lambda p_x, \lambda p_y)$ points in the same direction and the line spanned by this vector intersects the plane $x_0 = 1$ in the same point P. As in the previous case, we shorten notations by writing $P = \mathbf{p}\mathbb{R}$ as abbreviation for $\{\mathbf{p} \cdot \lambda \mid \lambda \in \mathbb{R}\}$. Once again, only the ratio $(1 : p_x : p_y)$ matters, and thus, we define $(1 : p_x : p_y)$ as the homogeneous coordinates of P. As indicated in Figure 5.15, there is an affine coordinate system in the plane $x_0 = 1$. The axes are labeled with x and y. Thus, (p_x, p_y) are the inhomogeneous coordinates of P.

On the other hand, we can extract the homogeneous coordinates of a point from the vector (v_0, v_1, v_2): We simply write $(v_0 : v_1 : v_2)$ in order to emphasize that these are homogeneous coordinates and only the ratio of these three values matters. If $v_0 \neq 0$, we can easily switch to a homogeneous coordinate representation of the point $V = \mathbf{v}\mathbb{R}$ determined by $\mathbf{v}$ with $(v_0 : v_1 : v_2) = (1 : v_1 v_0^{-1} : v_2 v_0^{-1})$.

There is one major advantage of homogeneous coordinates: Ideal points (points at infinity) can be described by homogeneous coordinates. The vectors $(0, f_1, f_2)$ are parallel to the plane $x_0 = 1$. To any of these vectors, we find a one-parameter family of parallels in the plane $x_0 = 1$. Thus, $(0 : f_1 : f_2)$ are the *homogeneous coordinates of an ideal point* in the plane $x_0 = 1$. The vector $\mathbf{f} = (0, f_1, f_2)$ is a representative of the common ideal point of all lines parallel to the vector $(f_1, f_2) \in \mathbb{R}^2$, considered as a vector in the plane $x_0 = 1$.

We call the triangle with vertices $(1 : 0 : 0)$, $(0 : 1 : 0)$, $(0 : 0 : 1)$ the *fundamental triangle* of one particular coordinatization, (cf. Figure 5.16). The point $(1 : 1 : 1)$ is called *unit point*. We will see that any four points that form a quadrangle can be chosen as the vertices of a fundamental triangle plus the unit point.

We know that instead of the standard basis $\{\mathbf{b}_1, \mathbf{b}_2, \mathbf{b}_3\}$ of $\mathbb{R}^3$ any three linearly independent vectors $\mathbf{a}$, $\mathbf{b}$, $\mathbf{c}$ can be used as a basis of a coordinatization. This leads to another system of homogeneous coordinates in $\mathbb{P}^2(\mathbb{R})$ with the fundamental points $A = \mathbf{a}\mathbb{R}$, $B = \mathbf{b}\mathbb{R}$, $C = \mathbf{c}\mathbb{R}$, and the unit point $E = (\mathbf{a} + \mathbf{b} + \mathbf{c})\mathbb{R}$ which is never collinear with any two of the points A, B, C. When in particular $\mathbf{a}$, $\mathbf{b}$, $\mathbf{c}$ are orthonormal in $\mathbb{E}^3$, we call the coordinates *homogeneous Cartesian coordinates*.

If we replace the fundamental points A, B, C by any other three non-collinear points A', B', C', then we replace the basis $\{\mathbf{a}, \mathbf{b}, \mathbf{c}\}$ by another basis, say $\{\mathbf{a}', \mathbf{b}', \mathbf{c}'\}$. From Linear Algebra, we know that the coordinates (x_0, x_1, x_2) are then replaced by the coordinates (x_0', x_1', x_2') which are given in matrix form by

$$\begin{pmatrix} x_0' \\ x_1' \\ x_2' \end{pmatrix} = \mathbf{T} \begin{pmatrix} x_0 \\ x_1 \\ x_2 \end{pmatrix}.$$

Therein, $\mathbf{T}$ is a regular 3×3-matrix with entries from $\mathbb{R}$. The columns of $\mathbf{T}$ are the 'new' coordinates of the previous basis vectors $\mathbf{a}$, $\mathbf{b}$, $\mathbf{c}$. Because of the homogeneity of the coordinates, the matrix $\mathbf{T}$ is unique only up to a non-zero real factor. When using the matrix notation, it makes sense to write the homogeneous coordinates as columns. Therefore, from now on we insert the symbol $^\mathrm{T}$ for 'transposed' when we write the coordinates of a column vectors in a line.

Because of (DE) it is easy to see that all this works in the same way when $\mathbb{R}$ is replaced by any commutative field $\mathbb{F}$, *e.g.*, by $\mathbb{C}$.

Computing with homogeneous coordinates

How to make calculations with homogeneous coordinates? Two points P and Q are represented by their homogeneous coordinates $(p_0 : p_1 : p_2)$ and $(q_0 : q_1 : q_2)$ or, equivalently, $P = \mathbf{p}\mathbb{F}$ and $Q = \mathbf{q}\mathbb{F}$. In the following, coordinates and scalars are taken from any commutative field $\mathbb{F}$. The points X on the line $l = [P, Q]$ spanned by P and Q can be parametrized with the help of *homogeneous parameters* $(\lambda : \mu)$. For the point $X = \mathbf{x}\mathbb{F}$, we have

$$\mathbf{x}(\lambda, \mu) = \{\mathbf{p} \cdot \lambda + \mathbf{q} \cdot \mu \mid (\lambda, \mu)^{\mathrm{T}} \in \mathbb{F}^2 \setminus \{(0,0)^{\mathrm{T}}\}\}.$$

Proportional pairs $(\lambda : \mu)$ represent the same point on l. Sometimes, we shall write $\mathbf{p}\mathbb{F} + \mathbf{q}\mathbb{F}$ instead of $\{\mathbf{p} \cdot \lambda + \mathbf{q} \cdot \mu \mid (\lambda, \mu)^{\mathrm{T}} \in \mathbb{F}^2 \setminus \{(0,0)^{\mathrm{T}}\}\}$. Note that $\mathbf{p}\mathbb{F} + \mathbf{q}\mathbb{F} \neq (\mathbf{p} + \mathbf{q})\mathbb{F}$.

For variable $(\lambda, \mu) \in \mathbb{F}^2$ the linear combination $\mathbf{x}(\lambda, \mu)$ is a *two-dimensional subspace of the vector space* $\mathbb{F}^3$. Thus, it can also be described by a linear equation of the form

$$\langle \mathbf{l}, \mathbf{x} \rangle := l_0 x_0 + l_1 x_1 + l_2 x_2 = 0, \quad (l_0, l_1, l_2) \neq (0, 0, 0). \tag{5.3}$$

We call $(l_0 : l_1 : l_2)$ the *homogeneous coordinates of the line* l since scaling the equation of l in (5.3) does not change its solutions. As we have done with points, we shall also write $l = \mathbf{l}\mathbb{F}$ and call $\mathbf{l}$ a *representative of* l.[5] Note that the incidence between the point $\mathbf{x}\mathbb{F}$ and the line $\mathbf{l}\mathbb{F}$ is expressed by $\langle \mathbf{l}, \mathbf{x} \rangle = 0$, which in $\mathbb{E}^3$ would mean the orthogonality $\mathbf{l} \perp \mathbf{x}$ between vectors $\mathbf{l}$ and $\mathbf{x}$.

Let $X = \mathbf{x}\mathbb{F}$ with $\mathbf{x} = (x_0, x_1, x_2)^{\mathrm{T}}$ be a point on $l = [P, Q]$, where $P = \mathbf{p}\mathbb{F}$ and $Q = \mathbf{q}\mathbb{F}$ (Figure 5.17). Then, the vectors $\mathbf{p}$, $\mathbf{q}$, and $\mathbf{x}$ are linearly dependent. Consequently, we have

$$\det(\mathbf{p}, \mathbf{q}, \mathbf{x}) = 0$$

which is again an equation of l. The latter determinant can be written in a different form as

$$\det(\mathbf{p}, \mathbf{q}, \mathbf{x}) = \langle \mathbf{p} \times \mathbf{q}, \mathbf{x} \rangle,$$

---

[5] In the case a of non-commutative field $\mathbb{F}$, there is a difference between left and right multiplication. Therefore, we have to differ between a *left* and a *right* vector space on $\mathbb{F}$. Consequently, lines should then be written as $\mathbb{F}\mathbf{l}$ if points are written as $\mathbf{p}\mathbb{F}$.

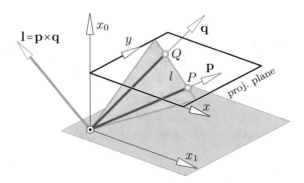

FIGURE 5.17. The line $l = [P, Q]$ through $P = \mathbf{p}\mathbb{F}$ and $Q = \mathbf{q}\mathbb{F}$ has the representative $\mathbf{l} = \mathbf{p} \times \mathbf{q}$ which satisfies $\langle \mathbf{l}, \mathbf{p} \rangle = \langle \mathbf{l}, \mathbf{q} \rangle = 0$.

where $\langle \mathbf{a}, \mathbf{b} \rangle = a_1 b_1 + a_2 b_2 + a_3 b_3$ is the *canonical scalar product* of vectors in $\mathbb{F}^3$ and $\mathbf{a} \times \mathbf{b} = (a_2 b_3 - a_3 b_2, a_3 b_1 - a_1 b_3, a_1 b_2 - a_2 b_1)$ is the induced *exterior product* of vectors in $\mathbb{F}^3$. Thus, a coordinate vector $\mathbf{l}$ of the line $l = [P, Q]$ can be found via

$$\mathbf{l} = \mathbf{p} \times \mathbf{q}. \tag{5.4}$$

Replacing $\mathbf{p}$ and $\mathbf{q}$ by non-zero multiples means changing the representatives of P and Q. The cross product of $\mathbf{p}$ and $\mathbf{q}$ then changes to a multiple of $\mathbf{l}$ and yields another representative of the same line.

Linearly dependent vectors $\mathbf{p}$ and $\mathbf{q}$ describe the same point and $\mathbf{p} \times \mathbf{q} = \mathbf{0}$ expresses the fact that there is no unique line joining a point with itself.

Obviously, in homogeneous Cartesian coordinates, the ideal line or line at infinity spanned by two (different) points $(0 : f_1 : f_2)$ and $(0 : g_1 : g_2)$ with $f_1 g_2 - f_2 g_1 \neq 0$ has the homogeneous coordinates $(1 : 0 : 0)$ and has, thus, an equation of the form $x_0 = 0$.

In the dual version, assume that two lines l and m are given by the respective homogeneous coordinates $\mathbf{l}$ and $\mathbf{m}$ (Figure 5.18). A coordinate vector $\mathbf{s}$ of the point S of intersection is given by

$$\mathbf{s} = \mathbf{l} \times \mathbf{m}. \tag{5.5}$$

This is clear since S is incident with both lines, and thus, we have

$$\langle \mathbf{l}, \mathbf{s} \rangle = 0 \quad \text{and} \quad \langle \mathbf{m}, \mathbf{s} \rangle = 0,$$

and consequently, $\mathbf{s}$ is a non-trivial scalar multiple of $\mathbf{l} \times \mathbf{m}$ provided that l and m are linearly independent. Linearly dependent vectors $\mathbf{l}$ and $\mathbf{m}$

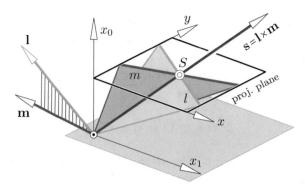

FIGURE 5.18. The point $S = \mathbf{s}\mathbb{F}$ of intersection of two lines $l = \mathbf{l}\mathbb{F}$ and $m = \mathbf{m}\mathbb{F}$ can be found as $\mathbf{s} = \mathbf{l} \times \mathbf{m}$.

are coordinates of the same line, and consequently, there is more than a single common point. This statement is expressed by $\mathbf{l} \times \mathbf{m} = \mathbf{0}$.

Lemma 5.3.1 *For any two different points $P = \mathbf{p}\mathbb{F}$ and $Q = \mathbf{q}\mathbb{F}$ in $\mathbb{P}^2(\mathbb{F})$, the connecting line is $[P, Q] = (\mathbf{p} \times \mathbf{q})\mathbb{F}$. For any two different lines $l = \mathbf{l}\mathbb{F}$ and $m = \mathbf{m}\mathbb{F}$, the point of intersection is $(\mathbf{l} \times \mathbf{m})\mathbb{F}$.*

Any line l in a projective plane is spanned by two points P and Q. Any point $X = \mathbf{x}\mathbb{F}$ on l can also be described by the two homogeneous coordinates $(x_0, x_1)^\mathrm{T} \in \mathbb{F}^2 \setminus \{(0,0)^\mathrm{T}\}$ on the line l, if there are representatives $\mathbf{p}$ and $\mathbf{q}$ such that $\mathbf{p} \cdot x_0 + \mathbf{q} \cdot x_1 = \mathbf{x}$.

This also holds for lines in a pencil: Any point $S = \mathbf{s}\mathbb{F}$ is the carrier of a pencil of lines spanned by two different lines $l = \mathbf{l}\mathbb{F}$ and $m = \mathbf{m}\mathbb{F}$ through S. If $u = \mathbf{u}\mathbb{F}$ is any line through S, then there are representatives $\mathbf{l}$ and $\mathbf{m}$ such that $\mathbf{l} \cdot u_0 + \mathbf{m} \cdot u_1 = \mathbf{u}$. Then, $(u_0 : u_1)$ are the homogeneous coordinates of the line $u = \mathbf{u}\mathbb{F}$ in the pencil about S.

Affine ratios, cross ratios, and projective frames

In $\mathbb{E}^2$, the inhomogeneous coordinate x of a point X on a line $l = [0, E]$ with respect to the coordinate system with origin O and unit point E has been defined as the affine ratio $x = \mathrm{ar}(O, E, X)$ as given in (5.1). The name affine ratio indicates that $\mathrm{ar}(O, E, X)$ is invariant under affine transformations, *i.e.*, transformations that map any point X with inhomogeneous coordinate vector $\mathbf{x}$ to $\mathbf{x}' = \mathbf{A}\mathbf{x} + \mathbf{a}$. Here, $\mathbf{A}$ is a regular 2×2

matrix and $\mathbf{a} \in \mathbb{R}^2$ is a vector.[6] The invariance of $\mathrm{ar}(O, E, X)$ under affine transformations is easily checked: If $\mathrm{ar}(O, E, X) = \lambda = \overline{OX}/\overline{OE}$, then $\overline{OE} \cdot \lambda = \overline{OX}$, and equivalently, $(\mathbf{e} - \mathbf{o}) \cdot \lambda = \mathbf{x} - \mathbf{o}$ with $\mathbf{o}$, $\mathbf{e}$, and $\mathbf{x}$ being the inhomogeneous coordinate vectors of O, E, and X. Thus, $\mathbf{x}$ can be expressed in terms of $\mathbf{o}$ and $\mathbf{e}$ by

$$\mathbf{x} = \mathbf{o} \cdot (1 - \lambda) + \mathbf{e} \cdot \lambda$$

which is a so-called *affine combination*. Now, we apply the affine transformation to O, E, X which gives $\mathbf{o}' = \mathbf{a}$, $\mathbf{e}' = \mathbf{A}\mathbf{e} + \mathbf{a}$, and $\mathbf{x}' = \mathbf{A}\mathbf{e} \cdot \lambda + \mathbf{a}$. Then, we compute

$$\mathbf{o}' \cdot (1 - \lambda) + \mathbf{e}' \cdot \lambda = \mathbf{a} \cdot (1 - \lambda) + (\mathbf{A}\mathbf{e} + \mathbf{a}) \cdot \lambda = \mathbf{A}\mathbf{e} \cdot \lambda + \mathbf{a} = \mathbf{x}'$$

which shows that the affine images O', E', and X' define the same affine ratio as O, E, and X, i.e., $\mathrm{ar}(O, E, X) = \mathrm{ar}(O', E', X')$.

Now, we extend $\mathbb{E}^2$ to the real projective plane $\mathbb{P}^2(\mathbb{R})$. Let O, U, and E be three points on a line l. We assign homogeneous coordinates to these points on the line l by setting $O = (1 : 0)$ and $U = (0 : 1)$. Without loss of generality, we can assume that $E = (1 : 1)$. This is possible since the homogeneous coordinate vectors of three collinear points are always linearly dependent. However, sometimes the homogeneous representatives $\mathbf{o}$, $\mathbf{u}$ of O, U have to be rescaled in order to fulfill $\mathbf{o} + \mathbf{u} = \mathbf{e}$. If X with representative $\mathbf{x}$ is a further point that is collinear with O, U, and E, then it has homogeneous coordinates $(x_0 : x_1)$ if $\mathbf{x} = \mathbf{v}x_0 + \mathbf{u}x_1$. All this works in $\mathbb{P}^2(\mathbb{F})$ in the same way.

We define the *cross ratio* $\mathrm{cr}(A, B, C, D) \in \mathbb{F} \cup \{\infty\}$ *of four collinear points* (on a line l) $A = \mathbf{a}\mathbb{F}$, $B = \mathbf{b}\mathbb{F}$, $C = \mathbf{c}\mathbb{F}$, and $D = \mathbf{d}\mathbb{F}$ with $A \neq D$ and $B \neq C$ in $\mathbb{P}^2(\mathbb{F})$ by

$$\mathrm{cr}(A, B, C, D) := \frac{\det(\mathbf{a}, \mathbf{c}) \cdot \det(\mathbf{b}, \mathbf{d})}{\det(\mathbf{a}, \mathbf{d}) \cdot \det(\mathbf{b}, \mathbf{c})}. \tag{5.6}$$

If $A = D$ or $B = C$ and no other two points coincide, we set $\mathrm{cr}(A, B, C, D) = \infty$. Rescaling of vectors, i.e., changing the representatives of points, does not alter the cross ratio. When we deal with projectivities (cf. page 210) and projective collineations (cf. page 233), we shall see that the cross

---

[6]The definition of an affine transformation in the n-dimensional space $\mathbb{F}^n$ over any field $\mathbb{F}$ is analogous: Take some regular $n \times n$ matrix $\mathbf{A}$ and some vector $\mathbf{a} \in \mathbb{F}^n$.

ratio is invariant under projective transformations. Thus, it is defined in a useful and proper way.

We observe that special values of $\mathrm{cr}(A, B, C, D)$ appear if D attains special positions with respect to the points A, B, C, e.g., $\mathrm{cr}(A, B, C, A) = \infty$, $\mathrm{cr}(A, B, C, B) = 0$, and $\mathrm{cr}(A, B, C, C) = 1$. Note that 0 and 1 always exist independently of the underlying field $\mathbb{F}$.

The base points of our coordinatization on l, $O = (1 : 0)$ and $U = (0 : 1)$ together with $E = (1 : 1)$ and $X = (x_0 : x_1)$ have the cross ratio $\mathrm{cr}(O, U, E, X) = \frac{x_0}{x_1}$. We call $F := (O, U; E)$ a *projective frame* or *projective coordinate system* on a line. The cross ratio $x_0 x_1^{-1}$ formed by O, U, E, and X can be interpreted as a coordinate of X in the given projective frame. In this sense, $\mathrm{cr}(A, B, C, D)$, as defined in (5.6) can be viewed as projective coordinate of D with respect to the frame $(A, B; C)$.

When O, U, E are replaced with three other points, then the basis $\{\mathbf{o}, \mathbf{u}\}$ of the two-dimensional subspace is replaced by another basis $\{\mathbf{o'}, \mathbf{u'}\}$. We know from Linear Algebra that the coordinates $(x_0 : x_1)$ for points X on the line l are replaced with $(x_0' : x_1')$ satisfying

$$\begin{pmatrix} x_0' \\ x_1' \end{pmatrix} = \mathbf{T} \begin{pmatrix} x_0 \\ x_1 \end{pmatrix}$$

with a regular 2×2-matrix $\mathbf{T} \in \mathbb{F}^{2 \times 2}$ which is unique only up to a non-zero scalar from $\mathbb{F}$. Now it is easy to prove that for the new coordinates $\mathbf{a'}$, $\mathbf{b'}$, $\mathbf{c'}$, $\mathbf{d'}$ of the four points A, B, C, D the formula (5.6) delivers the same result since $\det(\mathbf{a'}, \mathbf{c'}) = \det \mathbf{T} \cdot \det(\mathbf{a}, \mathbf{c})$ and $\det T \neq 0$.

Any four ordered collinear points define a unique cross ratio. On the other hand, the cross ratio does not uniquely define a quadruple of collinear points. However, it is surprising that the 24 permutations of four collinear points yield only six different cross ratios: If $\delta = \mathrm{cr}(A, B, C, D)$, then the five other values of the cross ratio are $1 - \delta$, δ^{-1}, $(1 - \delta)^{-1}$, $\delta(\delta - 1)^{-1}$, and $(\delta - 1)\delta^{-1}$. For example, $\mathrm{cr}(B, A, C, D) = \mathrm{cr}(A, B, C, D)^{-1}$, due to (5.6). Another consequence of (5.6) is the *product rule for cross ratios* concerning five collinear points A, B, C, D, E:

$$\mathrm{cr}(A, B, C, D) = \mathrm{cr}(A, E, C, D) \cdot \mathrm{cr}(E, B, C, D). \tag{5.7}$$

Suppose that in the real projective plane $\mathbb{P}^2(\mathbb{R})$ the point U is the ideal point of the line $[O,E]$. Then, for any finite point $X = \mathbf{x}\mathbb{R}$, we have $\mathbf{x} = \mathbf{o}x_0 + \mathbf{u}x_1 = x_0^{-1}(\mathbf{o} + x_0^{-1}x_1\mathbf{u})$ with $\mathbf{u} = \mathbf{o} - \mathbf{e}$. Therefore, we have

$$x_0^{-1}x_1 = \mathrm{ar}(O,E;X) = \mathrm{cr}(O,U,E,X)^{-1} = \mathrm{cr}(U,O,E,X). \qquad (5.8)$$

Let A, B, C, D be four finite points on the line $[O,E]$. Then, by virtue of (5.8) and (5.7), we find

$$\mathrm{cr}(A,B,C,D) = \mathrm{cr}(A,U,C,D) \cdot \mathrm{cr}(U,B,C,D) = \frac{\mathrm{ar}(B,C,D)}{\mathrm{ar}(A,C,D)}. \qquad (5.9)$$

Finally, it should be said that cross ratios can also be defined for quadruples of concurrent lines. It is easy to prove that the cross ratio of four concurrent lines a, b, c, d equals the cross ratio of the four points $l \cap a$, $l \cap b$, $l \cap c$, $l \cap d$ for any line l which is not incident with the common point of a, b, c, and d. With the help of the cross ratio formula (5.6), one can show that the cross ratio of four concurrent lines a, b, c, and d equals

$$\mathrm{cr}(a,b,c,d,) = \frac{\sin(\sphericalangle a,c) \cdot \sin(\sphericalangle b,d)}{\sin(\sphericalangle a,d) \cdot \sin(\sphericalangle b,c)}. \qquad (5.10)$$

Theorem 5.3.1 *The cross ratio of four collinear points (four concurrent lines) is invariant under perspectivities and projectivities.*

A projectivity $\pi : l \to m$ that maps points A, B, C to A', B', C' maps a point X to a point X' if, and only if, $\mathrm{cr}(A,B,C,X) = \mathrm{cr}(A',B',C',X')$.

Proof: We show that $\mathrm{cr}(A,B,C,D)$ is invariant under perspectivities. Let $A = (1:0:0)$, $B = (0:1:0)$, $C = (1:1:0)$, $D = (d_0:d_1:0)$, *i.e.*, the four points are lying on the line l with homogeneous equation $x_2 = 0$, see Figure 5.19. Without loss of generality, we can use $Z = (0:0:1)$ as the center of a perspectivity from l to the line m with homogeneous equation $x_1 - \lambda x_2 = 0$ with $\lambda \in \mathbb{F}$. The choice of m means no restriction. Due to the product rule (5.7), we can express each cross ratio in terms of cross ratios where $A = l \cap m$ is involved.

The perspectivity $l \overset{Z}{\barwedge} m$ sends A, B, C, and D to the points

$$A' = (1:0:0), \quad B' = (0:\lambda:1), \quad C' = (\lambda:\lambda:1), \quad D' = (\lambda d_0:\lambda d_1:d_1).$$

We introduce homogeneous coordinates on the line m and let $A' = (1:0)$ and $B' = (0:1)$. Then, $C' = (\lambda:1)$ and $D' = (\lambda d_0:d_1)$. With (5.6), we find $\mathrm{cr}(A',B',C',D') = \frac{d_0}{d_1} = \mathrm{cr}(A,B,C,D)$.

Each projectivity is a finite sequence of perspectivities (cf. Definition 5.2.2). Consequently, the cross ratio of four points (lines) is not altered by projectivities.

It is also clear that the cross ratio between the lines in a pencil is invariant under projectivities because the coordinatization on the line m is equivalent to the aforementioned coordinatization in the pencil about Z.

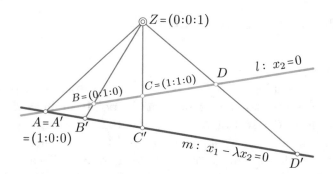

FIGURE 5.19. Proper choice of a coordinate system, cf. proof of Theorem 5.3.1

For given mutually distinct collinear points A, B, C and a given cross ratio $\delta = \mathrm{cr}(A, B, C, D)$, the fourth point D is uniquely determined. This follows from (5.6), since the given δ leads to a linear homogeneous equation for the coordinates of D. If now A', B', C' are the images of A, B, C, under a projectivity, then there is a unique point D' satisfying $\mathrm{cr}(A, B, C, D) = \delta = \mathrm{cr}(A', B', C', D')$. Since projectivities are bijective, D' is the image of D. $\blacksquare$

Let $\pi : l \to m$ be a projectivity between two lines l and m that maps the points A, B, C to the points A', B', C'. If a point X has homogeneous coordinates $(x_0 : x_1)$ with respect to the frame $F = (A, B; C)$, then its image point $\pi(X) = X'$ has homogeneous coordinates $(x_0' : x_1')$ with respect to the frame $F' = (A', B'; C')$. Since cross ratios are invariant under projectivities, we have $\mathrm{cr}(A, B, C, X) = \mathrm{cr}(A', B', C', X')$, and thus,

$$x_1 : x_0 = x_1' : x_0'.$$

If $A = \mathbf{a}\mathbb{F}$, $B = \mathbf{b}\mathbb{F}$, $C = \mathbf{c}\mathbb{F}$, $X = \mathbf{x}\mathbb{F}$, $A' = \mathbf{a}'\mathbb{F}$, $B' = \mathbf{b}'\mathbb{F}$, $C' = \mathbf{c}'\mathbb{F}$, $X' = \mathbf{x}'\mathbb{F}$ and $\mathbf{a} + \mathbf{b} = \mathbf{c}$, $\mathbf{a}' + \mathbf{b}' = \mathbf{c}'$, then in the case $\mathbf{x} = \mathbf{a}x_0 + \mathbf{b}x_1$, we can set $\mathbf{x}' = \mathbf{a}'x_0 + \mathbf{b}'x_1$ which shows the linearity of the induced mapping of vectors.

5.4 Harmonic quadruples

The harmonic conjugate D of a point C with respect to a pair of points (A, B) is closely related to reflections in points or lines. If not stated otherwise, from this section on we assume that FANO's axiom holds true, or in terms of algebra, that the characteristic of the field $\mathbb{F}$ is different from 2. We shall see that harmonic quadruples of collinear points can be easily characterized and the construction of a harmonic conjugate, *i.e.*, the construction of the "fourth harmonic point" is linear. First, we give a purely synthetic definition:

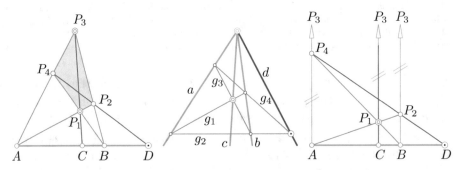

FIGURE 5.20. Left: Construction of a the fourth harmonic point D for given A, B, and C. Middle: Harmonic quadruple of lines obtained by the dualized construction. Right: Constructing the fourth harmonic point using an ideal point P_3 simplifies the construction in $\mathbb{P}^2(\mathbb{R})$.

Definition 5.4.1 *Let A, B, C, D be four collinear points. The pair (C, D) lies harmonic to the pair (A, B) if A and B are opposite vertices of a quadrilateral while C and D lie on the remaining diagonals. We denote harmonic quadruples by $H(A, B, C, D)$.*

How to find the "fourth harmonic point" if A, B, and C are given? Assume the three given points are mutually distinct. Then, we choose two points P_1 and P_3 which are collinear with C, as can be seen in Figure 5.20 (left). The lines $[A, P_1]$ and $[B, P_3]$ intersect in P_2. Finally, the lines $[A, P_3]$ and $[B, P_1]$ intersect in P_4. Now, we have a quadrangle $P_1 P_2 P_3 P_4$, and obviously, A and B are two of its diagonal points. The point C lies on the diagonal $[P_1, P_3]$ of the quadrangle. Then, $D := [A, B] \cap [P_2, P_4]$ is the harmonic conjugate of C with respect to (A, B).

In order to find the harmonic conjugate of a line for three concurrent lines, we dualize the construction of the harmonic conjugate point (see Figure 5.20, middle). It will turn out that the harmonic conjugate D does not depend on the choice of P_1, P_3 as long as C, P_1, and P_3 are mutually distinct because (DE) holds. Sometimes, things become simpler if P_3 is chosen as an ideal point (cf. Figure 5.20).

A closer look at Figure 5.20 shows that $H(A, B, C, D)$ is equivalent to $H(A, B, D, C)$, and further, $H(A, B, C, D)$ implies $H(B, A, C, D)$. Moreover, $H(A, B, C, D)$ also implies $H(C, D, A, B)$. Consequently, there are eight permutations of the four points A, B, C, D which do not harm harmonic quadruples.

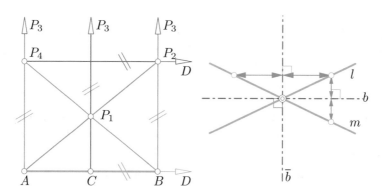

FIGURE 5.21. Left: C is the midpoint of AB and the ideal point D is the harmonic conjugate of C with respect to (A, B). Right: Two lines l and m together with their interior and exterior angle bisectors b and $\bar{b}$ form a harmonic quadruple.

In $\mathbb{P}^2(\mathbb{R})$, harmonic conjugates are closely related to midpoints of lines. Assume C is the midpoint of AB (see Figure 5.21): If C is the midpoint of a line segment AB, then the line $[P_2, P_4]$ that meets $[A, B]$ in the harmonic conjugate D is parallel to $[A, B]$. Therefore, D is the ideal point of the line $[A, B]$.

The pair $(b, \bar{b})$ of angle bisectors of two non-parallel lines l and m is the set of all points which are equidistant to the lines l and m. The lines b and $\bar{b}$ intersect at $l \cap m$ (for this point has zero distance to both lines l and

m), and b and $\bar{b}$ are orthogonal. Any line parallel to b or $\bar{b}$ meets l and m in a pair of points which are equidistant to $\bar{b}$ or b. Thus, $H(l, m, b, \bar{b})$ holds for any pair (l, m) of non-parallel lines and the pair $(b, \bar{b})$ of angle bisectors.

From Theorem 5.2.3 we know that there exists a projectivity from a line l onto itself with $A \mapsto B$, $B \mapsto A$, $C \mapsto C$, and $D \mapsto D$ for any four mutually distinct points A, B, C, and D. Therefore, $\delta = \mathrm{cr}(A, B, C, D) = \mathrm{cr}(B, A, C, D) = 1/\mathrm{cr}(B, S, C, D) = 1/\delta$. Harmonic position implies $\delta^2 = 1$, hence $\delta = \pm 1$. However, $\delta = +1$ would mean $C = D$ which is impossible in a Fano plane. We obtain

$$H(A, B, C, D) \implies \mathrm{cr}(A, B, C, D) = -1.$$

Also the converse is valid, since the cross ratio defines D uniquely and there is also just one harmonic conjugate.

Lemma 5.4.1 *Four mutually different collinear points A, B, C, D and four mutually different concurrent lines a, b, c, d are harmonic if, and only if, $\mathrm{cr}(A, B, C, D) = \mathrm{cr}(a, b, c, d) = -1$.*

This shows that the harmonic position is invariant under projectivities, and the construction of the fourth harmonic conjugate, as shown in Figure 5.20, gives a unique result independent on the choice of the points P_1 and P_3.

Note that permutations of A, B, C, and D yield only three different values for the cross ratio: -1, 2, $\frac{1}{2}$. From the algebraic point of view, we can say: A harmonic quadruple of points or lines can be characterized by the fact that three pairs of cross ratios (on any choice of ordering in the quadruple) are equal, *i.e.*, $\delta = \delta^{-1}$, $1 - \delta = (\delta - 1)/\delta$, and $(1 - \delta)^{-1} = \delta/(\delta - 1)$. The case $\delta = (\delta - 1)/\delta$ leads to $\delta = \frac{1}{2} \pm i\frac{\sqrt{3}}{2}$. These complex values characterize *equi-anharmonic* quadruples. These do not play an important role in the theory of conics, but in the theory of cubics.

If $\overline{AB} = b$, $\overline{AC} = c$, and D is the fourth harmonic, then we have

$$\overline{AD} = \frac{bc}{2c - b}$$

which can easily be shown with (5.6) or (5.9). This shows that $d = \infty$ if $c = \frac{b}{2}$, *i.e.*, the point at infinity D is the harmonic conjugate of the midpoint C of AB. Of course, this can be concluded from Figure 5.21.

■ **Example 5.4.1** On a line in the complex projective plane $\mathbb{P}^2(\mathbb{C})$, let four points with homogeneous coordinates $(1:x)$, $(1:-x)$, $(1:ix)$, and $(1:-ix)$ with $i^2 = -1$ be given. Show that they form a harmonic quadruple.

These four points (like any other) can be considered as points in the complex plane $\mathbb{C} \cong \mathbb{R}^2$. We identify a point (x_0, x_1) with the complex number $x_0 + ix_1$. Then, $(1:\pm x)$ yields $1 \pm ix$ and $(1:\pm ix)$ leads to $1 \mp x$. These four points lie on a circle with the equation $(x_0 - 1)^2 + x_1^2 = 1$. It can be shown that four points in the Euclidean plane $\mathbb{E}^2$ considered as the complex plane $\mathbb{C}$ are *concyclic* if, and only if, their cross ratio is real.

■ **Example 5.4.2** Harmonic quadruples in some finite projective planes. In the Fano plane there are only three points on each line and the same number of lines incident with each point. Thus, a fourth point or line that completes a harmonic quadruple does not exist. The cross ratio of three points (lines) in the Fano plane can attain only the values 0, 1, ∞ (provided that we have used an analytic model, such as $\mathbb{P}^2(\mathbb{Z}_2)$).

In the projective plane of order three, each line carries four points (and each pencil consists of four concurrent lines). The choice of any three points on a line (lines in a pencil) yields the fourth point (line) as the harmonic conjugate independent from the ordering of points (lines). Synthetically and without any computation, this can be verified with the help of the incidence diagram given in Figure 5.1. If one prefers the analytic approach, then we build the vector space model $\mathbb{P}^2(\mathbb{Z}_3)$. There, $\mathbb{Z}_3$ consists of the elements 0, 1, 2. In order to meet the requirements of the definition of the cross ratio, we add ∞ to the field $\mathbb{Z}_3$. For any four mutually distinct points, the cross ratio has to be different from 0, 1, and ∞. Thus, it has to be 2. Actually, $2 \equiv -1$ (in $\mathbb{Z}_3$). Moreover, we have $2 \equiv -1 \equiv (2)^{-1}$. Consequently, any four (mutually distinct) points (lines) form a harmonic quadruple.

Harmonic conjugate lines and points - Triangle polarity

In Projective Geometry as well as in Triangle Geometry there appears a mapping which is sometimes called the *triangle polarity*.

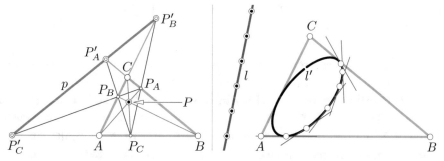

FIGURE 5.22. Left: The line p is the polar line of P with regard to the base triangle ABC. The line p carries the three harmonic conjugates P'_A, P'_B, and P'_C of P's projections P_A, P_B, and P_C of P from the triangle's vertices to the opposite lines. Right: The triangle polarity maps the points of a line l to the tangents of a conic l'.

We start with a triangle ABC in $\mathbb{P}^2(\mathbb{F})$. It means no restriction to assume that $A = (1 : 0 : 0)$, $B = (0 : 1 : 0)$, and $C = (0 : 0 : 1)$. Let further $P = (\xi : \eta : \zeta)$ be a point in the plane with $P \notin [A, B] \cup [B, C] \cup [C, A]$.

We project P from each vertex onto the opposite side and obtain P_A, P_B, and P_C. Then, the harmonic conjugates P_A', P_B', and P_C' with $H(B, C, P_A, P_A')$, $H(C, A, P_B, P_B')$, and $H(A, B, P_C, P_C')$ are located on a line p which is called the *polar line of p w.r.t. the triangle ABC* (Figure 5.22).

It is left as an exercise to the reader to verify that the triangle polarity maps the point $P = (\xi : \eta : \zeta)$ to the line according to

$$P = (\xi : \eta : \zeta) \mapsto p = (\eta\zeta : \zeta\xi : \xi\eta). \tag{5.11}$$

Obviously, this mapping is not linear in the coordinates of the point P and the name triangle polarity is misleading since in Section 7.1, we shall see that polarities (with regard to conics) allow a linear representation in terms of homogeneous coordinates.

Later, we shall see that the triangle polarity is not a polarity as defined in Section 7.1. However, we shall learn in Section 7.5 that it is a quadratic Cremona transformation.

Analytic representation of projectivities

Later, when we derive the equation and a parametrization of a conic, we need to have an analytic description of a projectivity. The case of a projectivity $\pi : l \to m$ is already described on page 205. It is sufficient to deal with projectivities between two ranges of points. The dual case does not differ very much.

In the following, we treat the case of a projectivity $\pi : l \to l$. We have already learned that, due to Theorem 5.3.1, for all points $X \in l$ the homogeneous coordinates of $X' = \pi(X)$ in the frame $(A', B'; C')$ equal that of X in the frame (A, B, C). Therefore, in any fixed frame on l, the coordinates of X' can be obtained from the coordinates of X by multiplication with a regular matrix. The following theorem clarifies how to obtain this matrix:

Theorem 5.4.1 *In any projective coordinate frame on a line l, each projectivity $\pi : l \to l$ is described by a bijective linear mapping, and conversely, each bijective linear map defines a projectivity on l.*

Proof: We can represent the given points A, B, C and A', B', C' in $\mathbb{P}^2(\mathbb{F})$ by homogeneous coordinates as $A = \mathbf{a}\mathbb{F}$, $B = \mathbf{b}\mathbb{F}$, $C = \mathbf{c}\mathbb{F}$ and $A' = \mathbf{a}'\mathbb{F}$, $B' = \mathbf{b}'\mathbb{F}$, $C' = \mathbf{c}'\mathbb{F}$.

Since scalar multiples do not matter when dealing with homogeneous coordinates, we can rescale the representatives of all given points. Therefore, we can replace $\mathbf{a}$ and $\mathbf{b}$ by $\widetilde{\mathbf{a}} = \mathbf{a} \cdot \alpha$ and $\widetilde{\mathbf{b}} = \mathbf{b} \cdot \beta$ such that $\widetilde{\mathbf{a}} + \widetilde{\mathbf{b}} = \mathbf{c}$. The scaling factors α and β can be found as the solutions of a system of three inhomogeneous linear equations. This system is guaranteed to have a unique solution since A, B, C are different but collinear points, and thus, only one representative is linearly dependent from the others. We also rescale the representatives of the image points such that $\widetilde{\mathbf{a}'} + \widetilde{\mathbf{b}'} = \mathbf{c}'$. Hence, $\pi : l \to l$ is represented by the linear map $\widetilde{\mathbf{a}}x_0 + \widetilde{\mathbf{b}}x_1 \mapsto \widetilde{\mathbf{a}'}x_0' + \widetilde{\mathbf{b}'}x_1'$ for all $(x_0, x_1) \in \mathbb{F}^2$.

In any frame $F = (P_0, P_1; P)$ on l, the given points can be represented by vectors $\mathbf{a}_F, \dots,$ $\mathbf{c}_F' \in \mathbb{F}^2$ obeying $\mathbf{a}_F + \mathbf{b}_F = \mathbf{c}_F$ and $\mathbf{a}_F' + \mathbf{b}_F' = \mathbf{c}_F'$. When we write these vectors as columns in matrices, the projectivity π can be represented as

$$\begin{pmatrix} y_0' \\ y_1' \end{pmatrix} = \mathbf{T} \begin{pmatrix} y_0 \\ y_1 \end{pmatrix} \quad \text{with} \quad \mathbf{T} = \begin{pmatrix} \mathbf{a}_F' & \mathbf{b}_F' \end{pmatrix} \begin{pmatrix} \mathbf{a}_F & \mathbf{b}_F \end{pmatrix}^{-1}. \tag{5.12}$$

The coordinates of $X = \widetilde{\mathbf{a}}x_0 + \widetilde{\mathbf{b}}x_1$ and $X' = \widetilde{\mathbf{a}'}x_0' + \widetilde{\mathbf{b}'}x_1'$ in the frame $F = (P_0, P_1; P)$ are

$$\begin{pmatrix} y_0 \\ y_1 \end{pmatrix} = \begin{pmatrix} \mathbf{a}_F & \mathbf{b}_F \end{pmatrix} \begin{pmatrix} x_0 \\ x_1 \end{pmatrix} \quad \text{and} \quad \begin{pmatrix} y_0' \\ y_1' \end{pmatrix} = \begin{pmatrix} \mathbf{a}_F' & \mathbf{b}_F' \end{pmatrix} \begin{pmatrix} x_0 \\ x_1 \end{pmatrix}.$$

Obviously, the multiplication of $(y_0, y_1)^{\mathrm{T}}$ with the matrix $\mathbf{T}$ as given in (5.12) yields the requested result $(y_0', y_1')^{\mathrm{T}}$ for all $(x_0, x_1) \in \mathbb{F}^2$. ∎

■ **Example 5.4.3** Projective mappings on a line in the real projective plane. We look at some projective mappings on a line. Their actions can be seen in Figure 5.23. First, we shall write these mappings in affine coordinates as functions over the real numbers in order to convince ourselves that these are elementary and well-known functions. After rewriting these mappings in terms of homogeneous coordinates, we shall see that we are dealing with projective mappings. Along the way, we can introduce some new vocabulary and study some special types of projective mappings:

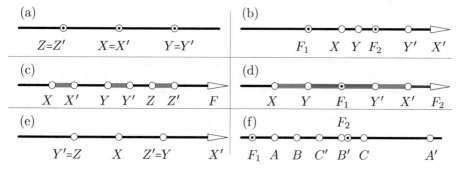

FIGURE 5.23. The projectivities from Example 5.4.3: (a) the identity mapping, (b) an inversion, (c) a translation, (d) the reflection in F_1, (e) the anti-inversion, (f) a further hyperbolic projectivity.

The identity mapping $\text{id} : \mathbb{R} \to \mathbb{R}$ is given by $x \mapsto x$ for any $x \in \mathbb{R}$. We introduce homogeneous coordinates, and thus, any point with inhomogeneous *coordinate* x is now represented by $(1 : x)$. Since x is sent to x via id, the homogeneous representation is given by

$$(1 : x) \mapsto (1 : x) \quad \Longleftrightarrow \quad \begin{pmatrix} p_0 \\ p_1 \end{pmatrix} \mapsto \begin{pmatrix} 1 & 0 \\ 0 & 1 \end{pmatrix} \begin{pmatrix} p_0 \\ p_1 \end{pmatrix}.$$

Thus, the identity mapping on a projective line is described by the identity matrix $\mathbf{I}_2$ or by one of its non-trivial scalar multiples, for multiples of representatives do not change the points.

Any point F that coincides with its image is called a fixed point. It is obvious that each point is a fixed point under id.

On page 247, we shall see how to find fixed points by ruler and compass, even if the projectivity is not the identity mapping. Geometrically speaking, the compass is replaced by an arbitrary conic.

Inversion. Now, we study the mapping $x \mapsto \frac{1}{x}$ which, considered as a mapping over the real or complex numbers, causes troubles at $x = 0$. We shall see that on the projective line, there are no exceptional points. In terms of homogeneous coordinates, we write

$$(1 : x) \mapsto \left(1 : \frac{1}{x}\right) = (x : 1) \quad \Longleftrightarrow \quad \begin{pmatrix} p_0 \\ p_1 \end{pmatrix} \mapsto \begin{pmatrix} 0 & 1 \\ 1 & 0 \end{pmatrix} \begin{pmatrix} p_0 \\ p_1 \end{pmatrix}.$$

From this, we learn that the point $x = 0$ is mapped to the point with homogeneous coordinates $(0 : 1)$ which is the ideal point of the line.

The computation of fixed points is equivalent to the determination of the eigenvectors of the matrix. We find $\mathbf{e}_1 = (1, -1)$ and $\mathbf{e}_2 = (1, 1)$ which correspond to the points $x = -1$ and $x = 1$. Thus, the projectivity has *two fixed points* and is, therefore, called *hyperbolic*. The mapping $x \mapsto \frac{1}{x}$ is also called *inversion*. It is the one-dimensional analogue to the *reflection in a circle*. Note that the center of inversion, *i.e.*, the point $x = 0$ and the ideal point are interchanged.

It is worth saying that the inversion is *involutive*, *i.e.*, the mapping itself is not the identity, but applying it twice yields the identity. Involutive mappings are sometimes called *involutions* and play an important role in the theory of conics.

Translation. Let us now consider the mapping $x \mapsto x + 1$. In fact, this is a *translation*. All points on the line are translated about one unit. If we use homogeneous coordinates, we can describe this mapping with the help of a linear mapping, though a translation is not linear considered as a mapping on the real line. We find

$$(1 : x) \mapsto (1 : x + 1) \quad \Longleftrightarrow \quad \begin{pmatrix} p_0 \\ p_1 \end{pmatrix} \mapsto \begin{pmatrix} 1 & 0 \\ 1 & 1 \end{pmatrix} \begin{pmatrix} p_0 \\ p_1 \end{pmatrix}.$$

We immediately see that this linear mapping has only one eigenvector, and thus, the projective mapping has only *one fixed point*; the point $(0 : 1)$ which corresponds to the ideal point $x = \infty$. Projectivities with precisely one fixed point/line are called *parabolic*.

Reflection in a point. The *reflection in a point* can be given by $x \mapsto -x$. Without loss of generality, we have chosen the point $x = 0$ for the *center of the reflection*. The corresponding linear mapping is given by

$$(1 : x) \mapsto (1 : -x) \quad \Longleftrightarrow \quad \begin{pmatrix} p_0 \\ p_1 \end{pmatrix} \mapsto \begin{pmatrix} 1 & 0 \\ 0 & -1 \end{pmatrix} \begin{pmatrix} p_0 \\ p_1 \end{pmatrix}.$$

Again, we find two fixed points: $(1 : 0)$ and $(0 : 1)$ or, in terms of inhomogeneous coordinates, $x = 0$ and the ideal point $x = \infty$, and thus, the reflection is a hyperbolic projectivity.

There is an obvious advantage in considering a line to be projectively closed. Otherwise, we would not have seen that a reflection has two fixed points.

The reflection in X is involutive which is clear from its definition. On the other hand, this property can also be seen from the matrix representation: The square of the transformation matrix is a scalar multiple of $\mathbf{I}_2$.

Anti-inversion. The mapping $x \mapsto -\frac{1}{x}$ is a composition of an inversion and the reflection in a point. In the older literature, one can find the name *anti-inversion*. The matrix representation of this projectivity is given by

$$(1:x) \mapsto \left(1:-\frac{1}{x}\right) = (x:-1) \quad \Longleftrightarrow \quad \begin{pmatrix} p_0 \\ p_1 \end{pmatrix} \mapsto \begin{pmatrix} 0 & 1 \\ -1 & 0 \end{pmatrix} \begin{pmatrix} p_0 \\ p_1 \end{pmatrix}.$$

This mapping has no real fixed point, since $(1,i)$ and $(1,-i)$ are the two linearly independent eigenvectors of the linear mapping. Because of the absence of real fixed points, the anti-inversion is called an *elliptic* projectivity. If we consider this mapping as mapping over the complex number field, then it would be hyperbolic.

Some projective mapping. In Figure 5.23 (f), there is one more projectivity. After fixing an origin $x = 0$ and further a unit point $x = 1$ of an affine coordinate system, we can assign homogeneous coordinates to all points given there. Suppose, we have

$$A = (1:-2), \quad B = (1:-1), \quad C = (1:2),$$
$$A' = (1:\ 6), \quad B' = (1:\ 1), \quad C' = (1:0).$$

As we have shown above, we rescale the coordinate vectors of A and B so that they sum up to a representative of C; and we do the same with A', B', and C': Since $-3 \cdot (1,-2) + 4 \cdot (1,-1) = (1,2) = C$ and $-1 \cdot (1,6) + 6 \cdot (1,1) = (5,0) = 5 \cdot C'$, we arrive at

$$\widetilde{\mathbf{a}} = (-3:6), \quad \widetilde{\mathbf{b}} = (4:-4), \quad \widetilde{\mathbf{a}}' = (-1:-6), \quad \widetilde{\mathbf{b}}' = (6:6),$$

and with (5.12) this yields

$$\mathbf{T} = \begin{pmatrix} -1 & 6 \\ -6 & 6 \end{pmatrix} \begin{pmatrix} -3 & 4 \\ 6 & -4 \end{pmatrix}^{-1} \cong \begin{pmatrix} -1 & 6 \\ -6 & 6 \end{pmatrix} \begin{pmatrix} 4 & 4 \\ 6 & 3 \end{pmatrix} \cong \begin{pmatrix} 16 & 7 \\ 6 & -3 \end{pmatrix}.$$

We have used the symbol $\cong$ in order to indicate that the left-hand side and right-hand side are equal up to a non-zero factor. It is left to the reader as an exercise to show that there are two fixed points $F_1 = (7:2)$ and $F_2 = (1:-3)$.

Theorem 5.4.1 relates linear mappings and projectivities. From Linear Algebra we know that the regular linear mappings of a vector space $\mathbb{F}^2$ form the *general linear group* $GL(\mathbb{F}, 2)$. This gives rise to the following result:

Theorem 5.4.2 *The set of projective mappings on a line l in $\mathbb{P}^2(\mathbb{F})$ together with the composition of projectivities forms a group $PGL(l)$. The group $PGL(l)$ is isomorphic to the factor group $GL(\mathbb{F}, 2)/D(\mathbb{F})$ where $D(\mathbb{F})$ is the group of dilations formed by the set of scalar multiples of the unit matrix $\mathbf{I}_2$.*

Proof: According to Theorem 5.4.1 we can formulate everything in terms of matrices: Two projectivities α, β correspond to two regular matrices $\mathbf{A}, \mathbf{B} \in \mathbb{F}^{2 \times 2}$. The matrices are unique up to a non-zero factor from $\mathbb{F}$ and, provided that $\alpha \neq \beta$, the matrices $\mathbf{A}$ and $\mathbf{B}$ are not proportional. Now, we only have to verify the laws (except the commutativity) of a group as given in the footnote on page 182. The proof of the isomorphy $PGL(l) \cong GL(\mathbb{F}, 2)/D(\mathbb{F})$ can be found in any textbook on algebra, e.g., [54], because it is the content of the homomorphism theorem for groups: If $f : A \to B$ is a homomorphism with kernel $\ker f$, then $A/\ker(f)$ is isomorphic to the image $f(A)$. ∎

The cross ratio can be used to define an invariant of projective mappings:

Theorem 5.4.3 *If $\pi : l \to l$ is a projectivity with two fixed points F_1 and F_2, then the cross ratio $\delta = \mathrm{cr}(F_1, F_2, X, \pi(X))$ is called the* characteristic cross ratio *of the hyperbolic projectivity. It is independent of X.*

Proof: Without loss of generality, we may assume that $F_1 = (1 : 0)$ and $F_2 = (0 : 1)$ are the fixed points of π. Then, the transformation matrix $\mathbf{T}$ of π is a diagonal matrix, and we may set

$$\mathbf{T} = \begin{pmatrix} \delta & 0 \\ 0 & 1 \end{pmatrix}$$

with $\delta \neq 1$. For any point $X = (x_0 : x_1) \neq F_1, F_2$ we have $\pi(X) = (\delta x_0 : x_1)$, and thus, $\mathrm{cr}(F_1, F_2, X, \pi(X)) = \delta$. ∎

The projectivity π is involutive if, and only if, $\delta = -1$. Therefore, a hyperbolic projectivity is involutive, *i.e.*, $\pi^2 = \mathrm{id}_l$ and $\pi \neq \mathrm{id}_l$ if, and only if, $H(F_1, F_2, X, \pi(X))$ for any X. According to (FT), an involutive projectivity (or *involution* in brief) π (on a line) is uniquely determined by prescribing two pairs (X, X') and (Y, Y') (which are four mutually distinct points) since $\pi(X') = X$ and (X', X) is the third pair of assigned elements. By virtue of Theorem 5.2.3 (with $A = X$, $B = X'$, $C = Y$, $D = Y'$), π also satisfies the condition $\pi(Y') = Y$.

A hyperbolic involutive projectivity is uniquely determined, if we prescribe the two fixed points $F_1 \neq F_2$ because, for any pair (X, X') of assigned points, we have $H(F_1, F_2, X, X')$, *i.e.*, the points F_1, F_2, X, and X' form a harmonic quadruple.

Parabolic projectivities have only one fixed point, and instead of a characteristic cross ratio we can state:

Lemma 5.4.2 *If $\pi : l \to l$ is a parabolic projectivity on a line l with fixed point F, then we have*

$$H(F, \pi(X), X, \pi(\pi(X)))$$

for all X.

Proof: We assume that $F = (1 : 0)$ is the only fixed point of π. Therefore, the transformation matrix $\mathbf{T}$ of π is an upper triangular matrix of the form

$$\mathbf{T} = \begin{pmatrix} 1 & c \\ 0 & 1 \end{pmatrix}$$

with $c \neq 0$. If $X = (x_0 : x_1) \neq F$, then $\pi(X) = (x_0 : x_0 + cx_1)$ and $\pi(\pi(X)) = (x_0 : x_0 + 2cx_1)$. So, we have $\mathrm{cr}(F, \pi(X), X, \pi(\pi(X))) = -1$.   ∎

■ **Example 5.4.4** The trace of $\mathbf{T}$. Here and in the following, $\mathrm{tr}\mathbf{T}$ denotes the *trace* of the matrix $\mathbf{T}$, *i.e.*, the sum of the diagonal elements:

$$\mathrm{tr}\mathbf{T} = \mathrm{tr}\begin{pmatrix} t_{00} & t_{01} \\ t_{10} & t_{11} \end{pmatrix} = t_{00} + t_{11}.$$

Lemma 5.4.3 *A projectivity* $\pi : l \to l$, $\pi \neq \mathrm{id}_l$ *with* $\mathbf{x} \mapsto \mathbf{T}\mathbf{x}$ *is involutive if, and only if,* $\mathrm{tr}\mathbf{T} = 0$.

Proof: First, let us assume that π is involutive. Then, $\pi \circ \pi = \mathrm{id}_l$, and consequently, $\mathbf{T}^2 = c\mathbf{I}_2$. The square of $\mathbf{T}$ equals

$$\mathbf{T}^2 = \begin{pmatrix} t_{00}^2 + t_{01}t_{10} & t_{01}(t_{00} + t_{11}) \\ t_{10}(t_{00} + t_{11}) & t_{01}t_{10} + t_{11}^2 \end{pmatrix}.$$

The right upper entry of $\mathbf{T}$ has to be zero, and thus, $t_{01}(t_{00} + t_{11}) = 0$. Now, we have two subcases: If $t_{00} + t_{11} = \mathrm{tr}\mathbf{T} = 0$, the square of $\mathbf{T}$ becomes a scalar diagonal matrix with entry $c = t_{00}^2 + t_{01}t_{10} \neq 0$ and is, thus, a scalar multiple of $\mathbf{I}_2$. The second case corresponds to $t_{01} = 0$. In this case, $\mathbf{T}^2$ is a lower triangular matrix

$$\begin{pmatrix} t_{00}^2 & 0 \\ t_{10}(t_{00} + t_{11}) & t_{11}^2 \end{pmatrix}$$

which becomes diagonal if, and only if, either $t_{10} = 0$ or $t_{11} + t_{00} = 0$. If now $t_{10} = 0$, then $\mathbf{T}^2$ is a scalar multiple of $\mathbf{I}_2$ and $\mathbf{T}$ describes an involutive projectivity if, and only if, $t_{00}^2 = t_{11}^2$, *i.e.*, in case $t_{00} = t_{11}$ or $t_{00} = -t_{11}$. The case $t_{00} = t_{11}$ is excluded by assumption ($\pi \neq \mathrm{id}_l$) and the case $t_{00} = -t_{11}$ is equivalent to $\mathrm{tr}\mathbf{T} = 0$.   ∎

A comparison with Lemma 5.4.2 reveals that involutions can never be parabolic in Fano planes, *i.e.*, if $\mathrm{char}\mathbb{F} \neq 2$

Commuting involutions.

Lemma 5.4.4 *Assume* $\alpha, \beta : l \to l$ *are involutive projectivities on a line* l *with respective transformation matrices* $\mathbf{A}$ *and* $\mathbf{B}$, *and* $\alpha \neq \beta$.

The two involutive projectivities $\alpha, \beta : l \to l$ *commute, i.e., they fulfill* $\alpha \circ \beta = \beta \circ \alpha$ *if, and only if, their product is involutive:*

$$\mathbf{A}\mathbf{B} = \mathbf{B}\mathbf{A} \iff (\mathbf{A}\mathbf{B})^2 = c\mathbf{I}_2, \quad c \in \mathbb{F} \setminus \{0\}.$$

Proof: First, we assume that α and β commute: $\mathbf{A}\mathbf{B} = \mathbf{B}\mathbf{A}$. We have to show that this implies $(\mathbf{A}\mathbf{B})^2 = c\mathbf{I}_2$ with some $c \in \mathbb{F}$.

Since $\mathbf{A}$ and $\mathbf{B}$ are matrices of involutions, we have $\mathbf{A}^2 = a\mathbf{I}_2$ and $\mathbf{B}^2 = b\mathbf{I}_2$, and consequently, $\mathbf{A} = a\mathbf{A}^{-1}$ and $\mathbf{B} = b\mathbf{B}^{-1}$ with some $a, b \in \mathbb{F} \setminus \{0\}$. Now, we compute $(\mathbf{A}\mathbf{B})^2 = \mathbf{A}\mathbf{B} \cdot \mathbf{A}\mathbf{B} =$

$\mathbf{AB} \cdot \mathbf{BA}$, where the latter makes use of the assumption. Now, we use the fact that α and β are involutive: $\mathbf{ABBA} = \mathbf{A}b\mathbf{I}_2\mathbf{A} = b\mathbf{AA} = ab\mathbf{I}_2$.

Conversely, we assume that $(\mathbf{AB})^2 = c\mathbf{I}_2$ and show that α and β commute:

We start with $(\mathbf{AB})^2 = c\mathbf{I}_2$ and multiply with the inverses of $\mathbf{B}$ and $\mathbf{A}$ from the right and find $\mathbf{ABA} = c\mathbf{B}^{-1}$ and $\mathbf{AB} = c\mathbf{B}^{-1}\mathbf{A}^{-1}$. Since α and β are involutive, we have $\mathbf{A} = a\mathbf{A}^{-1}$ and $\mathbf{B} = b\mathbf{B}^{-1}$ and find $\mathbf{AB} = a^{-1}b^{-1}c\mathbf{BA}$. ∎

● Exercise 5.4.1 Prove the following statement: Two involutions commute if the fixed points of one involution are corresponding under the second involution.

6 Projective conics

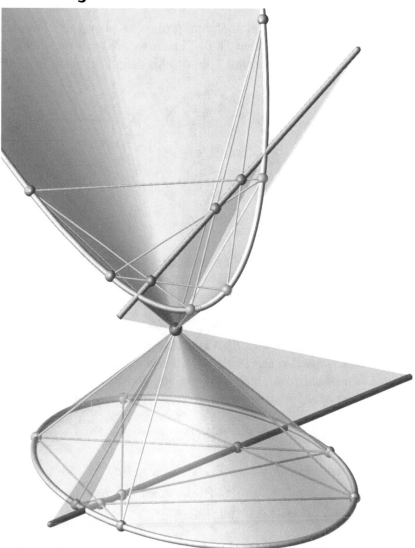

Studying conics in the framework of Projective Geometry leads to a much deeper understanding of their properties. The results are independent on the choice of the model as is the case for example with PASCAL's theorem stating that any six points on a conic define a Pascal axis. The distinction between the three affine types (ellipse, parabola, hyperbola) is no longer necessary.

6.1 Steiner's definition of a conic

We will stick to the convention that our projective planes are Pappian so that the fundamental theorem (FT) (Theorem 5.2.1 on page 190) is always valid. Following J. STEINER,[1] we use his definition given in 1832:

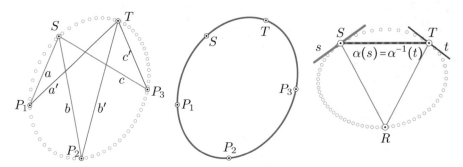

FIGURE 6.1. STEINER's definition of a conic: Left: Five points in an admissible position define a conic. Middle: the unique conic on five points. Right: Two line elements plus one point also define a unique conic.

Definition 6.1.1 STEINER's **definition:** *Assume we are given two pencils of lines with respective carriers $S \neq T$. Let $\sigma : S \barwedge T$ be a projectivity that cannot be reduced to a single perspectivity from the pencil S to the pencil T. The set of all points of intersection*

$$X = l \cap \sigma(l)$$

of all lines l through S with their σ-images is a conic.

A line t that contains only one point P of a conic c is called a tangent of c at P and the pair (P, t) is called a line element.

The points S and T are sometimes referred to as the *base points of the generation* of c. However, the term base point will also appear in a different meaning when we treat pencils of conics in Section 7.3.

From Definition 6.1.1 we can conclude that there is no restriction to the base points on a conic:

---

[1] JAKOB STEINER (1796–1863) was a Swiss mathematician.

Theorem 6.1.1 *Any two points on a conic c can serve as the base points of a projective generation of c according to* STEINER's *definition.*

Proof: The theorem is valid if lines in the underlying plane contain 3, 4, or 5 points, since then conics carry as many points.

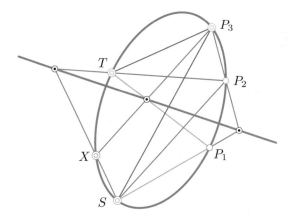

FIGURE 6.2. The base points of a conic's projective generation can be chosen freely, cf. Theorem 6.1.1.

Let now S and T be the base points of a projective generation of a conic c. Further, let P_1, P_2, P_3 be three mutually distinct and non-collinear points which, together with S and T, meet the requirements of Definition 6.1.1, compare Figure 6.2. The generating projectivity α is defined by $[S, P_i] \mapsto [T, P_i]$ with $i \in \{1, 2, 3\}$, and we call c the unique conic on S, T, P_1, P_2, and P_3.

Now, replace S with P_3 (for example) and show that the projectivity from the pencil P_3 to the pencil T generates a conic $\bar{c}$ which contains all the points of c, and thus, $\bar{c} = c$. For the technical details, we refer to [11]. ∎

In principle, the above results enables us to show the following:

Theorem 6.1.2 *A conic is defined by five points no three of which are collinear.*

Proof: We specify two of the five points as the base points S and T. Then, we use the three remaining points P_1, P_2, P_3 in order to define two times three lines for the projectivity σ (cf. Figure 6.1): Both triplets $\{[S, P_i] | i \in \{1, 2, 3\}\}$ and $\{[T, P_i] | i \in \{1, 2, 3, \}\}$ consist of three mutually distinct lines. Thus, the projectivity σ is defined by $[S, P_i] \mapsto [T, P_i]$ with $i \in \{1, 2, 3\}$. According to Theorem 6.1.1, the choice of the base points doesn't matter.

If P_1, P_2, and P_3 are collinear, then σ is a perspectivity with axis $[P_1, P_2]$ and there is no conic at all. ∎

6.2 Pascal, Pappus, Brianchon

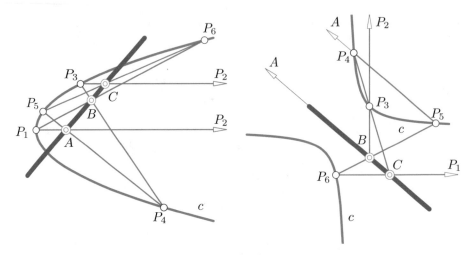

FIGURE 6.3. In the real projective plane $\mathbb{P}^2(\mathbb{R})$, PAPPUS's theorem may appear in different forms: Left: The ideal line is tangent to c at P_2. Right: The line $[P_1, P_2]$ coincides with the ideal line.

PAPPUS's theorem (cf. Figure 6.3) gives an incidence geometric, and thus, a simple criterion for six points to lie on a conic:

Theorem 6.2.1 *Let P_1, P_2, P_3, P_4, P_5, and P_6 be six points on a conic, then the three points*

$$A := [P_1, P_2] \cap [P_4, P_5], \quad B := [P_2, P_3] \cap [P_5, P_6], \quad C := [P_3, P_4] \cap [P_5, P_6]$$

are collinear.

But even the converse is true, and so we state:

Theorem 6.2.2 *Six points P_1, P_2, P_3, P_4, P_5, P_6, no three of which are collinear, lie on a conic if, and only if, the three points*

$$A := [P_1, P_2] \cap [P_4, P_5], \quad B := [P_2, P_3] \cap [P_5, P_6], \quad C := [P_3, P_4] \cap [P_5, P_6]$$

are collinear.

Proof: The synthetic proof consists of two parts:

1. First, we assume that the six points lie on a conic c, and we show that A, B, and C, as defined above, are collinear.

We use the notation from Figure 6.4 and relabel two of the six points by letting $S := P_1$ and $T := P_3$. The projectivity α from the pencil about S to the pencil about T generates c. The pencil about S is perspective to the line $g = [P_4, P_5]$, and the pencil about T is perspective to the line $h = [P_5, P_6]$. Consequently, the two ranges g and h are linked by a projectivity

$$\beta: \ g \barwedge S \overset{\alpha}{\barwedge} T \barwedge h$$

which turns out to be a perspectivity since $\alpha(P_5) = P_5$, according to Theorem 5.2.2. The projectivity β sends $A = [P_1, P_2] \cap g$ to $B = [P_2, P_3] \cap h$, for $\alpha([P_1, P_2]) = [P_2, P_3]$.

Since β is a perspectivity, A and $\beta(A) = B$ have to be collinear with the center C of the perspectivity. Let now $Q := [P_1, P_6] \cap g$ and $R := [P_3, P_4] \cap h$. Since $\alpha([P_1, P_6]) = [P_3, P_6]$, we have $\beta(Q) = P_6$. Further, $\beta([P_1, P_4]) = [P_3, P_4]$, and therefore, $\beta(P_4) = R$. Thus, the center of perspectivity is found as $C = [P_3, P_4] \cap [P_6, P_1]$ and is collinear with A and B.

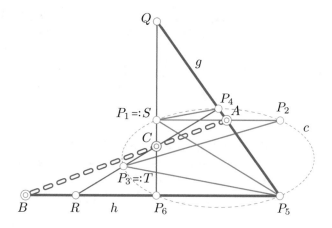

FIGURE 6.4. The synthetic proof of Theorem 6.2.2.

2. Now, assume that the six points $P_1, \ldots, P_6$ lie such that $A := [P_1, P_2] \cap g$, $B := [P_2, P_3] \cap h$, and $C := [P_3, P_4] \cap [P_6, P_1]$ are well defined and collinear. We have to show that the six points belong to one, and only one, conic.

The chain β of perspectivities from the pencil about S to the line g and further to the line h (center C) and then to the pencil about T is a projectivity, for it is a finite chain of perspectivities. It can be shown by contradiction that β is well defined. It acts in the following way: $[S, P_4] \mapsto [T, P_4]$, $[S, P_5] \mapsto [T, P_5]$, and further $[S, P_2] \mapsto [T, P_2]$, for A, B, and C are collinear. Therefore, this chain is a factorization of the projectivity α and $[S, P_6]$ is mapped to $[T, P_6]$, and thus, the point P_6 also lies on the same conic c on P_1, $\ldots$, P_5.

So far we have used P_1 and P_3 as the vertices of the pencils generating the conic. If we replace P_1 with P_5, the line $[A, B]$ remains unchanged und the generation of c is independent of the choice of the base points. For details we refer to [11].   ∎

There is also an analytic proof of PASCAL's or PAPPUS's theorem which is given in Example 6.4.3. However, the synthetic proof seems to be more powerful since it covers also the converse statement.

The dual of a conic defined by Definition 6.1.1 is a set of lines. It can be proved that in Fano planes this *dual conic* consists of the tangents of a conic. For a proof, we again refer to the literature (*e.g.* [11]). Later, in Section 7.1 on polarities this result becomes obvious.

The Theorems 6.2.1 and 6.2.2 have their dual versions. The dual version of the theorem by PAPPUS is usually referred to as BRIANCHON's theorem. It is due to the French mathematician CHARLES JULIEN BRIANCHON (1783–1864).

Theorem 6.2.3 *Six lines l_1, l_2, l_3, l_4, l_5, l_6, no three of which are concurrent, are tangents of one conic c if, and only if, the three lines*

$$a := [l_1 \cap l_2, l_4 \cap l_5], \quad b := [l_2 \cap l_3, l_5 \cap l_6], \quad c := [l_3 \cap l_4, l_6 \cap l_1]$$

are concurrent.

Both theorems, PAPPUS's and BRIANCHON's, are still valid if some points or lines happen to coincide. We formulate these special cases for the case of PAPPUS's theorem. The special cases of BRIANCHON's are nothing else but the dual counterparts, illustrated in Figure 6.6.

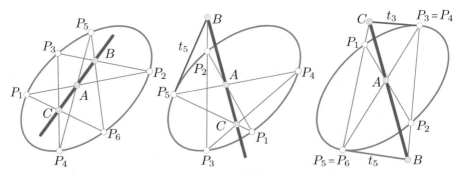

FIGURE 6.5. PAPPUS's theorem and its special forms: Left: the original version. Middle: The point P_6 has moved towards P_5. Now, the tangent t_5 at P_5 plays the role of the former line $[P_5, P_6]$. Right: A further coincidence, say $P_3 = P_4$, replaces the pair (P_3, P_4) by the line element (P_3, t_3) and PAPPUS's theorem still holds.

Figure 6.5 (on the left) shows a PAPPUS figure in the generic case. All six points on the conic c are mutually distinct, and in the closed polygon $P_1 \ldots P_6$ there are the three collinear crossing points A, B, and C. Assume now that the point P_6 is moving on c towards P_5. The theorem still remains valid if we replace the pair of points (P_5, P_6) by the line element consisting of the point P_5 and the tangent t_5 to c at P_5, see Figure 6.5 (in the middle). With the point $B := [P_2, P_3] \cap [P_5, P_6]$, we have $P_5 = [P_2, P_3] \cap t_5$. Now, we assume that a further pair of points on c coincides: Let $P_3 = P_4$ and replace the pair (P_3, P_4) by the line element (P_3, t_3) with the tangent t_3 at P_3. Now, $B = [P_2, P_3] \cap t_5$, and additionally, the point C which was defined by $C = [P_3, P_4] \cap [P_6, P_1]$ is found as $C = t_3 \cap [P_6, P_1]$. However, the three points A, B, and C are still collinear.

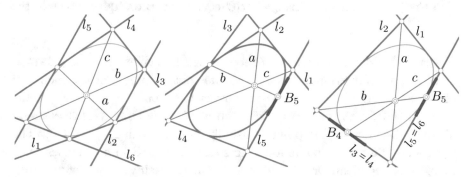

FIGURE 6.6. Some versions of the theorem by BRIANCHON, cf. Theorem 6.2.3: Left: the generic case with six mutually distinct tangents. Middle: One pair of coinciding tangents somehow "converges" to a line element. Right: Two pairs of tangents are replaced by two line elements.

The dual and degenerate versions of the PAPPUS configurations are illustrated in Figure 6.6. Note that pairs of lines are again to be replaced by line elements, *i.e.*, pairs (P, t) consisting of a tangent t and the point of contact P.

Both theorems, PAPPUS's and BRIANCHON's, apply to the most degenerate configurations where three pairs of points or three pairs of lines collapse as shown in Figure 6.7. The configuration on the left-hand side of Figure 6.7 is self-dual. The lines connecting the vertices of the tangent triangle of the conic with the opposite contact points are concurrent in the BRIANCHON point B.

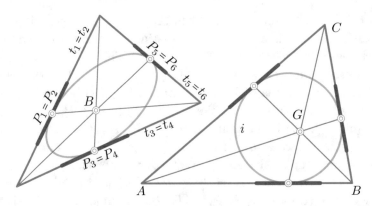

FIGURE 6.7. The theorems of PAPPUS and BRIANCHON in the most degenerate form: Left: The self-dual configuration of three line elements of a conic. Right: The Gergonne point of a triangle ABC equals the BRIANCHON point of the incircle i.

The right-hand side of Figure 6.7 shows a well-known example of a BRIAN-CHON point from triangle geometry. The contact points of the incircle i of a triangle ABC define three concurrent Cevians. The locus of concurrency is the *Gergonne point* named after JOSEPH DIAZ GERGONNE. In Triangle Geometry, this point is usually labeled with X_7, see [40]. The Gergonne point can also be found in some Cayley-Klein geometries, *e.g.*, in pseudo-Euclidean or MINKOWSKI-geometries since it is defined in a projective way. For these generalizations, we refer to [50].

■ **Example 6.2.1** Find the intersection of a conic and a line through a given point of the conic. The construction of the common points of a line x and a conic c is a quadratic problem, *i.e.*, algebraically speaking, one has to solve a quadratic equation in order to find the points of intersection. If one of the common points is already known, then the problem reduces to the search of the remaining intersection. This construction is linear. It is the solution of a linear equation.

Assume we are given five points and a line x through one of them, say S. We want to find the point of intersection of the conic on the five given points and the line x without drawing the conic itself. In the first step, we define a projectivity from the pencil of lines about S to the pencil about one of the remaining four points, say T (cf. Figure 6.8). The remaining points shall be labeled with 1, 2, and 3. The projectivity α from the pencil about S to the pencil about T is uniquely defined by

$$g_1 := [S, 1] \mapsto [g_1' := T, 1], \quad g_2 := [S, 2] \mapsto g_2' := [T, 2], \quad g_3 := [S, 3] \mapsto g_3' := [T, 3].$$

In the next step, we construct the center P_α of the projectivity α. Following Theorem 5.2.4, we have

$$P_\alpha = [[g_1 \cap g_2', g_2 \cap g_1'], [[g_2 \cap g_3'], [g_3 \cap g_2']]].$$

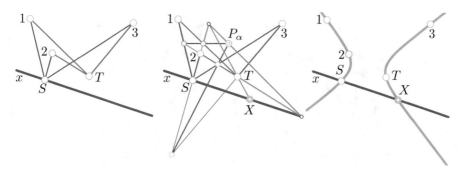

FIGURE 6.8. Construction of the remaining point of intersection of a line x with a conic. Left: The initial situation shows five points and a line through one of them. Middle: The generating projectivity is determined and the center P_α of the projectivity is constructed. Then, the line x through S is mapped to $x' = \alpha(x)$ through T, and according to STEINER's definition of a conic, $x \cap' x = X$ is the desired point of intersection. Right: The conic on the given five points, the line x, and the point X.

The third and last step is the construction of $\alpha(x)$. Again, we apply Theorem 5.2.4. Thus, we have

$$X = [[x \cap g_2', P_\alpha] \cap g_2, T] \cap x.$$

Note that every step of the construction is linear.

By constructing the remaining point of intersection, we have also shown how to find further points on a conic that is given by five of its points. Following the Definition 6.1.1, we just have to turn x around the one *a priori* known point of intersection and then construct the second one. In this way, we can find infinitely many further points of c beside the given ones.

■ **Example 6.2.2** Elementary generation of conics in the projective extension of $\mathbb{E}^2$.

Assume $S \neq T$ are proper (finite) vertices of two pencils of lines in $\mathbb{P}^2(\mathbb{R})$, and let further P_1 be a point which is not collinear with S and T. Now, we define a projectivity α in the following way: Let $l \neq [S, P_1]$ be a line through S. The image $m = \alpha(l)$ of l under α runs through T and it satisfies $\angle([P_1, S], l) = \angle([P_1, T], m)$. Take care of the orientation of angles! If the line l traverses the pencil about S, then we see that the points $l \cap m$ trace the circumcircle of STP_1 because the angles measured at S and T are the *angles of circumference* (Figure 6.9, left).

The pencils of lines about S and T can be mapped via a Euclidean congruence transformation onto each other such that each line in the pencil about S coincides with its image under alpha in the pencil about T. Therefore, we call these pencils a pair of *congruent pencils* of lines. In other words: A circle can be generated with the help of congruent pencils of lines. For the sake of completeness, one should show that this congruence transformation is a projectivity: We use Theorem 5.3.1 and (5.10), and thus, $cr([S, P_1], [S, P_2], [S, P_3], [S, X]) = cr([T, P_1], [T, P_2], [T, P_3], [T, X'])$.

What happens if we change the orientation of the angles? With the same prerequisites, assume now that the image of a line l is defined by $\angle([P_1, S], l) = - \angle([P_1, T], m)$. In Figure 6.9 (right), we can see the resulting curve. However, here we also have to show that the mapping $\alpha : l \to m$ is a projectivity. We can be sure that if congruences are projectivities, then their

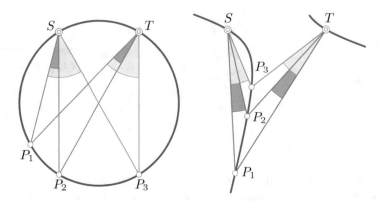

FIGURE 6.9. Elementary generation of conics: Left: A circle is generated by "congruent pencils of lines". Right: An equilateral hyperbola results if the pencils are congruent but the orientation of angles is reversed as described in Example 6.2.2.

compositions with reflections are also projectivities. Or else, we can translate the point T into $\overline{T}$ such that S and $\overline{T}$ are mirrors with respect to a line.

Further, we can state and prove what is illustrated in Figure 6.1 (right):

Theorem 6.2.4 *Let (S,s) and (T,t) be two line elements with $S \notin t$, $T \notin s$, and let further $R \notin \{s \cup t\}$ be a point. Then, there is a unique conic c through the two line elements (S,s), (T,t), and R.*

Proof: If (S,s), (T,t), and R are in an admissible position, then there is a uniquely defined projective mapping α from the pencil of lines about S to the pencil of lines about T rendering $s \mapsto [S,T]$, $[S,T] \mapsto t$, and $[S,R] \mapsto [T,R]$.

According to the construction of α, the thus generated conic c passes through S, T, and R. We only have to show that s and t are tangents to c at S and T. We recall Definition 6.1.1: A line s is a tangent of c if it contains precisely one point of c. The pre-image under α of $[S,T]$ equals s and is unique. There is no further point of c, except S, on s. If there was such a point $X \in s$ with $X \neq S$, then $[S,X] = s$ would have two images under α: $[S,T]$ and $[X,T]$. This contradicts the bijectivity of α. Similar arguments apply to the line element (T,t). ∎

As already mentioned, the dualized Definition 6.1.1 yields a dual conic, *i.e.*, the set of tangents of a conic in Fano planes. In other words: The set of tangents of a conic can be generated by a projective mapping $\alpha: l \to m$ between two ranges of points l and m provided that α is not a perspectivity. Consequently, a conic is also uniquely defined if we prescribe five tangents $t_1, \ldots, t_5$ where no three are concurrent. A conic on five tangents is illustrated in Figure 6.10.

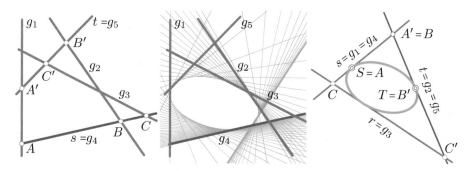

FIGURE 6.10. Applying the principle of duality: Five tangents in an admissible position define a unique conic (left) because they determine a projectivity. The conic on five tangents (middle). Two line elements plus one tangent also define one, and only one, conic if they lie in an admissible position (right).

Recall that a line element is self-dual for it is a line with an incident point. Applying the principle of duality to Theorem 6.2.4 shows that a conic is uniquely defined by prescribing two line elements, say (S, s), (T, t), and a further tangent r. Figure 6.10 shows the conic which is given that way. The generating projectivity $s \barwedge t$ is defined by $S = A \mapsto s \cap t = A'$, $s \cap a = B \mapsto T = B'$, and $s \cap r = C \mapsto t \cap r = C'$.

6.3 Equation and parametrization of a conic

In this section, we shall see that the conics defined via projectivities are algebraic curves of degree two. Therefore, we obtain a parametrization and an equation.

From Theorem 6.2.4, we know that two line elements (S, s) and (T, t) together with a point $R \notin \{s \cup t\}$ define a unique conic c. The projectivity α that generates c is defined by

$$s \mapsto [S, T], \quad [S, T] \mapsto t, \quad [R, S] \mapsto [R, T]. \tag{6.1}$$

We can impose a projective frame on the given points and lines as indicated in Figure 6.11: $S = (1 : 0 : 0)$, $T = (0 : 0 : 1)$, and $R = (1 : 1 : 1)$. Further, it is no restriction to assume that the tangents s and t meet at the point $(0 : 1 : 0)$, and therefore, s and t have the homogeneous equations $s: x_2 = 0$ and $t: x_0 = 0$. In our setting, the line $[S, T]$ has the equation $x_1 = 0$, and finally, the lines $[S, R]$ and $[T, R]$ have the equations $x_1 - x_2 = 0$ and $x_0 - x_1 = 0$.

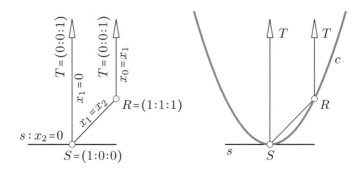

FIGURE 6.11. The proper choice of a coordinate frame simplifies the equation of the conic c. Left: the chosen frame. Right: the unique conic c on (S, s), (T, t), and R.

In order to find an analytic representation of α, we recall that the equations of the three pairs of lines determining α can be given in terms of their homogeneous coordinate vectors which are nothing but the coefficient vector of their equations. Note that non-trivial multiples of the equations stand for the same geometric objects. Consequently, (6.1) can be written either in terms of equations

$$-x_2 = 0 \mapsto -x_1 = 0, \quad x_1 = 0 \mapsto x_0 = 0, \quad x_1 - x_2 = 0 \mapsto x_0 - x_1 = 0$$

or in terms of homogeneous coordinate vectors

$$(0{:}0{:}{-}1) \mapsto (0{:}{-}1{:}0), \quad (0{:}1{:}0) \mapsto (1{:}0{:}0), \quad (0{:}1{:}{-}1) \mapsto (1{:}{-}1{:}0). \quad (6.2)$$

Here, some equations are multiplied by -1. This turns out to be useful, since $(0,0,-1) + (0,1,0) = (0,1,-1)$ and $(0,-1,0) + (1,0,0) = (1,-1,0)$. If we impose a frame on both pencils, we can assign the homogeneous coordinates $(1:0)$ to s and $[S,T]$ and the coordinates $(0:1)$ to the lines $[S,T]$ and t in the second pencil. Therefore, the mapping α from the pencil about S to the pencil about T is described via equal coordinates. This means that a line g through S with coordinate vector $\mathbf{g}(\lambda,\mu) = \lambda(0,0,-1)+\mu(0,1,0) = (0:\mu:-\lambda)$ is mapped to the line g with coordinate vector $\mathbf{g}'(\lambda,\mu) = \lambda(0,-1,0) + \mu(1,0,0) = (\mu,-\lambda,0)$ where $(\lambda,\mu) \neq (0,0)$.

The points of c are the intersection of the lines $\mathbf{g}\mathbb{F}$ and $\mathbf{g}'\mathbb{F}$. With formula(5.5) we have

$$\mathbf{c}(\lambda:\mu) = (0,\mu,-\lambda)^{\mathrm{T}} \times (\mu,-\lambda,0)^{\mathrm{T}} \cong (\lambda^2 : \lambda\mu : \mu^2) \qquad (6.3)$$

Summarizing these calculations yields:

Corollary 6.3.1 *In a suitable coordinate frame, all points of a conic c in the projective plane $\mathbb{P}^2(\mathbb{F})$ can be parametrized by a homogeneous parameter $(\lambda,\mu) \neq (0,0)$ as $(\lambda^2 : \lambda\mu : \mu^2)$. The homogeneous parameter ranges in a projective line.*

The homogeneous parameter (λ,μ) in (6.3) can be replaced by an *affine parameter* $t = \mu\lambda^{-1}$ provided that $\lambda \neq 0$. Thus, we obtain the "standard parametrization" of a conic which reads

$$\mathbf{c}(t) = (1,t,t^2)\mathbb{F}.$$

In this parametrization, we have lost one point of c, *i.e.*, the point corresponding to $t = \infty$. Since the first coordinate function is constant and equals 1, we choose in $\mathbb{P}^2(\mathbb{R})$ the lines $x_0 = 0$ as the ideal line and arrive at a representation of the affine part of c in terms of affine coordinates

$$\mathbf{c}(t) = (t,t^2).$$

From the parameter representation of c, we can deduce an equation in terms of homogeneous coordinates. With $x_0 = \lambda^2$, $x_1 = \lambda\mu$, and $x_2 = \mu^2$, we find

$$\lambda^2 \cdot \mu^2 - (\lambda\mu)^2 = x_0 x_2 - x_1^2 = 0. \tag{6.4}$$

Conversely, for any point $(x_0 : x_1 : x_2)$ with $x_0 \neq 0$ satisfying (6.4), we can set $(x_0, x_1) = (\lambda, \mu)$ and obtain $x_2 = \mu^2\lambda^{-1}$, hence $(x_0 : x_1 : x_2) = (\lambda : \mu : \mu^2\lambda^{-1})$. For $x_0 = 0$, only $(0 : 0 : 1)$ solves (6.4) which corresponds to $(\lambda : \mu) = (0 : 1)$.

Here, we should emphasize that a change of the frame does not change the algebraic degree of the curve: This change is performed by multiplying coordinates with a suitable matrix, *i.e.*, the change of coordinates is linear. So, we have:

Corollary 6.3.2 *The coordinates of the points of a conic in the projective plane $\mathbb{P}^2(\mathbb{F})$ satisfy a quadratic equation.*

On page 239, we shall show that there is only one type of conic in any projective plane.

■ **Example 6.3.1** The circle, rational parametrization, and THALES's theorem.

Assume $S = (1 : -1 : 0)$ and $T = (1 : 1 : 0)$ are the base points of a STEINER generation in the projective extension of $\mathbb{E}^2$. Further, let $s : x_0 + x_1 = 0$ and $t : x_1 - x_0 = 0$ be the tangents at the base points. Finally, the point $R = (1 : 0 : 1)$ helps to define the projectivity. In terms of Cartesian coordinates, these points are given by $(-1, 0)$, $(1, 0)$, and $(0, 1)$. Now, it is clear that S, T, and R lie on the unit circle with the inhomogeneous equation $x^2 + y^2 = 1$, and the lines s and t are tangents (cf. Figure 6.12, left).

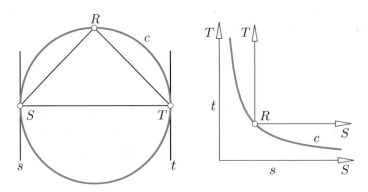

FIGURE 6.12. The conics from Example 6.3.1.

The projectivity α from the pencil about S to the pencil about T is defined by the following three lines and their respective images:

$$
\begin{aligned}
x_0 \ + \ x_1 \qquad\qquad &= \ 0 \ \mapsto \qquad\qquad\qquad\qquad x_2 \ = \ 0, \\
- \ x_2 \ &= \ 0 \ \mapsto \ -x_0 \ + \ x_1 \qquad\quad = \ 0, \\
x_0 \ + \ x_1 \ - \ x_2 \ &= \ 0 \ \mapsto \ -x_0 \ + \ x_1 \ + \ x_2 \ = \ 0.
\end{aligned}
$$

Since the equations of these lines are scaled such that the sum of the first two equals the third, we have a mapping via equal coordinates, and thus,

$$l: \ \lambda(x_0 + x_1) + \mu(-x_2) = 0 \ \text{ is mapped to } \ l': \ \lambda x_2 + \mu(-x_0 + x_1) = 0. \tag{6.5}$$

We intersect corresponding lines, *i.e.*, lines with equal coordinates

$$(\lambda : \lambda : -\mu) \cap (-\mu : \mu : \lambda) = (\lambda^2 + \mu^2 : \mu^2 - \lambda^2 : 2\lambda\mu)$$

with $\lambda : \mu \neq 0 : 0$. If we replace the homogeneous parameter $(\lambda : \mu)$ by the affine parameter $t = \lambda^{-1}\mu$ and switch to inhomogeneous coordinates by setting $x = x_1 x_0^{-1}$ and $y = x_2 x_0^{-1}$, then we arrive at

$$\left(\frac{1 - t^2}{1 + t^2}, \frac{2t}{1 + t^2} \right). \tag{6.6}$$

The latter is a rational parametrization of the unit circle, as we expected. Moreover, this parametrization is usually found via the *stereographic projection* from the line $x = 0$ to the unit circle c as shown in Figure 6.13: We parametrize the line $x = 0$ (y-axis of a Cartesian frame) by $P = (0, t)$ and project the points P from $T = (1, 0) \in c$ onto c. This means that the lines $(1 + \lambda, -t\lambda)$ have to be intersected with c. This yields exactly the parametrization of the unit circle as given in (6.6).

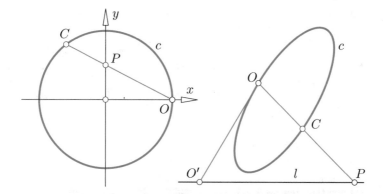

FIGURE 6.13. Stereographic projection with center O: Left: Deriving the rational parametrization of the unit circle. Right: A stereographic projection from an arbitrary conic c to a line l maps any point $C \in c$ to a unique and well-defined point P on the line l. One hole is to be filled: The stereographic image of the center $O \in c$ is the intersection O' of c's tangent at O with the line l.

The construction of the unit circle as the locus of all points $l \cap l'$ (as given in Equation (6.5)) shows the following which is called THALES's theorem (THALES OF MILETUS, Greek philosopher, 624 BC–546 BC):

Corollary 6.3.3 *In $\mathbb{E}^2$, the locus of all points from which a line segment ST is seen at right angles is a circle.*

Proof: We use homogeneous Cartesian coordinates with $S = (1 : -1 : 0)$ and $T = (1 : 1 : 0)$. Coefficient vectors of the equations of l and l' are $(\star : \lambda : -\mu)$ and $(\star : \mu : \lambda)$. The normal vectors to these lines are given by the second and third entry, so the first entry does not matter. The Euclidean scalar product of the normal vectors vanishes:

$$\left\langle \begin{pmatrix} \lambda \\ -\mu \end{pmatrix}, \begin{pmatrix} \mu \\ \lambda \end{pmatrix} \right\rangle = 0,$$

and consequently, l and l' are orthogonal. Exactly the same coordinates were used before. Therefore, we obtain the homogeneous equation $-x_0^2 + x_1^2 + x_2^2 = 0$. ∎

We can generalize the stereographic projection from a line to the unit circle in the following way (Figure 6.13): The circle c can be replaced with any conic in the projective plane $\mathbb{P}^2(\mathbb{F})$, and the center O may be an arbitrary point on c. The line l can be chosen freely, however, it is not allowed to coincide with c's tangent t_O at O. The mapping $c \to l$ is one-to-one and onto if we additionally define $O \mapsto O' := l \cap t_O$ (see Figure 6.13, right). Later, we shall learn that there are n-dimensional analogues of the planar stereographic projection.

From the definition of a stereographic projection, we obtain:

Corollary 6.3.4 *There are as many points on a conic as there are lines in a pencil or points on a line.*

Equilateral hyperbola.

Now, we return to the projective extension of $\mathbb{E}^2$. Let's generate a conic with ideal base points $S = (0 : 1 : 0)$ and $T = (0 : 0 : 1)$. If we interpret $x_0 = 0$ as the ideal line, then S and T are the ideal points of the axis of a Cartesian frame. We cannot choose a further ideal point since no three points on a conic are collinear. Let $R = (1 : 1 : 1)$, *i.e.*, the unit point in the frame, and assume further that $s : x_2 = 0$ and $t : x_1 = 0$ are the tangents at the base points S and T, as shown in Figure 6.12, right.

Similar to the example above, the projectivity α from the pencil about S to the pencil about T is defined by the following three lines and their respective images:

$$
\begin{aligned}
- \quad x_2 \quad &= 0 \mapsto \quad x_0 \qquad\qquad = 0, \\
x_0 \qquad\qquad &= 0 \mapsto \quad - \quad x_1 \quad = 0, \\
x_0 \quad - \quad x_2 \quad &= 0 \mapsto \quad x_0 \quad - \quad x_1 \quad = 0.
\end{aligned}
$$

Again, the mapping between the pencils is described via equal coordinates and so the lines $l = (\mu : 0 : -\lambda)$ and their images $l' = (\lambda, -\mu, 0)$ (under α) have to be intersected. This yields

$$l \cap l' = (\lambda\mu : -\lambda^2 : -\mu^2)$$

with $(\lambda, \mu) \neq (0, 0)$. The coordinates of the points on this conic satisfy the homogeneous equation $x_1 x_2 - x_0^2 = 0$. In terms of inhomogeneous coordinates, we have $xy - 1 = 0$ which is the well-known graph of the rational function $x \mapsto \frac{1}{x}$. The resulting conic is an *equilateral hyperbola*, *i.e.*, a hyperbola with orthogonal asymptotes.

• **Exercise 6.3.1** Hexagrammum mysticum.

According to Theorem 6.2.1, any six points on a conic define a Pascal line. What happens if we fix the points but change their labels? At first glance, we would expect to find $6! = 720$

FIGURE 6.14. The 60 Pascal lines of an unlabeled hexagon inscribed in a conic.

Pascal lines, *i.e.*, for each permutation a new Pascal line. However, this is not the case. There are only 60 of them (Figure 6.14).

The 60 Pascal lines of a hexagon inscribed in a conic intersect three at a time in 20 Steiner points, and also three at a time in 60 Kirkman points. Each Steiner point lies together with three Kirkman points on 20 Cayley lines. Between the 60 Kirkman points and the 60 Pascal lines there is a relation that seems to be a duality. The 20 Cayley lines generated by a hexagon inscribed in a conic pass four at a time through 15 points known as Salmon points. There is a dual relationship between the 15 Salmon points and the 15 Plücker lines.

Find out which are the essential permutations of $P_1, \ldots, P_6$. Without loss of generality, the conic $(1 : t : t^2)$ (or $x_0 x_2 - x_1^1 = 0$) can be used and it means no restriction to assume $P_1 = (1 : 0 : 0)$, $P_2 = (1 : 1 : 1)$, and $P_3 = (0 : 0 : 1)$. Let $u, v, w \neq 0, 1, \infty$ be the parameters that correspond to the points P_3, P_4, P_5. For further details see [63].

Perspective and projective collineations

Perspectivities and projectivities between ranges of points and pencils of lines are the basic building blocks in Projective Geometry. However, there are further interesting and important mappings between projective planes, namely *collineations*. In this section, we shall describe these mappings and show their relations to conics and their action on the set of conics. We make use of

Definition 6.3.1 *A collineation from one projective plane to another projective plane is a mapping that sends points to points and satisfies the following two axioms:*

(K1) *Collinear points are mapped to collinear points.*
(K2) *The mapping is one-to-one and onto.*

We call a collineation *projective* if its restriction to any range of points or pencil of lines is a projectivity. It can be shown that in the real projective plane $\mathbb{P}^2(\mathbb{R})$ every collineation is projective.

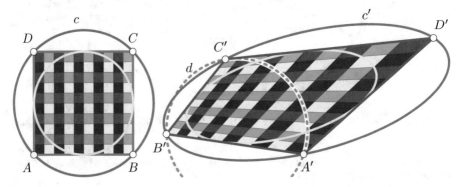

FIGURE 6.15. A projective collineation in $\mathbb{P}^2(\mathbb{R})$: In general, parallels are not preserved and even orientations may change. The circumcircle c of the square $ABCD$ is not mapped to the circumcircle of $A'B'C'$ but to some curve c'.

We have seen that, in a Pappian projective plane, a projective map between lines or pencils is uniquely determined if we prescribe three elements with their images (cf. page 191). A projective collineation is uniquely determined if we prescribe a labeled quadrangle and its labeled image. We say that projective collineations act sharply transitive on labeled quadrangles. Figure 6.15 shows the action of a collineation in $\mathbb{P}^2(\mathbb{R})$. We observe some phenomena:

- Parallel lines can be mapped to non-parallel lines.
- Affine ratios are not preserved: As can be seen in Figure 6.15, the square $ABCD$ is mapped to an arbitrary quadrangle $A'B'C'D'$ which has no center, in the elementary sense.

- The circumcircle c of $ABCD$ is not mapped to the circumcircle d of ABC. It is mapped to some curve c'.

A point F that is mapped to itself under a collineation κ is called a *fixed point of the collineation*. A line f can be fixed *pointwise* under κ, i.e., all points on f are fixed points of κ. Such a line is called an *axis of the collineation*. If a line is fixed and not all points on it are fixed, we call it a *fixed line*. A point C is called a *center of a collineation* if it is a fixed point and every line through C is a fixed line but not necessarily an axis.

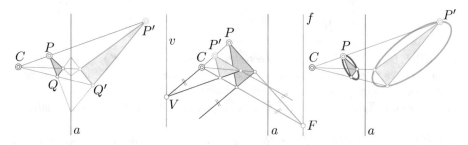

FIGURE 6.16. Left: A perspective collineation is defined by prescribing a center C, an axis a, and a pair (P, P') of assigned points such that C, P, P' are collinear. We can also see that a Desargues figure and a perspective collineation determine each other mutually. Middle: In the real plane $\mathbb{P}^2(\mathbb{R})$ we find the *vanishing line* v as the set of images of ideal points under the collineation. The line f is mapped to the ideal line. Right: Conics are mapped to conics via perspective collineations, as we shall see later.

A collineation with a center is called *perspective collineation*. Any perspective collineation has exactly one center and one axis. The center C of a perspective collineation κ may be incident with the axis a. In this case, we call κ an *elation*, otherwise a *homology*. Perspective collineations will be the key to the definition of osculating and hyperosculating conics in Section 7.3, especially in Definition 6.4.1. Figure 6.16 shows the action of three homologies in $\mathbb{P}^2(\mathbb{R})$. The left-hand side of Figure 6.16 illustrates the close relation of Desargues configurations and perspective collineations. In the middle of Figure 6.16, the vanishing line v as well as the image f of the line at infinity are displayed. The right-hand side of Figure 6.16 shows a circumconic of a triangle and its image.

Perspective collineations are uniquely determined by prescribing a center C, an axis a, and a pair (P, P') of assigned points such that C, P, P' are

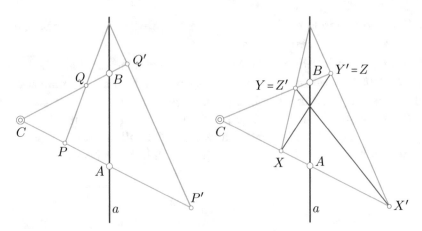

FIGURE 6.17. The characteristic cross ratio $\delta = \mathrm{cr}(C, A, P, P') = \mathrm{cr}(C, B, Q, Q')$ of a homology is, indeed, a characteristic of the collineation and independent of the choice of assigned points P and P' (left). Harmonic homologies are involutive, and $\mathrm{cr}(C, A, X, X') = \mathrm{cr}(C, B, Y, Y') = -1$, *i.e.*, $X \mapsto X'$ and $X' \mapsto X$ (right).

collinear and $P, P' \notin a$. Therefore, in case of a homology, we can assign a *characteristic cross ratio* δ to a perspective collineation. If $A = [P, P'] \cap a$, then $\delta = \mathrm{cr}(C, A, P, P')$ and is independent of the choice of the assigned pair of points, see Figure 6.17. A perspective collineation is involutive if, and only if, the characteristic cross ratio equals -1, *i.e.*, in this case, we have $H(C, A, P, P')$ for all $P, P' \neq C$ and $P, P' \notin a$. This can easily be read off from Figure 6.17. In Fano planes, harmonic perspective collineations are homologies. Harmonic homologies are also called *projective reflections* since they generalize the concept of the elementary reflection in a point or in a line. Projective reflections are important in connection with conics: There exist projective reflections that leave conics invariant (Figure 6.18).

In $\mathbb{P}^2(\mathbb{F})$, the projective collineations are represented by linear mappings $\mathbb{F}^3 \to \mathbb{F}^3$. Therefore, in any coordinate frame, we can describe them by regular 3×3 matrices.

The analytic representation of collineations is given with the help of matrices. The computation of the transformation matrix $\mathbf{T}$ of a projective collineation κ for given quadruples of corresponding points is similar to that of a projectivity as described on page 210.

Assume $A = \mathbf{a}\mathbb{F}$, $B = \mathbf{b}\mathbb{F}$, $C = \mathbf{c}\mathbb{F}$, and $D = \mathbf{d}\mathbb{F}$ are four given points. Further, the points $A' = \mathbf{a}'\mathbb{F}$, $B' = \mathbf{b}'\mathbb{F}$, $C' = \mathbf{c}'$, and $D' = \mathbf{d}'\mathbb{F}$ shall be the respective κ-images. Then, we rescale the representatives **a**, **b**, **c** such that they sum up to a multiple of the representative **d**. We also do this for the image points. Now, the matrix $\mathbf{T} \in \mathbb{F}^{3\times3}$ is given by

$$\mathbf{T} = \left(\widetilde{\mathbf{a}'}\ \widetilde{\mathbf{b}'}\ \widetilde{\mathbf{c}'}\right) \cdot \left(\widetilde{\mathbf{a}}\ \widetilde{\mathbf{b}}\ \widetilde{\mathbf{c}}\right)^{-1} \tag{6.7}$$

where $\widetilde{}$ indicates that the column vectors are rescaled. This formula is the analogon to the analytic description of projectivities (5.12) and confirms that a projective collineation is uniquely determined by two labeled quadrangles.

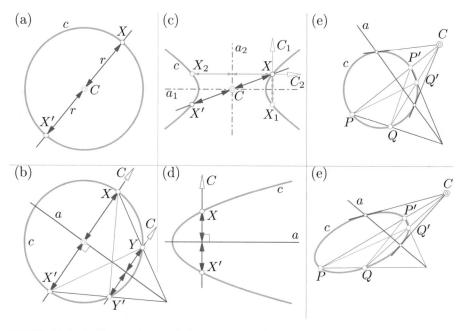

FIGURE 6.18. Projective and elementary reflections which leave conics invariant: (a) The reflection in the center C of a circle c fixes the circle. The axis of this harmonic homology is the line at infinity. (b) The reflection in a diameter a of a circle c is a harmonic homology with axis a. The center C of the collineation is the ideal point of lines orthogonal to a. (c) A conic c with center C (ellipse or hyperbola) is invariant under reflections in the axis and under the reflection in the center C. (d) A parabola c has no center but allows the reflection in its axis a. (e) There are more harmonic homologies that fix the circle c. (f) For each conic c, we can find harmonic perspective collineations that leave the conic c fixed as a whole but not pointwise.

Finally, we point out that any collineation κ of points also determines a mapping of lines $\kappa^\star$. The matrix $\mathbf{U}$ acting on the coordinate vectors of lines can be derived from $\mathbf{T}$. Assume $X = \mathbf{x}\mathbb{F}$ is a point on the line $u = \mathbf{u}\mathbb{F}$, i.e., $\langle \mathbf{u}, \mathbf{x} \rangle = \mathbf{u}^{\mathrm{T}}\mathbf{x} = 0$. If now $\mathbf{x}' = \mathbf{T}\mathbf{x}$ and $\mathbf{u}' = \mathbf{U}\mathbf{u}$ with $\mathbf{U} \in \mathbb{F}^{3\times3}$, then $\langle \mathbf{u}', \mathbf{x}' \rangle = 0$ for κ preserves incidences. Thus, we have $\langle \mathbf{u}', \mathbf{x}' \rangle = \mathbf{u}'^{\mathrm{T}}\mathbf{x}' = \mathbf{u}^{\mathrm{T}}\mathbf{U}^{\mathrm{T}}\mathbf{T}\mathbf{x} = 0$ for all $\mathbf{x}, \mathbf{u} \in \mathbb{F}^3$ with $0 = \mathbf{u}^{\mathrm{T}}\mathbf{x} = \mathbf{u}^{\mathrm{T}}\mathbf{I}_3\mathbf{x}$. This yields

$$\mathbf{U}^{\mathrm{T}}\mathbf{T} = \lambda \mathbf{I}_3 \quad \Longleftrightarrow \quad \mathbf{U} = \lambda \left(\mathbf{T}^{\mathrm{T}}\right)^{-1} = \lambda \left(\mathbf{T}^{-1}\right)^{\mathrm{T}} \qquad (6.8)$$

for any $\lambda \in \mathbb{F}\backslash\{0\}$. Since $\left(\mathbf{T}^{\mathrm{T}}\right)^{-1} = \left(\mathbf{T}^{-1}\right)^{\mathrm{T}}$ holds, transposing and inverting a matrix commute. Therefore, we simply write $\mathbf{T}^{-\mathrm{T}}$ for the transposed inverse of a regular matrix $\mathbf{T}$. We summarize:

Theorem 6.3.1 *Every projective collineation $\kappa : \mathbb{P}^2(\mathbb{F}) \to \mathbb{P}^2(\mathbb{F})$ with a commutative field $\mathbb{F}$ can be described by a linear mapping $f_\kappa : \mathbb{F}^3 \to \mathbb{F}^3$. Conversely, any linear mapping $f : \mathbb{F}^3 \to \mathbb{F}^3$ defines a projective collineation $\kappa_f : \mathbb{F}^3 \to \mathbb{F}^3$. For any given coordinate frame, the transformation matrix $\mathbf{T}$ describing κ_f is unique only up to scalar multiples with factors taken from $\mathbb{F}$.*

It is useful to have the following result:

Lemma 6.3.5 *Perspective collineations $\kappa : \mathbb{P}^2(\mathbb{F}) \to \mathbb{P}^2(\mathbb{F})$ with center $C = \mathbf{c}\mathbb{F}$ and axis $a = \mathbf{a}\mathbb{F}$ can be written in the form*

$$\mathbf{x} \mapsto \mathbf{x}' = \mathbf{x} + \mathbf{c}\langle \mathbf{a}, \mathbf{x} \rangle (\delta - 1) \qquad (6.9)$$

where $\delta \neq 1$ is the characteristic cross ratio.

Proof: Any two corresponding points X and $X' = \kappa(X)$ in a perspective collineation are collinear with the center C. The characteristic cross ratio is defined as $\mathrm{cr}(C, A, X, X') = \mathrm{cr}(X, X', C, A)$ with $A = a \cap [X, C]$. For $A = \mathbf{b}\mathbb{F}$ we have $\mathbf{b} = (\mathbf{x} \times \mathbf{c}) \times \mathbf{a} = \mathbf{c}\langle \mathbf{a}, \mathbf{x} \rangle - \mathbf{x}\langle \mathbf{c}, \mathbf{a} \rangle$. Thus, the points C, A, and X on the line $[C, X]$ have the homogeneous coordinates $(1 : 0)$, $(\langle \mathbf{a}, \mathbf{x} \rangle : \langle \mathbf{c}, \mathbf{a} \rangle)$, and $(0 : 1)$. Now, we determine $X' = (\xi : \eta)$ on $[C, X]$ such that $\mathrm{cr}(C, A, X, X') = \delta$. With (5.6) we find $\xi : \eta = \langle \mathbf{a}, \mathbf{x} \rangle (\delta - 1) : \langle \mathbf{a}, \mathbf{c} \rangle$ which gives $X' = \mathbf{x}'\mathbb{F}$ with

$$\mathbf{x}' = \mathbf{x}\langle \mathbf{c}, \mathbf{a} \rangle + \mathbf{c}\langle \mathbf{a}, \mathbf{x} \rangle (\delta - 1).$$

We can replace the representatives of C and a such that $\langle \mathbf{c}, \mathbf{a} \rangle = 1$ and we are done. ∎

Especially for harmonic homologies we have

$$\mathbf{x}' = \mathbf{x} - 2\mathbf{c}\langle \mathbf{a}, \mathbf{x} \rangle \langle \mathbf{a}, \mathbf{c} \rangle^{-1}. \qquad (6.10)$$

6.4 There is only one conic in a projective plane

From STEINER's definition of a conic it is clear that projective collineations map conics to conics. For any two conics in $\mathbb{P}^2(\mathbb{F})$ with mutually
different points $A, B, D \in c$ and $A', B', D' \in c'$ there is a unique collineation
with $A \mapsto A'$, $B \mapsto B'$, and $D \mapsto D'$.

Though there is a huge variety of conics from the Euclidean point of view.
In projective planes, everything simplifies. Now, we can show:

Theorem 6.4.1 *For any two conics c, c' in $\mathbb{P}^2(\mathbb{F})$ with mutually different
points $A, B, D \in c$ and $A', B', D' \in c'$ there is a unique collineation with
$A \mapsto A'$, $B \mapsto B'$, and $D \mapsto D'$.*

All conics c are collinear images of the conic given in (6.4).

Proof: A projective collineation is uniquely defined by prescribing a labeled quadrangle $ABCD$
and its image, the labeled quadrangle $A'B'C'D'$.

We recall the projective generation used on page 228 in order to compute a parametrization
and an equation of any given conic c. Now, we relabel the points used there: $S = (1:0:0) =: A$,
$T = (0:0:1) =: B$, $s \cap t = (0:1:0) =: C$, and $R = (1:1:1) =: D$ (Figure 6.19).

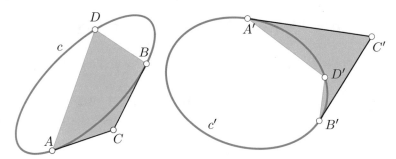

FIGURE 6.19. Up to collineations, there is only one conic in $\mathbb{P}^2(\mathbb{F})$: There is
exactly one projective collineation κ mapping the conic c to the conic c' with
$A \mapsto A'$, $B \mapsto B'$, $C \mapsto C'$, and $D \mapsto D'$. The proper choice of $A, B, C, D \in c$ and
A', B', C', D' has some degrees of freedom.

Similarily we define the points A', B', C', D' related to c'. The uniquely defined collineation
with $A \mapsto A'$, $B \mapsto B'$, $C \mapsto C'$, and $D \mapsto D'$ maps the conic c onto the conic c'. The second
statement of Theorem 6.4.1 follows if c' is chosen as the conic with equation (6.4). ∎

It turns out to be useful to write the simple equation of the conic given in
(6.4) in matrix form. First, we observe that $2(x_0 x_2 - x_1^2) = 0$ is an equation

of the same conic, since $\mathrm{char}\mathbb{F} \neq 2$. Thus, we have

$$0 = 2(x_0 x_2 - x_1^2) = (x_0, x_1, x_2) \begin{pmatrix} 0 & 0 & 1 \\ 0 & -2 & 0 \\ 1 & 0 & 0 \end{pmatrix} \begin{pmatrix} x_0 \\ x_1 \\ x_2 \end{pmatrix}.$$

This is a matrix equation of the form

$$0 = \mathbf{x}^{\mathrm{T}} \mathbf{A} \mathbf{x}, \tag{6.11}$$

where $\mathbf{A} \in \mathbb{F}^{3 \times 3}$ is a symmetric and regular matrix and $\mathbf{x} = (x_0, x_1, x_2)^{\mathrm{T}}$.
Now, we show that the equation of any conic c in $\mathbb{P}^2(\mathbb{F})$ can be written in
this form. Let κ be a projective collineation described by a regular 3×3-
matrix $\mathbf{T} \in \mathbb{F}^{3 \times 3}$ and points $X = \mathbf{x}\mathbb{F}$ are transformed to points $X' = \mathbf{x}'\mathbb{F}$
with $\mathbf{x}' = \mathbf{T}\mathbf{x}$. Consequently, $\mathbf{x} = \mathbf{T}^{-1}\mathbf{x}'$, and therefore, we have

$$\mathbf{x}^{\mathrm{T}} \mathbf{A} \mathbf{x} = (\mathbf{T}^{-1}\mathbf{x}')^{\mathrm{T}} \mathbf{A} \mathbf{T}^{-1} \mathbf{x}' = \mathbf{x}'^{\mathrm{T}} \mathbf{T}^{-\mathrm{T}} \mathbf{A} \mathbf{T}^{-1} \mathbf{x}' = \mathbf{x}'^{\mathrm{T}} \mathbf{B} \mathbf{x}'.$$

Since $\mathbf{A}$ is symmetric, $i.e.$, $\mathbf{A}^{\mathrm{T}} = \mathbf{A}$, we find that $\mathbf{B}$ is also symmetric:

$$\mathbf{B}^{\mathrm{T}} = (\mathbf{T}^{-\mathrm{T}} \mathbf{A} \mathbf{T}^{-1})^{\mathrm{T}} = \mathbf{T}^{-\mathrm{T}} \mathbf{A}^{\mathrm{T}} \mathbf{T}^{-1} = \mathbf{T}^{-\mathrm{T}} \mathbf{A} \mathbf{T}^{-1} = \mathbf{B}.$$

Hence, the general homogeneous equation of a conic in $\mathbb{P}^2(\mathbb{F})$ reads

$$Q(\mathbf{x}, \mathbf{x}) = a_{00}x_0^2 + 2a_{01}x_0x_1 + 2a_{02}x_0x_2 + a_{11}x_1^2 + 2a_{12}x_1x_2 + a_{22}x_2^2 = 0,$$

or in matrix form

$$(x_0, x_1, x_2) \begin{pmatrix} a_{00} & a_{01} & a_{02} \\ a_{01} & a_{11} & a_{12} \\ a_{02} & a_{12} & a_{22} \end{pmatrix} \begin{pmatrix} x_0 \\ x_1 \\ x_2 \end{pmatrix} = 0 \tag{6.12}$$

with $a_{ik} = a_{ki}$ and a regular matrix (a_{ik}). It is a quadratic form, $i.e.$, it is
homogeneous, and thus, $Q(\mathbf{x} \cdot \lambda, \mathbf{x} \cdot \lambda) = \lambda^2 Q(\mathbf{x}, \mathbf{x}) = 0$ for any $\lambda \in \mathbb{F} \setminus \{0\}$.
So, with $\mathbf{x}$ also the multiples of $\lambda\mathbf{x}$ also satisfy the equation $Q(\mathbf{x}, \mathbf{x}) = 0$,
and therefore, it is really the equation of a set of points in the projective
plane $\mathbb{P}^2(\mathbb{F})$.

Note that, conversely, not each such form defines a conic. $E.g.$, in $\mathbb{P}^2(\mathbb{R})$
there are no points satisfying $x_0^2 + x_1^2 + x_2^2 = 0$.

● **Exercise 6.4.1** Rational parametrization.

Show that for each conic $c \subset \mathbb{P}^2(\mathbb{F})$ (char$\mathbb{F} \neq 2$) there is a homogeneous rational parametrization

$$\begin{pmatrix} x_0 \\ x_1 \\ x_2 \end{pmatrix} = \mathbf{T} \begin{pmatrix} \lambda^2 \\ \lambda\mu \\ \mu^2 \end{pmatrix}$$

with $\mathbf{T} \in \mathbb{F}^{3\times3}$ and $\det \mathbf{T} \neq 0$, and conversely, each such parametrization defines a conic.

● **Exercise 6.4.2** Conic on five or six points.

Assume $P_i = (p_{i0}, p_{i1}, p_{i2})$ with $i = 1, \ldots, 5$ are five points. There is a uniquely defined conic c on these five points according to Definition 6.1.1 unless no three of the given points are collinear. The equation of c

$$c: \quad a_{00}x_0^2 + 2a_{01}x_0x_1 + 2a_{02}x_0x_2 + a_{11}x_1^2 + 2a_{12}x_1x_2 + a_{22}x_2^2 = 0$$

is satisfied by the coordinates of any of the P_is. On the other hand, the coefficients of c's equation can be determined from a system of linear equations. The values a_{ij} (with $i, j \in \{0, 1, 2\}$) are uniquely determined (up to a common non-vanishing factor) as the kernel of

$$\begin{pmatrix} p_{10}^2 & 2p_{10}p_{11} & 2p_{10}p_{12} & p_{11}^2 & 2p_{11}p_{12} & p_{12}^2 \\ \vdots & & & & & \vdots \\ p_{50}^2 & 2p_{50}p_{51} & 2p_{50}p_{52} & p_{51}^2 & 2p_{51}p_{52} & p_{52}^2 \end{pmatrix}.$$

Show that

$$\det \begin{pmatrix} p_{10}^2 & 2p_{10}p_{11} & 2p_{10}p_{12} & p_{11}^2 & 2p_{11}p_{12} & p_{12}^2 \\ p_{20}^2 & 2p_{20}p_{21} & 2p_{20}p_{22} & p_{21}^2 & 2p_{21}p_{22} & p_{22}^2 \\ p_{30}^2 & 2p_{30}p_{31} & 2p_{30}p_{32} & p_{31}^2 & 2p_{31}p_{32} & p_{32}^2 \\ p_{40}^2 & 2p_{40}p_{41} & 2p_{40}p_{42} & p_{41}^2 & 2p_{41}p_{42} & p_{42}^2 \\ p_{50}^2 & 2p_{50}p_{51} & 2p_{50}p_{52} & p_{51}^2 & 2p_{51}p_{52} & p_{52}^2 \\ p_{60}^2 & 2p_{60}p_{61} & 2p_{60}p_{62} & p_{61}^2 & 2p_{61}p_{62} & p_{62}^2 \end{pmatrix} = 0 \qquad (6.13)$$

is a condition on six points $P_i = (p_{i0} : p_{i1} : p_{i2})$ with $i \in \{1, \ldots, 6\}$ to lie on a conic and compare it with the synthetic equivalent in form of the Pappus theorem. Clarify what happens in the singular and degenerate cases. How does (6.13) change when it becomes the condition of four points to be concyclic?

We continue with examples in the projective extension of $\mathbb{E}^2$:

■ **Example 6.4.1** Contact of a conic and a line - contact conditions.

An ellipse/hyperbola shall be given by the equation

$$c: \quad \frac{x^2}{a^2} \pm \frac{y^2}{b^2} = 1.$$

We ask for conditions on a and b such that c touches a line $l: y = kx + d$ with $k, d \in \mathbb{R}$. For that purpose, we insert $y = kx + d$ into c's equation and find

$$x^2(b^2 \pm a^2k^2) \pm 2a^2dkx \pm a^2(d^2 - b^2) = 0.$$

By Definition 6.1.1, a tangent has exactly one point with the conic c in common. Therefore, the obtained quadratic equation has one solution with multiplicity two if l is a tangent of c. This yields

$$a^2k^2 \pm b^2 = d^2.$$

The case of the asymptotes of a hyperbola is also covered, since there $k = \frac{b}{a}$, $d = 0$, and thus, $a^2 \left(\frac{b}{a} \right)^2 - b^2 = 0$.

A parabola p with the equation

$$p : y^2 - 2qx = 0, \quad q \in \mathbb{R} \setminus \{0\},$$

touches the line l if, and only if,

$$2dk = q.$$

■ **Example 6.4.2** Conics as images of circles under collineations.

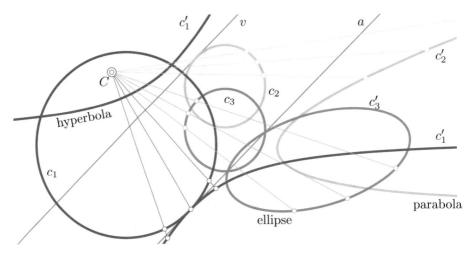

FIGURE 6.20. A perspective collineation with center C, axis a, and the pre-image v of the line at infinity maps circles to ellipses, or parabolas, or hyperbolas if the circles do not intersect v, or touch v, or intersect v in two different (real) points.

Many problems in connection with conics allow a graphical solution with ruler and compass. For that purpose, perspective collineations turn out to be very helpful since any conic in the real plane can be mapped to a circle and then most of the constructions are very simple. This is a good reason to have a closer look at the perspective collineations that transform a circle into the three affine types of conics, i.e., into an ellipse, a parabola, or a hyperbola. The result depends on the position of the circle c relative to the pre-image $v = \kappa^{-1}(\omega)$ of the line ω at infinity. If $v \cap c = \varnothing$, then $c' = \kappa(c)$ has no ideal points, and therefore, it is an ellipse. The intersection $v \cap c$ consists of two points, we find a hyperbola c'. The conic c' is a parabola if v is a tangent of c (see Figure 6.20).

Later, we shall use the following result:

Theorem 6.4.2 *If two conics c and d in a projective plane share a line element (P,t), i.e., they have the common point P and the same tangent* *t there, then there exists exactly one perspective collineation α with center P that maps c to d.*

Proof: Assume l and m are two lines through P and $l, m \neq t$. Then, l intersects c and d in P and two further points, say C and C' (Figure 6.21). The line m intersects c and d in D and D', besides P. Then, PCD and $P'C'D'$ are two triangles. Referring to the proof of Theorem 6.4.1, there exists a unique collineation κ with $\kappa(P) = P$, $\kappa(C) = C'$, and $\kappa(D) = D'$. Consequently, $\kappa(l) = l$, $\kappa(m) = m$, $\kappa(t) = t$. Thus, the restriction of κ to the pencil of lines about P is the identity mapping and P turns out to be a center of κ which is now recognized as a perspective collineation. ∎

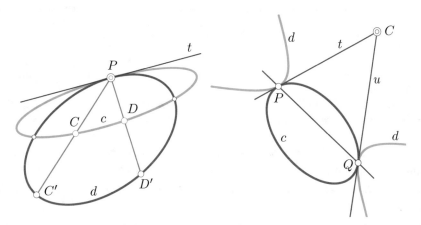

FIGURE 6.21. Left: The perspective collineation κ maps c to d. Right: A common mistake: Two conics c and d with two common line elements (P, t) and (Q, u) are images of each other under perspective collineations, but the center differs from the point $t \cap u$ and the axis is different from the line $[P, Q]$.

We know from Theorem 6.4.1 that there is one prototype of conic in the projective plane, and therefore, we can simplify many proofs. If a theorem is valid for a conic with a simple equation or a simple parametrization, then it is valid for all conics. In this sense, we can prove the theorem by PAPPUS or PASCAL (Theorem 6.2.1, page 220) by means of analytic geometry with the most simple equation and parametrization (cf. Example 6.4.3).

▪ **Example 6.4.3** Analytic proof of Theorem 6.2.1.

The analytic proof uses the fact that the conic given in (6.4) or in (6.3) can be used instead of any conic according to Theorem 6.4.1.

Further, it is no restriction to assume that the points P_1, P_2, and P_3 have the homogeneous coordinates

$$P_1 = (1:0:0), \quad P_2 = (0:0:1), \quad P_3 = (1:1:1),$$

i.e., they correspond to the affine parameter values $t = 0, \infty, 1$ in (6.3). The remaining three points shall be defined by the values $t = u, v, w$ of the affine parameter. Consequently,

$$P_4 = (1:u:u^2), \quad P_5 = (1:v:v^2), \quad P_6 = (1:w:w^2).$$

In order to make the computations traceable, we give the homogeneous coordinates of all lines needed:

$$[P_1, P_2] = (0:1:0), \qquad [P_4, P_5] = (uv:-u-v:1),$$

$$[P_2, P_3] = (1:-1:0), \qquad [P_5, P_6] = (vw, -v-w, 1),$$

$$[P_3, P_4] = (u:-u-1:1), \qquad [P_6, P_1] = (0:w:-1).$$

Note that factors of the form $u - v$ or $u - 1$ (likewise for other combinations of u, v, and w) can be canceled out, for $u = v$ means $P_4 = P_5$ and $u = 1$ causes $P_4 = P_1$. The three presumably collinear points A, B, and C have the following homogeneous coordinates

$$A = (1:0:-uv), \quad B = (1:1:v+w-vw), \quad C = (1+u-w:u:uw).$$

The coordinate vectors of A, B, and C are linearly dependent, and therefore, these three points are collinear.

The limiting cases of Theorem 6.2.2 as described on page 223 can also be treated analytically. We only have to be careful when canceling out factors of homogeneous coordinates.

Definition 6.4.1 *Two conics c and d with a common point P and a common tangent t at P are said to osculate each other at P if there exists a perspective collineation α with axis t and a center $C \neq P$ on t such that $\alpha(c) = d$.*

Two conics c and d with a common point P and a common tangent t at P are said to hyperosculate each other at P if there is a perspective collineation α with axis t and center P such that $\alpha(c) = d$.

The first part of Definition 6.4.1 could be slightly modified without changing the meaning substantially: The conics c and d with a common point P and a common tangent t at P are osculating each other if there is a perspective collineation β with center P and an axis a through P such that $\beta(c) = d$. The left-hand side of Figure 6.22 shows the osculating conics c and d being each others collinear images.

Anyway, the center of the collineation mentioned in the first version has to lie on t, and the axis a mentioned in the second version of the definition has to pass through the two common points of c and d. Hence, both collineations are elations. In Figure 6.22 (left), we can see the centers and axes of both perspective collineations: C and t are the center and the axis of α; P and a are the center and axis of β.

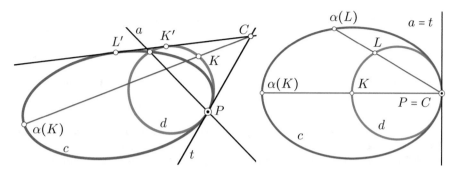

FIGURE 6.22. Left: osculating conics. Right: hyperosculating conics.

In $\mathbb{P}^2(\mathbb{R})$, a special example of an osculating pair of conics is given by a conic and one of its oculating circles. Especially the "osculating circles" at the vertices of a conic are hyperosculating circles. The right-hand side of Figure 6.22 shows a pair of hyperosculating conics. The circle k hyperosculates the ellipse c at the right principal vertex P. The perspective collineation with center P and axis t transforms c into d, or likewise, α^{-1} transforms d into c.

■ **Example 6.4.4** The osculating circle o of an ellipse e as a collinear image of e.

As shown in Figure 6.23, we can use a the perspective collineation in order to find the osculating circle o_P of an ellipse e at any point P. The vertices are somehow special, and therefore, we exclude them in this example.

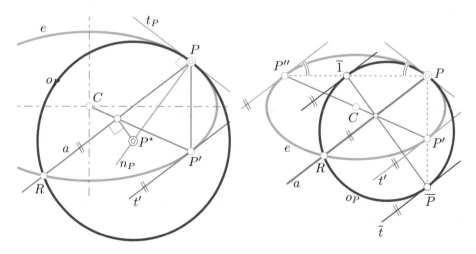

FIGURE 6.23. The osculating circle o_P of an ellipse e as collinear image of e.

According to Definition 6.4.1, the axis a of this elation passes through P. Now, we have to determine a such that tangents of e parallel to a are mapped to parallel tangents of o. Because of the symmetry of o_P with respect to any diameter, we reflect P in the principal and auxiliary axes or e and we obtain P' and P''. The tangents of c at P' and P'' are parallel to the wanted axis. Thus, the pre-image of the center $P^\star$ of o_P is known (Figure 6.23, left). In Section 3.2, we learned that there is also a differential geometric approach to the osculating circle.

■ **Example 6.4.5** Intersection of a line and a conic in $\mathbb{P}^2(\mathbb{R})$ with ruler and compass.

At this point, we should note that the perspective collineation α that maps a conic c into one of its hyperosculating circles can be used in order to intersect a conic and a line graphically (by straightedge and compass) with a minimum number of lines. Such questions seem to be academic in nature, but line-saving drawings are useful even when using CAD systems.

• In the first case, we construct the intersection of a given line l with a hyperbola c. The construction is shown in detail in Figure 6.24, left.

We are given the hyperbola c and its hyperosculating circle d at the vertex P. Now, we use the perspective collineation κ that maps c to d. Its center is the point P of hyperosculation and the axis coincides with the common tangent t of c and d at P. The asymptote a of c carries the ideal point $A \in c$ which is mapped to $A' \in d$ under κ. Thus, we have found the vanishing line v. The line $l' = \kappa(l)$ is now determined by its fixed point $l \cap t$ and its vanishing point which is the intersection of v and the parallel to l through P. The points T_1 and T_2 are the intersections of l' and d. The inverse collineation κ^{-1} sends them to the wanted intersection points S_1 and S_2 on l.

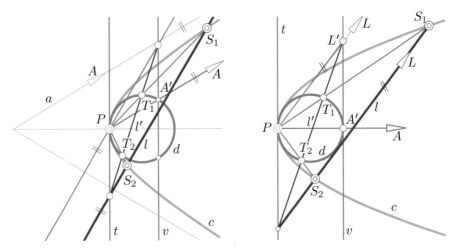

FIGURE 6.24. Line-saving construction of the intersection of a line l and a hyperbola (left) or a parabola (right).

• The intersection of a parabola c and a line l (see Figure 6.24, right) is relatively simple, compared to the previous case. The vanishing line v of the perspective collineation κ (with center P and axis t) has to be that tangent of d which is parallel to t. The line $l' = \kappa(l)$ is again determined by its fixed point $l \cap t$ and its vanishing point L'. Since $\kappa(c) = d$, we find the points of intersection of c and l as the collinear images of $d \cap l'$ under κ.

• In the case of an ellipse c, we avoid the construction of the vanishing line. The line l intersects the tangent of c at the principal vertex A that is opposite to P in a point 1. Since the latter tangent is parallel to the tangent t at P, its κ-image is a parallel tangent of d. Thus, $1'$ and the fixed point $l \cap t$ determine l', and the common points of d and l' are the κ^{-1}-images of $l' \cap d$ (see Figure 6.25).

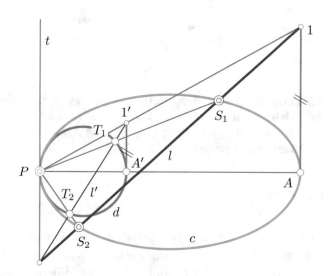

FIGURE 6.25. Line-saving construction of the intersection of a line l and an ellipse c.

Fixed points of projectivities and conics

The stereographic projection, as described in Example 6.3.1 and depicted in Figure 6.13, is a bijective mapping between a conic c and a line l or a pencil about some point. The connection of each point on c with the center of the projection $C \in c$ is a perspectivity between the pencil about C and the conic. In analogy to Definitions 5.2.1 and 5.2.2, we use:

Definition 6.4.2 *Let c be a conic and let $C \in c$ be a point. We call the mapping α that maps points $X \in c \setminus \{C\}$ on c to the lines $[X, C]$ and C to the tangent t_C a perspectivity $c \,\overline{\wedge}\, C$. The inverse α^{-1} of α is also called a perspectivity. A mapping β from a conic c to a conic d is called a projectivity if it is the composition of finitely many perspectivities of the form*

$$c \,\overline{\wedge}\, C \,\overline{\wedge}\, \ldots \,\overline{\wedge}\, D \,\overline{\wedge}\, d$$

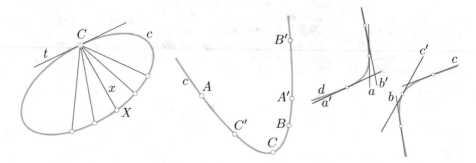

FIGURE 6.26. Perspectivities and projectivities on conics: Left: the perspectivity $c(X) \barwedge C(x)$ between a conic c and the pencil about $C \in c$. Middle: A projectivity $\alpha : c \mapsto c$ is uniquely defined by $A \mapsto A'$, $B \mapsto B'$, and $C \mapsto C'$. Right: The dual version is a projectivity between the tangents of a conic d.

with $C \in c$ and $D \in d$.

According to Theorem 6.1.1, any two points of a conic can serve as base points of a Steiner generation, as explained in Definition 6.1.1. Thus, Definition 6.4.2 makes sense: If C and D are two different points on a conic c, then c is perspective to both, the pencil C and the pencil D, and so we have a projectivity between the two pencils, *i.e.*, $C \barwedge c \barwedge D \implies C \barwedge D$.

As a consequence of the axiom (FT), a projectivity α between two conics c and d is uniquely defined by prescribing three pairs (P_i, P_i') of assigned points with $\alpha(P_i) = P_i'$, $P_i \in c$, $P_i' \in d$, and $i \in \{1,2,3\}$. Dual to that, we have to define a mapping between the sets of tangents of two conics by prescribing three pairs (t_i, t_i') of assigned tangents (cf. Figure 6.26). Further, if $c = d$, then $\alpha : c \mapsto c$ is an *automorphism* of c. Figure 6.19 and Theorem 5.4.2 reveal that each projectivity $\alpha : c \to c$ can be extended to a projective collineation which mas c onto itself. Conversely, the restriction of each such projective collineation onto c is a projectivity on c. We call such collineations *projective automorphisms of c* or *automorphic collineations of c* and claim:

Theorem 6.4.3 *In $\mathbb{P}^2(\mathbb{F})$, the projective automorphisms of a conic form a group which is isomorphic to the group $\mathrm{PGL}(\mathbb{F}, 2)$ of projectivities on a line in $\mathbb{P}^2(\mathbb{F})$.*

The fact that conics are isomorphic to lines and pencils can be used for the determination of fixed points of projectivities. Let $\alpha : l \to l$ be a projectivity on a line l with fixed points G_1 and G_2. Projecting l from some point $C \in c$ (but $C \notin l$) to c yields a projectivity $\beta : c \to c$, and the fixed points G_1 and G_2 of α are mapped to fixed points F_1 and F_2 of β.

An important tool for the constructive approach to projectivities between lines was given in Theorem 5.2.4. There, we have shown that a projectivity $\alpha : l \to m$ between lines $l \neq m$ defines an axis p_α such that for any two points $X, Y \in l$ with $X \neq Y$ and $X' \neq Y' \in m$ with $\alpha(X) = X'$ and $\alpha(Y) = Y'$ the point $[X, Y'] \cap [X', Y]$ lies on p_α. Theorem 6.2.2 was a generalization of PAPPUS's theorem to conics. These results lead to

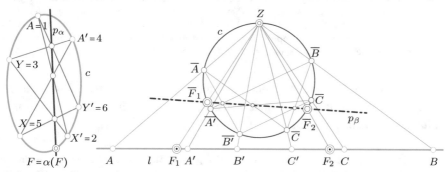

FIGURE 6.27. Left: A projectivity α on a conic c and the axis p_α, cf. Theorem 6.4.4. Right: The construction of the fixed points F_1 and F_2 of a projectivity on a line l as described in Example 6.4.6. The conic c and the point $Z \in c$ can be chosen freely as long as $Z \notin l$.

Theorem 6.4.4 *Let $\alpha : c \to c$ with $X \mapsto X'$ and $Y \mapsto Y'$ be a projectivity (different from the identity) on a conic c. Then, there exists an axis p_α and the lines $[X, Y']$ and $[X', Y]$ intersect in points of p_α for all $X, Y \in c$ with $X \neq Y$.*

Proof: Assume that the projectivity $\alpha : c \to c$ is not the identity mapping. Then, there exists a point A with $A \neq \alpha(A) =: A'$ and α induces a projectivity σ from the pencil A' to the pencil A which equals the following chain of perspectivities and projectivities:

$$A' \barwedge c(X') \overset{\alpha}{\barwedge} c(X) \barwedge A.$$

Since the line $[A', A]$ is assigned to itself, the above chain is a perspectivity. Thus, pairs of assigned lines meet on the axis p_α, i.e., $[A', X] \cap [A, X']$ and $[A', Y] \cap [A, Y']$ intersect at p_α. Let $A =: 1$, $X' =: 2$, $Y =: 3$, $A' =: 4$, $X =: 5$, and $Y' = 6$, see Figure 6.27. Now, Theorem 6.2.2 applies and, conversely, also $[X', Y] \cap [X, Y']$ is a point on p_α. If two points coincide that have to be connected by a line in the construction, then the line is to be replaced by the tangent to c at this point. Obviously, $F = \alpha(F)$ is equivalent to $F \in c \cap p_\alpha$. ∎

■ **Example 6.4.6** Fixed points of a projectivity on a line in $\mathbb{P}^2(\mathbb{R})$.

The right-hand side of Figure 6.27 shows a line l and the points A, B, C, A', B', C' on it. According to (FT), there is a unique projectivity α defined by $\alpha(A) = A'$, $\alpha(B) = B'$, and $\alpha(C) = C'$. We want to construct the fixed points of α.

First, we use a perspectivity from l to an arbitrarily chosen conic c and project all given points on l according to $l \barwedge Z \barwedge c$. The center $Z \in c$ with $Z \notin l$ of the projection can also be chosen freely. The projection results in the points $\overline{A}$, $\overline{B}$, $\overline{C}$, $\overline{A'}$, $\overline{B'}$, and $\overline{C'}$. Now, a projectivity $\beta : c \to c$ is uniquely defined by $\beta(\overline{A}) = \overline{A'}$, $\beta(\overline{B}) = \overline{B'}$, and $\beta(\overline{C}) = \overline{C'}$. Here, we can apply Theorem 6.4.4 and construct the axis p_β of β. The fixed points $\overline{F}_1$ and $\overline{F}_2$ of β are found as the points of intersection of p_β and c. The projection of the latter points from Z to l yields the fixed points F_1 and F_2 as shown in Figure 6.27. The conic c as well as any other conic with the same properties that can be used to find the fixed points of a projectivity is called *Steiner conic*. Constructions simplify if c is chosen as a circle which is then referred to as *Steiner circle*. Steiner conics are named after J. STEINER (cf. page 218).

Theorem 6.4.5 *Let c be conic and let $Z \notin c$ be a point. There exists a unique harmonic homology ϱ with center Z which is automorphic for c. The axis a of ϱ is not tangent to c.*

Proof: There are at most two tangents u, v of c passing through Z. Besides their points of contact U and V, there are two further points A and C on c which are not collinear with the center Z. Hence, $[A, Z] \cap c = \{A, B\}$ and $[C, Z] \cap c = \{C, D\}$. Then, $ABCD$ is a quadrangle with Z as a diagonal point. Since the respective collineation ϱ keeps $[Z, A]$ and $[Z, C]$ fixed, it interchanges A and B as well as C and D. The axis a must be the line carrying the remaining diagonal points: $[A, C] \cap [B, D]$ and $[A, D] \cap [B, C]$. On the other hand, only a harmonic perspective collineation with center Z and axis a interchanges A with B and C with D.

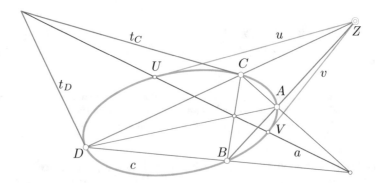

FIGURE 6.28. The harmonic homology ϱ (center Z, axis a) from Theorem 6.4.5.

It remains to show that ϱ maps c onto itself. Let t_C and t_D be the tangents of c at C and D. We look at the projectivity σ from the pencil C to the pencil D that generates c. It maps

$$[C,A] \mapsto [D,A], \quad [C,B] \mapsto [D,B], \quad [C,D] \mapsto t_D, \quad t_C \mapsto [D,C].$$

Since $[D,A]$, $[D,B]$, $[D,C]$, t_D are mutually distinct, Theorem 5.2.3 (page 191) applies, and there exists a projectivity τ in the pencil D which interchanges $[D,A]$ with $[D,B]$ and $[D,C]$ with t_D. Therefore, the composition $\tau \circ \sigma$ is a projectivity $C \barwedge D$ with

$$[C,A] \mapsto [D,B], \quad [C,B] \mapsto [D,A], \quad [C,D] \mapsto [D,C], \quad t_C \mapsto t_D.$$

Furthermore, we look at the restriction ϱ to the pencil about C: It maps $[A,C] \mapsto [B,D]$, $[B,C] \mapsto [A,D]$, $[C,D] \mapsto [D,C]$. Therefore, $\sigma \circ \tau$ acts like ϱ if restricted to the pencil about C, and thus, $\varrho(t_C) = t_D$. Consequently, c and $\varrho(c)$ share four points A, B, C, D, and the tangent t_D at D. According to Theorem 6.1.2, we have $c = \varrho(c)$.

Since ϱ is involutive, its restriction to the conic c is an involutive projectivity. The line a is also the axis of ϱ_c in the sense of Theorem 6.4.4, since $[A, \varrho(C)] \cap [\varrho(A), C] = [A, D] \cap [C, B] \in a$, and further $[A, \varrho(D)] \cap [\varrho(A), D] = [A, C] \cap [B, D] \in a$, according to Theorem 6.2.2. In Fano planes, involutive projectivities on lines as well as conics are never parabolic. Thus, a is never tangent to c.   ∎

■ **Example 6.4.7** Center of an involution.

As we have seen in the proof of Theorem 6.4.5, the restriction of a harmonic automorphic collineation ϱ of a conic c to c is an involutive projectivity $\eta : c \to c$. Any point $X \in c$ is collinear with its image $X' = \eta(X)$ and the center Z of the collineation ϱ. Usually, the point Z is called the *center of the involution*. Sometimes we denote it by I_η in order to indicate that I_η is the center of the involution η. If t is a tangent of c at T with $I_\eta \in t$, then T is a fixed point of η. Conversely, if $T \in$ is a fixed point of $\eta : c \to c$, then the tangent t of c at T passes through I_η.

It is not hard to show that, conversely, for each involution η on c its extension to an involutive automorphic collineation ϱ of c is a harmonic homology, and hence, a perspective collineation. This can be concluded from Figure 6.28 and the fact that the axis of η must be fixed pointwise. Therefore, each involution η has a center $I_\eta \notin c$.

Corollary 6.4.1 *Any involutive projectivity on a conic c can be extended to a harmonic homology leaving c invariant.*

Elliptic involution, interior and exterior of a conic.

An involutive projectivity $\eta : c \to c$ is either hyperbolic or elliptic. In the first case, the center I_η of the involution lies *outside* the conic, *i.e.*, it is called an *outer point* of c. In the second case, I_η is an *interior point*. Any outer point of a conic c sends two tangents to the conic, and there are no tangents through interior points.

The existence of elliptic involutive projectivities on a conic c causes the existence of interior points, no matter, if the projective plane (or equivalently, its algebraic model) is finite or not. There are no elliptic involutions on a conic in $\mathbb{P}^2(\mathbb{C})$, and consequently, there are no interior points of c in $\mathbb{P}^2(\mathbb{C})$.

Non-involutive projective mappings.

The construction of fixed points or fixed lines of involutive projectivities on lines or in pencils is a little less work than in the case of a non-involutive projectivity, cf. Example 6.4.6.

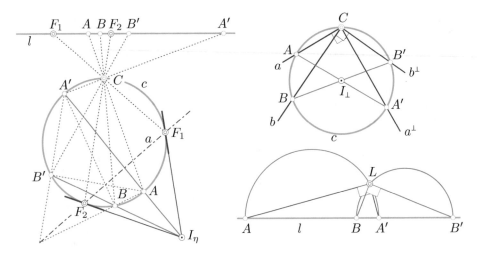

FIGURE 6.29. Left: Fixed points F_1 and F_2 of a hyperbolic involution on a line l constructed on a conic c. Right, top: An elliptic involution in a pencil of lines, the involution of right angles. The center of involution is the center of the circle c, and thus, there are no fixed points on c, and consequently, no fixed lines in the pencil about C. Right, bottom: An elliptic involution on a line l and one of its Laguerre points L.

In Figure 6.29, we treat the following examples in $\mathbb{P}^2(\mathbb{F})$:

• On the left-hand side, we are given an involution $\eta : l \to l$ on a line l that maps $A \mapsto A'$ and $B \mapsto B'$. The projectivity η is lifted to a conic c which is a circle just in order to simplify the constructions. Here, c is a Steiner circle. A point $C \in c$ can be chosen freely as a center of the stereographic projection $l \to c$. Note that C may not coincide with points of intersections of c and l if there are some. All stereographic images of points on l are labeled in the same way, though we know that these are different points. Now, we apply Theorem 6.4.5 and find the center I_η of the involution η on c. The tangents of c through I_η touch c in the fixed points F_1 and F_2. The subsequent stereographic projection yields the fixed points F_1 and F_2 of η considered as an involution on l.

• On the right-hand side of Figure 6.29 (top), we can see some lines of a pencil with vertex C. There is an involutive projectivity that maps the lines a and b to the uniquely defined orthogonal lines $a^\perp$ and $b^\perp$ ($a^\perp, b^\perp \ni C$). We call this mapping the *involution of right angles* at C. Without knowledge from Projective Geometry, we would have recognized this mapping as involutive: Applying it twice to a line, say l, yields the line $(l^\perp)^\perp = l$. We also would guess that this mapping (if we knew that it is a projectivity) is elliptic, since there are no self-orthogonal lines in the real projective plane.

The constructive treatment of the involution of right angles can be done on a conic c. However, we take circle c through the point C (cf. Figure 6.29), not only to simplify the construction. According to the theorem by THALES (cf. the footnote on page 56 and Corollary 6.3.3), the lines a and $a^\perp$ meet c in C and in a pair (A, A') of anti-podal points. This is also the case for the lines b and $b^\perp$. Therefore, the center $I_\perp$ is the center of the circle which is an interior point of c. Thus, once again, the involution of right angles is recognized as an elliptic involution.

We should add a little comment on the analytic treatment of the involution of right angles. Assume that C is given by its homogeneous coordinates $(1:0:0)$. Then, we can represent the lines in the pencil about C by their direction vectors which are at the same time homogeneous coordinates of the lines in this pencil. Without loss of generality, $\mathbf{a} = (1:0)$, $\mathbf{a}^\perp = (0:1)$, $\mathbf{b} = (\cos\varphi : \sin\varphi)$, and $\mathbf{b}^\perp = (-\sin\varphi : \cos\varphi)$ with $0 < \varphi < \frac{\pi}{2}$. Then, the involution is described by the matrix

$$\mathbf{T} = \begin{pmatrix} 0 & -1 \\ 1 & 0 \end{pmatrix}.$$

The eigenvectors of $\mathbf{T}$ are the coordinate vectors of the fixed lines. We use the complex extension $\mathbb{P}^2(\mathbb{C})$ of $\mathbb{P}^2(\mathbb{R})$ and find $\mathbf{f}_1 = (1:i)$ and $\mathbf{f}_2 = (1:-i)$ which correspond to the two complex conjugate lines with Cartesian coordinates

$$y = \pm ix.$$

Indeed, these two lines are fixed under Euclidean rotations with any angle of rotation. These lines are called *isotropic lines of Euclidean Geometry* through C. Through any point in the (complex extended) Euclidean plane there exists such a pair of isotropic lines. The common ideal points $(0:1:\pm i)$ are called *absolute points* or *absolute circle points* of Euclidean Geometry.

On the line $y = ix$, the Euclidean measurement of lengths degenerates: Let $a, b \in \mathbb{R}$ be two different values. Then, the two points $A = (a, ia)$ and $B = (b, ib)$ have the Euclidean distance

$$\overline{AB} = \sqrt{(a-b)^2 + (ia - ib)^2} = \sqrt{(a-b)^2 - (a-b)^2} = 0$$

although $a \neq b$ and $A \neq B$.

• Any elliptic involution η on a line l gives rise to an elliptic involution in a pencil of lines by simply inserting a perspectivity in between, while the property of being elliptic is not violated. The right-hand side of Figure 6.29 (bottom) shows that for any elliptic and involutive projectivity we can find a pair of points such that pairs (A, A') of corresponding points can be seen under right angles. (Two pairs (A, A') and (B, B') of corresponding points determine η.) Figure 6.29 shows only one such point L; the second one is the reflection of L in l. The point L is usually called *Laguerre point* named after the French mathematician EDMOND LAGUERRE (1834–1886). L can be found as the intersection of the two Thales circles over the segments AA' and BB'.

The Laguerre point is useful when we have to find the image X' of a point X under an elliptic involution on l: Draw the normal to $[X, L]$ at L and intersect it with L, i.e., $[X, L]^\perp \cap l = X'$.

Euclidean circles.

The name *circle point* can also be justified in the following way: The equation of the Euclidean circle c with center $C = (m, n)^\mathrm{T} \in \mathbb{R}^2$ and radius $r \in \mathbb{R}^+$ reads either (in terms of inhomogeneous Cartesian coordinates)

$$x^2 + y^2 - 2mx - 2ny + m^2 + n^2 - r^2 = 0$$

or (in terms of homogeneous coordinates)

$$x_1^2 + x_2^2 - 2mx_0x_1 - 2nx_0x_2 + (m^2 + n^2 - r^2)x_0^2 = 0.$$

From the latter we can find the intersection of c with the ideal line $x_0 = 0$ (cf. page 200) of the projectively closed Euclidean plane as

$$I = (0:1:i) \quad \text{and} \quad \overline{I} = (0:1:-i).$$

On the other hand, any conic that passes through I and $\overline{I}$ has an equation of the (inhomogeneous) form

$$a_{11}x^2 + a_{22}y^2 + 2a_{01}x + 2a_{02}y + a_{00} = 0$$

which can be verified by simply inserting the coordinates of I and $\overline{I}$ into the most general form (6.12) (see page 240) $\sum_{i,j=0\ldots2} a_{ij}x_i x_j = 0$ of a conic's equation. Summarizing, we can say:

Corollary 6.4.2 *Any conic in $\mathbb{P}^2(\mathbb{F})$ that passes through the absolute points $I, \overline{I}$ of Euclidean geometry is a Euclidean circle, and* vice versa.

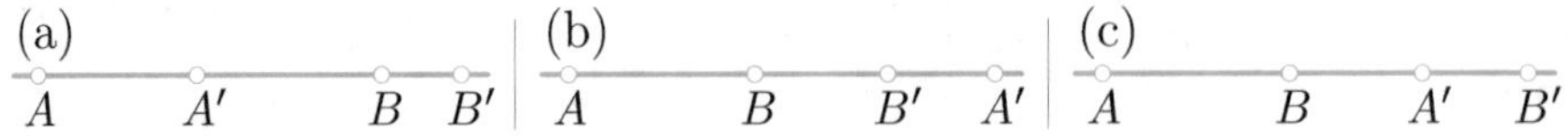

FIGURE 6.30. (a) A hyperbolic involution determined by two separated pairs (A, A') and (B, B') of corresponding points. (b) A hyperbolic involution can also be determined by two pairs (A, A') and (B, B') where one pair encloses the other pair (completely). (c) If the pairs (A, A') and (B, B') of corresponding points are entangled, then $A \mapsto A'$, $B \mapsto B'$ determines a unique elliptic involution.

The ordering of four different collinear points A, A', B, and B' that determine a unique involution η via $A \mapsto A'$ and $B \mapsto B'$ decides whether η is elliptic or hyperbolic. In Figure 6.30, we can see the three main cases. Examples (a) and (b) show pairs (A, A') and (B, B') of corresponding points of hyperbolic involutions. In the first case, the segments AA' and BB' are separated; in the second case the segment AA' encloses the segment BB'. The elliptic involution in the example (c) shows two entangled pairs of points, the two pairs separate each other. Any two pairs of corresponding points in an elliptic involution are entangled. These configurations can easily be found by means of a stereographic projection from a circle. In the cases (a) and (b) the cross ratio $\mathrm{cr}(A, A', B, B')$ is positive, in (c) negative.

When we are given two involutions on a line, then it is hard to decide whether there are pairs of points that correspond in either involution or not. It is easier to transfer the involutions to a conic c. Then, any involution is represented by its center which is an interior or outer point of c depending on whether the involution is elliptic or hyperbolic.

The line joining the two centers of involutions meets the conic c in those two points which are corresponding in both involutions. In Figure 6.31, we have illustrated all relevant cases. In two cases, (a) and (b), the existence of a common pair is guaranteed. Though in case (d) there is a common pair of corresponding points, we cannot be sure that any two hyperbolic involutions have a common pair. This is only the case as long as the line

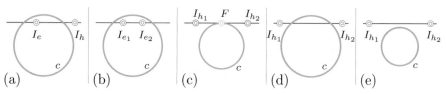

FIGURE 6.31. Common pairs of two involutions and common fixed points: I_e and I_h are the centers of an elliptic and a hyperbolic involution on the conic c. (a) and (b): There is a common pair of corresponding points if at least one involution is elliptic. (c): A common fixed point F of two involutions can only appear if both involutions are hyperbolic. (d): In case of two hyperbolic involutions, a common corresponding pair either *can* appear, or it does not exist like in (e): If the line $[I_e, I_h]$ connecting the two centers does not intersect c.

through the centers of involutions intersects c. A common fixed point of two involutions can only appear if both involutions are hyperbolic, or equivalently, if the line through the centers of involutions is tangent to c as displayed in Figure 6.31, (c).

■ **Example 6.4.8** Interior and outer points of a conic in a finite projective plane.

In a *finite projective plane of order* N, any line carries $N + 1$ points. There are as many lines through any point. Since a conic is equipotent to a line, there are $N + 1$ points on the conic. Since there are $\binom{N+1}{2} = \frac{1}{2}N(N + 1)$ pairs of points on c, we have $\frac{1}{2}N(N + 1)$ outer points of c, *i.e.*, this equals the number of intersections of the tangents at these points.

The number of points in a plane of finite order N equals $N^2 + N + 1$: On any line (range of points) find $N + 1$ points. Let l be such a line. There exists a point P off the line l. The $N + 1$ lines connecting P with all $N + 1$ points of the line l carry P (that is one point), the $N + 1$ points on l, and $N + 1 - 2$ further points different from P and those on l. Thus, we have

$$\underbrace{1}_{P} + \underbrace{(N + 1)}_{\text{on } l} + \underbrace{(N + 1)(N + 1 - 2)}_{\text{the others}} = N^2 + N + 1$$

points, and according to an earlier result, $N^2 + N + 1$ lines in the plane. Consequently, the number of interior points of c equals $N^2 + N + 1 - (N + 1) - \frac{1}{2}N(N + 1) = \frac{1}{2}N(N - 1) = \binom{N}{2}$.

Different appearances of centers of involutions

Some theorems on conics seem to be surprising and hard to prove at first. With a little bit of knowledge from Projective Geometry, they appear more or less trivial. As such, there is a theorem ascribed to FRÉGIER (see [2]):

Theorem 6.4.6 *Let P be a point on a conic c in $\mathbb{P}^2(\mathbb{R})$. All chords of c seen from P at right angles are concurrent in one point F.*

Proof: Assume $[X, X']$ is a chord of c seen at a right angle from P (Figure 6.32, left). Then, the lines $[P, X]$ and $[P, X']$ correspond to each other in the involution of right angles in the pencil of lines about P. Since c passes through P the involution in the pencil induces an involution on c, and according to Example 6.4.7 there exists a center of the involution where the chords joining corresponding points meet. ∎

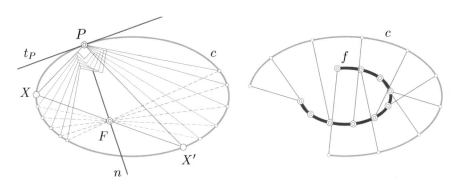

FIGURE 6.32. Left: FRÉGIER's theorem. Right: The points F trace a conic f if P traces c.

FRÉGIER's theorem is shown in Figure 6.32. The chords meet in a point F that lies on c's normal n at P. If c is a circle, then F equals the center of c and FRÉGIER's theorem covers THALES's theorem (cf. Corollary 6.3.3) as a special case. On the right-hand side of Figure 6.32 we can see that the points F trace a conic f if P traces c.

● **Exercise 6.4.3** FRÉGIER transforms of ellipses, hyperbolas, and parabolas.

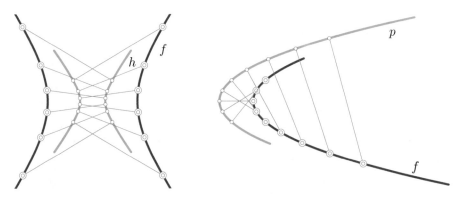

FIGURE 6.33. The traces of centers of the involution of right angles lifted to a hyperbola (left) and parabola (right): The hyperbola h produces a homothetic hyperbola f. The parabola f is a translated copy of p.

Assume that an ellipse e, a hyperbola h, and a parabola p are given by

$$e, h: \quad \frac{x^2}{a^2} \pm \frac{y^2}{b^2} = 1 \quad \text{and} \quad p: \ y^2 - 2qx = 0$$

with $a, b, q \in \mathbb{R} \setminus \{0\}$. Lift the involution of right angles to these conics by projecting the orthogonal and corresponding rays from a point P (on the conic) to the conic. Assume further that P traces the entire (affine part) of the conic and show that the curve f (as shown in figures 6.32 and 6.33) is either a homothetic ellipse/hyperbola or a translated copy of the parabola with the equations

$$f_e, f_h: \quad \frac{x^2}{a^2} \pm \frac{y^2}{b^2} = \frac{a^2 + b^2}{a^2 - b^2} \quad \text{and} \quad f_p: \ y^2 - 2qx + 4q^2 = 0$$

Fixed points of a collineation and the common points of two conics

A projective collineation that is not a perspective collineation κ has at most three fixed points and at most three fixed lines. This is clear from the synthetic point of view: A quadrangle consisting of fixed points characterizes the identity mapping. Any restriction of κ to a line l is a projective mapping $l \barwedge \kappa(l)$. If κ_l has three fixed points, then $\kappa_l = \mathrm{id}_l$ and l is an axis in contradiction to κ is not perspective. Hence, there are no three collinear fixed points. The dual statement holds as well.

The existence and determination of fixed points of a projective collineation is closely related to conics:

Theorem 6.4.7 *The determination of fixed points of a projective and non-perspective collineation is equivalent to the determination of the remaining common points of two conics c and d that share a point A without touching each other at A. Through any fixed point F there exist no two fixed lines such that F is the only fixed point on them. If there exists a fixed point, then there also exists a fixed line.*

Proof:

1. Assume X is not a fixed point of κ and let $X' = \kappa(X)$. If the restriction $\kappa|_X$ of κ to the pencil of lines through X is not perspective, i.e., $[X, X']$ is not a fixed line of κ, then $\kappa|_X$ generates a conic according to Definition 6.1.1. Since κ (as well as its dual) has at most three fixed points (fixed lines), there exists at least one point $A \neq A' = \kappa(A)$ that is not incident with any fixed line.

   Then, A, A', and $\kappa(A') = A''$ form a triangle. The projectivities $\kappa|_A$ and $\kappa|_{A'}$ generate conics c and $d = \kappa(c)$, see Figure 6.34. The line $a = [A, A']$ is tangent to d at A; whereas $a' = [A', A''] \neq a$ is tangent to c at A'. Any fixed point of κ is on c and on $\kappa(c)$, since $\kappa: \ [A, F] \mapsto [A', F] \mapsto [A'', F]$. On the other hand, each $F \in (c \cap \kappa(c)) \setminus \{A'\}$ must be fixed, since $F = [A, F] \cap [A', F] \mapsto [A', F] \cap [A'', F] = F$.

2. The tangent a/a' of d/c at A' intersects c/d in a point A/A''. Any line p through A' different from a and a' meets c in a second point P and d in P'. Now, there exists exactly

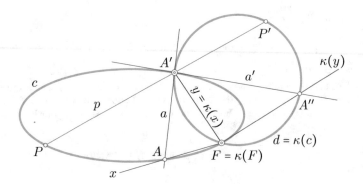

FIGURE 6.34. The two conics mentioned in the proof of Theorem 6.4.7.

one projective collineation κ with $\kappa(A) = A'$, $\kappa(A') = A''$, $\kappa(P) = P'$, and $\kappa(c) = d$. The collineation κ is not perspective since A, A', and A'' are not collinear.

The projective mapping $\kappa|_c : c \to d$ defines a projective mapping $\lambda_{A'}$ in the pencil A'. The lines a, a', and p are fixed lines of $\lambda_{A'}$. Thus, $\lambda_{A'}$ is the identity mapping in the pencil. Therefore, the projective mapping $\kappa|_A$ generates c which according to 1. carries all fixed points of κ. Since $\lambda_{A'}$ is the identity mapping, the fixed points of κ are precisely the common points of c and $d = \kappa(c)$ different from A'.

3. Let $F \in c \cap d$ be a fixed point and $f \ni F$ a fixed line which meets c in a point $G \neq F$. Then, $G \neq A'$ and $G = f \cap [G, A] \mapsto \kappa(G) = f \cap [G, A'] = G$. Consequently, f must be the connection of F with another fixed point G of κ. Therefore, if f is a fixed line through F with the only incident fixed point F, then f is the uniquely determined tangent to c at F.

4. If F is fixed point on g with $g' = \kappa(g) \neq g$, then $g'' = \kappa(g') \neq g'$ and $F = g \cap g' = g' \cap g''$. The projective mappings $\kappa|_g : g \to g'$ and $\kappa|'_g : g' \to g''$ are perspective because F corresponds to itself. The centers Z and $Z' = \kappa(Z)$ deliver a fixed line $[Z, Z']$ in case $Z \neq Z'$, otherwise $[F, Z]$ is fixed.

It is clear that the dual statements hold as well.   ■

If the projective collineation is given analytically by its transformation matrix $\mathbf{T}$ according to Theorem 5.4.1, then the coordinate vectors of the fixed points are eigenvectors of $\mathbf{T}$.

7 Polarities and pencils

A hyperbolic pencil of circles gives rise to a hyperbolic pencil of spheres by simply revolving it about its axis. Pencils of circles are special families of conics which are classified by means of a projectivity induced on the circles' common diameter line.

7.1 Polar system of a conic

In this section, we study a mapping that is associated with any conic: the *polar system* or *polarity*. Moreover, one can use a polarity as a starting point for the theory of conics. Later, in Section 7.2, we shall see that a conic can be defined as the set of self-conjugate points in a hyperbolic polarity.

We still presuppose that the underlying projective plane is Pappian and the Fano axiom is fulfilled. With Theorem 6.4.5 in mind, we give:

Definition 7.1.1 *Let c be a conic in $\mathbb{P}^2(\mathbb{F})$. Then, there exists a mapping*

$$\pi_c : \mathcal{P} \to \mathcal{L}$$

with the following two properties:

1. *For any point $P \notin c$, the line $\pi_c(P)$ is the axis of the unique automorphic harmonic homology ϱ_P of c with center P.*
2. *For any $Q \in c$, the line $\pi_c(Q)$ coincides with the tangent to c at Q.*

The mapping π_c is called polar system of c *or the* polarity with respect to *c. The line $p := \pi_c(P)$ is called the* polar line *of P w.r.t. c; and the point P is called the* pole *of p w.r.t. c.*

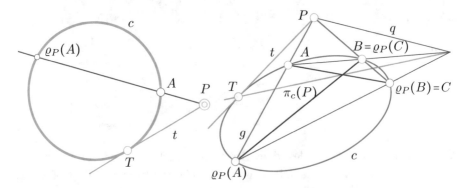

FIGURE 7.1. Left: The polarity π_c maps P to the axis p of ϱ_P. Right: The action of ϱ_P and π_c.

If the point $P \notin c$ lies on the tangent to c at T, then $T \neq P$ is a fixed point of ϱ_P. Consequently, T has to be a point on $\pi_c(P)$. Conversely, $T \in p \cap c$ remains fixed under ϱ_P. Hence, $[P,T]$ must be tangent to c.

If Q is any point of $p = \pi_c(P)$ and $Q \notin c$, then Q is the center of an automorphic harmonic homology ϱ_Q. Since $\varrho_P(Q) = Q$, also the axis $q = \pi_c(Q)$ must be fixed under ϱ_P. Hence, Q must pass through P, since in the case $q = p$, the axis q of the homology ϱ_Q would pass through the center of ϱ_Q.

Lemma 7.1.1 *The polarity w.r.t. any conic c has the property that $Q \in \pi_c(P)$ implies that $P \in \pi_c(Q)$ for all points P and Q.*

This lemma implies that π_c is bijective since the inverse mapping exists: If any line p is given, whether a tangent of c or not, choose two different points $Q_1, Q_2 \in p$. The corresponding polars $\pi_c(Q_1)$ and $\pi_c(Q_2)$ share a point which must be the pole of p.

Any line $g \neq t$ through P that intersects c in a point, say A, intersects c also at $\varrho_P(A)$ with $H(P, g \cap \pi_c(P), A, \varrho_P(A))$ (cf. Figure 7.1). Hence, on the chords $[A, \varrho_P(A)]$ through P, the polar line p is the locus of the fourth harmonic conjugates of P w.r.t. the pair $(A, \varrho_P(A))$.

Let $A, B \in c \setminus \{P\}$ such that A, B, and P are not collinear. Then, $A \varrho_P(A) B \varrho_P(B)$ is a quadrangle where P is one diagonal point. The other two diagonal points span the polar line p of P (Figure 7.1).

Analytic description of a polarity

In Section 6.4, we have shown that the equation of a conic in terms of homogeneous coordinates of points is a quadratic form $\mathbf{x}^T \mathbf{A} \mathbf{x} = 0$ with a regular symmetric matrix $\mathbf{A}$, cf. (6.11). Now, we study the linear mapping $\lambda \colon \mathbb{F}^3 \to \mathbb{F}^3$ that maps points $X = \mathbf{x}\mathbb{F}$ to lines $l = \mathbf{l}\mathbb{F}$ via

$$\mathbf{l} = \mathbf{A}\mathbf{x} \quad \text{with } \mathbf{A}\mathbf{A}^T = \mathbf{I}_3, \ \det \mathbf{A} \neq 0. \tag{7.1}$$

In other words, we interpret the result of the multiplication of the vector $\mathbf{x}$ with a regular symmetric matrix $\mathbf{A}$ from the left as the coordinate vector of a line. If $X = \mathbf{x}\mathbb{F}$ is a point on the conic c with the equation $\mathbf{x}^T \mathbf{A} \mathbf{x} = 0$, then we find

$$\mathbf{x}^T \mathbf{A} \mathbf{x} = \langle \mathbf{A}\mathbf{x}, \mathbf{x} \rangle = \langle \mathbf{l}, \mathbf{x} \rangle = 0$$

which tells us that $X \in c$ is incident with its λ-image. For any point $X \notin c$, we have $\lambda(X) \not\ni X$. We recall (6.10) for the vector representation of a harmonic homology and insert $\mathbf{c}$ as a representative of the center C and

$\lambda(C) = \mathbf{Ac}$ for the axis. Thus, the harmonic homology ϱ_C with center C and axis $\lambda(C)$ has the vector representation

$$\mathbf{x} \mapsto \mathbf{x'} = \mathbf{x}\langle \mathbf{Ac}, \mathbf{c}\rangle - \mathbf{c}\, 2\langle \mathbf{c}, \mathbf{Ax}\rangle, \tag{7.2}$$

where fractions can be circumvented since we are dealing with homogeneous coordinates. Then, we observe that the quadratic form giving the equation of the conic does not change, apart from non-vanishing factors:

$$
\begin{aligned}
\mathbf{x'}^{\mathrm{T}}\mathbf{Ax'} &= \left(\mathbf{x}\langle \mathbf{Ac}, \mathbf{c}\rangle - \mathbf{c}\,2\langle \mathbf{c}, \mathbf{Ax}\rangle\right)^{\mathrm{T}}\mathbf{A}\left(\mathbf{x}\langle \mathbf{Ac}, \mathbf{c}\rangle - \mathbf{c}\,2\langle \mathbf{c}, \mathbf{Ax}\rangle\right) = \\
&= \left(\mathbf{x}^{\mathrm{T}}(\mathbf{c}^{\mathrm{T}}\mathbf{Ac}) - \mathbf{c}^{\mathrm{T}}\,2\left(\mathbf{c}^{\mathrm{T}}\mathbf{Ax}\right)\right)\mathbf{A}\left(\mathbf{x}(\mathbf{c}^{\mathrm{T}}\mathbf{Ac}) - \mathbf{c}\,2\left(\mathbf{c}^{\mathrm{T}}\mathbf{Ax}\right)\right) = \\
&= \mathbf{x}^{\mathrm{T}}\mathbf{Ax}(\mathbf{c}^{\mathrm{T}}\mathbf{Ac})^2 \quad \text{where } \mathbf{c}^{\mathrm{T}}\mathbf{Ac} \neq 0.
\end{aligned}
$$

Note that $\langle \mathbf{Ay}, \mathbf{x}\rangle = \mathbf{y}^{\mathrm{T}}\mathbf{Ax} = \mathbf{x}^{\mathrm{T}}\mathbf{Ay} = \langle \mathbf{Ax}, \mathbf{y}\rangle$ since the matrix $\mathbf{A}$ is symmetric. Therefore, the harmonic homology with center $C = \mathbf{c}\mathbb{F}$ and axis $a = \lambda(C)$ leaves the conic c invariant. Now, we have:

Theorem 7.1.1 *If $\mathbf{x}^{\mathrm{T}}\mathbf{Ax} = 0$ is the equation of a conic c with a regular symmetric $\mathbf{A} \in \mathbb{F}^{3\times 3}$, then $\mathbf{x} \mapsto \mathbf{Ax}$ describes the polarity π_c w.r.t. to c.*

Since $\mathbf{A}$ is regular by assumption, we can say:

Corollary 7.1.2 *1. Any polarity $\pi_c : \mathcal{P} \to \mathcal{L}$ with $\pi_c : \mathbf{x} \mapsto \mathbf{l} = \mathbf{Ax}$ in a projective plane $\mathbb{P}^2(\mathbb{F})$ is a collineation from $\mathbb{P}^2$ onto its dual plane $\mathbb{P}^2(\mathbb{F})^\star$.*
2. Any polarity $\pi_c : \mathcal{P} \to \mathcal{L}$ defines a mapping $\pi_c^\star : \mathcal{L} \to \mathcal{P}$ by $X \in l \Leftrightarrow \pi_c^\star(l \in \pi_c(X))$ for all X and l. The mapping $\pi_c^\star$ is described by $\mathbf{A}^{-1}$ if π is described by $\mathbf{A}$ and $\pi_c^{\star-1}$ is called the adjoint *mapping.*

Proof: 1. Let the polarity π_c be described by a symmetric matrix $\mathbf{A}$. Then, $\mathbf{A}$ is also the coordinate matrix of a collineation according to Theorem 6.3.1. The coordinates $\mathbf{l}$ can also be interpreted as homogeneous coordinates of points in a projective plane $\mathbb{P}^2(\mathbb{F})$. So, π can be seen as a collineation from $\mathbb{P}^2(\mathbb{F})$ to $\mathbb{P}^{2\star}(\mathbb{F})$.
2. According to (6.8), if the transformation of points $\mathbf{x}\mathbb{F}$ is given by the multiplication with the matrix $\mathbf{A}$ from the left, then the lines $\mathbf{l}\mathbb{F}$ are transformed by the multiplication of $\mathbf{l}$ with $\mathbf{A}^{-\mathrm{T}}$ from the left. In the case of a polarity, $\mathbf{A}$ is symmetric, and therefore, $\mathbf{A}^{-1}$ is too. Thus, $\kappa^\star$ is described by $\mathbf{A}^{-1}$. ∎

If $\mathbf{A}$ is singular, then the polarity defined by $\mathbf{x} \mapsto \mathbf{Ax}$ is *singular* or *degenerate* and so is the curve $c: \mathbf{x}^{\mathrm{T}}\mathbf{Ax} = 0$ of degree 2. In this case, the conic c is either a *pair of lines* (which, in the complex extension of $\mathbb{P}(\mathbb{R}^3)$ may be real or conjugate complex), or c is a *repeated line*. A separate

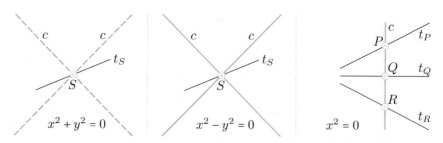

FIGURE 7.2. Tangents to degenerate curves of degree 2 in the real projective plane: Left: One of many tangents at the only real point S of $x^2 + y^2 = 0$ from the pencil S. Middle: One of many tangents at the singular point S of $x^2 - y^2 = 0$. Right: Every line is a tangent of the repeated line $x^2 = 0$.

discussion is necessary if we work in $\mathbb{P}(\mathbb{F}^3)$ with an arbitrary field. The tangent of a conic c is a line which meets with multiplicity two. We extend this definition for degenerate curves by calling a line t a *tangent* also if t is a component of c. This may be possible in a single point, in case of a pair of lines, or in any point in the case of a repeated line, cf. Figure 7.2.

Since π_c and $\pi_c^\star$ are described by inverse matrices $\mathbf{A}$ and $\mathbf{A}^{-1}$, we have $\pi_c : x \mapsto 1 = \mathbf{A}x$ and $\pi_c^\star : 1 \mapsto x = \mathbf{A}^{-1}x$, and therefore, π equals the inverse of $\pi^\star$. We call π_c *self-adjoint*. Lemma 7.1.1 is a direct consequence of this fact.

Now, we generalize the the term polarity w.r.t. a conic by:

Definition 7.1.2 *Each mapping* $\pi : \mathcal{P} \to \mathcal{L}$ *with* $x\mathbb{F} \mapsto 1 = \mathbf{A}x$ *where* $\mathbf{A}^{\mathrm{T}} = \mathbf{A}$ *and* $\det \mathbf{A} \neq 0$ *is called a* polarity.

This definition is independent of whether $x^{\mathrm{T}}\mathbf{A}x = 0$ is a conic or empty. Lemma 7.1.1 is still valid for all polarities.

In the following we use:

Definition 7.1.3 *Let* $\pi : \mathcal{P} \to \mathcal{L}$ *be a polarity in a Pappian projective plane with the Fano property.*

1. *All points* $Y \in \pi(X)$ *are called* conjugate *(with respect) to* X. *All lines* $l \ni \pi^\star(m)$ *are called* conjugate *to* $M = \pi^\star(m)$.

2. *A point* X *which is incident with its polar* $x = \pi(X)$ *is called* self-conjugate. *A line which contains its pole is also called* self-conjugate.

3. *A polarity is called* elliptic *if there are no self-conjugate points.*

4. A polarity is called hyperbolic *if there exist self-conjugate as well as not self-conjugate points.*

Figure 7.3 illustrates the contents of Definition 7.1.3.

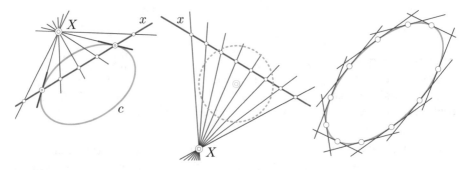

FIGURE 7.3. Left: The points on $x = \pi_c(X)$ are conjugate (w.r.t. the conic c) to the point X. All lines through X are conjugate to x. Since there is a conic (with real points), we have a hyperbolic polarity. Middle: Conjugate points and lines in an elliptic polarity. Right: Some of the self-conjugate points and lines of a hyperbolic polarity: the points and tangents of a conic.

With the analytic description of a polarity we can easily show:

Theorem 7.1.2 *The conjugacy of points and lines w.r.t. polarity π is a symmetric relation: If the point P (line l) is conjugate to the point Q (line m) w.r.t. π, then Q (m) is conjugate to P (l).*

Proof: Assume that P and Q are given by the homogeneous coordinates $\mathbf{p}$ and $\mathbf{q}$ taken from $\mathbb{F}^3$. According to Definition 7.1.2, the polarity π is described by a symmetric matrix $\mathbf{A} \in \mathbb{F}^{3 \times 3}$. Following Definition 7.1.3, P is conjugate to Q if, and only if, $P \in \pi(Q)$. In terms of coordinates this reads $\langle \mathbf{p}, \pi(\mathbf{q}) \rangle = 0$. The symmetry of $\langle \cdot, \cdot \rangle$ and $\mathbf{A}$ lead to

$$\langle \mathbf{p}, \pi(\mathbf{q}) \rangle = \mathbf{p}^T \mathbf{A} \mathbf{q} = \mathbf{q}^T \mathbf{A}^T \mathbf{p} = \mathbf{q}^T \mathbf{A} \mathbf{p} = \langle \mathbf{q}, \pi(\mathbf{p}) \rangle$$

which proves the symmetry of the conjugacy for points. An analogous computation shows the symmetry of the conjugacy of lines. ∎

■ **Example 7.1.1** Some polarities.

Elliptic polarity. The matrix

$$\mathbf{M} = \begin{pmatrix} 1 & 0 & 0 \\ 0 & 1 & 0 \\ 0 & 0 & 1 \end{pmatrix}$$

is the coefficient matrix of the quadratic form

$$Q(\mathbf{x}, \mathbf{x}) = x_0^2 + x_1^2 + x_2^2 = \langle \mathbf{x}, \mathbf{x} \rangle.$$

The conic $Q(\mathbf{x}, \mathbf{x}) = 0$ carries no points if considered as a conic in the real projective plane $\mathbb{P}^2(\mathbb{R})$. On the other hand, the matrix $\mathbf{M}$ is symmetric. Thus, the matrix $\mathbf{M}$ defines an elliptic polarity ι considered as a polarity in the real projective plane. Note that ι as well as $\iota^\star$ are described by the identity matrix and $\iota \neq \mathrm{id}_{\mathbb{P}^2}$.

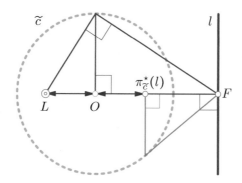

FIGURE 7.4. The particular elliptic polarity ι can be realized as the composition of the polarity w.r.t. the circle $\widetilde{c}$ and the reflection in the common center O of the elliptic polarity and $\widetilde{c}$. A line l is mapped to its anti-pole L, *i.e.*, the reflection of $\pi^\star_{\widetilde{c}}$ in O. The anti-inversion w.r.t. $\widetilde{c}$ sends the pedal point F of l to L.

We can visualize ι by assuming that the underlying coordinate frame is Cartesian. The pole of the ideal line $(1:0:0)$ is the point $(1:0:0)$ which has inhomogeneous coordinates $O = (0,0)$ and be the origin of a Cartesian frame. The point O is called the *center of the polarity*.

A line l with inhomogeneous equation $x = d$ (and homogeneous coordinates $(-d:1:0)$) is mapped to the point $L = (-d:1:0)$, *i.e.*, $L = (-\frac{1}{d}, 0) =: (d', 0)$ which gives

$$dd' = -1.$$

Consequently, L can be found be constructing a right angled triangle as shown in Figure 7.4. Let F denote the pedal point of l w.r.t. O. Then, the mapping $F \mapsto L$ is an elliptic and involutive projectivity just as we have seen in Example 5.4.3 (on page 213) and in Figure 5.23.

The reflection M of L in O satisfies $\overline{OM} \cdot \overline{OL} = 1$. Further, M is the pole of the line l under the polarity η with the coordinate matrix $C = \mathrm{diag}(-1, 1, 1)$. Since $\mathbf{x}^{\mathsf{T}} \mathbf{C} \mathbf{x} = 0$ is the equation of the unit circle $\widetilde{c}$, M is the pole of l w.r.t. $\widetilde{c}$. Therefore, ι can be realized as the composition of the reflection in O and the subsequent polarity with respect to the unit circle. In the older literature, the mapping ι is called an *anti-polarity*.

The mapping that sends the point F to L is called *anti-inversion*. The restriction of the anti-inversion to lines through the center O is that kind of projectivity that is shown in Example 5.4.3 on page 213. Here, the unit circle $\widetilde{c}$ takes the role of the Laguerre points and is sometimes called the *real representative* for the conic $\langle \mathbf{x}, \mathbf{x} \rangle = 0$ without real points.

Pairs of conjugate points on the ideal line ω are seen from the center O under a right angle. It is easy to see that each elliptic polarity in $\mathbb{P}^2(\mathbb{R})$ is the composition of the polarity in an ellipse $\widetilde{c}$ and the reflection in the center of $\widetilde{c}$.

A hyperbolic polarity - orthogonality and the Euclidean unit circle. *The Euclidean unit circle c* with its Cartesian equation $x^2 + y^2 = 1$ can be given an equation in homogeneous coordinates by letting $x = x_1 x_0^{-1}$ and $y = x_2 x_0^{-1}$ which yields

$$-x_0^2 + x_1^2 + x_2^2 = (x_0, x_1, x_2) \begin{pmatrix} -1 & 0 & 0 \\ 0 & 1 & 0 \\ 0 & 0 & 1 \end{pmatrix} \begin{pmatrix} x_0 \\ x_1 \\ x_2 \end{pmatrix} = \mathbf{x}^{\mathrm{T}} \mathbf{M} \mathbf{x} = 0.$$

Note that $\mathbf{M}^{-1} = \mathbf{M}$.

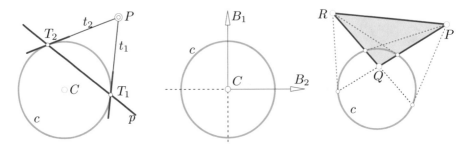

FIGURE 7.5. Left: The polar line p of P w.r.t. c intersects c in the points T_1 and T_2. These are the points of contact of c's tangents t_1 and t_2 through P. Middle: The pair $([C, B_1], [C, B_2])$ of axes is a pair of conjugate diameters of c. The triangle CB_1B_2 is a polar triangle of c. Right: Any vertex of a polar triangle PQR is the pole of the opposite side line. In $\mathbb{P}^2(\mathbb{R})$, each polar triangle has precisely one interior vertex.

We define the *center of the circle c* as the center of the polarity π_c. In this case, it is the point $C = (1:0:0)$. A line d through the center of the conic is called a *diameter*. Here, all lines $a_1 x_1 + a_2 x_2 = 0$ with $(a_1 : a_2) \neq (0:0)$ are diameters.

In general, the *diameters* of a conic are the polar lines of all ideal points. The *center of the conic* is the pole of the ideal line w.r.t. the conic. A pair (d_1, d_2) of diameters is called *conjugate pair* or a *pair of conjugate diameters* if, and only if, the two diameters (or equivalently their ideal points) are conjugate w.r.t. c.

A pair (d_1, d_2) of conjugate diameters of a circle is a pair of orthogonal lines: Assume $\mathbf{u} = (u_0, u_1, u_2)^{\mathrm{T}}$ and $\mathbf{v} = (v_0, v_1, v_2)^{\mathrm{T}}$ are the homogeneous coordinates of two different diameters u and v of a circle centered at the origin of a Cartesian coordinate system. Then, $u_0 = v_0 = 0$ and the conjugacy of $\mathbf{u}$ and $\mathbf{v}$ can be expressed in terms of coordinates as

$$(0, u_1, u_2) \mathbf{M}^{-1} \begin{pmatrix} 0 \\ v_1 \\ v_2 \end{pmatrix} = u_1 v_1 + u_2 v_2 = \left\langle \begin{pmatrix} u_1 \\ u_2 \end{pmatrix}, \begin{pmatrix} v_1 \\ v_2 \end{pmatrix} \right\rangle = 0$$

which means that the canonical scalar product of the normal vectors of the lines u and v vanishes. Therefore, the diameters are orthogonal (see Figure 7.5, middle). There, we can see the special pair $([C, B_1], [C, B_2])$ with $B_1 = (0:1:0)$ and $B_2 = (0:0:1)$.

On the right-hand side of Figure 7.5, we can see a *polar triangle* of c. In general, a triangle PQR is called a *polar triangle* of a polarity or a conic, if each vertex is the pole of the opposite side line. In case of a hyperbolic polarity, precisely one vertex of the polar triangle is an interior point of the conic c. In case of an elliptic polarity, we cannot distinguish between interior and outer points.

The construction of a polar triangle within a given polarity π (no matter, whether it is elliptic or hyperbolic) starts with the choice of a point P which is not self-conjugate. Then, we determine the polar line $p = \pi(P)$ of P. Finally, we choose a point $Q \in p$ which implies $P \in \pi(Q)$, by virtue of Lemma 7.1.1. The third vertex of the polar triangle equals $R = p \cap \pi(Q)$.

Looking back to Figure 7.4 accompanying the previous example of an elliptic polarity, we see that the points $F = (1 : d : 0)$ and $M = (d : 1 : 0)$ satisfy $\overline{OF} \cdot \overline{OM} = d \cdot \frac{1}{d} = 1$. They are said to lie *inverse w.r.t. the unit circle* $\widetilde{c}$. The mapping $F \mapsto M$ is called *inversion w.r.t. the unit circle* or *reflection in the unit circle*. In Section 7.5, we shall see that the inversion in a circle is a special case of a birational quadratic transformation. We speak briefly of a conjugate pair of points and lines.

Polar forms

Each quadratic form $\mathbf{x}^T \mathbf{A} \mathbf{x} = 0$ with $\mathbf{A} = \mathbf{A}^T$ defines an associated symmetric bilinear form

$$p(\mathbf{x}, \mathbf{y}) = \mathbf{y}^T \mathbf{A} \mathbf{x} \qquad (7.3)$$

called the *polar form*. Independently of whether $\mathbf{A} \in \mathbb{F}^{3 \times 3}$ is regular or not, the polar form characterizes conjugate points (lines): $\mathbf{x}$ is conjugate to $\mathbf{y}$ w.r.t. the polarity defined by $\mathbf{A}$ if, and only if, $p(\mathbf{x}, \mathbf{y}) = 0$. In the case of a singular matrix $\mathbf{A}$ we speak of a *singular polarity* which is no longer bijective.

■ **Example 7.1.2** Polar forms of conics in the Euclidean plane. We assume that

$$c : \quad \frac{x^2}{a^2} \pm \frac{y^2}{b^2} = 1$$

is an ellipse (or hyperbola) with semiaxis lengths a and b. Thus, $\mathbf{A} = \mathrm{diag}(-1, a^{-2}, \pm b^{-2})$. For $\mathbf{x} = (1, x, y)$ and $\mathbf{y} = (1, x', y')$, we obtain

$$p(\mathbf{x}, \mathbf{y}) = -1 + \frac{xx'}{a^2} \pm \frac{yy'}{b^2}$$

which gives the equation of the tangent t_P to c at $P = (x', y')$ if $P \in c$, *i.e.*, if $-1 + \frac{x'^2}{a^2} \pm \frac{y'^2}{b^2} = 0$. A parabola may be given by

$$x^2 - 2qy = 0 \quad \Longleftrightarrow \quad \mathbf{x}^T \begin{pmatrix} 0 & 0 & -q \\ 0 & 1 & 0 \\ -q & 0 & 0 \end{pmatrix} \mathbf{x} = 0.$$

With $\mathbf{x} = (1, x, y)$ and $\mathbf{y} = (1, x', y')$, we find the polar form as

$$p(\mathbf{x}, \mathbf{y}) = xx' - q(y + y') = 0.$$

Involution of conjugate points/lines

Assume c is a conic in a Pappian projective plane with the Fano property and π_c is the polarity w.r.t. c. What happens to the polar line if a point P

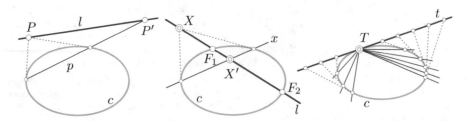

FIGURE 7.6. Involutions induced by the polar system of a conic c: Left: The involution of conjugate points on the line l is elliptic if $l \cap c = \emptyset$. Middle: The fixed points of the involution of conjugate points on the line l are the common points of l and c. Right: There is no involution on a tangent t. The point T is conjugate to all points of t.

traverses a line l? The polar line p goes through the pole $P = \pi_c(P)$, and we get $P' := p \cap l$ for the point P' which is conjugate to P. The mapping $\lambda : P \mapsto P'$ is linear in terms of homogeneous coordinates of points, and bijective, if l is not tangent to c. Thus, it is a projectivity. Moreover, λ is involutive because $\pi(P') \ni P$, according to Lemma 7.1.1. The mapping λ is called *involution of conjugate points* (see Figure 7.6).

If the involution λ of conjugate points on a line l is elliptic, then there are no common points of c and l. On the other hand, if there exists a fixed point F_1 of λ, then it is a common point of l and c. In this case, there exists a further fixed point F_2 since involutions are never parabolic in Fano planes. Hence, l meets c in these two points (cf. Figure 7.6).

However, the case of a tangent t of c is different: Any point on t is conjugate to the point T of contact of c and t, since the polar lines of all points on t are passing through T (see Figure 7.6). So, there is no involutive projectivity on t induced by π_c.

As the dual counterpart of the involution of conjugate points, we find the involution of conjugate lines about a point P, provided that P is not self-conjugate, *i.e.*, $P \notin c$. Figure 7.7 shows pairs (l, l') and (m, m') of lines conjugate w.r.t. a conic c. In the left-hand figure, the common point of these four lines lies in the interior of c, and thus, the involution of conjugate lines is elliptic. The right-hand figure shows the hyperbolic case. The point P is an outer point of c. Then, the fixed lines of the involution of conjugate lines through the point P are the tangents t_1 and t_2 of c through P.

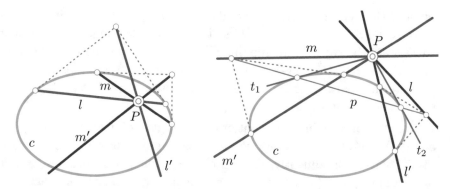

FIGURE 7.7. Involution of conjugate lines: Left: The involution of conjugate lines through an interior point P of c is elliptic. Right: The involution of conjugate lines through an outer point P of c is hyperbolic and the fixed lines are the tangents t_1 and t_2 from P to c.

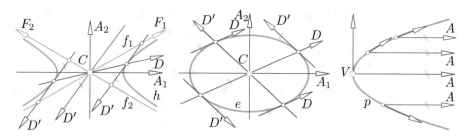

FIGURE 7.8. Conjugate diameters: Left: The two fixed lines of the involution of conjugate diameters of a hyperbola h are the asymptotes of h. Middle: An ellipse e has no asymptotes. Like the hyperbola h, the ellipse has two axes being the only pair of orthogonal conjugate diameters. Right: The parabola p with its only axis and the vertex V has parallel diameters.

As defined in Example 7.1.1, in the projective extension of $\mathbb{E}^2$, the center of a conic is the pole of the ideal line provided that this point is not self-conjugate. Thus, the ellipse and the hyperbola have a center and the parabola is a conic without center.

Now, we assume that c is a conic with center C, $i.e.$, it is either a hyperbola or a parabola as shown in Figure 7.8. We look at the involution η of conjugate lines in the pencil C. If η is hyperbolic, $i.e.$, it has two fixed lines f_1 and f_2, then there exist two tangents from the center C to the conic c. The tangents f_1 and f_2 touch c at the ideal points F_1 and F_2. Since F_1 and F_2 are ideal points on c, the conic c is a hyperbola and the lines f_1 and f_2 are its asymptotes. The center of a hyperbola is an outer point of the hyperbola. Any other diameter $d = [C, D]$ is the line spanned by the center C and an arbitrary ideal point D. The diameter d' conjugate to d passes through the ideal point D' which is the conjugate of D on the ideal line w.r.t. D. Consequently, the tangents at the endpoints of any diameter are parallel to the conjugate diameter. Note that conjugacy is a symmetric relation.

In the case of an elliptic involution of conjugate diameters in the pencil about the center, the conic c is an ellipse since it has no self-conjugate ideal points. Consequently, the center of an ellipse c is an interior point of c. Again, pairs (d, d') of conjugate diameters show the characteristic behavior: The tangents at the endpoints of d and d' form a parallelogram (cf. Figure 7.8).

In both cases, ellipse and hyperbola, we defined the *axes of the conic* as the orthogonal pair (a_1, a_2) of conjugate diameters. Thus, (a_1, a_2) is a pair of assigned elements in both involutions, the involution of right angles and the involution of conjugate diameters. Since the involution of right angles is elliptic, there exists one pair, no matter if the involution of conjugate diameters is elliptic or hyperbolic (cf. page 255).

The points of intersection of an ellipse or a hyperbola with its axes are the vertices. Since the center C is an interior point of an ellipse c, there are four vertices, because both axes meet c in two points. In the case of a hyperbola, there are only two (real) vertices.

In the case of a parabola, there is exactly one ideal point A that is self-conjugate since the parabola touches the ideal line. Therefore, the parabola has no center. It is impossible to view a parabola as the limit of an infinite sequence of ever stretching ellipses. The lines through A are the diameters of the parabola (see Figure 7.8). The parabola has only one axis which is that diameter of the parabola which is perpendicular to the tangent at its finite point of intersection with the parabola. So, if we want to find the axis of a parabola, we first have to find the point V on c whose tangent is orthogonal to the diameters. The point V is the only vertex of the parabola.

■ **Example 7.1.3** Vertex of a parabola on two line elements.

A parabola p shall be given by two line elements, say (A, a) and (B, b). Assuming that neither $A \in b$, nor $B \in a$, nor $a \parallel b$, we can be sure that there is a unique parabola on these two line elements. We shall construct the vertex of the parabola.

As shown in Figure 7.9, we are looking for the parabola's ideal point P in the first step. On the chord $[A, B]$, the ideal point I and midpoint M of AB are conjugate w.r.t. p. Since $T = a \cap b$ is the pole of $[A, B]$, the line $[T, M]$ carries p's ideal point P.

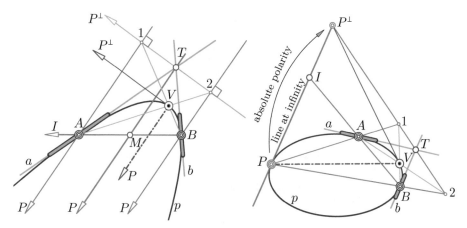

FIGURE 7.9. The vertex of a parabola on two line elements (A, a) and (B, b) is constructed with the help of only six lines (left). The construction is based on three harmonic quadruples (right).

Once we have found the direction of p's axis, *i.e.*, the ideal point P, we know that the orthogonal direction $P^\perp$ is the direction of the tangent at the vertex V.

The second step is the construction of the vertex V being a point of p with a tangent $[V, P^\perp]$ orthogonal to the diameters: We intersect the diameters through A and V with the orthogonal line $[T, P^\perp]$ and obtain the points 1 and 2. Now, we claim that $V = [1, B] \cap [2, A]$ is the vertex. Proof: The point 1 is the center of a harmonic homology ϱ_1 which is automorphic for p. We have $\varrho_1 : A \mapsto P$, and T on the tangent a at A is mapped to $P^\perp$ on the line at infinity which is tangent at P. The point 2 as the fourth harmonic point to the center 1 w.r.t. $(T, P^\perp)$ must be a point of the axis of ϱ_1, and therefore, remains fixed. Hence ϱ_1 maps $B \in [2, P] \cap p$ to $V = [2, A] \cap [1, B] \in p$ and the tangent $b = [B, T]$ to the tangent $[V, P^\perp]$ at V to p which confirms that V is the vertex. ∎

If $P^\perp$ is replaced with another ideal point, then this construction gives the point of contact of the second tangent drawn from this point to p.

■ **Example 7.1.4** A proof of Seydewitz's theorem.

Theorem 7.1.3 *Let A, B, C be three mutually distinct points on a conic c. Any line l conjugate to one side of the triangle ABC intersects the other two lines in a pair (P, Q) of points conjugate w.r.t. c.*

Proof: It means no restriction to assume that l is conjugate to $[A, B]$ (see Figure 7.10). Therefore, l passes through the pole S of $[A, B]$. Without loss of generality, we let $c : x_0 x_2 - x_1^2 = 0$ and $A = (1 : 0 : 0)$, $B = (0 : 0 : 1)$, and $C = (1 : 1 : 1)$. Consequently, $S = (0 : 1 : 0)$ and the line l has an equation of the form $\mu x_0 - \lambda x_2 = 0$. The intersections of l with $[A, C]$ and $[B, C]$

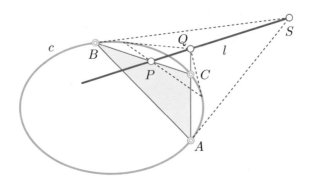

FIGURE 7.10. SEYDEWITZ's theorem.

are thus $P = (\lambda : \mu : \mu)$ and $Q = (\lambda : \lambda : \mu)$. Since the coeffcient matrix of c (which is also the coordinate matrix of c's polarity) equals

$$\mathbf{A} = \begin{pmatrix} 0 & 0 & 1 \\ 0 & -2 & 0 \\ 1 & 0 & 0 \end{pmatrix},$$

the points $P = \mathbf{p}\mathbb{F}$ and $Q = \mathbf{q}\mathbb{F}$ turn out to be conjugate w.r.t. c, since $\mathbf{p}^T\mathbf{A}\mathbf{q} = 0$.   ∎

● **Exercise 7.1.1** Conjugate pairs in a quadrilateral. **Prove the following result:**
If, at any quadrilateral, two pairs of opposite vertices are conjugate w.r.t. any polarity π, then the same holds for the third pair.

Hint: Let $q_1q_2q_3q_4$ be the quadrilateral and $P_ik = q_i \cap q_k$. Set $P_{12} = \mathbf{a}\mathbb{F}$, $P_{23} = \mathbf{b}\mathbb{F}$, $P_{34} = \mathbf{c}\mathbb{F}$, and $P_{41} = (\mathbf{a} + \mathbf{b} + \mathbf{c})\mathbb{F}$.

● **Exercise 7.1.2** Commuting involutions. **Prove the following statement:**
Two involutions on a conic c commute if, and only if, the centers of the two involutions are conjugate w.r.t. c.

Equation of the dual conic

We have seen that the set of points of a conic c can be described as the zero set of a quadratic form

$$\mathbf{x}^T\mathbf{A}\mathbf{x}$$

with a regular symmetric 3×3-matrix with entries from an arbitrary commutative field $\mathbb{F}$, char$\mathbb{F} \neq 2$. Since $\mathbf{l} = \mathbf{A}\mathbf{x}$ represents the polar line l of $X = \mathbf{x}\mathbb{F}$, we can give a condition on the coordinates $\mathbf{u} = (u_0, u_1, u_2)$ of c's tangents. To put it in another way: We can derive an equation of the *dual curve* $c^\star$.

First, we observe $\mathbf{x} = \mathbf{A}^{-1}\mathbf{u}$. Then, we write

$$\mathbf{x}^{\mathrm{T}}\mathbf{A}\mathbf{x} = (\mathbf{A}^{-1}\mathbf{u})^{\mathrm{T}}\mathbf{A}\mathbf{A}^{-1}\mathbf{u} = \mathbf{u}^{\mathrm{T}}\mathbf{A}^{-1}\mathbf{u},$$

where we used the fact that the inverse of a symmetric matrix is again symmetric. So, we can say:

Corollary 7.1.3 *Let* $c:$ $\mathbf{x}^{\mathrm{T}}\mathbf{A}\mathbf{x} = 0$ *with a regular symmetric* 3×3-*matrix with entries from an arbitrary commutative field* $\mathbb{F}$ *(*$\mathrm{char}\mathbb{F} \neq 2$*) be a conic in* $\mathbb{P}^2(\mathbb{F})$. *Then, the equation of the dual curve* $c^{\star}$ *equals*

$$c^{\star} : \ \mathbf{u}^{\mathrm{T}}\mathbf{A}^{-1}\mathbf{u} = 0. \tag{7.4}$$

■ **Example 7.1.5** Dual curve of a circle. In $\mathbb{P}^2(\mathbb{R})$, the equation of the conic $c: -r^2x_0^2 + x_1^2 + x_2^2 = \mathbf{x}^{\mathrm{T}}\mathbf{A}\mathbf{x} = 0$ with $r \in \mathbb{R}$ and $r > 0$ equals the equation of its dual curve if, and only if, $r = 1$ since then

$$\mathbf{A} = \begin{pmatrix} -r^2 & 0 & 0 \\ 0 & 1 & 0 \\ 0 & 0 & 1 \end{pmatrix} \quad \text{and} \quad \mathbf{A}^{-1} = \begin{pmatrix} -r^{-2} & 0 & 0 \\ 0 & 1 & 0 \\ 0 & 0 & 1 \end{pmatrix}$$

are proportional matrices.

Dual parabola. What does the dual of a parabola look like? Assume we are given the parabola $p: 2ax_0x_2 - x_1^2 = 0$ with positive a. Here, we have

$$\mathbf{A} = \begin{pmatrix} 0 & 0 & a \\ 0 & -1 & 0 \\ a & 0 & 0 \end{pmatrix} \quad \text{and} \quad \mathbf{A}^{-1} = \begin{pmatrix} 0 & 0 & a^{-1} \\ 0 & -1 & 0 \\ a^{-1} & 0 & 0 \end{pmatrix}.$$

Thus, the dual curve has the equation

$$p^{\star} : \ 2u_0u_2 - au_1^2 = 0.$$

It depends on the choice of the ideal line in the dual plane if the curve $p^{\star}$ is a parabola or not.

Focal points

In Section 2.1, we have defined conics in $\mathbb{E}^2$ as sets of points satisfying certain distance relations. There, the focal points appeared first. Now, we shall see that the focal points of a conic can be defined within the framework of Projective Geometry. Later, in Section 7.5, we shall treat focal points once again, but in a different way.

We define a focal point of a conic c in the real projective plane as follows:

Definition 7.1.4 *Let* c *be a conic in the real projective plane. A point* F *is called a* focus *or* focal point *if the involution of conjugate lines (w.r.t.* c*) in the pencil about* F *equals the involution of right angles.*

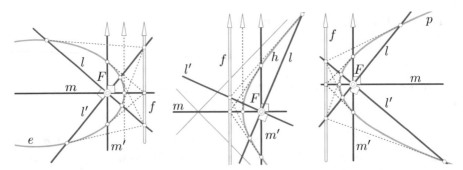

FIGURE 7.11. The involution of conjugate lines w.r.t. the conic equals the involution of right angles exactly in the pencils about the focal points. For the case of an ellipse e (left), a hyperbola h (center), and a parabola p (right), two pairs (l, l') and (m, m') of conjugate and orthogonal lines are displayed.

Figure 7.11 shows two pairs of corresponding orthogonal lines through a focal point of an ellipse, hyperbola, and through the focus of a parabola.

The involution of right angles is elliptic and its fixed lines are isotropic (cf. Example 6.4.7, page 253). Thus, a point F is a focal point of a conic, if the tangents from F to c are isotropic lines. These isotropic tangents a_1, a_2 connect F with one of the absolute points of Euclidean geometry, i.e., either with $I = (0 : 1 : i)$ or $\overline{I} = (0 : 1 : -i)$.

Assume we are given the ellipse

$$c: \frac{x^2}{a^2} + \frac{y^2}{b^2} - 1 = (1 \; x \; y) \, \mathbf{M} \begin{pmatrix} 1 \\ x \\ y \end{pmatrix} = 0.$$

with $\mathbf{M} = \mathrm{diag}\left(-1, a^{-2}, b^{-2}\right)$ where $a > b$ means no restriction. We want to find the focal points of c. Let $F = \mathbf{f}\mathbb{F}$ with $\mathbf{f} = (f_0, f_1, f_2)$ be a point in the complex extension $\mathbb{P}^2(\mathbb{C})$. The homogeneous coordinates of the isotropic lines $a_1 = [F, I]$ and $a_2 = [F, \overline{I}] = \overline{a_1}$ are

$$\mathbf{a}_1 = (if_1 - f_2 : -if_0 : f_0) \quad \text{and} \quad \mathbf{a}_2 = (-if_1 - f_2 : if_0 : f_0) = \overline{\mathbf{a}_1}.$$

The lines a_1 and a_2 are tangent to c if, and only if, they are self-conjugate w.r.t. c. In other words, a_1 and a_2 have to be points of the dual curve $c^\star$. Thus, the coordinate vectors of a_1 and a_2 fulfill

$$\mathbf{a}_1^{\mathrm{T}} \mathbf{M}^{-1} \mathbf{a}_1 = f_1^2 + 2if_1 f_2 - f_2^2 - f_0^2(a^2 - b^2) = 0,$$
$$\mathbf{a}_2^{\mathrm{T}} \mathbf{M}^{-1} \mathbf{a}_2 = f_1^2 - 2if_1 f_2 - f_2^2 - f_0^2(a^2 - b^2) = 0.$$

The difference of the latter equations yields $f_1 f_2 = 0$ which tells us that the focal points are located on an axis of c. The sum of the latter equations yields $f_1^2 - f_2^2 - f_0^2(a^2 - b^2) = 0$ which gives

$$f_2 = \pm i\sqrt{a^2 - b^2}\, f_0 \quad \text{if} \quad f_1 = 0,$$
$$f_1 = \pm \sqrt{a^2 - b^2}\, f_0 \quad \text{if} \quad f_2 = 0.$$

Without loss of generality, we may assume that $f_0 = 1$. Thus, we find the four focal points of the ellipse c, according to Definition 7.1.4, as

$$F_1 = (1 : \sqrt{a^2 - b^2} : 0), \quad F_2 = (1 : -\sqrt{a^2 - b^2} : 0),$$
$$F_3 = (1 : 0 : i\sqrt{a^2 - b^2}), \quad F_4 = (1 : 0 : -i\sqrt{a^2 - b^2}).$$

A simple construction of the focal points of an ellipse is shown in Figure 7.12. The points F_3 and F_4 are sometimes called the *anti-foci* of the ellipse.

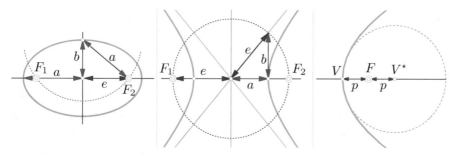

FIGURE 7.12. Construction of the focal points of the ellipse (left), the hyperbola (center), and the parabola (right).

● Exercise 7.1.3 Focal points of a hyperbola. Compute the focal points of the hyperbola

$$h: \quad \frac{x^2}{a^2} - \frac{y^2}{b^2} = 1$$

in the way shown above. Prove that the construction shown in Figure 7.12 is a proper way of finding the focal points of a hyperbola.

The parabola's only focal point. Show that the point $F = (1 : q : 0)$ is the only focus of the parabola $p: \ y^2 - 4qx = 0$. The point F is the midpoint of the segment between the vertex $V = (1 : 0 : 0)$ and its center $V^\star = (1 : q : 0)$ of curvature.

Conjugacy between focal points and directrices. Show that a focal point and the associated directrix of a conic c are polar w.r.t. c, *i.e.*, a pair consisting of a pole and a polar line.

7.2 Definition of a conic according to von Staudt

In Section 7.1 we have seen that any conic c defines its polarity π_c. It turned out that the self-conjugate points of π_c are exactly the points of c. Now, we can ask ourselves if any polarity (as described in Definition 7.1.2) defines a conic. In fact, this is not the case. We just have to recall that elliptic polarities do not have self-conjugate points (cf. Definition 7.1.3).

For historical reasons, we recall an alternative to STEINER's generation of a conic (cf. Definition 6.1.1). The following definition was given by the German mathematician KARL GEORG CHRISTIAN VON STAUDT (1798–1867).

Definition 7.2.1 VON STAUDT's **definition:**
The set of self-conjugate points of a hyperbolic polarity in $\mathbb{P}^2(\mathbb{F})$ (with commutative $\mathbb{F}$ and $\mathrm{char}\mathbb{F} \neq 2$) is called a conic.

Each polarity π in $\mathbb{P}^2(\mathbb{F})$ is described by a symmetric matrix $\mathbf{P} \in \mathbb{F}^{3 \times 3}$, i.e., $\mathbf{P} = \mathbf{P}^{\mathrm{T}}$. The polar line $l = \mathbf{l}\mathbb{F}$ of the point $X = \mathbf{x}\mathbb{F}$ is given by $\mathbf{l} = \mathbf{Px}$. Self-conjugate points of π are incident with their image lines. Hence, they satisfy

$$\langle \mathbf{l}, \mathbf{x} \rangle = \langle \mathbf{Px}, \mathbf{x} \rangle = (\mathbf{Px})^{\mathrm{T}}\mathbf{x} = \mathbf{x}^{\mathrm{T}}\mathbf{Px} = 0$$

which is the equation of a conic (cf. (6.11)) since it has non-zero solutions.

In projective planes over finite fields or even over the real number field we can find *empty conics*: They are given by a regular symmetric matrix from $\mathbb{F}^{3 \times 3}$ but there is no triplet from $\mathbb{F}^3$ that satisfies the corresponding quadratic equation. In the case of the real projective plane, the coefficient matrix of empty conics is definite.

Conics from a given polar triangle

According to Definition 6.1.1 (page 218), a conic is uniquely determined by five points where no three of them are collinear. An analogous definition uses only the terms pole, polar line, and polar triangle. We can say that a polarity is uniquely defined by prescribing a polar triangle $P_1P_2P_3$ (definition on page 266) and a pair of corresponding elements (Q, q) with Q being on neither side of the triangle $P_1P_2P_3$ and q not passing through any of its vertices.

An example is illustrated in Figure 7.13: The pole Q and its polar line q help to define involutions of conjugate points on the side lines of the polar

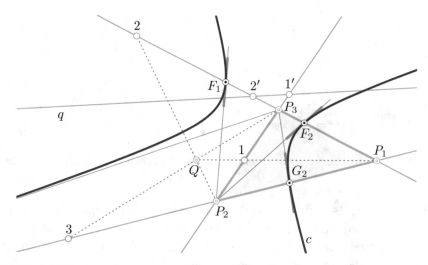

FIGURE 7.13. A conic constructed from the polar triangle $P_1P_2P_3$ and a pair (Q, q) consisting of a pole with its polar line.

triangle $P_1P_2P_3$. These involutions are the restriction of the polarity to the side lines of the polar triangle. The fixed points of these involutions are self-conjugate points of the polarity, and therefore, they are points of the underlying conic, provided that there exist fixed points.

The involution η_1 acting on the line $[P_2, P_3]$ maps P_2 to P_3, and *vice versa*; further $1 = [Q, P_1] \cap [P_2, P_3]$ is mapped to $1' = q \cap [P_2, P_3]$. In the version of Figure 7.13, the pair $(1, 1')$ is separated by the pair (P_2, P_3), and thus, η_1 is elliptic and there are no self-conjugate points on this line. On the line $[P_3, P_1]$ we find the corresponding pair of points $2 = [Q, P_2] \cap P_3, P_1$ and $2' = q \cap [P_3, P_1]$. The involution η_2 on $[P_3, P_1]$ is hyperbolic, and its fixed points F_1 and F_2 are points on the conic c. Note that the tangents to c at F_1 and F_2 pass through P_2 for it is the pole of $[P_3, P_1]$. Furthermore, η_3 acting on $[P_1, P_2]$ yields two more fixed points in a similar way. This results in two more line elements of the desired conic. So, the unique solution c is more than well-determined by four line elements. For the sake of completeness, it has to be checked that these four line elements fit to only one conic. This is easily done by studying the actions of the harmonic collineations with any vertex of the polar triangle as center and the opposite side line as axis.

Note that the polar triangle $P_1P_2P_3$ together with the pair (Q, q) (pole and polar line) may define an elliptic polarity. In such a case, the involu-

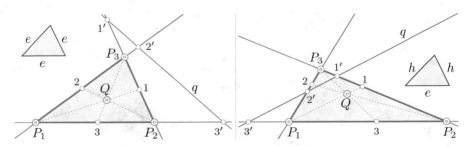

FIGURE 7.14. The sequence of elliptic (e) and hyperbolic (h) involutions on a polar triangle is either (e,h,h) (left) or (e,e,e) (right).

tions on all three side lines of the polar triangle are elliptic. Then, there are no self-conjugate points, and therefore, there is no conic. Figure 7.14 shows the two possible distributions of elliptic and hyperbolic involutions induced on the side lines of a polar triangle.

● **Exercise 7.2.1** A conic from a polar triangle and a line element. How does the construction of the above example change if the pair (Q, q) is replaced by an incident (self-conjugate) pair $Q \in q$, i.e., a point Q plus the tangent q at Q.

Common polar triangles

Assume $\mathbf{P}, \mathbf{Q} \in \mathbb{F}^{3 \times 3}$ are symmetric and not proportional matrices. They determine two different polarities π_P, π_Q in $\mathbb{P}^2(\mathbb{F})$. A triangle $A_1 A_2 A_3$ is a *common polar triangle* of π_P and π_Q if it is a polar triangle with respect to π_P and π_Q at the same time.

Let $A_i = \mathbf{a}_i \mathbb{F}$ and further define the lines $b_i = [A_j, A_k]$ with coordinates $\mathbf{b}_i \mathbb{F}$ where $i \in \{1, 2, 3\}$. Without loss of generality, we may assume that $\mathbf{a}_1 = (1, 0, 0)$, $\mathbf{a}_2 = (0, 1, 0)$, $\mathbf{a}_3 = (0, 0, 1)$. Thus, $\mathbf{b}_i = \mathbf{a}_i$ for $i \in \{1, 2, 3\}$.

Then, according to the definition of a polar triangle (cf. Example 7.1.1) and with proper scaling of $\mathbf{P}$'s and $\mathbf{Q}$'s rows, we have

$$\mathbf{P} \mathbf{a}_i = \mathbf{a}_i, \quad \mathbf{Q} \mathbf{a}_i = \mathbf{a}_i \quad \text{with } i \in \{1, 2, 3\}.$$

Therefore, $\mathbf{P}$ and $\mathbf{Q}$ have to be diagonal matrices. Thus, we can say:

Lemma 7.2.1 *The determination of common polar triangles of two polarities is equivalent to the simultaneous diagonalization of the corresponding symmetric matrices.*

From Linear Algebra we know that $\mathbf{P}$ and $\mathbf{Q}$ can be diagonalized simultaneously if, and only if, the corresponding endomorphisms commute.

The simultaneous diagonalization requires the computation of eigenvectors and eigenvalues of the (symmetric) matrix $\mathbf{R} = \lambda\mathbf{P} + \mu\mathbf{Q}$ ($\lambda : \mu \neq 0 : 0$) since $\mathbf{R}\mathbf{a}_i = \mathbf{a}_i$ tells us that $\mathbf{a}_i$ are eigenvectors of $\mathbf{R}$. On the other hand, the solutions $\lambda : \mu$ of $\det\mathbf{R} = 0$ yields the three (algebraically counting) singular polarities in the *pencil of polarities* spanned by π_P and π_Q. In Section 7.3, we shall consider π_P and π_Q as polarities of conics.

Polar transform of conics

It is worth to be noted that the image of a conic c under a polarity π is a conic $\pi(c)$, whether the polarity π is elliptic or not. We assume that π is desribed by the regular and symmetric coordinate matrix $\mathbf{P} \in \mathbb{F}^{3\times3}$ and the conic is given by $\mathbf{x}^T\mathbf{A}\mathbf{x} = 0$ with $\mathbf{A} \in \mathbb{F}^{3\times3}$ and $\mathbf{A} = \mathbf{A}^T$. We compute the π-transform of c by first mapping points of c to the tangents of the dual conic $\pi(c)^\star$. Since points $X = \mathbf{x}\mathbb{F}$ are mapped to lines $l = \mathbf{l}\mathbb{F}$ with $\mathbf{l} = \mathbf{P}\mathbf{x}$, we arrive at the equation of the dual of $\pi(c)^\star$ which reads

$$\pi(c)^\star : \ \mathbf{l}^T\mathbf{P}^{-1}\mathbf{A}\mathbf{P}^{-1}\mathbf{l} = 0.$$

Note that $(\mathbf{P}^{-1})^T = \mathbf{P}^{-1}$ if $\mathbf{P}^T = \mathbf{P}$. In other words: The inverse of a polarity is also a polarity. Now, we recall (7.4) and with $\mathbf{B} := \mathbf{P}^{-1}\mathbf{A}\mathbf{P}^{-1}$ we find the equation of $\pi(c)$ as

$$\pi(c) : \ \mathbf{x}^T\mathbf{B}^{-1}\mathbf{x} = \mathbf{x}^T\mathbf{P}\mathbf{A}^{-1}\mathbf{P}\mathbf{x} = 0. \tag{7.5}$$

■ **Example 7.2.1** The polar transform of a circle.

Let us start in $\mathbb{P}^2(\mathbb{R})$ with the polarity w.r.t. the circle $p: \ x^2 + y^2 = r^2$ centered at $(0,0)$. The circle to be transformed shall be given by $c: \ (x-m)^2 + (y-n)^2 = R^2$, i.e., its center equals (m,n) and the radius equals R. In terms of homogeneous coordinates, the polarities w.r.t. both circles can be described by linear mappings $\mathbb{R}^3 \to \mathbb{R}^{3\star}$ with the coordinate matrices

$$\mathbf{P} = \begin{pmatrix} -r^2 & 0 & 0 \\ 0 & 1 & 0 \\ 0 & 0 & 1 \end{pmatrix} \quad \text{and} \quad \mathbf{A} = \begin{pmatrix} m^2 + n^2 - R^2 & -m & -n \\ -m & 1 & 0 \\ -n & 0 & 1 \end{pmatrix}.$$

Inserting $\mathbf{P}$ and $\mathbf{A}$ into (7.5), we arrive at the coordinate matrix of the conic d that is polar to c w.r.t. p:

$$\mathbf{P}\mathbf{A}^{-1}\mathbf{P} = \begin{pmatrix} -r^4 & r^2 m & r^2 n \\ r^2 m & R^2 - m^2 & -mn \\ r^2 n & -mn & R^2 - n^2 \end{pmatrix}.$$

Figure 7.15 shows the images d of three different circles c under the polarity π_p. The common points of d and p transform to the common tangents of p and d, while the common tangents of c and p map to the common points of c and d.

Now, it is easy to show that the conic d has the center

$$\frac{r^2}{m^2 + n^2 - R^2}(m,n)$$

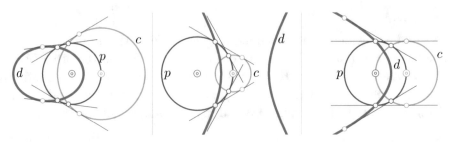

FIGURE 7.15. The polar transform of a circle c w.r.t. a circle p yields a conic d: Common points of c and p are mapped to common tangents of p and d; common tangents of c and q are mapped to common points of p and d.

if $m^2 + n^2 \neq R^2$ and its principal axis encloses the angle

$$\varphi = \frac{1}{2} \arctan \frac{2mn}{m^2 - n^2} \quad \text{or} \quad \varphi = \arccos \frac{m}{\sqrt{m^2 + n^2}}$$

with the x-axis. The polar image $d = \pi_P(c)$ is an ellipse if $R^2 > m^2 + n^2$, i.e., there are no real tangents from P's center to c. The conic d is a hyperbola if $R^2 < m^2 + n^2$, or equivalently, there are two real tangents from P's center to c.

If $R^2 = m^2 + n^2$, then c contains the center of p and d is a parabola. The angle φ enclosed by the axis of the parabola and the x-axis of the coordinate system is still given by $\tan 2\varphi = \frac{2mn}{\sqrt{m^2 + n^2}}$.

This classification of the polar image of a conic remains valid if the hyperbolic polarity π_P from Example 7.2.1 is replaced by an elliptic polarity since an elliptic polarity also has a center. The center is just the pole of the line at infinity.

The results of this section can also can be used in order to solve some conic problems by means of a graphical construction.

■ **Example 7.2.2** Find conics on points or lines with a given focus.

The generic case. Assume that we are given three non-collinear points P_1, P_2, P_3 and a further point F. We are looking for all conics passing through P_1, P_2, P_3 and having F for a focus.

Applying a polarity to a conic interchanges points and lines. A focus F, for example, is the meet of two isotropic tangents. When polarized w.r.t. the circle f with the center F, they map to the points of contact, i.e., the absolute points of Euclidean geometry. The given points map to three lines: $\pi(P_i) = p_i$, $i \in \{1, 2, 3\}$. By assumption, the given points form a triangle, and therefore, their polars p_i are the side lines of a triangle (see the light grey triangle in Figure 7.16). The conics we are looking for, are now transformed to conics e_i (with $i \in \{1, 2, 3, 4\}$) that touch the side lines p_i of the triangle and pass through the absolute points of Euclidean geometry.

Thus, the conics e_i are the incircle together with the excircles of the triangle. Applying the polarity once again yields the four solutions l_i.

In general, there are four conics on three points and a given focus. These four conics are either four hyperbolas or three hyperbolas plus an ellipse or parabola depending on whether the

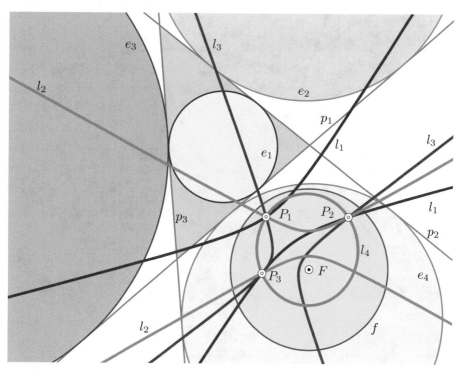

FIGURE 7.16. The construction of conics on three points P_1, P_2, P_3 with a given focus F is equivalent to the construction of all circles tangent to three lines. The four solutions $l_1, \ldots, l_4$ are the polar images of the four tri-tangent circles $e_1, \ldots, e_4$ of the triangle all of whose sides lines p_1, p_2, p_3 are the polars of P_1, P_2, P_3 w.r.t. any circle f centered at F.

given focus F is an interior point of one of the tritangent circles of the lines p_i or lies on one of them. F cannot be the interior point of more than one circle e_i because the tritangent circles e_i of the lines p_i are the incircle together with the excircles of the triangle built by the lines p_1, p_2, p_3. These circles are known to be disjoint.

Homofocal parabolas. We try a variation of the previous example. We are looking for all parabolas on two points P_1, P_2 and a given focus F (Figure 7.17).

Again, we apply the polarity w.r.t. to a circle f centered at F: The given points P_1, P_2 are mapped to lines p_1, p_2 and possible solutions (which should pass through P_1 and P_2) are, thus, transformed into conics tangent to p_1 and p_2. The focus considered as the intersection of a pair of isotropic tangents to the solutions maps to the absolute points of Euclidean geometry, and therefore, the polar transforms of the desired conics are circles tangent to p_1, p_2. Further, the ideal line is mapped to F, because the center of f and the ideal line correspond to each other in the polarity.

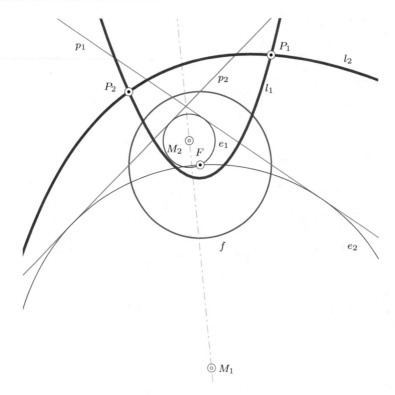

FIGURE 7.17. The two parabolas l_1 and l_2 on two points with a common focus are found in the following way: Find the circles e_1 and e_2 which are tangent to the polar lines p_1, p_2 of P_1, P_2 w.r.t. f. Then, apply the polarity w.r.t. f to e_1, e_2.

Hence, we are looking for the circles tangent to p_1, p_2 passing through F. In the generic case, there are two solutions e_1 and e_2, as shown in Figure 7.17. Their polar transforms are the parabolas l_1 and l_2.

7.3 Pencils of conics

Pencils of conics are special one-parameter families of conics. The projective classification of the pencils of conics leads to five different types. All these types can be classified by means of the set of singular conics included in the pencil. The computation as well as the construction of common points of two conics out of this pencil can be reduced to the case where one of these curves of degree two is reducible. This is interesting at least from the theoretical point of view, since in general it is not possible to find the common points of two conics by ruler and compass.

In this section, the term 'conic' means a curve of degree two, no matter if it is irreducible (regular) or reducible (singular).

An introductory example

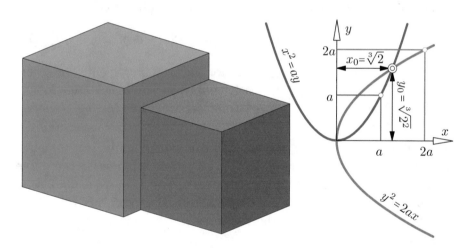

FIGURE 7.18. The Delian cube duplication problem: The volume of the two cubes on the left have ratio $2 : 1$. How can one construct the scaling factor of the edge length?

Let us look at the following classical problem. A cube of side length $a = 1$ is to be duplicated, *i.e.*, the volume shall be doubled. The question is: What is the side length of the new cube with twice the volume?

This problem is usually called the *cube duplication problem*. The name *Delian problem* was also given to this question: It is reported that the citizens of the Greek island Delos once asked the oracle how to defeat the plague. The oracle wanted the Delians to build a new altar of twice the volume of the old altar. Unfortunately, the altar was a regular cube, and so, the Delians had to duplicate a cube. The ancient Egyptians and Indians also knew this type of mathematical problem.

In principle, one has to find or construct $\sqrt[3]{2}$ in order to solve the Delian problem. Sad to say, but ruler and compass are not sufficient to construct the cube root of two, as shown in 1837 by the French mathematician PIERRE LAURENT WANTZEL (1814–1848).

However, the Greek geometers discovered a way to construct $\sqrt[3]{2}$ using two parabolas, as shown in Figure 7.18. Assume that

$$p_1 : y^2 = 2ax \quad \text{and} \quad p_2 : ay = x^2 \tag{7.6}$$

are the two parabolas. Obviously, (7.6) is a system of two quadratic equations in the unknowns x and y. The solutions of (7.6) are

$$(x,y) = (0,0), \quad (x,y) = (\sqrt[3]{2}a, \sqrt[3]{4}a). \tag{7.7}$$

The first solution is of no importance for the Delian problem. The second one with $a = 1$ gives $x = \sqrt[3]{2}$, precisely that what the Delians were looking for. If we allow a tool that draws parabolas, then the problem is solvable in a constructive way.

We have solved a system of two quadratic equations in two variables. An equivalent problem is to search for all points common to two curves of degree two (cf. Section 6.3). In this particular case, we could extract the solutions by mere elementary operations. However, this will not be sufficient when it comes to the more general case with two quadratic equations.

On the other hand, how do the configurations of two conics look like? It can easily be imagined that there may occur more than two common points, see Figure 7.19.

FIGURE 7.19. Some configurations of pairs of conics with different numbers of common points.

Pencils of conics in the projective plane

In the following we assume that the field $\mathbb{F}$ is algebraically closed, *i.e.*, any non-constant polynomial with coefficients in $\mathbb{F}$ has at least one root in $\mathbb{F}$. This allows us to generalize and extend the case of a pair of circles which is somehow special: Any two circles share the circle points / absolute points (but only after the projective closure and complex extension of the Euclidean plane). Their radical axis l joins the two common proper points S_1 and S_2 whether they are real or not.

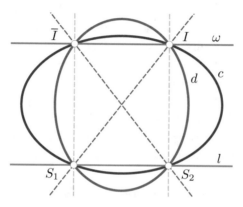

FIGURE 7.20. A complex and projective view on two circles: There are four common points spanning three pairs of lines.

The four common points of the two "circles" c and d, as illustrated in Figure 7.20, can be thread up by three pairs of lines, one real pair (magenta), two of them consisting of a pair of complex lines (dashed in cyan and violet): The real pair is $(l\,m)$ with $l = [I,\overline{I}]$ and $m = [S_1,S_2]$. The two complex pairs are $([I,S_1],[\overline{I},S_2])$ and $([I,S_2],[\overline{I},S_1])$.

These pairs of lines can be viewed as *singular conics* each sharing the same four points with c and d.

Consequently, it is natural to treat pairs of conics in the projective plane. Whenever we deal with equations and coordinates, we prefer the homogeneous representation and extend the underlying field $\mathbb{F}$ properly. As we shall see, a purely synthetic approach to pencils of conics is also possible.

Definition of a pencil of conics

From Section 5.1, we already know that in $\mathbb{P}^2(\mathbb{F})$ a conic is uniquely defined by prescribing five points such that no three of them are collinear.

On the other hand, five lines considered as the tangents of a conic also define a unique conic if no three of these lines are concurrent.

What happens if we remove one point or line? In the following, we use

Definition 7.3.1 *The family of all regular conics through the vertices of a quadrangle $B_1B_2B_3B_4$ is called a pencil (of conics) of the first kind. The points B_i with $i \in \{1, 2, 3, 4\}$ are called base points of the pencil.*

Some conics in a pencil of the first kind are displayed in Figure 7.21.

How many conics are in this pencil? If we choose an arbitrary point P which is not collinear with any two of the given base points, then we know that there is a unique conic through $B_1, \ldots, B_4$, and P. So, it seems that there are as many conics in the pencil as there are points in the plane (besides the one which are collinear with the base points). However, there is only one degree of freedom in a pencil of conics, and the pencil is a fibration of the projective plane. This can easily be seen with (6.13) by writing down a conic's equation in terms of homogeneous coordinates

$$a_{00}x_0^2 + 2a_{01}x_0x_1 + \ldots + a_{22}x_2^2 = 0.$$

Inserting the homogeneous coordinates of the four given base points, we obtain a system of four linear equations in the six unknown coefficients a_{ij} with $i, j \in \{0, 1, 2\}$. The solution of this system is a two-dimensional subspace of $\mathbb{F}^6$ provided that the rank of the coefficient matrix is 4. Hence, there are as many conics in a pencil as there are points on a projective line.

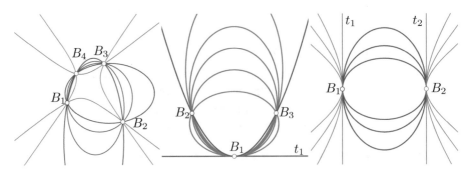

FIGURE 7.21. Definition of pencils of conics: from left to right; the first three kinds.

There are some other types of pencils of conics which differ from the viewpoint of synthetic geometry but not from the algebraic point of view:

Definition 7.3.2 *Assume that B_1, B_2, and B_3 are the vertices of a triangle and t_1 is a line through B_1 that does neither contain B_2 nor B_3. Then, the family of (regular) conics through B_1, B_2, B_3 that touch t_1 at B_1 is called a pencil of the second kind.*

The pencil of the second kind can be seen as a pencil of the first kind: One base point of a pencil of the first kind, say B_4, has moved infinitely close to B_1. In the limit, the two points become one point plus the tangent there. An example of a pencil of the second kind can be seen in Figure 7.21 (middle). Naturally, the second type of pencil contains a one-parameter family of conics as is the case for the pencil of the first kind.

The transition from the pencil of the first kind to that of the second kind gives a rough idea of how to proceed in order to find the remaining kinds of pencils of conics. Assume that we have two such limiting procedures: Let B_4 move towards B_1, and B_3 move towards B_2. Then, we obtain:

Definition 7.3.3 *Assume B_1 and B_2 are two (different) points, and t_1 and t_2 are two straight lines such that $B_1 \in t_1$, $B_2 \in t_2$, and either t_i is different from the line $[B_1, B_2]$. The family of regular conics through B_1 and B_2 that touch t_1 at B_1 and t_2 at B_2 is called a pencil of the third kind.*

The pencil of the third kind consists of *doubly touching conics* and is a one-parameter family of conics. Some of the conics in a pencil of the third kind can be seen in Figure 7.21 (right).

There are two further types of pencils. We can modify a pencil of the second kind as described in Definition 7.3.1 and assume that the third base point B_3 is moving towards the point B_1. Algebraically speaking, we have three points infinitely close together, and therefore, any two conics in this pencil intersect at B_3 with multiplicity three.

However, there is a more geometric way to define new pencils of conics. For that purpose, we use Definition 6.4.1 and recall the perspective collineation α between any two conics in the pencil.

From the definition of osculation and hyperosculation, it is clear that the family of conics which are osculating a given conic k at B_1 and share a further point B_2 with k is one-parametric: We can vary the center C of α on t_1. It is also obvious that the family of conics hyperosculating a given conic k at a point B_1 is one-parametric: On a line $[B_1, K]$ with $K \in k$ through the center B_1 of $\alpha : k \mapsto l$, we have one degree of freedom for the choice of the α-image L of K. Note that K is not allowed to lie on t_1 (Figure 6.22). So we have the natural

Definition 7.3.4 *The family of regular conics osculating a given conic k at a point B_1 and passing through a further point $B_2 \in k$ is called pencil of the fourth kind.*

The family of conics hyperosculating a given conic k at a point B_1 is called a pencil of the fifth kind.

Figure 7.22 shows some conics in a pencil of the fourth and fifth kind. The base points and some special conics in the pencil are highlighted.

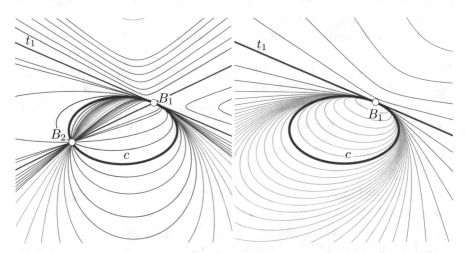

FIGURE 7.22. Pencils of conics in $\mathbb{P}^2(\mathbb{R})$: Left: The pencil of the fourth kind with base points B_1, B_2, and the base tangent t_1 consists of all conics that osculate a conic c at B_1. Right: The pencil of the fifth kind consists of all conics that hyperosculate c at B_1. In both cases, c is also an element of the pencil.

Analytic representation

The analytic representation of pencils of conics in $\mathbb{P}^2(\mathbb{F})$ enables us to extend the previously defined pencils to singular conics. The following result is necessary for the characterization of singular conics by means of the rank of the coefficient matrix.

Lemma 7.3.1 *In $\mathbb{P}^2(\mathbb{F})$ with char$\mathbb{F} \neq 2$, the curves of degree two with the equation $\mathbf{x}^T\mathbf{K}\mathbf{x} = 0$ where $\mathbf{K} \in \mathbb{F}^{3\times3}$ with $\mathbf{K}^T = \mathbf{K}$ is*
1. *a conic or the empty set if, and only if, rk$\mathbf{K} = 3$,*
2. *a repeated line if, and only if, rk$\mathbf{K} = 1$,*
3. *a pair of lines or a single point if, and only if, rk$\mathbf{K} = 2$.*

Proof: We have already learned that a regular symmetric matrix defines a polarity where the set of self-conjugate points is either empty or a conic. In the case of det $\mathbf{K} = 0$ the singular points $S = \mathbf{s}\mathbb{F}$ are solving the system of homogeneous linear equations $\mathbf{K}\mathbf{s} = \mathbf{0}$. Singular points have the property that with any other point $P = \mathbf{p}\mathbb{F}$ of the curve, i.e., with $\mathbf{p}^T\mathbf{K}\mathbf{p} = 0$, all points of the line $[S, P] = (\lambda\mathbf{s} + \mu\mathbf{p})\mathbb{F}$ with $(\lambda, \mu) \in \mathbb{F}^2 \setminus \{(0,0)\}$, satisfy the quadratic equation. In the case of rk$\mathbf{K} = 1$ there is a line of singular points. If there exists another point $\mathbf{p}$ satisfying the quadratic equation, all points in the plane satisfy this equation and $\mathbf{K} = \mathbf{0}$. ∎

We assume that two conics in $\mathbb{P}^2(\mathbb{F})$ have the following equations

$$k: \mathbf{x}^T\mathbf{K}\mathbf{x} = 0 \quad \text{and} \quad l: \mathbf{x}^T\mathbf{L}\mathbf{x} = 0$$

with symmetric matrices $\mathbf{K}, \mathbf{L} \in \mathbb{F}^{3\times3}$. Then, we can state:

Theorem 7.3.1 *If the pencil of conics of any kind is spanned by the conics $k: \mathbf{x}^T\mathbf{K}\mathbf{x} = 0$ and $l: \mathbf{x}^T\mathbf{L}\mathbf{x} = 0$, then all conics of the pencil are included in the family of curves satisfying any non-trivial linear combination*

$$\kappa\mathbf{x}^T\mathbf{K}\mathbf{x} + \lambda\mathbf{x}^T\mathbf{K}\mathbf{x} = \mathbf{x}^T(\kappa\mathbf{K} + \lambda\mathbf{L})\mathbf{x} = 0, \quad (\kappa, \lambda) \in \mathbb{F}^2 \setminus \{(0,0)\}. \quad (7.8)$$

Proof: The coordinates of each base point B_i annihilate the equations of k and l, and therefore, also each linear combination. Thus, all conics defined by (7.8) pass through the base points.

If any two points P, Q are conjugate w.r.t. k and l, then they are conjugate w.r.t. each curve given in (7.8), since $\mathbf{p}^T\mathbf{K}\mathbf{q} = \mathbf{p}^T\mathbf{L}\mathbf{q} = 0$ implies $\mathbf{p}^T(\kappa\mathbf{K} + \lambda\mathbf{L})\mathbf{q} = 0$. If, therefore, k and l share the line element (B_i, t_i), then each $P \in t_1$ is conjugate to B_i w.r.t. k and l. Hence, all other conics in (7.8) contain this line element.

If k and l are hyperosculating at B_1, then each $P \in t_1$ has the same polar line w.r.t. k and l. Consequently, all other conics in (7.8) hyperosculate k and l at B_1.

If finally k and l span a pencil of the fourth kind, then l share (B_1, t_1) and another point $B_2 \notin t_1$ with k such that no other point or the tangent at B_2 is common to k and l. This condition holds for all other conics in the pencil as well as for each conic in (7.8).

Hence, for all pencils of any kind, each included conics satisfies also (7.8). Conversely, if any regular conic satisfying (7.8) is given, choose any point Q different from the base points. The pencil as well as the linear family (7.8) send just one conic through this point Q, and the two conics must coincide. Thus, the set of regular conics in (7.8) is a pencil.   ∎

Note that either conic in the pencil can be seen as a point in a five-dimensional space. Then, a pencil of conics is a line in this space.[1]

The result of Theorem 7.3.1 is a good reason to extend the Definitions 7.3.1, 7.3.2, 7.3.3, and 7.3.4 as given below.

Definition 7.3.5 *The set of regular or singular conics which satisfy any linear combination of the equations of two different conics k and l, is called a pencil of conics. In this sense, we extend all kinds of pencils of conics by the respective singular conics.*

■ **Example 7.3.1** Conics of a pencil through certain points.

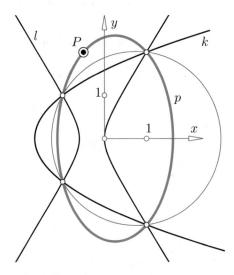

FIGURE 7.23. Conics of a pencil and the conic p (blue) through P.

In $\mathbb{P}^2(\mathbb{R})$, we are looking for the conic p through $P = (-\frac{1}{2}, 2)$ in the pencil of conics spanned by $k: \ 2y^2 - 3x - 5 = 0$ and $l: \ 5x^2 - 2y^2 + 3x = 0$. The equations of all conics in the pencil are given by

$$\kappa(2y^2 - 3x - 5) + \lambda(5x^2 - 2y^2 + 3x) = 0 \quad \text{with} \ \ \kappa : \lambda \neq 0 : 0.$$

If a certain curve from that family runs through P, then the ratio $\kappa : \lambda \neq 0 : 0$ is to be determined such that the latter equation is fulfilled. Therefore, we substitute the first and

---

[1]The idea of treating conics as points in five-dimensional projective space is due to the Italian mathematician Giuseppe Veronese, 1854–1917.

second coordinates of P for x and y and find

$$3\kappa - 11\lambda = 0 \quad \Longleftrightarrow \quad \kappa : \lambda = 11 : 6.$$

The conic p through P (as displayed in Figure 7.23) has the equation

$$p: \ 6x^2 + 2y^2 - 3x - 11 = 0.$$

We can look for special types of conics in a given pencil. We show this by means of examples:

■ **Example 7.3.2** Parabolas in a pencil of conics.

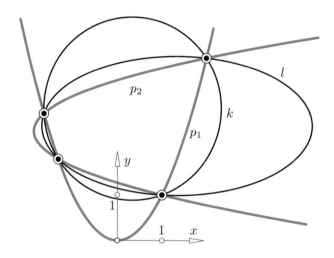

FIGURE 7.24. The parabolas p_1 and p_2 in the pencil of conics spanned by k and l, cf. Example 7.3.2.

In the projectively extended Euclidean plane let two conics k and l be given by their equations

$$k: \ 9x^2 + 9y^2 - 6x - 52y - 40 = 0, \ l: \ 83x^2 + 333y^2 - 222x - 1674y + 1480 = 0.$$

We homogenize the equations and write the pencil in the form given in (7.8). The intersection with the ideal line u is given by $x_0 = 0$. In the matrix representation of (7.8), we have to remove the first column and the first row, and we arrive at the quadratic form

$$(9\kappa + 83\lambda)x^2 + (9\kappa + 333\lambda)y^2 = 0.$$

The zeros of this quadratic form correspond to the ideal points of the conics in the pencil. In order to find parabolas, we have to determine $\kappa : \lambda$ such that the conics touch the ideal line. In other words: We have to look for double solutions. The existence of a double solution of a quadratic form is equivalent to the vanishing of the determinant of the coefficient matrix. In our example, we find

$$\kappa : \lambda = -83 : 9 \quad \text{and} \quad \kappa : \lambda = -37 : 1.$$

Now, the equations $\kappa \cdot k + \lambda \cdot l = 0$ are the equations of the two parabolas in the pencil and read

$$p_1: \ 9y^2 - 43y - 6x + 40 = 0 \quad \text{and} \quad p_2: \ x^2 - y = 0.$$

Figure 7.24 shows four conics in the pencil: k, l and the two parabolas p_1 and p_2.

From this example, we learn that *there are at most two parabolas in a generic pencil of conics* in $\mathbb{P}^2(\mathbb{R})$. This fits to Theorem 7.4.1, as we shall see later. However, there exist pencils of conics which contain only parabolas.

■ **Example 7.3.3** Circles in a pencil of conics.

We would like to find the circles in a pencil of conics if there are any. The pencil in question shall be spanned by two circles k and l with their Cartesian equations

$$k: \ 3x^2 + yx + 6y^2 - 14y - 12 = 0 \quad \text{and} \quad l: \ 3x^2 + 4yx + 15y^2 - 38y - 12 = 0.$$

Now, we have to find $\kappa : \lambda$ in (7.8) such that the coefficient matrix of the quadratic form in $x : y$ is a scalar multiple of the 2×2 unit matrix I_2. Therefore, we extract the matrices of the quadratic forms by removing the first rows and columns of the coefficient matrices of the homogeneous equations and arrive at

$$\overline{\mathbf{K}} = \begin{pmatrix} 6 & 1 \\ 1 & 12 \end{pmatrix} \quad \text{and} \quad \overline{\mathbf{L}} = \begin{pmatrix} 3 & 2 \\ 2 & 15 \end{pmatrix}$$

where $\overline{\mathbf{K}}$ is multiplied by 2. From $\kappa\overline{\mathbf{K}} + \lambda\overline{\mathbf{L}} = \alpha I_2$ with $\alpha \in \mathbb{R}$ we find

$$6\kappa + 3\lambda = 12\kappa + 15\lambda, \quad \kappa + 2\lambda = 0, \quad \kappa + 2\lambda = 0$$

which is a system of three homogeneous linear equations in two unknowns, namely κ and λ. Normally, we cannot expect a non-trivial solution. Since $\overline{\mathbf{K}}$ and $\overline{\mathbf{L}}$ are symmetric, the third equation equals the second one and can, thus, be canceled. Since the first equation simplifies to $\kappa + 2\lambda = 0$, it is equivalent to the second, and we end with one homogeneous equation with solutions $\kappa : \lambda = -2 : 1$.

Consequently, there is one circle c in the pencil. It has the equation

$$c: \ -2 \cdot (6x^2 + 2yx + 12y^2 - 28y - 24) + 1 \cdot (3x^2 + 4yx + 15y^2 - 38y - 12) =$$
$$= -9x^2 - 9y^2 + 18y + 36 = -9(x^2 + y^2 - 2y - 4) = 0.$$

Figure 7.25 shows the conics k and l as well as the circle c. The circle c is centered at $(0, 1)$, and the radius equals $\sqrt{5}$.

Singular conics in a pencil

In the pencil of conics given in (7.8), we always find *singular conics, i.e.*, degenerate conics. In the sequel we discuss the pencils of all five kinds.

Following Lemma 7.3.1, a necessary and sufficient condition for a conic to be singular is the singularity of the coefficient matrix. The singularity of the 3×3-matrix is equivalent to

$$\det(\kappa\mathbf{K} + \lambda\mathbf{L}) = 0. \tag{7.9}$$

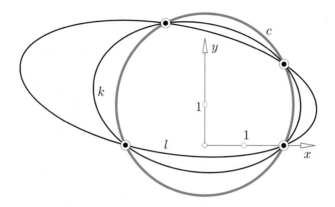

FIGURE 7.25. There is only one circle in the pencil of conics spanned by k and l (cf. Example 7.3.3).

From that we infer that there are at most three singular conics in the pencil since this determinant is a cubic form in $\kappa : \lambda$, *i.e.*, a homogeneous cubic polynomial. We shall emphasize that we are not interested in the trivial solution $\kappa : \lambda = 0 : 0$ for it corresponds to the zero matrix which is not the coefficient matrix of a conic.

It turns out that the singular conics can serve as a basis of a pencil in three types. As we shall see, there are two kinds of pencils of conics that cannot be spanned by taking only singular conics.

In the following, we assume that $\mathbf{K}$ and $\mathbf{L}$ are regular. Later we shall see what happens if we drop this assumption.

Remark 7.3.1 For any root $\kappa : \lambda$ of (7.8) the quotient κ/λ can be seen as generalized eigenvalue of $\mathbf{K}$ w.r.t. $\mathbf{L}$. We will meet these eigenvalues again in (9.24) in Section 9.5.

Case 1 - conics through the vertices of a quadrangle

First, we are dealing with the case of three different roots of (7.9). Any of these roots has multiplicity one, and thus, we have three different singular conics s_1, s_2, s_3 in the pencil. The singular conics s_1, s_2, and s_3 are pairs of lines through four common points B_1, B_2, B_3, and B_4. This is a pencil of conics of the first kind. All conics in the pencil, including the singular ones, share the four base points B_1, ..., B_4. Figure 7.26 shows some conics from a pencil of the first kind. The three singular conics are pairs of lines displayed in pink.

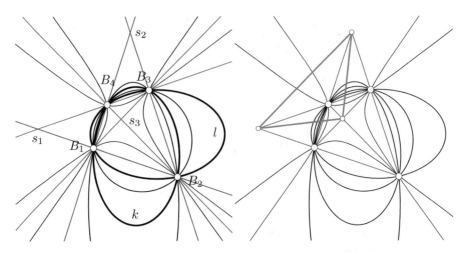

FIGURE 7.26. The first kind of a pencil of conics: Left: some conics, the four real base points B_1, ..., B_4, and the singular conics s_1, s_2, and s_3. Right: the common polar triangle (blue).

The diagonal triangle of the quadrilateral of base points B_1, ..., B_4 is the common polar triangle of all regular conics in the pencil. Figure 7.26 (right) also shows this polar triangle.

More than one pencil of the first kind is involved in the following result, called the *three-conics-theorem*:

Theorem 7.3.2 *Assume that three conics c_1, c_2, c_3 share two points S_1 and S_2. Then, the lines connecting the remaining two intersections of either pair of conics are concurrent.*

Figure 7.27 shows the three conics mentioned in Theorem 7.3.2 as well as its dual counterpart.

● **Exercise 7.3.1** The three-conics-theorem. Give a proof of Theorem 7.3.2. Why don't we have to give a proof for the dual statement? What happens if S_1 and S_2 are the absolute points of Euclidean geometry?

Show that Theorem 7.3.2 (together with its dual version) is still valid even if some conics degenerate into pairs of lines.

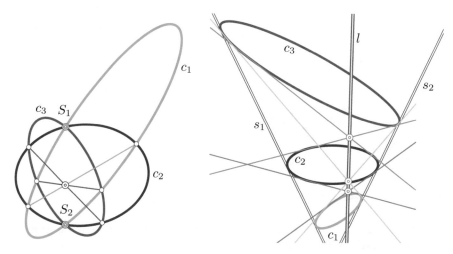

FIGURE 7.27. Left: Three conics with two common points S_1, S_2 and the three concurrent chords. Right: In the dual version, the intersections of the pairs of remaining common tangents are collinear.

Closely related to Theorem 7.3.2 is the following strange result, the *four-conics-theorem*[2]

Theorem 7.3.3 *If two points of intersection of each pair of three conics c_1, c_2, and c_3 lie on a conic c, then the lines joining the remaining two intersections of each pair are concurrent.*

Figure 7.28 illustrates Theorem 7.3.3 and its dual counterpart.

• Exercise 7.3.2 Give a precise formulation of the dual version of Theorem 7.3.3. It is illustrated in Figure 7.28 (right).

Cases 2 and 3 - conics in contact

If the cubic form given in (7.9) has two different zeros with respective multiplicities one and two, we find precisely two singular conics s_1 and s_2 in the pencil.

The singular conic s_1 that corresponds to the double root may be a pair of lines or a repeated whereas the singular conic s_2 corresponding to the simple root is always a pair of distinct lines. In the first case, we have

---

[2]Theorem 7.3.2 and Theorem 7.3.3 can be found among many other results in C.J.A. EVELYN, G.B. MONEY-COUTTS, AND J.A. TYRRELL: The Four-Conics Theorem. Stacey International, London, 1974.

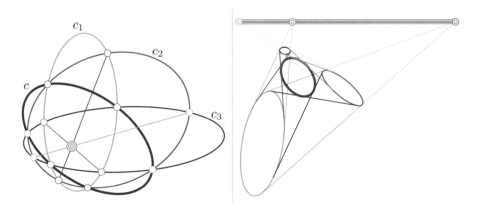

FIGURE 7.28. Left: the four conics theorem in its original version. Right: the dual version of the four conics theorem.

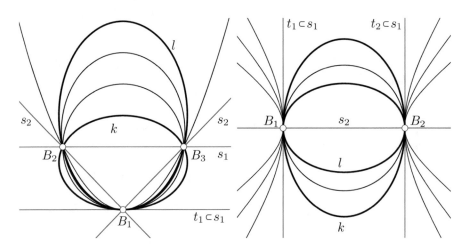

FIGURE 7.29. Left: pencil of the second kind with base points B_1, B_2, B_3 and singular conics s_1 and s_2. Right: pencil of the third kind with base points B_1, B_2 and singular conics s_1 and s_2.

a pencil of the second kind. All conics in the pencil pass through three different points one of which has multiplicity two. In other words: The conics of the pencil of the second kind share two different points B_2, B_3, and a line element (B_1, t_1). Figure 7.29 shows some of the conics in a pencil of the second kind besides the singular conics.

Now, there is no common polar triangle to all the conics in the pencil. Only B_1 and the point $[B_2, B_3] \cap t_1$ have equal polars w.r.t. all conics in the pencil.

If the singular conic s_1 corresponding to the double root of (7.9) turns out to be a *repeated line* (sometimes called a *double line*), we have a pencil of the third kind. Now, there are only two different base points, say B_1 and B_2, each of multiplicity two. Thus, the conics in the pencil of the third kind touch each other at B_1 and B_2. The line $s_1 = [B_1, B_2]$ joins these contact points. The singular conic s_2 is the union of the tangents t_1 at B_1 and t_2 at B_2. The right-hand side of Figure 7.29 shows some conics in a pencil of doubly touching conics as well as the respective singular conics.

Case 4 and 5 - osculating and hyperosculating conics

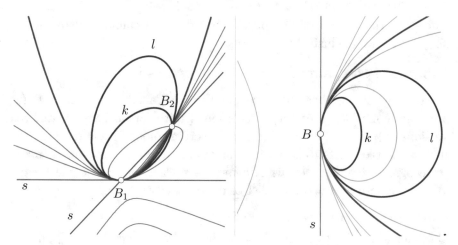

FIGURE 7.30. Pencils of conics: Left: pencil of the fourth kind with base points B_1, B_2 and singular conic s. Right: pencil of the fifth kind with base point B and singular conic s.

Finally the cubic form given in (7.9) may have one root with multiplicity three. In this case, there is only one singular conic s in the pencil. Depending on whether s is a pair of lines $(\text{rk}(\kappa \mathbf{K} + \lambda \mathbf{L}) = 2)$ or a repeated line $(\text{rk}(\kappa \mathbf{K} + \lambda \mathbf{L}) = 1)$, we have a pencil of the fourth or fifth kind.

In the case of a pencil of the fourth kind, all conics of the pencil share two points B_1 and B_2. At B_1 they have a common tangent t which is one component of the singular conic s. The second component of s is the line joining B_1 and B_2. Any pair of conics in the pencil intersects at the point B_1 with multiplicity three and at B_2 with multiplicity one. Therefore, any two conics of the pencil of the fourth kind *osculate* each other. Sometimes we say that the pencil of the fourth kind is a pencil of *osculating conics*. On the left-hand side of Figure 7.30 we can see some conics of a pencil of the fourth kind. The only parabola in the depicted pencil is shown in red, and the common osculating circle is shown in orange.

If the one and only singular conic s in the pencil is a repeated line, then there is only one base point B. Any two conics in the pencil of the fifth kind intersect at B with multiplicity four. This is also the case for the singular conics, and therefore, s is the common tangent to all conics in the pencil. Any two conics in the pencil intersect with multiplicity four at B, *i.e.*, they are *hyperosculating* at B, and thus, the pencil of the fifth kind is also called *pencil of hyperosculating conics*.

Note that the pencils of the fourth and fifth kind cannot be spanned by singular conics exclusively.

Different appearances of pencils of conics in $\mathbb{P}^2(\mathbb{R})$

Depending on whether the base points of a pencil of conics in $\mathbb{P}^2(\mathbb{R})$ are real or not, whether they are proper (finite) or not, we see different versions of pencils of conics. We shall describe some of them here in order to make the reader familiar with this fact and in order to enable the reader to recognize different pencils of conics as such at hand of some images. We do not aim at a complete list of different cases.

Case 1 - Four common points, the generic case

Consider a pencil of the first kind with two base points, say B_3 and B_4, at infinity. From the viewpoint of Euclidean geometry, the conics in this pencil are hyperbolas if B_3 and B_4 are real points. If B_3 and B_4 are complex conjugate, then the conics in the pencil are ellipses. In Figure 7.31 (left), we can see the conics of a pencil of the first kind with a pair of proper real base points and a pair of ideal base points. In this particular example, we have $B_3 = (0:1:0)$ and $B_4 = (0:0:1)$. The remaining base points are $B_1 = (1:1:1)$ and $B_2 = (1:-1:-1)$. Note that all conics in this pencil are equilateral hyperbolas with parallel asymptotes. This pencil can be spanned by

$$k:\ xy - 1 = 0 \quad \text{and} \quad l:\ xy - x + y - 1 = 0.$$

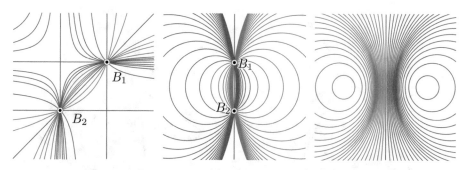

FIGURE 7.31. Some versions of pencils of the first kind in $\mathbb{P}^2(\mathbb{R})$. Left: Two real ideal base points and two real proper base points. Middle: A pair of complex conjugate ideal base points together with a real proper pair. The absolute points of Euclidean geometry are the two ideal base points, and thus, the conics are circles. Right: Two pairs of complex conjugate base points.

Note that these curves are frequently considered to be the graphs of rational functions.

The middle of Figure 7.31 shows an example of a pencil of the first kind with two real proper base points and a pair of complex conjugate ideal base points. The proper points could be placed at

$$B_1 = (1:0:1) \quad \text{and} \quad B_2 = (1:0:-1)$$

whereas the ideal points could be chosen as *absolute points of Euclidean geometry*, i.e.,

$$B_3 = (0:1:i) \quad \text{and} \quad B_4 = (0:1:-i).$$

Because of the latter choice, the regular conics in this pencil are circles.

Replacing B_1 and B_2 with a complex conjugate pair of points, we find a pencil of conics which is still of the first kind and contains only circles. The right-hand side of Figure 7.31 shows some of the circles in this pencil. We shall treat pencils of circles in more detail in Section 7.4 (cf. page 320).

■ **Example 7.3.4** A pencil of equilateral hyperbolas. We assume that the four base points of a pencil of the first kind are the vertices A, B, C of a triangle Δ in the Euclidean plane together with Δ's orthocenter O as shown in Figure 7.32. As a matter of fact, the three singular conics in this pencil are the three pairs of lines $([A, B], [C, O])$, $([B, C], [A, O])$, and $([C, A], [B, O])$, i.e., any side line of Δ together with the altitude through the opposite vertex. Any of these pairs can be viewed as a limiting case of an equilateral hyperbola with principal axis equal to zero. Any regular conic in this pencil is an equilateral hyperbola. This can be seen as follows: We impose a Cartesian frame on Δ such that $A = (p, 0)$, $B = (q, 0)$, and $C = (0, r)$ with $p \neq q$

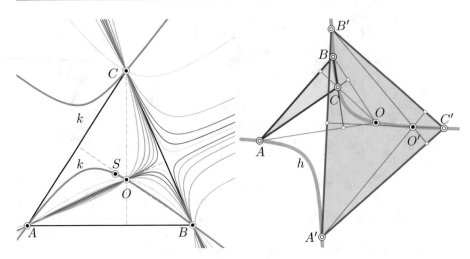

FIGURE 7.32. Left: The pencil of conics with base points A, B, C, and O contains three pairs of lines and only equilateral hyperbolas. KIEPERT's hyperbola k passes through the centroid of Δ. Right: The orthocenter of any triangle whose vertices are chosen on an equilateral hyperbola h is also located on h.

and $p, q, r \neq 0$. Then, $O = (0, -pqr^{-1})$ and the equation of the conics in the pencil are linear combinations of the equations of any two singular conics:

$$xy\lambda + (qr - rx - qy)(qx - ry - pq)\mu = 0$$

with $(\lambda, \mu) \neq (0, 0)$. The ideal points of all conics in this pencil are real and belong to orthogonal directions

$$\mathbf{v}_{1,2} = (\mu(q^2 - r^2) - \lambda \pm \sqrt{\mu^2(q^2 + r^2)^2 + 2\lambda\mu(r^2 - q^2) + \lambda^2}, \ 2qr\mu)$$

since $\langle \mathbf{v}_1, \mathbf{v}_2 \rangle = 0$.

Among the hyperbolas shown in Figure 7.32 (left), we find the KIEPERT hyperbola h (named after the German mathematician FRIEDRICH WILHELM AUGUST LUDWIG KIEPERT (1846–1934)). The hyperbola h also contains the centroid S of the base triangle Δ.

From Example 7.3.4, we can deduce:

Theorem 7.3.4 *Let A, B, C three non-collinear points chosen on an equilateral hyperbola h. Then, the orthocenter of the triangle ABC is a point on h.*

The contents of Theorem 7.3.4 are illustrated in Figure 7.32 for two different triangles ABC and $A'B'C'$ with vertices on one equilateral hyperbola h.

Case 2 - One point of contact

In the case of a pencil of conics sharing one line element and two arbitrary points, the different appearances in $\mathbb{P}^2(\mathbb{R})$ can be distinguished by the relative position of the line element and the ideal line and the reality of the two further base points.

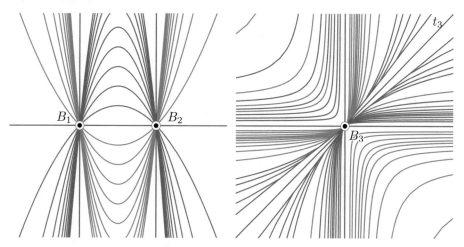

FIGURE 7.33. Some versions of pencils of the first kind in $\mathbb{P}^2(\mathbb{R})$. Left: A pencil containing only parabolas which could also be considered the graphs of all quadratic functions $y = a(x^2 - 1)$ with zeros -1 and 1; $a \in \mathbb{R}$. Right: A pencil of equilateral hyperbolas. These are the graphs of rational functions of the form $y = \frac{x}{1-ax}$ with $a \in \mathbb{R}$.

Let us consider the following two examples which are illustrated in Figure 7.33. The pencil spanned by

$$k: \ x^2 - y - 1 = 0 \quad \text{and} \quad l: \ x^2 + y - 1 = 0$$

is a pencil of the second kind, consists of parabolas only, and has the three real base points

$$B_1 = (1 : -1 : 0), \ B_2 = (1 : 1 : 0), \ B_3 = (0 : 0 : 1).$$

At the ideal point B_3, all the conics in the pencil share the tangent $x_0 = 0$ which is the ideal line. The singular conics in the pencil are two pairs of lines: a pair of parallels and the union of the ideal line and the line joining B_1 and B_2.

Another variant of a pencil of the second kind can be seen on the right-hand side of Figure 7.33. We have chosen the base points

$$B_1 = (0 : 1 : 0), \ B_2 = (0 : 0 : 1), \ B_3 = (1 : 0 : 0),$$

and the line $x - y = 0$ for the common tangent of the curves at B_3. Therefore, we see a pencil of equilateral hyperbolas all of whose orthogonal pairs of asymptotes are parallel and the curves in the pencil touch at B_3. In this case, the singular conics are $xy = 0$ and the union of the ideal line with the line $x = y$.

Note that the left and right image in Figure 7.33 are collinear copies of each other, even from the viewpoint of real geometry.

Case 3 - double contact

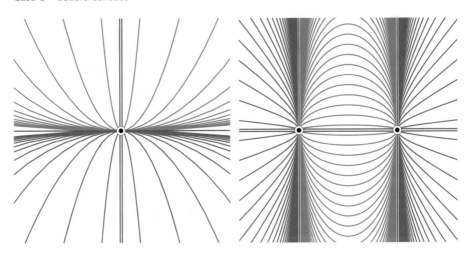

FIGURE 7.34. Some versions of pencils of the third kind. Left: The graphs of the family of functions $y = ax^2$ touch at $B_1 = (1 : 0 : 0)$ and at the ideal point $B_2 = (0 : 0 : 1)$. Right: Ellipses (blue) and hyperbolas (violet) in a pencil of the third kind.

The graphs of the quadratic functions $y = ax^2$ with variable $a \in \mathbb{R} \setminus \{0\}$ are parabolas which constitute a pencil of conics of the third kind with a proper base point $B_1 = (1 : 0 : 0)$ and an ideal base point $B_2 = (0 : 0 : 1)$. The image, as displayed in Figure 7.34 (left), shows some of the curves and the singular conics in the pencil. As the ideal line is a part of one of the singular conics, the singular conic consisting of the common tangents at the base points can only be seen partly. The repeated line, *i.e.*, the connection of B_1 and B_2, is drawn as a double line.

Next, we consider $B_1 = (1 : 1 : 0)$ and $B_2 = (1 : -1 : 0)$ to be the base points of a pencil of the third kind. Further, we let $(0 : 0 : 1)$ be the common point of the common tangents of all conics at B_1 and B_2. Now, the pencil of the third kind contains ellipses and hyperbolas as well (cf. Figure 7.34). The ellipses fill the strip in between the parallel tangents, whereas the hyperbolas cover the exterior of the strip. In this particular example in $\mathbb{P}^2(\mathbb{R})$, we have chosen the repeated line as the perpendicular to the tangents at B_1 and B_2.

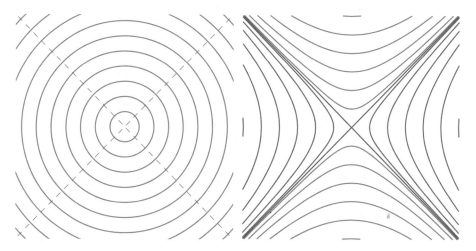

FIGURE 7.35. Some particular versions of pencils of the third kind. Left: Concentric circles form a pencil of doubly touching conics. Right: Concentric equilateral hyperbolas are concentric circles in pseudo-Euclidean geometry.

The base points of the pencil of the third kind can be chosen on the ideal line (Figure 7.35). The concentric circles displayed on the left form a pencil of conics of the third kind. The base points form a complex conjugate pair,

$$B_1 = (0 : 1 : i) \quad \text{and} \quad B_2 = (0 : 1 : -i),$$

and since these points are the absolute points of Euclidean geometry, the regular conics in the pencil are Euclidean circles. Thus, the repeated line in the pencil is the *real* ideal line which is spanned by B_1 and B_2. Note that the concentric circles touch at B_1 and B_2 along complex tangents meeting at the common center.

The pencil shown in Figure 7.35 (right) has the real base points

$$B_1 = (0:1:1) \quad \text{and} \quad B_2 = (0:1:-1).$$

It consists, therefore, of equilateral hyperbolas. The singular conics in the pencil are the repeated ideal line and the asymptotes of the hyperbolas. In pseudo-Euclidean geometry (see Section 10.2), we would interpret these as *concentric circles*, since, in the standard model, the absolute points can be chosen as B_1 and B_2.

Case 4 - osculating conics

A pencil of osculating conics can, for example, be spanned by

$$k: \ x^2 - x - y = 0 \quad \text{and} \quad l: \ x^2 + x - y = 0.$$

These parabolas intersect at $B_1 = (1:0:0)$ with multiplicity one and at $B_2 = (0:0:1)$ with multiplicity three. Therefore, these two curves osculate at B_2, and consequently, any two conics in the pencil osculate at B_2. The ideal line is the common tangent to all conics in the pencil. There is only one singular conic in the pencil. It is a pair of lines consisting of the ideal line and the line $[B_1, B_2]$. Some conics of this pencil are shown in Figure 7.36 (left).

Another example of a pencil of conics of the fourth kind is given by the one-parameter family of conics with the equations

$$\frac{1}{x} + \lambda - y = 0 \quad \text{with} \quad \lambda \in \mathbb{R}.$$

Again, the point of osculation lies at $(0:0:1)$. The conics in this pencil are equilateral hyperbolas sharing the asymptote $x = 0$. The complementary family of asymptotes is given by $y = \lambda$. Figure 7.36 (right) shows some of the conics in the pencil. In both pencils, the singular conics are a pair of lines containing the common tangent at the point of osculation and the ideal lines. In the displayed image, we see only "one half" of the singular conics.

Both examples show a further specialty: Any two parabolas/hyperbolas in the respective pencil differ only by a translation. This can easily be seen by direct computation, *i.e.*, by letting $x \mapsto x' = x + a$, $y \mapsto y' = y + b$.

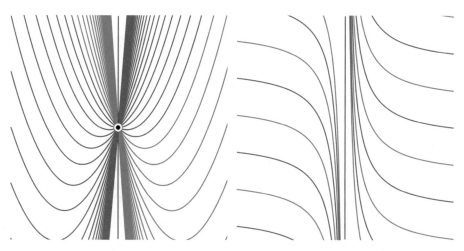

FIGURE 7.36. Some versions of pencils of osculating conics: Left: The parabolas $x^2 + \lambda \cdot x - y = 0$ with $\lambda \in \mathbb{R}$ osculate at $(0:0:1)$ and meet at the proper point $(1:0:0)$. Right: The hyperbolas $\frac{1}{x} + \lambda - y = 0$ with $\lambda \in \mathbb{R}$ also osculate at $(0:0:1)$ and share the ideal point $(0:1:0)$.

The equation $x^2 + \lambda \cdot x - y = 0$ of a parabola in the pencil changes to $x'^2 + x'(2a + \lambda) - y + a^2 + \lambda \cdot a - b = 0$. Obviously, only the coefficient of x changes. Since this coefficient is a linear function in the first parameter a of the translation, it can achieve any real value if λ does. However, we also have the possibility to argue without any computation. According to Definition 6.4.1, any pair of regular conics in the pencil of osculating conics is related via a perspective collineation. Assume k and l are (regular conics) taken from the pencil. They shall osculate at $B_2 = (0:0:1)$. Then, there exists a perspective collineation α with center B_2 and axis $[B_1, B_2]$ with $\alpha(k) = l$. The perspective collineation α is a translation for the center B_2 is an ideal point and the axis is the ideal line. In the case of the hyperbolas, the translation fixes the lines parallel to $x = 0$.

Case 5 - hyperosculating pair, intersection with multiplicity four

Finally we shall look at special examples of pencils of the fifth kind. First, let us assume that

$$\lambda \cdot x + \frac{1}{x} - y = 0 \quad \text{with } \lambda \in \mathbb{R}$$

are the equations of the conics in the pencil. Once again, we can also view these curves as the graphs of rational functions. Therefore, we have a pencil of hyperbolas. The left-hand side of Figure 7.37 shows some curves of the pencil. They are hyperosculating at $B = (0 : 0 : 1)$. The only singular conic in the pencil is the repeated line $x = 0$.

Another simple example in the projectively extended Euclidean plane is the family of parabolas given by

$$x^2 + \lambda - y = 0 \quad \text{with } \lambda \in \mathbb{R}$$

as shown on the right side of Figure 7.37. Note that these parabolas only differ by a translation in the direction of the common axes. All the curves in the pencil are congruent.

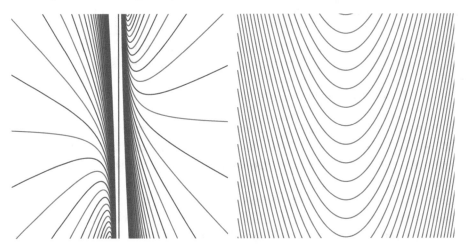

FIGURE 7.37. Some versions of pencils of hyperosculating conics: Left: A pencil of hyperbolas hyperosculating at the common ideal point. Right: Coaxial and congruent parabolas are hyperosculating at the common ideal point.

Singular pencils

A pencil of conics in the sense of Definition 7.3.5 (page 290) can be spanned by two singular curves in the pencil. However, spanning a pencil by two singular conics, *i.e.*, degenerate curves of degree two, may lead to a pencil that contains only singular conics. We call a pencil of conics *singular* if all conics in the pencil are singular. Figure 7.38 shows the bases of three types of singular pencils. The pencil $P(\lambda, \mu) = \lambda x^2 + \mu y^2 = 0$ is spanned by a pair of distinct repeated lines (Figure 7.38, left) and contains pairs of real lines if $\lambda\mu > 0$ and pairs of complex conjugate lines if

FIGURE 7.38. Singular pencils of conics spanned by three different pairs of singular conics.

$\lambda\mu > 0$. Changing the base by letting $P(\lambda,\mu) = \lambda x^2 + \mu(x^2 - y^2) = 0$ or $P(\lambda,\mu) = \lambda(x^2 - y^2) + \mu xy = 0$ gives rise to further examples of singular pencils (cf. Figure 7.38, middle and right).

• Exercise 7.3.3 Classification of singular pencils. Try to classify pencils of singular conics from the viewpoint of projective geometry. Are there distinctions to be made if the analytic model is based on a finite field $\mathbb{F}$ with char$\mathbb{F} \neq 0$? Is it possible to find pencils that are spanned by singular conics containing a finite number of non-singular pencils?

Flocks - The dual counter parts of pencils of conics

According to the principle of duality, any statement and any configuration of geometric objects in projective geometry has a dual version. A *flock* of conics of the first kind is the set of all conics tangent to four lines that form a quadrilateral, *i.e.*, no three of them are collinear. Note that in this particular case conics (of the flock) are considered as a set of lines. We have called these objects dual conics in Section 7.1.

Since line elements (lines with incident points) are self-dual, line elements are part of the base for flocks of conics of the second and third kind. A flock of conics of the third kind is the family of conics that share two line elements (A, a) and (B, b) in admissible position which means that $A \notin b$ and $B \notin a$. The regular conics in this flock belong also to a pencil of the third kind. However, the singular conics are different. The dual of a pair of lines is a pair of pencils of lines. The flock of conics of the second kind consists of all conics that share a line element (A, a) and two further tangents b and c with $A \notin b, c$ and $b \cap c \notin a$.

The flock of the fourth kind comprises the family of conics osculating a given conic c at a point $P \in c$ and sharing a tangent $t \not\ni P$. The flock of the fifth kind consists of all conics hyperosculating a given conic c at a point $P \in c$. Again, the regular conics belong also to a pencil of the fifth kind. However, the included singular dual conic is a repeated pencil of lines.

The synthetic treatment of flocks of conics does not differ too much from the treatment of pencils. Every construction is to be dualized. The same holds true for the analytical treatment: Point coordinates are replaced with line coordinates.

When we visualize the flocks of conics, we should be aware of the fact that the dual conic is a family of lines being the family of tangents of what we call a conic (note Figure 1.5). The images in Figure 7.39 show conics as sets of points which are just the envelopes of the dual conics. In Figure 7.39 we have displayed three types of flocks. The pencils of third and fifth kind are self-dual, *i.e.*, the base consisting either of a pair of line elements or a conic to be hyperosculated at some point stays the same when we apply a duality. In the case of a pencil of the first, second, or fourth kind the base changes to a quadrilateral of lines, a line element plus two tangents, or a conic to be osculated and a further tangent.

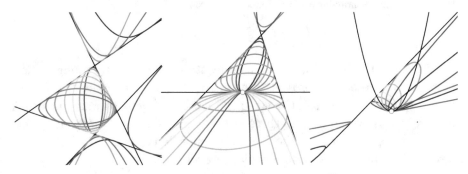

FIGURE 7.39. The three types of flocks of conics that differ from their dual counter parts: Left: A flock of the first kind consists of all conics tangent to a quadrilateral. Middle: A flock of the second kind consists of all conics with a common line element and two further common tangents. Right: The conics in a flock of the fourth kind share an osculating element (black parabola with a certain point of osculation) and one further tangent.

7.4 DESARGUES's involution theorem

First, we formulate the theorem of GÉRARD DESARGUES (first given in 1639) in the most general form. Afterwards, we treat the five cases of pencils of conics separately.

In Example 7.3.2, we have determined the parabolas in a pencil of conics. For that purpose, we have written the equations of the conics in the pencil in terms of homogeneous coordinates. Then, the intersection with the ideal line which are the solutions of a quadratic equation were depending on the homogeneous parameters (κ, λ) in the pencil. Consequently, we had to determine $(\kappa, \lambda) \neq (0,0)$ such that the quadratic equation has a double root. The geometric background is laid down in DESARGUES's involution theorem:

Theorem 7.4.1 *If the points S_1 and S_2 are the points of intersection between a curve k of any given pencil of conics and a given line l then S_1 and S_2 are corresponding in an involution δ_l, provided that l does neither pass through a base point of the pencil nor is a component of any curve in the pencil.*

Proof: Let $\mathbf{A} = (a_{ik})$ and $\mathbf{B} = (b_{ik})$ be the coefficient matrices of the two conics c and d which span the pencil, and let $x_0 = 0$ be the the the equation of the line l in $\mathbb{P}^2(\mathbb{F})$.

The points $X = (0 : x_1 : x_2)$ and $Y = (0 : y_1 : y_2)$ are conjugate w.r.t. c if, and only if,

$$a_{11}x_1y_1 + a_{12}(x_1y_2 + x_2y_1) + a_{22}x_2y_2 = 0. \tag{7.10}$$

The pairs (X, Y) constitute an involution if the determinant $a_{11}a_{22} - a_{12}^2 \neq 0$. Otherwise l is a tangent or, in the case $a_{11} = a_{12} = a_{22} = 0$ a component of l. There is an analogous bilinear form characterizing the conjugate position of X and Y w.r.t. d.

If (P, Q) is a pair of conjugate points w.r.t. c and d, then it is conjugate w.r.t. all conics in the pencil. Consequently, if fixed points S_1 and S_2 of such an involution of conjugate points exist, they are harmonic w.r.t. P and Q. Hence, the fixed points are pairs of an involution on l.

In the case of an algebraically closed field $\mathbb{F}$ the proof is already done. However, in any other case, the pair (P, Q) needs not exist. Therefore, we give a second proof which is valid for all comutative fields $\mathbb{F}$ with char$\mathbb{F} \neq 2$.

Let us project the points $X = (0 : x_1 : x_2)$ on l from the center $A = (1 : 0 : 0)$ onto the "standard conic" $s:\ x_0x_2 - x_1^2 = 0$. Then, X is mapped to $X' = (x_1^2 : x_1x_2 : x_2^2)$. The points X, Y which are conjugate w.r.t. c satisfying (7.10) are projected to points $X', Y' \in s$ which are collinear with $I_c = (a_{22} : -a_{12}; a_{11})$. This is either the center of an involution on s or the image T' of the point T of contact between c and l, if l is tangent to c. The verification of this statement is left to the reader as an exercise.

In the same way, the second conic d defines a point $I_d = (b_{22} : -b_{12} : b_{11})$. The corresponding centers I_k of all other conics k in the pencil are represented as linear combinations. Hence,

I_k, I_c, and I_d are collinear. Before confirming that $I_c \neq I_d$ let us finish the reasoning: If k intersects the line l at S_1 and S_2, their projections $S'_1, S'_2 \in s$ are fixed points of the involution on s with center I_k. Hence, S'_1 and S'_2 lie on the polar line p_k of I_k w.r.t. s. While I_k traverses the line $[I_c, I_d]$, the polars p_k pass through a fixed point I. If $I \in s$, then there is a point of intersection $B \in l$ with $B' = I$ which belongs to all conics in the pencil, and l would pass through any base point, but this was excluded. Therefore, all pairs (S'_1, S'_2) are aligned with a point $I \notin s$. They are corresponding in an involution on s which, after projection from s back to l confirms the statement.

Finally, we have to exclude the case $I_c = I_d$: Suppose $\alpha(a_{22}, -a_{12}, a_{11}) + \beta(b_{22}, -b_{12}, b_{11}) = (0,0,0)$ for any $(\alpha, \beta) \neq (0,0)$. Then, the conic with coefficient matrix $\alpha\mathbf{A} + \beta\mathbf{B}$ has the equation

$$(\alpha a_{00} + \beta b_{00})^2 x_0^2 + 2(\alpha a_{01} + \beta b_{01})x_0 x_1 + 2(\alpha a_{02} + \beta b_{02})x_0 x_2 = 0.$$

The conic is reducible and contains l as a component. This has been excluded as well. ∎

In order to get a deeper insight into DESARGUES's involution theorem, we add synthetic proofs for the different kinds of pencils.

The most simple case - pencils of the first kind

Proof: 1. Assume $P_1 P_2 P_3 P_4$ is the quadrangle of base points of the given pencil and c is a conic of the pencil, i.e., $P_i \in c$ for all $i \in \{1,2,3,4\}$. Let further l be a line which does not contain any P_i. Let S and S' be common points of c and l (Figure 7.40).

Then, the points $Q := [P_2, P_4] \cap l$, $R := [P_2, P_3] \cap$, S, and S' are four different points. Let α be the projectivity from the pencil P_1 to the pencil P_2 that generates c. If π_1 is the perspectivity from the pencil P_1 to l, and similar, π_2 is the perspectivity from the pencil P_2 to l, then $\pi_2^{-1} \circ \alpha \circ \pi_1 : g \to g$ is a projectivity with

$$[P_1, P_3] \cap l =: A \mapsto R, \quad [P_1, P_4] \cap l =: B \mapsto Q, \quad S \mapsto S, \quad S' \mapsto S'.$$

Now, there exists an involutive projectivity $\eta : l \to l$ with $Q \mapsto R \mapsto Q$ and $S \mapsto S' \mapsto S$. The

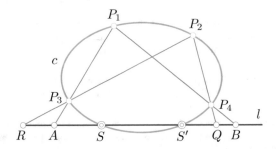

FIGURE 7.40. The Desargues involution for a pencil of the first kind on a line l, as stated in Theorem 7.4.1.

composition $\delta_l := \eta \circ \pi_2 \circ \alpha \circ \pi_1^{-1}$ maps

$$A \mapsto Q, \quad B \mapsto R, \quad S \mapsto S', \quad S' \mapsto S.$$

Since $S \neq S'$, δ_l is involutive. Since $A \neq B$, $Q \neq R$, $A \neq R$, $B \neq Q$, the involution δ_l is uniquely defined by $A \mapsto Q$ and $B \mapsto R$. This means that the quadrangle $P_1 P_2 P_3 P_4$ of base points of the pencil of the first kind uniquely determines δ_l and δ_l does not depend on the choice of the conic c.

2. If δ_l is the involutive projectivity determined by $A \mapsto Q$ and $B \mapsto R$, then $\eta \circ \delta_l : l \to l$ maps $A \mapsto R$ and $B \mapsto Q$ and fixes S and S'. Consequently, the projectivity $\alpha^{-1} := \pi_2^{-1} \circ \eta \circ \delta_l \circ \pi_1$ from the pencil P_1 to the pencil P_2 is identical with the projectivity α which generates the conic c, because $\alpha^{-1}([P_1, P_3]) = [P_2, P_3]$, $\alpha^{-1}([P_1, P_4]) = [P_2, P_4]$, and $\alpha^{-1}([P_1, S]) = [P_2, S]$. Hence, the point S' is also a point of c since $\alpha^{-1}([P_1, S']) = [P_2, S']$. ∎

In the case of a pencil of conics of the first kind, the Desargues involution δ_l is determined by the quadrangle $P_1P_2P_3P_4$ of base points. Also in the case of a pencil of the second or third kind, the base points and base lines determine the involutive projectivity δ_l on any line l that is in an admissible position w.r.t. the base points (see Figure 7.41), and the proof given above remains valid.

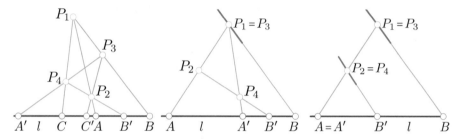

FIGURE 7.41. Assigned pairs of points of the Desargues involution δ_l for pencils of the first kind (left), the second kind (middle), and the third kind (right).

● **Exercise 7.4.1** Three corresponding pairs in the Desargues involution. Show the following: The three pairs of opposite side lines in a quadrangle $P_1P_2P_3P_4$ in a projective Pappian Fano plane intersect any line l which does not pass through a vertex of the quadrangle in three pairs of corresponding points of an involutive projectivity on l.

Hint: Study the action of $l \overset{P_1}{\barwedge} [P_2, P_3] \overset{P_4}{\barwedge}$.

If a line l contains one or even two diagonal points of the quadrangle $P_1P_2P_3P_4$ of base points, then these points are fixed points of δ_l. This is clear from the proof of Theorem 7.4.1. The pencils of the second determines fixed points of δ_l on special lines. In the case of a pencil of the third kind, one fixed point is known from the beginning. It is the point $A = A'$ (in Figure 7.41) that comes as the point of intersection of l with the line joining the only two base points of this type of pencil.

For fixed points of the Desargues involution δ_l which are not determined by the special choice of l there is a very important meaning. We can show:

Theorem 7.4.2 *Any fixed point of the Desargues involution δ_l on the line l either the point of contact of l with a regular conic out of the pencil or a singular point of a singular conic included in the pencil.*

Proof: Let F be a fixed point of δ_l which is not located on any side of $P_1P_2P_3P_4$. In the case of a pencil of the first kind, the base points P_1, P_2, P_3, and P_4 together with the point F determine a unique conic c. The line l is tangent and contains only the point F: Otherwise, l would also carry $\delta_l(F) \neq F$ (according to Theorem 7.4.1) which contradicts $\delta(F) = F$. If F lies on any side of $P_1P_2P_3P_4$, then it must be a diagonal point.

Conversely, if c and l are in contact at a point X with $X \neq \delta_l(X)$, then, according to Theorem 7.4.1, c passes through $\delta_l(X)$ which is not possible for a tangent of the conic c. ∎

■ **Example 7.4.1** Applications of DESARGUES's theorem. DESARGUES's involution theorem is useful for the construction of conics from certain given configurations of lines, points, and conics in the projective extension of $\mathbb{E}^2$. In this example we shall give a list of possible problems. This list is far from being complete.

Parabolas on four points. We continue Example 7.3.2 and look for the parabolas in a pencil of the first kind. A parabola is a conic in the real plane that touches the ideal line ω. Therefore, we have to find the fixed points A_1 and A_2 of the Desargues involution δ_ω, if there are some.

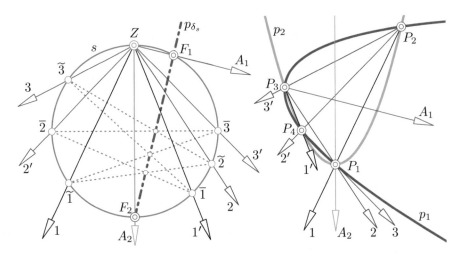

FIGURE 7.42. How to find the parabolas in a pencil of the first kind? The six lines through the base points P_1, P_2, P_3, P_4 have six ideal points which comprise three pairs of corresponding points in the Desargues involution δ_ω on the ideal line ω. The projection of ω to the conic s from $Z \in s$ yields an involutive projectivity on s whose fixed points F_1 and F_2 correspond to the fixed points of δ_ω. The lines $[Z, F_1]$ and $[Z, F_2]$ are parallel to the axes of the two wanted parabolas p_1 and p_2.

Assume we are given the four base points P_1, P_2, P_3, and P_4 of a pencil of the first kind, as shown in Figure 7.42. We are looking for the parabolas in the pencil, *i.e.*, we want to find the parabolas through the points P_1, P_2, P_3, and P_4, if there are some.

A parabola is a conic that touches the ideal line ω, and thus, we use Theorem 7.4.2 which says that the fixed points of δ_ω belong to those conics that touch ω.

In order to handle the involution δ_ω, we project the ideal points 1, 2, 3 (the ideal points of the lines $[P_1, P_2]$, $[P_1, P_3]$, $[P_1, P_4]$) as well as the ideal points $1'$, $2'$, $3'$ (the ideal points of the opposite lines $[P_3, P_4]$, $[P_2, P_4]$, $[P_2, P_3]$) to an arbitrary conic s from some of its points Z. Note that the ideal points of opposite sides in the quadrangle of base points are assigned points in δ_ω. For the sake of simplicity, we have chosen a circle s.

This yields a projectivity δ_s on s with $(\widetilde{1}, \widetilde{1})$, $(\widetilde{2}, \widetilde{2})$, and $(\widetilde{3}, \widetilde{3})$ for its assigned pairs. The projectivity δ_s is involutive since δ_ω is. Thus, already two pairs of assigned points determine δ_s. A priori, we do not know which pairs are useful for the construction of the fixed points. Therefore, we have projected all points onto s. Now, Theorem 6.4.4 comes into play: The axis p_{δ_s} carries the points

$$[\widetilde{1}, \widetilde{2}] \cap [\widetilde{1}, \widetilde{2}], \quad [\widetilde{1}, \widetilde{3}] \cap [\widetilde{1}, \widetilde{3}], \quad [\widetilde{2}, \widetilde{3}] \cap [\widetilde{2}, \widetilde{3}].$$

The fixed point F_1 and F_2 of δ_s are the points of intersection of s and p_{δ_s}.

Finally, we project the fixed points of δ_s to that of δ_ω by simply connecting Z with F_1 and F_2. The latter lines are parallel to the axes of the parabolas p_1 and p_2 in the pencil. In Figure 7.42, the parabolas p_1 and p_2 are shown though we do not describe all constructions in order to find the vertices and the precise position of the axes.

■ **Example 7.4.2** Conics in a pencil of the second kind touching a line l. The line l does not pass through any of the base points P_1, P_2, and P_3. The base points together with the tangent t_1 at P_1 determines the Desargues involution δ_l on l as shown in Figure 7.41. Thus, we obtain the following pairs of assigned points

$$[P_1, P_2] \cap l =: 1 \mapsto 1' := [P_1, P_3] \cap l, \quad [P_2, P_3] \cap l =: 2 \mapsto 2' := t_1 \cap l.$$

Since δ_l is involutive, we have $\delta(1') = 1$, and thus, we have at least three pairs of assigned points which are sufficient according to Theorem 5.2.1. According to Theorem 7.4.2, the fixed points of δ_l are the points of contact of conics in the pencil with the line l.

We choose a conic s and a point $Z \in s$, project the points 1, $1'$, 2, $2'$ from Z to s. With $1' = 3$ and $1 = 3'$ we can apply Theorem 6.4.4 and find the fixed points F_1 and F_2 of the projectivity δ_l lifted to s (as in the previous example). The fixed points are projected to the points C_1 and C_2 where the conics c_1 and c_2 touch l, see Figure 7.43.

■ **Example 7.4.3** Conics on three points and two tangents. Assume that three non-collinear points P_1, P_2, P_3 and two lines l and m are given, as shown in Figure 7.44. The lines l and m shall not pass through any of the given points. Now, we are looking for conics through P_1, P_2, and P_3 that touch both l and m.

If a conic c_1 is a solution of the given problem, then it touches l and m at the points C_1 and C_2. Thus, c_1 and the pair (l, m) span a pencil of the third kind with the two line elements (C_1, l) and (C_2, m) for its base. Let $g_{12} := [P_1, P_2]$ be the line through the given points P_1 and P_2. According to Theorem 7.4.2, the line g_{12} (which is not passing through a base point) intersects the conics as well as the lines l and m in pairs of assigned points in the Desargues involution δ_{12} on g_{12}. The assigned pairs are

$$l \cap g_{12} \mapsto m \cap g_{12}, \quad P_1 \mapsto P_2$$

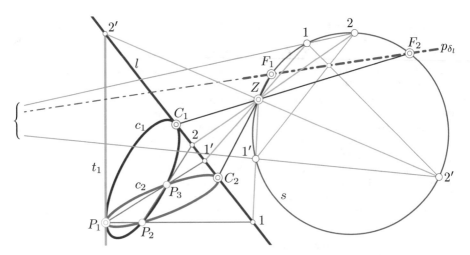

FIGURE 7.43. The conics c_1 and c_2 in a pencil of the second kind with base points P_1, P_2, P_3 and base line t_1 are uniquely determined if the contact points C_1 and C_2 with the line l are known. These contact points are the fixed points of the Desargues involution δ_l on l. The involution δ_l is projected from the center Z to the Steiner conic s where we find the fixed points. (The image points are labeled in the same way.) The axis p_{δ_l} intersects s at the fixed points F_1 and F_2 which are mapped to the desired contact points C_1 and C_2 via the projection from Z to l.

and δ_{12} is uniquely determined.

A fixed point F_{12} of δ_{12} is the point of intersection of g_{12} with a possible chord $[C_1, C_2]$. So, the connection of two fixed points of the Desargues involutions on different lines, $e.g.$, $g_{12} := [P_1, P_2]$ and $g_{13} := [P_1, P_3]$, is a chord of a possible conic c_1 and intersects l and m in the contact points C_1 and C_2.

As we can see in Figure 7.44, the six fixed points on the sides of the triangle gather on four lines, $e.g.$, the fixed points F_{12}, F_{13}, and F_{23} are collinear. (The proof of this fact is left as an exercise to the reader.) Consequently, there are four chords which are the double lines in the pencils of the third kind hidden in this configuration of conics and lines. Therefore, there are up to four (real) solutions.

■ Example 7.4.4 Conics that touch a given conic twice. In this example, we shall treat conics on three points P_1, P_2, P_3 that touch a given conic c twice. Like in the previous item, we are concerned with pencils of doubly touching conics. In the previous example, two lines l and m were given. Somehow, this pair of lines can be seen as a degenerate conic.

Again, we aim at a determination of the points of contact of the given conic c and the conics we are looking for. Just as in the example before, the Desargues involutions on the sides of $P_1 P_2 P_3$ are defined by their intersections with c and the pair of given points on it, $e.g.$, in the case of the line $[P_1, P_2]$, the involution δ_{12} is determined by the following two pairs of corresponding points

$$P_1 \mapsto P_2, \quad S_1 \mapsto S_2$$

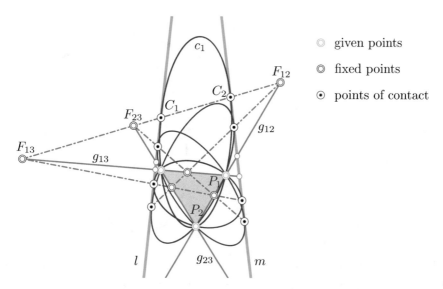

FIGURE 7.44. Conics on three points and two tangents: There is a Desargues involution on each side of $P_1P_2P_3$. Triples of fixed points are collinear and there are four such lines which meet the tangents l and m at points where one of the solutions touches.

where S_1 and S_2 are the common points of c and $[P_1, P_2]$, if they exist.

The fixed points of the three Desargues involutions, and thus, the points of contact of c and the solutions give rise to the same configuration of points and lines as in the previous example, cf. Figure 7.44.

The left-hand side of Figure 7.45 shows the configuration of given points, fixed points of Desargues involutions, and points of contact.

■ **Example 7.4.5** A conic on three real points and a pair of complex conjugate points. **How to** find a conic c on a pair $(K, \overline{K})$ of complex conjugate points and three real points A, B, C?

The pair $(K, \overline{K})$ of complex conjugate points can be given by its real join l and by the elliptic involution with the fixed points K and $\overline{K}$. If there exists a solution c, then the given involution is the involution α of points conjugate w.r.t. c.

Applying a perspective collineation κ with l as its vanishing line and the Laguerre point L_α for its center to A, B, C results in three points, say A', B', C'. The collineation α maps the pair $(K, \overline{K})$ to the pair of absolute points of Euclidean geometry. Thus, the κ image c' of c is the unique circle on A', B', C'. The κ^{-1}-image of c' gives the conic we are looking for.

In Section 4.4, we have seen a different approach to these kind of conic problems. However, the constructive methods and possibilities apply to many more kinds of constructive problems compared to the spatial interpretation.

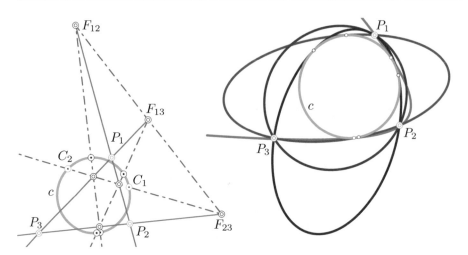

FIGURE 7.45. Left: The conic c, the three given points P_1, P_2, P_3, the fixed points of the Desargues involutions on the sides of $P_1P_2P_3$, and the points of contact of the solutions. Two points of contact of one solution are always collinear with one fixed point of each of the three involutions. Right: Four conics on the same three points in double contact with c.

DESARGUES's involution theorem for pencils of the fourth and fifth kind

In the previous section, we have learned that the *base figure* of a pencil of conics, *i.e.*, the base points together with base tangents, determines the Desargues involution. In the case of a pencil of the fourth or fifth kind, the base figure contains a non-degenerate conic. However, there is still an involution determined on any line, that neither contains a base point, nor coincides with a base line, nor passes through a common pole of the conics in the pencil.

We state and show:

Corollary 7.4.1 *Let (P_1, t_1, P_2, c) be the base of a pencil of conics osculating the conic c at P_1 with the common tangent t_1 at P_1 and passing through P_2. Then, all conics in the pencil meet any line l that is neither passing through P_1 nor through P_2 in corresponding pairs of an involutive projectivity.*

In case of a pencil of the fifth kind with hyperosculation at P_1, the tangent t_1 meets l in a fixed point of the involution.

Proof: According to Definition 6.4.1, any conic d in the pencil defined by the base (P_1, t_1, P_2, c) can be mapped onto c by an elation η with center P_1 and axis $a := [P_1, P_2]$. Assume now that

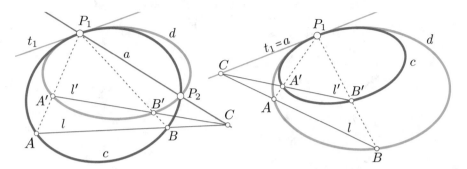

FIGURE 7.46. Pencils of conics osculating each other at a common point also define a Desargues involution on a line in an admissible position.

a line l with $P_1, P_2 \nmid l$) carries two different points A and B of a such conic d in the pencil (cf. Figure 7.46, left-hand side).

The line $l = [A, B]$ meets the axis a of the collineation η in a point C. Also the η-image of l meets a in C. Further, the points $A' = \eta(A)$ and $B' = \eta(B)$ are points on $\eta(d) = c$. Since $C \nmid c$, it is the center of an involutive projectivity $\alpha : c \to c$ that interchanges A' and B'. Let us use the projective mapping

$$\beta : l \ \overline{\barwedge} \ P_1 \ \overline{\barwedge} \ c$$

in order to define the projective mapping

$$\delta_l := \beta^{-1} \circ \alpha \circ \beta$$

that acts on the line l and is involutive, for it sends A to B and B to A. It is independent of the choice of the conic d in the pencil. Since η sends A to A', η is uniquely defined.

In case of a pencil of the fifth kind, the base consists of (P_1, t_1, c) where (P_1, t_1) is a line element of c. The axis of $\eta : d \to c$ now coincides with t_1 (Figure 7.46, right). Everything else of the previous proof remains valid. ∎

■ Example 7.4.6 Parabolas that osculate a hyperbola. Figure 7.47 shows a hyperbola h with the line element (P_1, t_1) and a further point P_2. We are looking for all parabolas that osculate h at P_1 and pass through P_2.

In order to determine the parabolas, we are searching for their points at infinity, i.e., the direction of the axis. The base of the pencil of osculating conics consist of the line element (P_1, t_1), the point P_2, and the conic h. Now, we look at the Desargues involution on the line at infinity: We find the two ideal points 1 and 1' of h as a pair of corresponding points. A further pair comes from the singular conic in the pencil. The tangent t_1 at P_1 has the ideal point 2' which corresponds to the ideal point 2 of the line $[P_1, P_2]$. In this particular example, we can simplify the construction of the fixed elements of the involution by using h as the Steiner conic (cf. Example 6.4.7). Furthermore, we use P_1 as the center of the perspectivity that maps the range of points on the ideal line to the points on h. In Figure 7.47, the images of points under the perspectivity $\omega \ \overset{P_1}{\overline{\barwedge}}$ are labeled $\overline{1}, \overline{2}, \overline{1'}$, and $\overline{2'}$. The construction of fixed points $\overline{F}_1$ and $\overline{F}_2$ follows the advice given in Example 6.4.7. In Figure 7.47, the center of the involution on h is the ideal point $I = 2$. The dashed lines $[P_1, \overline{F}_1]$ and $[P_1, \overline{F}_2]$ join P_1 with the ideal points being fixed points of the Desargues involution on the line at infinity, and thus, they are parallel to the axes of the parabolas p_1, p_2 we were looking for.

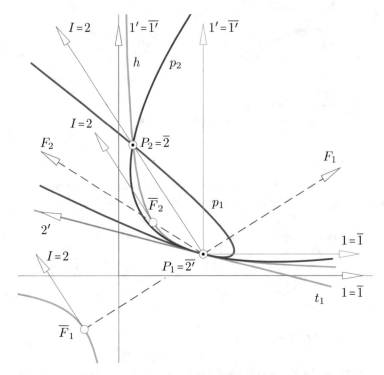

FIGURE 7.47. How to find the parabolas p_1 and p_2 (if there are some) that osculate the hyperbola h at P_1 and share a further point P_2 with h: According to Theorem 7.4.1, the Desargues involution on the line at infinity is uniquely determined. The hyperbola serves as Steiner conic and the fixed points of the involution determine the axes of the desired parabolas.

■ **Example 7.4.7** Conics that hyperosculate a given conic and pass through two given points. Figure 7.48 shows a parabola p and two further points A and B. We are looking for all conics that hyperosculate p somewhere and contain A and B if there are some.

The conics in a pencil of the fifth kind intersect a generic line l in pairs of corresponding points in an involution (see Corollary 7.4.1). A fixed point F of this involutive projective mapping on l is either a point of contact of l with a conic in the pencil or a point on the tangent at the point H of hyperosculation.

Therefore, we determine the fixed points of the involution on $l = [A, B]$. The points A and B form a pair of corresponding points as well as the two common points 1 and 2 of l and p do. As a matter of fact, one fixed point, say F_2, lies in the interior of p. The other fixed point F_1 is an outer point of p. From F_1 we can draw two real tangents to p that meet the given conic at potential points H_1 and H_2 of hyperosculation. The conics c_1 and c_2 appearing as solutions to the present problem, can be completed by applying the elations with centers H_1 / H_2 and the respective axes $[F_1, H_1]$ / $[F_1, H_2]$ to the given conic p (see Definition 6.4.1).

This problem can also be solved by means of spatial interpretation as explained in Section 4.4.

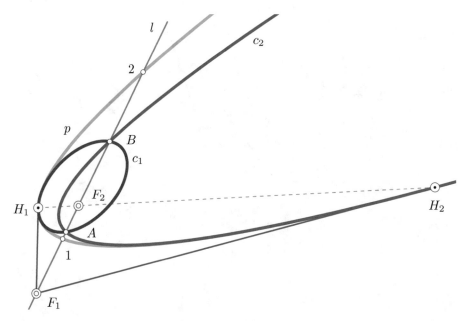

FIGURE 7.48. The conics c_1 and c_2 hyperosculate p at H_1 and H_2, respectively. According to Theorem 7.4.1, the tangents to p at the latter points pass through the fixed points of the Desargues involution involution on l.

DESARGUES's involution for flocks of conics

The principle of duality allows us to formulate the dual version of DE-SARGUES's involution Theorem 7.4.1. It reads:

Corollary 7.4.2 *The tangents drawn from a generic point P to the conics of a flock of any kind constitute corresponding pairs of an involutive projective mapping in the pencil P. The fixed lines in this involution are tangents of conics in the pencil that pass through P.*

The conics mentioned in Theorem 7.4.2 are to be considered as dual conics, *i.e.*, the set of tangents of a conic. Figure 7.49 shows how the base of a flock of conics defines the Desargues involution in the pencil of lines around a generic point. The depicted figures are obtained by dualizing the respective configurations shown in Figures 7.40 and 7.46.

● Exercise 7.4.2 A parabola tangent to three lines. Show the following result: The orthocenter O of each triangle built by three finite tangents of a parabola p lies on p's directrix.

Hint: Check DESARGUES's involution at O.

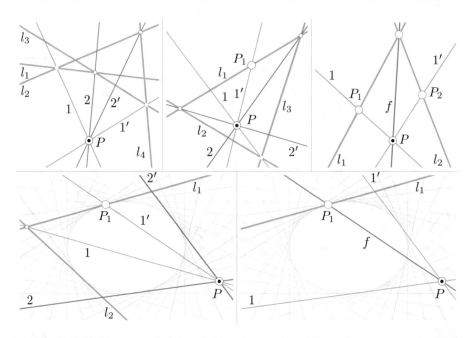

FIGURE 7.49. DESARGUES's involution in pencils of lines about a generic point P induced by the bases of flocks of conics of the first to fifth kind (from top-left to bottom-right): $(1, 1')$ an $(2, 2')$ are pairs of corresponding points, f indicates a fixed line. Note, that the bases of flocks of the fourth and fifth kind contain a dual regular conic.

■ **Example 7.4.8** Conics on two points and three tangents. Dual to Example 7.4.4, we briefly describe how to find conics on three given tangents a, b, c and two points P, Q. Figure 7.50 shows an example.

Now, we look at the Desargues involutions in the pencils of lines with vertices $a \cap b$, $b \cap c$, and $c \cap a$. Assume f_1 is a fixed line of $\delta_{a \cap b}$ and let f_2 be a fixed line of $\delta_{b \cap c}$, cf. Figure 7.50. For example, the involution $\delta_{a \cap b}$ is defined by

$$a \mapsto b, \quad [a \cap b, P] \mapsto [a \cap b, Q].$$

Dualizing the figure from the previous example, we find that $T_1 = f_1 \cap f_2$ is the common point of two tangents of a solution conic s. Consequently, we obtain a configuration of six fixed lines where three out of them pass through one such point T_i. There are four points T_i, and thus, there are four real solutions provided that all three Desargues involutions are hyperbolic.

Pencils of circles

Circles are special conics in many ways. From the viewpoint of geometry in the complex extended real projective plane, a circle is a conic through the absolute points $I = (0 : 1 : i)$ and $\bar{I} = (0 : 1 : -i)$ of Euclidean geometry (see Example 6.4.7 on page 253).

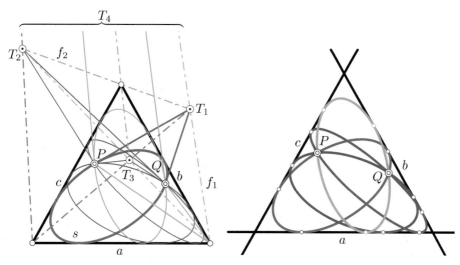

FIGURE 7.50. Conics on two points and three tangents. Left: The fixed lines of the Desargues involution in the pencils about $a \cap b$, $b \cap c$, and $c \cap a$ meet in the common points T_i (with $i \in \{1, 2, 3, 4\}$) of the tangents at P and Q of the desired conics. Right: In this case all three Desargues involutions are hyperbolic, and thus, we have four real solutions.

It is natural to distinguish pencils of circles by the number and type of proper base points. Thus, we have the following case (cf. Figure 7.51):

1. circles with two real common points - *elliptic pencil,*

2. circles with a pair of complex conjugate common points - *hyperbolic pencil,*

3. circles tangent to a common line element - *parabolic pencil,* and

4. *concentric circles.*

According to this list, the elliptic and the hyperbolic pencil of circles are pencils of conics of the first kind. The parabolic pencil is a pencil of conics of the second kind. Concentric circles form a pencil of conics of the third kind, because any two of the circle touch at the absolute points of Euclidean geometry. We shall see very soon where the names of the pencils originate. Thus, pencils of circles are pencils of conics whose bases contain the points I and $\bar{I}$.

A pencil of circles is an object of equiform geometry, *i.e.,* the four-parameter group of transformations in the Euclidean plane generated by Euclidean motions and uniform scalings.

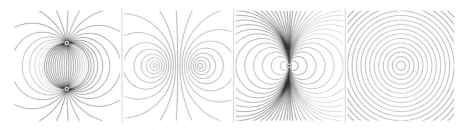

FIGURE 7.51. The four types of pencils of circles: an elliptic, a hyperbolic, a parabolic pencil, and a pencil of concentric circles.

The elliptic pencil - two real base points

Since the scale does not matter, it means no restriction to assume that the two common points B_1 and B_2 of all circles in the pencil have Cartesian coordinates

$$B_1 = (0,1) \quad \text{and} \quad B_2 = (0,-1).$$

Thus, the Cartesian equations of all circles in this elliptic pencil read

$$x^2 - 2xt + y^2 = 1 \quad \text{with } t \in \mathbb{R}. \tag{7.11}$$

It is clear and easily verified that in this pencil of conics (of the first kind), there are the following singular conics:

$$[I, \overline{I}] \cup [B_1, B_2], \quad [I, B_1] \cup [\overline{I}, B_2], \quad [I, B_2] \cup [\overline{I}, B_1].$$

The first is a pair of real lines; the remaining pairs consist of conjugate complex pairs of lines. The line $[B_1, B_2]$ is the *radical axis* of any pair of circles in the pencil. The centers of all circles lie on the bisector of the segment $B_1 B_2$. We call this bisector the *axis of the pencil*. The axis is a line of symmetry of the pencil as well as common line of symmetry for all circles in the pencil.

We choose two circles l and m of the elliptic pencil (see Figure 7.52). According to THALES's theorem (cf. 6.3.3), we can see the intersections 1, 1' and 2, 2' of l and m with the axis a at right angles from either of the two base points. However, the choice of l and m does not matter. Thus, the pair of intersections of any circle with the axis of the pencil is seen at right angles from the base points. Therefore, B_1 and B_2 are the Laguerre points of the elliptic involution δ_a on the axis a interchanging the intersection points of the circles in the pencil (see Example 6.4.7 in Section 6.4, page 253). We can summarize:

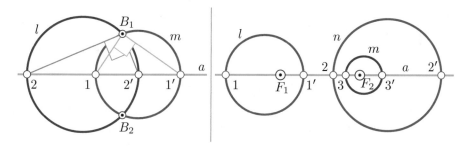

FIGURE 7.52. Left: An elliptic pencil is characterized by the elliptic involution on the axis a. The two real base points B_1 and B_2 are the Laguerre points of the Desargues involution on the axis a. Right: A hyperbolic pencil induces a hyperbolic involution on the line a of symmetry. The centers F_1 and F_2 of the two null circles in the pencil are the fixed points of the Desargues involution on the axis a. For any two circles in the hyperbolic pencil we can say that one circle is either completely inside or completely outside the other.

Theorem 7.4.3 *In the projective extension of* $\mathbb{E}^2$, *the circles in an elliptic pencil of circles meet the axis of the pencil in pairs of corresponding points in an elliptic involution with the two real base points for its Laguerre points.*

The hyperbolic pencil - no real base point

Again we make use of the fact that pencils of circles are objects of equiform geometry. Since the actual size and position is not of importance, we may choose the pair of complex conjugate base points as

$$B_1 = (0, i) \quad \text{and} \quad B_2 = (0, -i).$$

The regular circles in the hyperbolic pencil have the equations

$$x^2 - 2tx + y^2 = -1 \quad \text{with } t \in \mathbb{R} \setminus \{-1, 1\} \tag{7.12}$$

and the three singular conics in this pencil of conics of the first kind equal that in the previous case. The singular curves $(x \pm 1)^2 + y^2 = 0$ which are obtained for $t = \pm 1$ are called *null circles*. They split into two isotropic lines through their real centers $F_1 = (1, 0)$ and $F_2 = (-1, 0)$, respectively. These points are fixed under the Desargues involution δ_a induced on the axis a.

A circle $k(t)$ from the pencil meets the axis $a : y = 0$ in two points $1 = (t - \sqrt{t^2 - 1}, 0)$ and $1' = (t + \sqrt{t^2 - 1}, 0)$, see Figure 7.52. Exactly for the values $t = 1$ and $t = -1$, these two points coincide and we denote these *double points* by $F_1 = (1,0)$ and $F_2 = (-1,0)$ which can be considered as *null circles*, i.e., circles with radius zero. With (5.6), we can easily verify that

$$\mathrm{cr}(1, 1', F_1, F_2) = -1$$

independent of $t \in \mathbb{R} \setminus \{-1, 1\}$. Since the characteristic cross ratio of a hyperbolic involution equals -1 (cf. page 214), we can say:

Theorem 7.4.4 *The circles in a hyperbolic pencil of circles meet the axis a in pairs of corresponding points in a hyperbolic involution whose fixed points are the centers of the two included null circles.*

● **Exercise 7.4.3** Orthocircles of a hyperbolic pencil. Show that any circle (7.12) in the

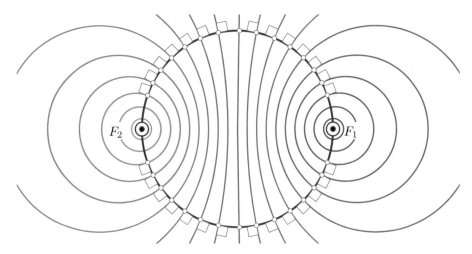

FIGURE 7.53. The Thales circle on the segment $F_1 F_2$ of fixed points of the hyperbolic involution in the hyperbolic pencil of circles is an orthocircle of all circles in the pencil.

hyperbolic pencil meets the Thales circle at right angles (see Figure 7.53). Hint: The radius of a circle (7.12) equals $\sqrt{t^2 - 1}$, the Thales circle has radius 1 and is centered at $(0,0)$, the distance of the centers equals t. Now, apply PYTHAGORAS's theorem.

● **Exercise 7.4.4** Pencils of orthogonal circles. We are given a hyperbolic pencil and an elliptic pencil of circles (see Figure 7.54) with the equations:

$$\begin{aligned} h: \quad & x^2 - 2tx + y^2 + 1 = 0 \quad \text{with } t \in \mathbb{R} \setminus \{-1, 1\}, \\ e: \quad & x^2 + y^2 - 2uy - 1 = 0 \quad \text{with } u \in \mathbb{R}. \end{aligned}$$

Show that for any (t, u) the circle h meets the circle e at right angles. Further, the base points of the elliptic pencil are the fixed points of the involution induced by the hyperbolic pencil.

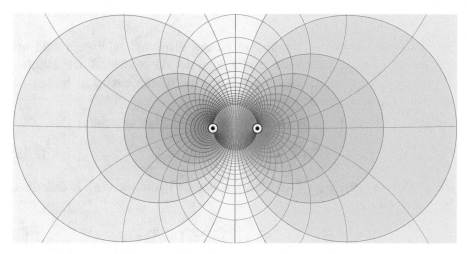

FIGURE 7.54. A hyperbolic pencil h and an elliptic pencil e of circles forming and orthogonal net of circles. This two-parameter family of circles is known as Apollonian circles.

The results from Exercises 7.4.3 and 7.4.4 can be summarized:

Theorem 7.4.5 *For any hyperbolic pencil of circles there exists an elliptic pencil of circles such that any circle of one family meets all circles of the other family at right angles. The axes of the pencils are orthogonal. The centers of the null circles of the hyperbolic pencil are the real base points of the elliptic pencil.*

The two pencils of circles mentioned in Theorem 7.4.5 are sometimes also called *conjugate* to each other.

Apollonian circles

The two one-parameter families of circles displayed in Figure 7.54 are frequently called *Apollonian circles*, due to APOLLONIUS OF PERGA. However, there is an elementary description of the circles in both pencils, the elliptic and the hyperbolic one.

Let $q \in \mathbb{R} \setminus \{1\}$ be any constant, and look at the set of points

$$c(q) = \{X | \overline{XF_1} : \overline{XF_2} = q\}. \tag{7.13}$$

Since $F_1 = (-1,0)$, $F_2 = (1,0)$, and $X = (x,y)$, the set $c(q)$ is described by the equation

$$(x^2 + y^2 + 1)(1 - q^2) + 2x(1 + q^2) = 0. \tag{7.14}$$

Now, substitute $q = \sqrt{\frac{t+1}{t-1}}$ and (7.14) changes to (7.12). Consequently, the Apollonius circles defined by (7.13) form a hyperbolic pencil of circles as long as $t \in \mathbb{R} \setminus \{-1, 1\}$.

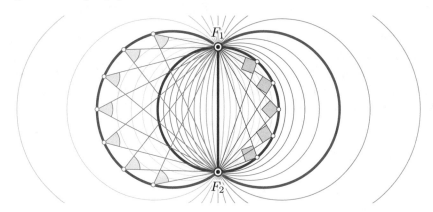

FIGURE 7.55. The family of isoptic curves of the line segment F_1F_2 consists of the circles of the elliptic pencil with base points F_1 and F_2. Any isoptic curve is a pair of circular arcs. Only in case of an optic angle equal to $\frac{\pi}{2}$ it is a unique circle: the Thales circle on F_1F_2.

Any circle of the elliptic pencil within the Apollonius circles is the locus of points where the segment F_1F_2 between the two base points is seen at a constant angle: on the arc from F_1 to F_2 at an angle of, say, φ and on the complementary arc at an angle of $\pi - \varphi$, according to the theorem of the angle of circumference. As can be seen in Figure 7.55, the *isoptic curve* of the segment $F_1 F_2$, *i.e.*, the locus of points from which the segment can be seen at constant angle φ, consists of parts of two circles and has a two-fold symmetry (see also Section 9.2).

The parabolic pencil - a common line element

The parabolic pencil of circles consists of all circles that share a line element. Though pencils of circle are objects of equiform geometry, there is only one parabolic pencil up to Euclidean motions. The parabolic pencil of circles is *self-similar*, *i.e.*, applying a similarity to the pencil may interchange the circles in the pencil but leaves the entire pencil unchanged. It is easy to check that the circles

$$l: \; x^2 - 2tx + y^2 = 0 \quad \text{with} \;\; t \in \mathbb{R}$$

share the point $O = (0,0)$ and touch the line $x = 0$ there, and thus, they form a parabolic pencil. The circles in the parabolic pencil

$$m: \; x^2 + y^2 - 2uy = 0 \quad \text{with} \;\; u \in \mathbb{R}$$

intersect all circles l at right angles, for all m touch $y = 0$ at O. Figure 7.56 shows the two parabolic pencils that form a rectangular grid of circles. We can also say: The pencil conjugate to a parabolic pencil of circles is again parabolic.

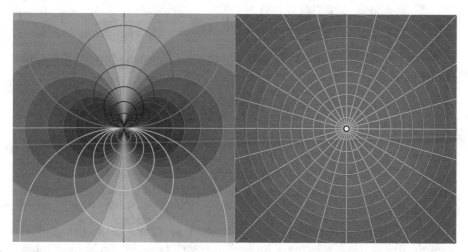

FIGURE 7.56. Left: A parabolic pencil of circles. Right: Concentric circles.

Concentric circles

Finally, the pencil of concentric circles builds an orthogonal net together
with the common diameters of all circles. This pencil is invariant under
rotations about the common center, for it is invariant under reflections
about all common diameters. On any diameter, the points of intersection
of the circles can be arranged in pairs of corresponding points in a hyperbolic involution, *i.e.*, the reflection in the common center, see Figure 7.56.

Bipolar coordinates

We can introduce curvilinear coordinates (u, v) in the plane $\mathbb{E}^2$ such that
the parameter lines u = const. and v = const. form the orthogonal net
consisting of the circles of an elliptic pencil and the conjugate hyperbolic
pencil. To this end, we choose points $F_1 = (e, 0)$ and $F_2 = (-e, 0)$, $e \in
\mathbb{R} \setminus \{0\}$, as the real base points. The circles of the two pencils satisfy the
respective equations

$$l : \ x^2 + y^2 - 2uy - e^2 = 0 \quad \text{and} \quad m : \ x^2 + y^2 - 2vx + e^2 = 0.$$

It is a matter of elementary computations to find the common points S_1
and S_2 of l and m. The coordinates of S_1 and S_2 are algebraic expressions
in terms of u and v. After a parameter substitution $(u, v) \to (U, V)$ with

$$u = \frac{e(1 - U^2)}{2U} \quad \text{and} \quad v = \frac{e(V^2 + 1)}{2V}$$

at least one of the points S_1 and S_2 can be given in terms of rational
coordinate functions as

$$\left(\frac{eV(1 + U^2)}{U^2 + V^2}, \frac{eU(V^2 - 1)}{U^2 + V^2} \right).$$

This is a rational parametrization of the Euclidean plane such that the
curves of U = const. or V = const. are circles of an elliptic and a hyperbolic
pencil forming an orthogonal net. Coordinates with these parameter lines
are called bipolar.

Usually, bipolar coordinates are expressed in terms of angles and the
computations involve hyperbolic and ordinary trigonometric functions.

7.5 Quadratic Cremona transformations

In this section, we shall deal with the *quadratic birational mappings*, frequently called CREMONA *transformations* (after ANTONIO LUIGI GAUDENZIO GIUSEPPE CREMONA, 1830–1903, Italian mathematician, structural engineer, and politician). Unlike collinear transformations, quadratic transformations are defined by homogeneous quadratic coordinate functions. Any such transformation is called *birational* if its inverse can also be given by homogeneous polynomial coordinate functions.

Quadratic birational transformations are important objects in algebraic geometry. According to EMMY AMALIE NOETHER (German mathematician, 1882–1935), any planar algebraic curve can be transformed into a planar algebraic curve with only ordinary singularities by a finite sequence of quadratic birational transformations. In this sense the quadratic transformations are the fundamental building blocks and generate the CREMONA group.

First, we classify the quadratic CREMONA transformations and study the three basic types. Then, we pay attention to some special mappings in $\mathbb{P}^2(\mathbb{R})$ such as the inversion (in conics in general and especially in circles), the transformations of doubly conjugate points or doubly conjugate lines, the transformation of conjugate normals, HIRST's inversion, and the pedal transformation.

Classification of birational quadratic mappings

Let

$$\begin{aligned} \Phi: \quad &\mathbb{P} := \mathbb{P}^2(\mathbb{F}) \to \mathbb{P}' = \mathbb{P}^2(\mathbb{F}) \text{ with} \\ &\mathbf{x}\mathbb{F} = (x_0 : x_1 : x_2) \mapsto (g_0(\mathbf{x}) : g_1(\mathbf{x}) : g_2(\mathbf{x})) \end{aligned}$$

be a rational mapping with homogeneous polynomials $g_i(x_0, x_1, x_2)$, $i = 1, 2, 3$, of degree N without any non-trivial common divisor. After excluding the points of the *exceptional set*

$$\mathcal{E} := \{\mathbf{x}\mathbb{F} \mid g_0(\mathbf{x}) = g_1(\mathbf{x}) = g_2(\mathbf{x}) = 0\},$$

i.e., the so-called *base points* or *exceptional points* of Φ, we obtain the mapping

$$\varphi: \quad \mathbb{P} \setminus \mathcal{E} \to \mathbb{P}' \text{ with } \mathbf{x}\mathbb{F} \mapsto \mathbf{x}'\mathbb{F} = (g_0(\mathbf{x}) : g_1(\mathbf{x}) : g_2(\mathbf{x})).$$

For the moment, we assume that the two involved projective planes $\mathbb{P}$ and $\mathbb{P}'$ are different, but nevertheless isomorphic, and equipped with their own coordinate frames.

The pre-image of any line l' : $u_0 x_0' + u_1 x_1' + u_2 x_2' = 0$ in $\mathbb{P}'$ with $(u_0 : u_1 : u_2) \in \mathbb{F}^3 \setminus \{(0,0,0)\}$ is the $cycle^3$ of degree N, given by

$$\varphi^{-1}(l') : \quad u_0 g_0(\mathbf{x}) + u_1 g_1(\mathbf{x}) + u_2 g_2(\mathbf{x}) = 0.$$

These cycles form a linear two-parameter family of curves, called the *net associated with* Φ. The net is spanned by the three cycles

$$g_0(\mathbf{x}) = 0, \quad g_1(\mathbf{x}) = 0, \quad g_2(\mathbf{x}) = 0,$$

where the cycle $g_i = 0$ is the pre-image of the line $x_i' = 0$ in terms of the homogeneous coordinates chosen in $\mathbb{P}'$. The base points of the net are the common points of the base curves, and therefore, they agree with base points of Φ in the exceptional set $\mathcal{E}$, as defined above.

In case of a *birational* mapping, any image point $\mathbf{x}'\mathbb{F}$ (off an exceptional set $\mathcal{E}'$ in $\mathbb{P}'$) has to have exactly one pre-image. Since $\mathbf{x}'\mathbb{F}$ can be considered as the intersection of two different lines l' and m', the two curves $\varphi^{-1}(l')$ and $\varphi^{-1}(m')$ have to intersect in precisely one point $\mathbf{x}\mathbb{F} \in \mathbb{P} \setminus \mathcal{E}$. This has to be true for any point in $\mathbb{P}' \setminus \mathcal{E}'$. It is not possible that two curves of the net are in contact since there exists one, and only one, curve of the net passing through a given line element in general position.

The *quadratic transformations* are obtained if $N = 2$. Thus, we are looking for all the nets of conics with the additional property that any two curves $\varphi^{-1}(l')$ and $\varphi^{-1}(m')$ intersect in exactly one point which is (in general) different from all base points. Furthermore, it is necessary that any point in $\mathbb{P} \setminus \mathcal{E}$ can be found as a remaining point of intersection between two net curves.

The two curves $\varphi^{-1}(l')$ and $\varphi^{-1}(m')$ span a pencil of conics. Depending on the type of pencil (cf. Section 7.3), these curves share 4, 3, 2, or 1 point. One of them is the point $\mathbf{x}\mathbb{F}$, while the remaining ones must be contained

---

[3] We differ between a cycle and a curve: An algebraic curve in the projective plane is the set of points whose homogeneous coordinates satisfy an irreducible homogeneous polynomial equation. A cycle is the union of a finite number of algebraic curves. Thus, the equation of a cycle is the product of irreducible homogeneous polynomials. Some of the factors may even have a multiplicity greater than one.

in the exceptional set $\mathcal{E}$. Since a contact of $\varphi^{-1}(l')$ and $\varphi^{-1}(m')$ at $\mathbf{x}\mathbb{F}$ is not allowed, a pencil of doubly touching conics (third kind) and a pencil of hyperosculating conics (fifth kind) cannot occur. Thus, there are three types of associated nets left, provided, the field $\mathbb{F}$ is algebraically closed, like $\mathbb{C}$:

1. The three base points P_1, P_2, P_3 are the vertices of a triangle.
2. There are two different base points P_1, P_2. The curves of the net share a line element (P_1, t_1) and the point $P_2 \notin t_1$.
3. There is only one base point P_1. The curves of the net osculate each other at P_1.

Note that in $\mathbb{P}^2(\mathbb{R})$ two of the three base points in type 1 can be complex conjugate.

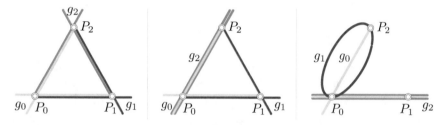

FIGURE 7.57. The three curves g_0, g_1, and g_2 span the associated net of conics. From left to right: the three types 1, 2, and 3.

Type 1: Three different base points

For the sake of simplicity, we choose the three base points of the quadratic transformation as the base points of the projective frame. Thus, $P_1 = (1 : 0 : 0)$, $P_2 = (0 : 1 : 0)$, and $P_3 = (0 : 0 : 1)$. The associated net now reads

$$g_0 : x_1 x_2 = 0, \quad g_1 : x_2 x_0 = 0, \quad g_2 : x_0 x_1 = 0.$$

Adapting the frame in $\mathbb{P}'$ such that these pairs of lines are exactly the images of the coordinate lines $x_0' = 0$, $x_1' = 0$, and $x_2' = 0$, we get the coordinate functions of the transformation Φ:

$$\Phi : (x_0 : x_1 : x_2) \mapsto (x_0' : x_1' : x_2') = (\lambda_0 x_1 x_2 : \lambda_1 x_2 x_0 : \lambda_2 x_0 x_1). \quad (7.15)$$

with coefficients $\lambda_i \in \mathbb{F} \setminus \{0\}$. Choosing the image point of $(1 : 1 : 1)$ as the unit point of the frame in $\mathbb{P}'$, we achieve $\lambda_0 = \lambda_1 = \lambda_2 = 1$.

Type 2: Two different base points

Again, we choose a proper coordinate system and let $P_1 = (1 : 0 : 0)$, $t_1 : X_2 = 0$, and $P_2 = (0 : 0 : 1)$. Then, the associated net is spanned by the curves

$$g_0 : x_1 x_2 = 0, \quad g_1 : x_0 x_2 = 0, \quad g_2 : x_1^2 = 0,$$

because these three curves pass through all the base points and touch t_1 at P_1, and they are linearly independet. (For tangents to a singular conic see page 263.)

With the suitable choice of the unit point, we arrive at the normal form for the second type of quadratic CREMONA transformation:

$$\Phi : (x_0 : x_1 : x_2) \mapsto (x_0' : x_1' : x_2') = (x_1 x_2 : x_0 x_2 : x_1^2). \tag{7.16}$$

Type 3: A single base point

We let $g_1 : x_1^2 - x_0 x_2 = 0$ be is a regular conic. It shall be osculated by all other conics of the associated net at $P_1 = (1 : 0 : 0)$. The singular conic $g_0 : x_1 x_2 = 0$ (a pair of lines) and g_1 span a pencil of osculating conics. The repeated line $g_2 : x_2^2 = 0$ and g_1 span a pencil of hyperosculating conics. Therefore, g_0, g_1, and g_2 span the associated net of the quadratic transformation with a single base point.

After choosing an appropriate unit point in $\mathbb{P}'$, the coordinate functions of the transformation read

$$\Phi : (x_0 : x_1 : x_2) \mapsto (x_0' : x_1' : x_2') = (x_1 x_2 : x_1^2 - x_0 x_2 : x_2^2). \tag{7.17}$$

The conics and pairs of lines, labeled with g_0, g_1, and g_2 which span the associated net of conics for all the three types of birational quadratic mappings, are illustrated in Figure 7.57.

We can summarize in the following theorem:

Theorem 7.5.1 *In projective planes $\mathbb{P}^2(\mathbb{F})$ over an algebraically closed field $\mathbb{F}$ there are three types of quadratic CREMONA transformations $\Phi : \mathbb{P} \to \mathbb{P}'$. Provided an appropriate choice of coordinate frames in both projective planes $\mathbb{P}$ and $\mathbb{P}'$, the coordinate functions of the three types are given by (7.15) with $\lambda_0 = \lambda_1 = \lambda_2 = 1$, (7.16), and (7.17). The respective* *inverse transformations read*

$$
\begin{aligned}
\varphi^{-1}: \ (x_0' : x_1' : x_2') &\mapsto (x_0 : x_1 : x_2) = (x_1'x_2' : x_2'x_0' : x_0'x_1'), \\
\varphi^{-1}: \ (x_0' : x_1' : x_2') &\mapsto (x_0 : x_1 : x_2) = (x_1'x_2' : x_0'x_2' : {x_0'}^2), \\
\varphi^{-1}: \ (x_0' : x_1' : x_2') &\mapsto (x_0 : x_1 : x_2) = ({x_0'}^2 - x_1'x_2' : x_0'x_2' : {x_2'}^2).
\end{aligned}
\tag{7.18}
$$

Proof: We have to show that the so far necessary conditions are sufficient.

In the case of type 1, we use $x_0'x_0 = x_0x_1x_2 = x_1'x_1 = x_2'x_2$ and find the system

$$
\begin{aligned}
x_0'x_0 \ \ -x_1'x_1 \ \ \ \ \ \ \ \ &= 0, \\
x_0'x_0 \ \ \ \ \ \ \ \ \ \ -x_2'x_2 &= 0
\end{aligned}
$$

of linear equations which can be solved for $(x_0 : x_1 : x_2)$.

A similar procedure for type 2, yields $x_0'x_0 = x_0x_1x_2 = x_1'x_1$ and $x_0'x_1 = x_1^2x_2 = x_2'x_2$, and consequently, the system of linear equations

$$
\begin{aligned}
x_0'x_0 \ \ -x_1'x_1 \ \ \ \ \ \ \ \ &= 0, \\
x_0'x_1 \ \ \ \ \ \ \ \ \ \ -x_2'x_2 &= 0.
\end{aligned}
$$

This system of equations can also be solved for (x_0, x_1, x_2).

In case of type 3, we have $x_0'x_2 = x_1x_2^2 = x_2'x_1$ and $x_0'x_1 - x_s'x_0 = x_1^2x_2 - x_2^2x_0 = (x_1^2 - x_0x_2)x_2 = x_1'x_2$. Thus, we find the system linear equations

$$
\begin{aligned}
-x_2'x_1 \ \ +x_0'x_2 \ \ \ \ \ \ \ \ &= 0, \\
x_2'x_0 \ \ -x_0'x_1 \ \ +x_1'x_2 &= 0.
\end{aligned}
$$

The solutions of the latter system are the coordinate functions of the inverse mapping. ∎

As an immediate consequence of the proof of Theorem 7.5.1, we have:

Theorem 7.5.2 *Each of the birational quadratic transformations listed in Theorem 7.5.1 can be determined by two bilinear forms, depending on the type of transformation:*

- *Type 1:* $\quad x_0x_0' - x_1x_1' = x_0'x_0 - x_2x_2' = 0,$
- *Type 2:* $\quad x_0x_0' - x_1x_1' = x_1'x_0 - x_2x_2' = 0,$
- *Type 3:* $\quad x_2x_0' - x_1x_2' = x_0x_2' - x_1x_0' + x_2x_1' = 0.$

Proof: We have to differ between the three types of transformations. The coordinate frames in $\mathbb{P}$ and $\mathbb{P}'$ are $(P_0, P_1, P_2; P)$ and $(P_0', P_1', P_2'; P')$, where P and P' are corresponding under φ.

- Type 1: The bilinear form $x_1'x_1 - x_2'x_2 = 0$ is equivalent to $(x_1' : x_2') = (x_2 : x_1)$. This is the analytic representation of a projective mapping π_0 from the pencil of lines about P_0 to the pencil of lines about P_0' (see Theorem 5.4.1). The mapping π_0 acts as follows:

$$
\pi_0: \ \mathcal{L}_{P_0} \barwedge \mathcal{L}_{P_0'} \quad \text{with} \quad
\begin{cases}
l_2 = [P_0, P_1]: \ x_2 = 0 \mapsto \pi_0(l_2) = [P_0', P_2']: \ x_1' = 0, \\
l_1 = [P_0, P_2]: \ x_1 = 0 \mapsto \pi_0(l_1) = [P_0', P_1']: \ x_2' = 0.
\end{cases}
$$

We can do the same with $x_0 x_0' - x_2 x_2' = 0$ and find $(x_0' : x_2') = (x_2 : x_0)$ which gives the analytic representation of a projective mapping β from the pencil about P_1 to the pencil about P_1':

$$\pi_1 : \mathcal{L}_{P_1} \barwedge \mathcal{L}_{P_1'} \quad \text{with} \quad \begin{cases} m_1 = [P_1, P_0] : x_2 = 0 \mapsto \pi_1(m_1) = [P_1', P_2'] : x_0' = 0, \\ m_2 = [P_1, P_2] : x_0 = 0 \mapsto \pi_1(m_2) = [P_1', P_0'] : x_2' = 0. \end{cases}$$

Finally, there exists a third projective mapping π_2 from the pencil about P_2 to the pencil about P_2' with

$$[P_2, P_0] \mapsto [P_2', P_1'], \quad [P_2, P_1] \mapsto [P_2', P_0'].$$

The points $X \notin \mathcal{E}$ can be transformed by mapping $[P_0, X]$ and $[P_1, X]$ to their respective images through P_0' and P_1'. The image point is $X' = \pi_0([P_0, X]) \cap \pi_1([P_1, X])$. Obviously, X' is also incident with $\pi_2([P_2, X])$.

Note the difference between the constructions of the image point under a birational quadratic mapping and under a collinear transformation. In case of a birational quadratic mapping of type 1, the line $[P_0, P_1]$ is not mapped to $[P_0', P_1']$, neither under π_0 nor under π_1.

All points $X \in [P_0, P_1] \setminus \{P_0, P_1\}$ are mapped to the same point $P_2' = \pi_0(l_1) \cap \pi_1(m_1)$. Thus, P_2' is an exceptional point for φ^{-1}, since its pre-image is not uniquely determined. With $\widetilde{\mathcal{E}} = [P_0, P_1] \cup [P_1, P_2] \cup [P_2, P_0]$ and $\widetilde{\mathcal{E}}' = [P_0', P_1'] \cup [P_1', P_2'] \cup [P_2', P_0']$, the restriction $\widetilde{\varphi}$ of φ as a mapping $\mathbb{P} \setminus \widetilde{\mathcal{E}} \to \mathbb{P}' \setminus \widetilde{\mathcal{E}}'$ is one-to-one and onto. It can be written in terms of homogeneous coordinates as

$$\widetilde{\varphi} : \mathbb{P} \setminus \widetilde{\mathcal{E}} \to \mathbb{P}' \setminus \widetilde{\mathcal{E}}', \quad (x_0 : x_1 : x_2) \mapsto (x_0' : x_1' : x_2') = \left(\frac{1}{x_0} : \frac{1}{x_1} : \frac{1}{x_2} \right),$$

since $x_0 x_1 x_2 \neq 0$ for any $\mathbf{x}\mathbb{F} \notin \widetilde{\mathcal{E}}$.

- Type 2: In this case there are only two projective mappings: π_0 from the pencil about P_0 to the pencil about P_0', and π_1 from the pencil about P_1 to the pencil about P_1'. The analytic representations of π_0 and π_1 read

$$\pi_0 : (x_0' : x_2') = (x_2 : x_1) \quad \text{and} \quad \pi_1 : (x_0' : x_1') = (x_1 : x_0).$$

Again, $\pi_0([P_0, P_1]) \neq [P_0', P_1']$ holds true, since $x_1 = 0$ causes $x_2' = 0$. Unlike in the previous case, $\pi_1([P_1, P_0]) = [P_0', P_1']$, because $x_1 = 0$ implies $x_0' = 0$.

- Type 3: In this case, there is only one projective mapping π_0 left, acting from the pencil about P_0 to the pencil about P_0'. Its analytical representation also follows form the coordinate functions of the birational quadratic mapping and reads

$$\pi_0 : (x_0' : x_2') = (x_1 : x_2).$$

The second bilinear form expresses the fact that the φ-image $\mathbf{x}'\mathbb{F}$ of the point $\mathbf{x}\mathbb{F}$ is a point on the polar line of $\mathbf{x}\mathbb{F}$ with regard to the correlation

$$\kappa : \begin{pmatrix} u_0' \\ u_1' \\ u_2' \end{pmatrix} = \begin{pmatrix} 0 & -1 & 0 \\ 0 & 0 & 1 \\ 1 & 0 & 0 \end{pmatrix} \begin{pmatrix} x_0 \\ x_1 \\ x_2 \end{pmatrix}$$

since

$$u_0' x_0' + u_1' x_1' + u_2' x_2' = -x_1 x_0' + x_2 x_1' + x_0 x_2' = 0.$$

With a cyclic shift of the coordinates $(y_0' : y_1' : y_2') := (x_2' : x_0' : x_1')$, the latter two bilinear forms (Theorem 7.5.2) change to

$$x_2 y_1' - x_1 y_0' = x_0 y_0' - x_1 y_1' + x_2 y_2' = 0.$$

The correlation κ is now replaced by

$$
\tilde{\kappa} : \begin{pmatrix} v'_0 \\ v'_1 \\ v'_2 \end{pmatrix} = \begin{pmatrix} 1 & 0 & 0 \\ 0 & -1 & 0 \\ 0 & 0 & 1 \end{pmatrix} \begin{pmatrix} x_0 \\ x_1 \\ x_2 \end{pmatrix} .
$$

Now, the coordinate matrix of $\tilde{\kappa}$ is symmetric. Identifying the coordinate frame of the pre-image with the new coordinate frame in the image plane, then $\tilde{\kappa}$ is the polarity w.r.t. the conic $p: \; {y'_0}^2 - {y'_1}^2 + {y'_2}^2 = 0$. The points $\varphi(X)$ is the intersection of $\pi_0([P_0, X])$ with the polar line of X with regard to p.

■

The following theorem will be useful, when we deal with special quadratic transformations. A proof can be found in books [13, 16]:

Theorem 7.5.3 *Under a birational quadratic transformation of type 1 any algebraic curve c of degree n is transformed to an algebraic curve of degree 2n, in general.*

If the exceptional points P_0, P_1, P_2 of the transformation are points of c with respective multiplicities $\mu(P_i) = m_i$ for $i = 0, 1, 2$, then the degree of the (proper) image curve equals $2n - m_0 - m_1 - m_2$. The multiplicities of the exceptional points P'_0, P'_1, P'_2 on c' in the image plane $\mathbb{P}'$ equal $n - m_1 - m_2$, $n - m_2 - m_0$, and $n - m_0 - m_1$, respectively (Figure 7.58). Moreover, if $R \neq P_1, P_2$ is a common point of c and $[P_1, P_2]$ and R is r-fold on c, then the line $\pi_0([P_0, R])$ is an r-fold tangent of c' at P'_0.

The contents of Theorem 7.5.3 will be of importance when we deal with special quadratic transformations, *e.g.*, the inversion which maps circles to circles.

Similar results hold for the transformations of type 2 and 3. For more details, we refer again to [13, 16].

In the beginning of the present section, we defined quadratic transformations by prescribing three homogeneous quadratic polynomials as the coordinate functions of the transformation. The geometric approach via pencils of conics showed that (from the viewpoint of Projective Geometry) there are only three types of birational quadratic transformation. However, is not possible to prescribe three arbitrarily chosen quadratic forms in order to define a quadratic birational transformation. The coefficients of the three underlying forms have to fulfill certain algebraic

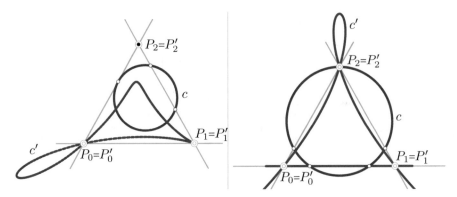

FIGURE 7.58. Quadratic Cremona transformation $\Phi : \mathbb{P} \to \mathbb{P}' = \mathbb{P}$ with canonical identification of base points. Left: A conic c touches the base line $[P_0, P_2]$, meets the line $[P_1, P_2]$ in two real points, and has no real point on $[P_0, P_1]$. Thus, the image curve c' has a cusp of the first kind at $P_1' = P_1$, a double point with two real tangents at $P_0' = P_0$, and an isolated double point at $P_2' = P_2$. Right: The conic c goes through the base point P_2. Thus, the image curve splits into a cubic c', being the proper image, and the line $[P_0', P_1'] = [P_0, P_1]$. Since c meets $[P_0, P_1]$ in two real points, c' has a double point with two real tangents at $P_2' = P_2$.

conditions[4] in order to make the transformation invertible and, thus, birational.

Special quadratic Cremona transformations

Transformation of doubly conjugate points

Assume we are given two different curves of degree two,

$$c : \mathbf{x}^\mathrm{T}\mathbf{A}\mathbf{x} = 0 \quad \text{and} \quad d : \mathbf{x}^\mathrm{T}\mathbf{B}\mathbf{x} = 0,$$

in any projective plane $\mathbb{P}^2(\mathbb{F})$. Herein, $\mathbf{A} = (a_{ik})$ and $\mathbf{B} = (b_{ik})$ with $i, k \in \{0, 1, 2\}$ are symmetric matrices in $\mathbb{F}^{3 \times 3}$. Let further $\mathcal{P}$ denote the pencil spanned by these two curves. The pencil should contain conics in order to exclude cases where both c and d are singular and share a line. We are going to show that the pairs (P, P') of conjugate points w.r.t. both curves c and d, are related in a birational quadratic transformation Φ depending on the type of the pencil. So these pairs of points are *doubly*

---

[4]The algebraic conditions on the coefficients on the quadratic forms are derived in: K. FLADT: *Die Umkehrungen der ebenen quadratischen Cremona Transformationen*. J. reine u. angew. Math. **170** (1933), 64–68.

conjugate. The resulting transformation Φ is involutive, *i.e.*, $\varphi = \varphi^{-1}$, since conjugacy is a symmetric relation (cf. Theorem 7.1.2).

Two points X and X' are conjugate w.r.t. c and d if their homogeneous coordinates $(x_0 : x_1 : x_2)$ and $(x_0' : x_1' : x_2')$ fulfill the two bilinear equations

$$\sum_{i,k=0}^{2} a_{ik}x_i x_k' = 0 \quad \text{and} \quad \sum_{i,k=0}^{2} b_{ik}x_i x_k' = 0.$$

They can be considered as linear homogeneous equations in the unknowns $(x_0' : x_1' : x_2')$ which yields

$$\sum_{i=0}^{2} a_{i0}x_i x_0' + \sum_{i=0}^{2} a_{i1}x_i x_1' + \sum_{i=0}^{2} a_{i2}x_i x_2' = 0,$$

$$\sum_{j=0}^{2} b_{j0}x_j x_0' + \sum_{j=0}^{2} b_{j1}x_j x_1' + \sum_{j=0}^{2} b_{j2}x_j x_2' = 0.$$

The solutions are, up to a common factor,

$$x_0' = \sum_{i,j=0}^{2} (a_{i1}b_{j2} - a_{i2}b_{j1})x_i x_j,$$

$$x_1' = \sum_{i,j=0}^{2} (a_{i2}b_{j0} - a_{i0}b_{j2})x_i x_j, \tag{7.19}$$

$$x_2' = \sum_{i,j=0}^{2} (a_{i0}b_{j1} - a_{i1}b_{j0})x_i x_j.$$

Obviously, we have found a rational mapping $\varphi : (x_0 : x_1 : x_2) \mapsto (x_0' : x_1' : x_2')$. It is of degree two, provided that the quadratic forms on the right-hand side of (7.19) do not have a non-constant common divisor. By construction, φ is involutive, and therefore, it is birational.

The mapping φ does not change if we replace the curves c and d with other curves from the pencil $\mathcal{P}$ because the conditions for conjugacy are linear combinations of the homogeneous equations above.

In order to make Φ a birational quadratic transformation, the existence of one, two, or three base points is necessary. These are points where Φ does not produce an unique image point, *i.e.*, in this case we are looking for points whose polar lines w.r.t. c and d agree or are even undetermined.

If $\mathcal{P}$ is a pencil of the third kind (two line elements) or a pencil of the fifth kind (pencil of hyperosculating conics), then there are infinitely many such exceptional points. In the first case, these are the points on the chord common to all conics in the pencil. In the second case, the points on the common tangent at the hyperosculation point play this particular role. It is easy to show that the quadratic forms on the right-hand side of (7.19) have a common linear factor if, *e.g.*, c is a repeated line. In this case, Φ becomes a collineation.

Consequently, the only cases that are left, originate from a pencil $\mathcal{P}$ of the first, second, or fourth kind. According to Theorem 7.5.1, the transformation Φ is then of type 1, 2, or 3. This type of quadratic CREMONA transformation is called *transformation of doubly conjugate points*.

The fixed points of this mapping are the self-conjugate points of c and d, *i.e.*, the base points of the pencil $\mathcal{P}$. The projective mappings π_0, π_1, π_2 in the pencils about the base points are involutive and their fixed lines pass through these fixed points.

Now, we can state:

Theorem 7.5.4 *Let c and d be two curves of degree two out of a pencil $\mathcal{P}$ in $\mathbb{P}^2(\mathbb{F})$ which is either of the first, or the second, or the fourth kind.*

1. *The transformation Φ of doubly conjugate points with respect to c and d is an involutive quadratic birational transformation of type 1, or 2, or 3 if the pencil $\mathcal{P}$ is of type 1, or 2, or 4. The mapping Φ remains unchanged if c and d are replaced by other curves from $\mathcal{P}$.*

2. *The exceptional points of Φ and its inverse are exactly those points whose polar lines with regard to c and d coincide. In the case of a pencil of the first kind, these are the vertices of the common polar triangle. The fixed points of Φ are the base points of the pencil $\mathcal{P}$. In the case of a quadratic transformation of the first type, the projective mappings in the pencils about the exceptional points are involutive projectivities whose fixed lines are the lines joining base points of Φ with the base points of $\mathcal{P}$.*

3. *The poles of lines $l \not\ni P_0, P_1, P_2$ with regard to all conics in $\mathcal{P}$ are contained in a conic l' that passes through all exceptional points of Φ. If $\mathcal{P}$ is of the first kind with base points B_1, B_2, B_3, and B_4, then l' contains on each line $[B_i, B_j]$ the fourth harmonic point L'_{ij} to $L_{ij} := [B_i, B_j] \cap l$ with respect to B_i and B_j, i.e., $H(B_i, B_j, L_{ij}, L'_{ij})$.*

Proof: The only thing that is left to show is the third part of the theorem. The locus of all image points P' of points $P \in l$ is an irreducible curve of degree two (cf. Theorem 7.5.3). The image l' passes through all exceptional points. On the other hand, the image point P' can be found as the intersection of the polar lines of P with regard to c and d which form pencils with the poles L_c and L_d of l w.r.t. c and d as carriers, when P traverses l. Thus, l' is generated by a projective mapping from the pencil L_c to the pencil L_d,

$$L_c(\pi_c(P)) \barwedge l(P) \barwedge L_d(\pi_d(P)),$$

and l' is passing through the pencils' vertices. Here and in the following, π_c and π_d denote the polarities with regard to c and d. Since any two different curves from $\mathcal{P}$ yield the same quadratic Cremona transformation, and thus, the same image curve l', it has to carry the poles of l with regard to all conics in the pencil. This is what has been stated.

If P is a point on a fixed line of the involution π_0, e.g., $P \in [B_1, B_2]$, then the conjugate point P' which lies on the polars $\pi_c(P)$ and $\pi_d(P)$, is the fourth harmonic point of P with respect to $[B_1, B_2]$, since $B_1, B_2 \in c, d$.   ∎

It is worth to be noted that in the special case of the projectively closed Euclidean plane, the line l can be chosen as the line at infinity and the common polar triangle can be chosen as a finite triangle. Then, l' carries the centers of all conics in the pencil $\mathcal{P}$, the vertices of the diagonal triangle of the base points of $\mathcal{P}$, and furthermore the midpoints of all sides of the base quadrangle of $\mathcal{P}$. This conic l' is called the *nine point conic*. This conic appears also in Section 9.1, cf. Corollary 9.1.1.

Doubly conjugate lines

The principle of duality (see Section 5.1, especially on page 182) guarantees that the results from this section so far have valid dual counter parts. The dual version of the transformation of doubly conjugate points is called the *transformation of doubly conjugate lines*. We denote this transformation by $\Phi^\star$.

$\Phi^\star$ acts on the set of line in $\mathbb{P}^2(\mathbb{F})$ as follows: Let $c^\star$ and $d^\star$ be two (dual) conics from a flock $\mathcal{F}$ of the first, or the second, or the fourth kind. Pairs (l, l') of lines are doubly conjugate if

$$l' \in \pi_{c^\star}(l) \text{ and } l' \in \pi_{d^\star}(l).$$

Again, $\pi_{c^\star}$ and $\pi_{d^\star}$ denote the polar systems with regard to $c^\star$ and $d^\star$. Moreover, the lines l and l' are doubly conjugate with respect to any two different curves out of $\mathcal{F}$.

The exceptional lines of $\Phi^\star$ are those lines whose poles with regard to $c^\star$ and $d^\star$ coincide. In the case of a flock of the first kind, these are the lines p_0, p_1, p_2 of the polar triangle common to all conics of $\mathcal{F}$. The fixed lines of $\Phi^\star$ are the base lines of the flock $\mathcal{F}$.

The points of intersection of corresponding lines l and l' with the exceptional lines are corresponding in involutive projective mappings π_0, π_1, π_2 acting on the base lines. The fixed points of these involutions are incident with the fixed lines of Φ^*, *i.e.*, they are the points of intersection of tangents common to c^* and d^*. Any two out of the three involutions are sufficient to determine the quadratic transformation Φ^*.

A pencil of lines about P_0 on the exceptional line p_0 is mapped to a pencil of lines about $\pi_0(P_0) \in p_0$, since the pencil that corresponds to p_0 with vertex $p_1 \cap p_2$ splits off. The points $p_0 \cap p_1$ and $p_0 \cap p_2$ are corresponding in this involution π_0, as well. On the other hand, pencils of lines whose vertices are not in $p_0 \cup p_1 \cup p_2$ are transformed to irreducible curves of class two containing the lines p_0, p_1, p_2.

Orthogonal conjugate lines

In the projectively extended Euclidean plane, we can study a special case of doubly conjugate lines. Assume that c^* and d^* are a pair of confocal conics with a center C. Thus, the four common tangents t_1, t_2, t_3, t_4 of c^* and d^* are isotropic lines, *i.e.*, two of them pass through $A_1 = (0 : 1 : i)$ while the others pass through $A_2 = (0 : 1 : -i)$. The three diagonal lines p_0, p_1, p_2 of the base quadrilateral, *i.e.*, the axes of c^* and d^* together with the line at infinity, form the exceptional set $\mathcal{E}^* = p_0 \cup p_1 \cup p_2$ of Φ^*.

The quadratic Cremona transformation Φ^* defined by c^* and d^* induces an involution π_2 on the ideal line $\omega : x_0 = 0$ leaving the absolute points A_1 and A_2 of Euclidean geometry fixed (cf. Example 6.4.7 on page 253). Therefore, any line $l \notin \mathcal{E}^*$ is orthogonal to its image $l' = \Phi^*(l)$. Φ^* is called *transformation of conjugate normals* with respect to any conic in the flock $\mathcal{E}$ of confocal conics spanned by c^* and d^*. Note that one conic is sufficient to determine this flock.

The involutions π_0 and π_1 on the axes are called the *focal involutions*. In the case of a conic c with center C, in both focal involutions the center C corresponds to the ideal point. The fixed points of the focal involutions π_1 and π_2 are the *focal points* of c and agree with those from Definition 7.1.4 in Section 7.1.

Each vertex V of c is mapped to its center of curvature V^* under the focal involution. This holds true for ellipses as well as hyperbolas as can be seen in Figure 7.59: In the case of the ellipse (left-hand side), we apply Φ^* to the line $[V_1, V_2]$ joining both vertices. The conjugate normal passes

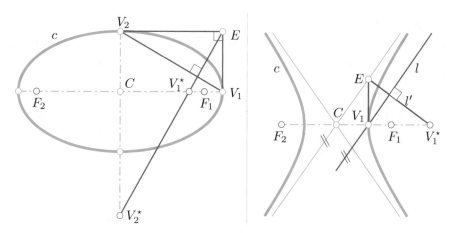

FIGURE 7.59. Centers of curvature at vertices: ellipse (left), hyperbola (right).

through E (pole of $[V_1, V_2]$ w.r.t. c) and is orthogonal to $[V_1, V_2]$. We know this construction from Corollary 3.2.1 (Figure 3.11, Section 3.2), where the proof was based on the comparison of similar right-angled triangles.

The construction of the centers of curvature at the vertices of the hyperbola shown on the right-hand side in Figure 7.59 is equivalent to the construction from Exercise 3.2.6 (cf. Figure 3.14). If $\Phi^\star$ is defined by a parabola, then the transformation of conjugate normals is a quadratic Cremona transformation of type 2. The focal involution on the parabola's axis is the reflection in the focal point.

There is a deeper reason for the fact that the focal involutions send vertices to the correspond centers of curvature, as shown below.

Conjugate normals and the evolute of a conic

The transformation $\Phi^\star$ of conjugate normals with respect to the conic c maps the set of tangents of c to the set of normals. From Section 3.2, we know that the envelope $c^\star$ of normals of a curve is its evolute. According to Theorem 7.5.3, the evolute of a conic is a curve of class four. This mirrors the results from Section 3.2. The axes of c and the ideal line ω are double tangents of the evolute. The vertices of c and the contact points with $c^\star$, $i.e.$, the centers of curvature at the vertices, are corresponding in the related focal involutions.

Evolutes of ellipses, parabolas, and hyperbolas are displayed in Figures 3.15, 3.18, and 3.20.

In the case of an ellipse and a hyperbola, we have learned in Exercise 3.2.8 that the construction of centers of curvature is equivalent to the transfer of an affine ratio which is based on (3.29). Since the quadratic transformation $\Phi^\star$ of conjugate normals preserves cross-ratios, on the tangent t_P the cross-ratio of $P \in c$ together with the points on the exceptional lines is equal to the cross-ratio of the respective image points on the image line n_P. One of the four points in question on t_P and its corresponding point on n_P are at infinity. Hence, by virtue of (5.8), the cross-ratio reduces to an affine ratio

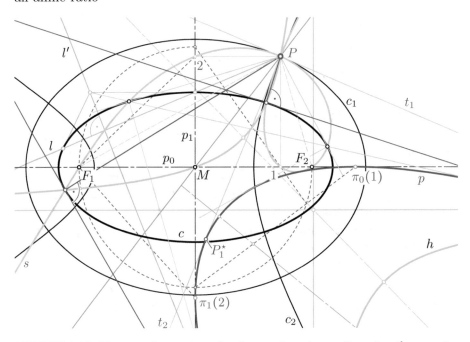

FIGURE 7.60. The transformation of orthogonal conjugate lines $l \mapsto l'$ w.r.t. the conic c maps the pencil P onto tangents of CHASLES's parabola p. The polarity in c sends p onto the Apollonius hyperbola h which intersects c at the pedal points w.r.t. P. These pedal points belong also to the strophoid s which is the pedal curve of p w.r.t. P.

■ **Example 7.5.1** Chasles's parabola and the Apollonian hyperbola. The transformation $\Phi^\star$ of conjugate normals w.r.t. the ellipse or hyperbola c maps the pencil of lines through a point P outside the exceptional lines onto the tangents of a conic p, which contacts all three exceptional lines (Figure 7.60). Therefore, p is a parabola which is named after the French mathematician MICHEL CHASLES, 1793–1880. On each axis, the contact point with p corresponds to the pedal point w.r.t. P in the related focal involution. The parabola p remains the same when c

is replaced by any other conic from the confocal family, e.g., by the ellipse c_1 and the hyperbola c_2 passing through P. Therefore p contacts the corresponding tangent lines t_1 and t_2 which bisect the angle $\sphericalangle F_1 P F_2$. The points of contact of these lines with p are the centers $P_1^\star, P_2^\star$ of curvature of P w.r.t. c_1 and c_2. It can be shown that the center $P_1^\star$ of curvature of c_1 is the pole of the tangent t_1 to c_1 with respect to c_2 (Figure 7.60), and *vice versa*.

The directrix of p (orthoptic curve of p) passes through P and the center M of c. Common tangents of p and c contact c at the pedal points of normals drawn from P onto the conic c. When p is polarized in c, we obtain an equilateral hyperbola $h = \pi_c(p)$ which contains P, the center M, the ideal points of the axes, and the at most four pedal points of normals on c w.r.t. P. It is the *Apollonian hyperbola* which we will meet again in Section 9.3. The pedal points in question are also located on the strophoid s which already has been depicted in Figure 2.27.

Inversions

In this section, we shall look at inversions from the projective point of view. This gives a more general concept and shows that the inversion from page 7.1.1 is just a special case and its properties are clear and obvious.

Projective Inversions

Let c be a conic in $\mathbb{P}^2(\mathbb{F})$ and let further $C \in \mathbb{P}^2(\mathbb{F})$ be some point. Two points X and X' in $\mathbb{P}^2(\mathbb{F})$ are said to be *inverse* if

$$
\begin{aligned}
&(1) \quad X, X', C \quad \text{are collinear and}\\
&(2) \quad X, X' \quad \text{are conjugate with regard to } c.
\end{aligned}
\tag{7.20}
$$

The mapping $X \mapsto X'$ is involutive, provided there exists a unique image point X' of X. We call such a mapping a *(projective) inversion*. The point C is called the *center of the inversion*, and the points of c remain fixed. Without loss of generality, we may assume that

$$c: \ x_0 x_2 - x_1^2 = 0$$

and

$$
\begin{aligned}
C &= P_1 = (0:1:0) \text{ if } C \notin c, \text{ and}\\
C &= P_0 = (1:0:0) \text{ if } C \in c.
\end{aligned}
$$

Then, in both cases we have two bilinear conditions. The first condition equals the polar form of c

$$x_0 x_2' + x_2 x_0' - 2 x_1 x_1' = 0,$$

and, depending on whether $C \notin c$ or $C \in c$, the second means

$$
\begin{aligned}
(x_0 : x_2) = (x_0' : x_2') &\iff x_0 x_2' - x_2 x_0' = 0 \quad \text{if } C \notin c,\\
(x_1 : x_2) = (x_1' : x_2') &\iff x_1 x_2' - x_2 x_1' = 0 \quad \text{if } C \in c.
\end{aligned}
$$

Solving the systems of linear equations yields the coordinate functions of the respective transformations

$$(x_0' : x_1' : x_1') = (x_0 x_1 : x_0 x_2 : x_1 x_2) \quad \text{if} \quad C \notin c,$$
$$(x_0' : x_1' : x_1') = (2x_1^2 - x_0 x_2 : x_1 x_2 : x_2^2) \quad \text{if} \quad C \in c.$$

Hence, this transformation is involutive, and therefore, birational.

Under $C \notin c$ the exceptional points of the inversion are $P_1 = C$ and the points P_0, P_2 of contact of the tangents drawn from C to c if they exist.

The projective mapping $\pi_0 : P_1 \barwedge P_1$ is the identity mapping. From the bilinear forms we can deduce the analytic description $(x_0' : x_1') = (x_2 : x_1)$ of the projective mapping $\pi_1 : P_0 \barwedge P_2$, and we have $\pi_2 = \pi_1^{-1}$. At this transformation, two base points are interchanged: $P_0 \mapsto P_2$ and $P_2 \mapsto P_0$. The conic c is generated by π_1 (or π_2) according to Definition 6.1.1, since all the points of c are fixed.

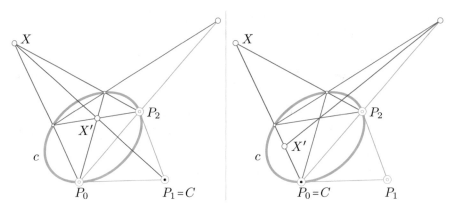

FIGURE 7.61. Projective inversions with the center C, either not on c (left) or on c (right).

We construct the image X' of X using the quadrangle on c which contains the points P_0 and P_2 and has X as a diagonal point (Figure 7.61). Since one of the remaining diagonal points lies on $[P_0, P_2]$ which is the polar line of P_1 w.r.t. c, the opposite side of the diagonal triangle passes through P_1 and, consequently, the third diagonal point coincides with X'.

If $C \in c$, then C is the only exceptional point and π_0 is the identity mapping. Therefore, this involutive quadratic transformation is of type 3.

Summarizing this, we can say:

Theorem 7.5.5 *The projective inversion in c with center C is an involutive quadratic (birational) transformation either of type 1 or of type 3, depending on whether $C \notin c$ or $C \in c$.*

In the first case, the base triangle consists of the center C of inversion and the points T_1, T_2 of contact of the two tangents drawn from C to c. The projective mapping in the pencil about C is the identity mapping. The conic c is generated by the projectivities $\pi_1 = \pi_2^{-1} : T_1 \barwedge T_2$. Any point on $c \setminus \{T_1, T_2\}$ is left fixed.

In the case $C \in c$, C is the only base point.

Inversion in a circle

Assume that c is a circle, and the center C of inversion coincides with the circle's center. Then the projective inversion in c with center C is called the *inversion* or *reflection in* the circle c which we have seen in Example 7.1.1 on page 267. The exceptional points of the inversion are C and the absolute points $I = (0 : 1 : i)$ and $\overline{I} = (0 : 1 : -i)$ of Euclidean geometry. If the radius of c equals r, then any pair (P, P') of inverse points satisfies

$$\overline{CP} \cdot \overline{CP'} = r^2.$$

According to Theorem 7.5.3, any line l (not incident with a base point) is mapped under the inversion to an irreducible curve l' of degree two passing through all three exceptional base points C, I, and $\overline{I}$. Thus, l' is a circle through C (see Figure 7.62).

A curve k of degree two that does not contain any of the base points is mapped to a quartic curve k' with double points at the base points C, I, and $\overline{I}$. Curves with double points at the absolute points of Euclidean geometry are called *bicircular*. If k is a circle with $C \notin k$, then the isotropic lines $[C, I]$ and $[C, \overline{I}]$ split off, and the proper image curve k' is a circle. Therefore, the inversion maps circles to circles, provided that lines are counted as circles. The center of a circle is in general not mapped to the center of the image circle.

■ **Example 7.5.2** Center of the inverse circle.

1. In order to derive the coordinate representation of the inversion in terms of Cartesian coordinates, we assume $c : x^2 + y^2 = r^2$ and $C = (0,0)$. Let further $X = (\xi, \eta) \neq (0,0)$ be an arbitrary point. Then, the polar lines of X with regard to c has the equation $p_X : x\xi + y\eta = r^2$, and the line $[X, C]$ has the equation $x\eta - y\xi = 0$. Computing $[X, C] \cap p_X$,

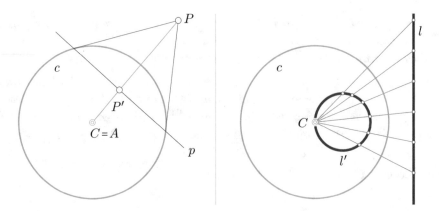

FIGURE 7.62. Left: The inversion in a circle c with center C (of the inversion) maps a point P to P' on P's polar line with regard to c. Right: A straight line l is mapped to a circle l' through C.

we obtain the coordinate functions of the inversion as

$$(\xi, \eta) \mapsto \left(\frac{r^2 \xi}{\xi^2 + \eta^2}, \frac{r^2 \eta}{\xi^2 + \eta^2} \right). \tag{7.21}$$

2. The image of a circle k in general position is a circle k', as we have deduced above. But where is the center of k'? We may assume that $k : (x - d)^2 + y^2 = R^2$ is the circle to be inverted in c. This means no restriction because the coordinate system can always be rotated about C such that the center of k lies on the x-axis. Inserting the coordinate functions of the inversion (and using x and y instead of ξ and η), we find

$$r^2 - 2dr^2 x + (d^2 - R^2)(x^2 + y^2) = 0$$

which is the equation of the image circle. The center can be found by completing to full squares and equals

$$\left(\frac{dr^2}{d^2 - R^2}, 0 \right).$$

Figure 7.63 shows that there is a simple construction of the center of the inverse circle.

Circles that intersect c at right angles are fixed as a whole, but not point-wise since in this case r^2 equals the power of C with respect to l.

Quadratic Cremona transformations (indeed any rational mapping) preserve the contact order of curves, provided, the contact takes place outside the exceptional set. To be precise: If the multiplicity $\mu_P(c, d)$ of two curves c, d at such a common point P is m, then the intersection multiplicity $\mu_{P'}(c', d')$ at P' of the transformed curves c' and d' also equals m. In the special case $m = 2$ we obtain that curves in contact are mapped to curves in contact. Therefore, these mappings are so-called *contact transformations*. If a curve $c \in \mathbb{P} \setminus \mathcal{E}$ and a circle (or line) l are in second or third

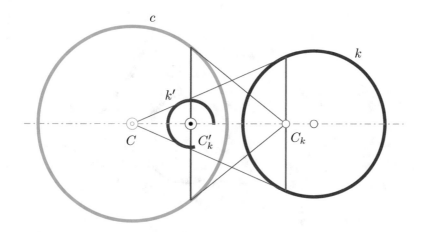

FIGURE 7.63. How to find the center C'_k of the inverse k' of a circle k? First, invert C in k, and second, invert the obtained point C_k in c.

order contact at P, i.e., $m = 3$ or $m = 4$, then c' and l' are in second or third order contact at P', and l' is again a circle or line. Therefore, the inversion maps osculating circles of c to such of c'. Moreover, vertices of c, i.e., points with stationary curvature, are mapped to vertices of c'.

The inversion in a circle is conformal but not orientation preserving. The inversion in c (with center C) maps the tangent t to a curve k at P to a circle t' that touches the image curve k' at P'. The two angles that are enclosed by t' and $[C, P]$ have equal measures but different orientations and the angle at C is congruent to the angle $\sphericalangle\,(k, [C, P]$ at P (Figure 7.64). We can summarize the results of this subsection:

Theorem 7.5.6 *In $\mathbb{E}^2$, the inversion in a circle preserves circles and is orientation reversing conformal at any point different from the center of inversion. Osculating circles are again mapped onto osculating circles, and vertices remain vertices.*

The inversion can also be performed when c is an empty circle with the equation $x^2 + y^2 = -r^2$. In this case, the inversion in c equals the product of the inversion in the circle $\widetilde{c}:\ x^2 + y^2 = r^2$ and the reflection in the center of c and $\widetilde{c}$.

■ **Example 7.5.3** Inversion and the complex plane. The Cartesian coordinates (x, y) in the Euclidean plane can be identified with complex numbers z by setting $z = x + iy$. We replace

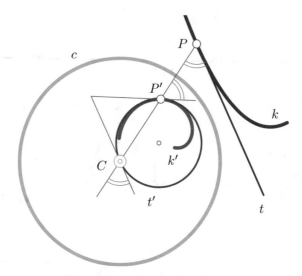

FIGURE 7.64. The inversion maps tangent lines to contacting circles. Angles are preserved, but their orientation is changed.

in (7.21) ξ by x and η by y and rewrite the coordinate functions of the inversion in a circle in terms of complex numbers as

$$\frac{x}{x^2 + y^2} + i\,\frac{y}{x^2 + y^2} = \frac{x + iy}{x^2 + y^2} = \frac{z}{z\overline{z}} = \frac{1}{\overline{z}}$$

with $\overline{z}$ being the complex conjugate number of z. The function $f(z) = \overline{z}^{-1}$ is anti-holomorphic, and thus, conformal but orientation reversing.

The function $f(z) = \overline{z}^{-1}$ is a special form of a more general linear rational mapping $g(z): \mathbb{C} \to \mathbb{C}$ which can also be viewed as projective mapping of the line $\mathbb{P}^1(\mathbb{C})$ onto itself since

$$g(z) = \frac{\alpha\overline{z} + \beta}{\gamma\overline{z} + \delta} \quad \Longleftrightarrow \quad g(z) = \begin{pmatrix} \alpha & \beta \\ \gamma & \delta \end{pmatrix} \begin{pmatrix} 1 \\ \overline{z} \end{pmatrix} \tag{7.22}$$

with $(\alpha, \beta, \gamma, \delta) \in \mathbb{C}^4 \setminus \{(0,0,0,0)\}$ and $\alpha\delta - \beta\gamma \neq 0$. The choice $\alpha : \beta : \gamma : \delta = 0 : 1 : 1 : 0$ yields the inversion in a circle given in (7.21). The complex conjugation in (7.22) is an automorphism of the field $\mathbb{C}$, and thus, $g(z)$ is an *anti-projectivity*.

The set of all $g(z)$ with all their compositions constitutes the three-parametric *Möbius group*.

● **Exercise 7.5.1** Hirst's inversion - inversion in a pair of lines. So far we have dealt with inversions in regular conics. However, we can replace the conic by a pair of lines $c = l_1 \cup l_2$ such that the center C of the inversion is not incident with any of the lines.

The mapping η which is called HIRST's inversion[5] after the British geometer THOMAS ARCHER HIRST (1830–1892), sends a point $X \notin \{c \cup C\}$ to a point X' collinear with C and X and satisfies $H(A_1, A_2, X, X')$, where the points A_1, A_2 are the intersections of $[C, X]$ with l_1, l_2.

---

[5]T.A. HIRST: *On the Quadric Inversion of Plane Curves.* Proc. R. Soc. of London, Phil. Trans. R. Soc. **14**, 91–106 (1865).

A projective frame can always be chosen such that $l_1 :\ x_0 - x_1 = 0$, $l_2\ :\ x_0 + x_1 = 0$, and $C = (1 : 0 : 0)$ hold. Derive the coordinate functions of the mapping η. Which type of quadratic Cremona transformation do we have? Assume further that we work in the projectively extended Euclidean plane and $x_0 = 0$ is the line at infinity. Which conics that are mapped to conics of the same affine type?

What happens if (l_1, l_2) is replaced by a pair (m_1, m_2) of complex conjugate lines, say $m_1 :\ x_0 - ix_1 = 0$ and $m_2 :\ x_0 + ix_1 = 0$?

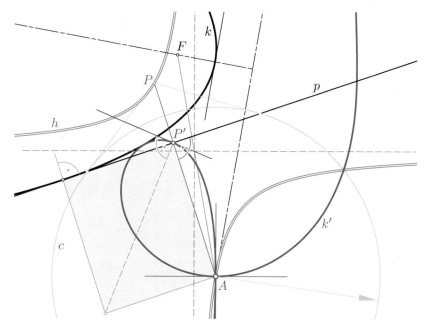

FIGURE 7.65. The pedal curve k' of the parabola k w.r.t. a point A of its directrix is a strophoid. The equilateral hyperbola h is polar to k w.r.t. the circle c centered at A, and therefore, invers to k' w.r.t. c.

Pedal transformation

Let A be any proper point in the Euclidean plane. The *pedal transformation* assigns to each line l the pedal point P' of l w.r.t A (compare with Remark 2.2.1 on page 34). The pedal transformation is a quadratic line-to-point transformation. It is the composition of a polarity and the inversion in a circle c centered at A (see Figure 7.62 left).

The *pedal curve* k' of a given curve k also admits a kinematic generation: One leg of a moving right angle remains tangent to k while the other leg is gliding through the fixed point A. Therefore, the tangent to k' at P' is a path tangent and, as such, orthogonal to the line joining P' with the instantaneous pole [66, 21]. Conversely, the curve k is called the *negative pedal curve* of k'.

Let k be a curve of class n which does neither contain the ideal line nor touch the isotropic lines through A. The dual curve c^* with respect to c is of degree n and does not contain any of the exceptional points of the inversion in c. Therefore, and according to Theorem 7.5.3, the pedal curve k' of k with respect to A is of degree $2n$ with n-fold points at the absolute points of Euclidean geometry (Figure 7.65).

8 Affine Geometry

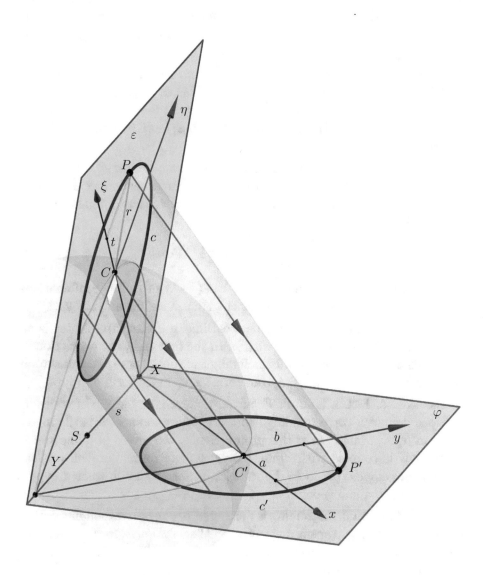

An ellipse can be the image of a circle under a parallel projection. The terms ellipse, parabola, hyperbola are typical notions of affine geometry.

In any projective plane $\mathbb{P}^2$, we can specify a single line ω and call it the *ideal line*. Points on ω are called *ideal points*, lines $l, m \neq \omega$ which are concurrent with ω are called *parallel*. The set of *proper points*, *i.e.*, the points not on ω, together with the set of lines $\neq \omega$ is called an *affine plane*.

If at any coordinatization of $\mathbb{P}^2(\mathbb{F})$ the line ω is represented by $x_0 = 0$, then for proper points we may set $x_0 = 1$ and switch to inhomogeneous coordinates $(x, y) \in \mathbb{F}^2$ obeying $(1 : x : y) = (x_0 : x_1 : x_2)$. Such a coordinate frame in the corresponding affine plane $\mathbb{A}^2(\mathbb{F})$ is uniquely determined by the origin $O = (0, 0)$ and the unit points $E_1 = (1, 0)$ and $E_2 = (0, 1)$, provided that OE_1E_2 is a triangle.

In $\mathbb{P}^2$, any collineation which fixes ω can be restricted to $\mathbb{A}^2$. Then, it is called an *affine transformation*. Projective collineations induce projective affine transformations, or *affinities* in brief. In $\mathbb{A}^2(\mathbb{F})$, affinities can be expressed as

$$\alpha : \ \mathbf{x} \mapsto \mathbf{x}' = \mathbf{A}\mathbf{x} + \mathbf{a} \quad \text{with } \mathbf{a}, \mathbf{x}, \mathbf{x}' \in \mathbb{F}^2, \ \mathbf{A} \in \mathbb{F}^{2 \times 2}, \ \det \mathbf{A} \neq 0.$$

Affinities map affine combinations $\mathbf{p} = \lambda \mathbf{x} + (1 - \lambda)\mathbf{y}$ to affine combinations $\alpha(\mathbf{p}) = \lambda \alpha(\mathbf{x}) + (1 - \lambda)\alpha(\mathbf{y})$ without changing $\lambda \in \mathbb{F}$ for all $\mathbf{p}, \mathbf{x}, \mathbf{y} \in \mathbb{F}^2$.

A perspective collineation fixes the ideal line ω if either its center is an ideal point or its axis coincides with ω. In the first case, the restriction to $\mathbb{A}^2$ is called a *perspective affinity* including the special case of a translation if the axis equals ω. The affine transformations in $\mathbb{A}^2$ form a group.

In the sense of FELIX KLEIN's Erlangen program, any group of transformations defines a corresponding geometry whose main goal is the investigation of invariants w.r.t. this group. Hence, the group of affine transformations is coupled with *affine geometry*. The terms parallel, convex, affine ratio, ellipse, parabola, hyperbola, center of an ellipse or a hyperbola belong to affine geometry.

If, additionally, an *orthogonality* is defined in $\mathbb{A}^2$, the corresponding orthogonality preserving transformations are called *equiform transformations*. The equiform transformations also constitute a group and the corresponding geometry is called *equiform geometry*.

Euclidean planes are equiform planes equipped with a *distance function* such that PYTHAGORAS's theorem holds.

Below, we provide selected affine properties of conics, often mixed with Euclidean properties.

We do not report about the various applications and chracterizations of ellipses in the theory of convexity. The interested reader is referred to [47].

8.1 Conjugate diameters of ellipses and hyperbolas

Ellipses as affine images of circles - de La Hire's construction

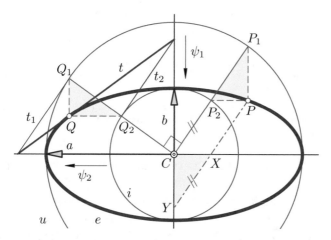

FIGURE 8.1. De La Hire's construction of an ellipse.

We recall from Figure 2.6 in Section 2.1 the elegant and simple construction of points and tangents of an ellipse which is ascribed to the French mathematician PHILIPPE DE LA HIRE. The construction of points of the ellipse e starts with the circumcircle u and incircle i of the ellipse, *i.e.*, the concentric circles with the semimajor and semiminor axes (lengths) for their radii (Figure 8.1).

We consider two perspective affinities ψ_1 and ψ_2 with the axes of the ellipse e for their axes that, respectively, map the circumcircle u and the incircle i to e. Thus, in both cases the pencil of fixed lines is orthogonal to the axis of the affine mapping.

The mapping $\psi_2 : i \to e$ maps any point $Q_2 \in i$ to a point $Q \in e$. The tangent t_2 of i at Q_2 is mapped to $t \ni Q$. The second affine mapping $\psi_1^{-1} : e \to u$ maps $Q \in e$ to a point $Q_1 \in u$. The tangent t at Q is mapped to the tangent t_1 of u at Q_1. The product $\alpha := \psi_1 \circ \psi_2^{-1}$ of these two affine mappings is still an affine mapping with the center C of e as a fixed point. Further, $\alpha(Q_1) = Q_2$ and α is central similarity. This shows that the construction by DE LA HIRE (as illustrated in Figure 8.1) yields points and tangents of an ellipse:

Draw the two concentric circles $u = (C; a)$ and $i = (C; i)$ with the semi-major axis a and semiminor axis b for their radii. A common radius of either circle meets u and i in points Q_1 and Q_2. The parallels through these points to the axes of the ellipse meet in a point $Q \in c$.

■ Example 8.1.1 Show that the perspective and affine mappings

$$\psi_1 : \ x' = x, \quad y' = \frac{b}{a}y \quad \text{and} \quad \psi_2 : x' = \frac{a}{b}x, y' = y$$

are the two mappings used in the above explanation of DE LA HIRE's construction. Show further that the circles u and i are mapped to e by deriving the equation of e. This proves the following statement:

Let $k \in \mathbb{R}^+$ be a constant scaling factor. Scale the segments on all chords perpendicular to a diameter of a circle while fixing their midpoints on the diameter. The endpoints of the scaled segments form an ellipse.

In Figure 8.1, we observe that the two right triangles QQ_1Q_2 and PP_2P_1 are congruent. If we draw the line parallel to $[C, P_1]$ through P, a further congruent right triangle appears: The parallel line meets the axes of the ellipse in X and Y. The triangle CXY is congruent to PP_2P_1. Obviously, $\overline{CP_1} = \overline{PY} = a$ and $\overline{P_1P_2} = \overline{XY} = a - b$, and therefore, $\overline{XP} = b$. This yields another construction of the points of an ellipse which is frequently called the *trammel construction*:

Corollary 8.1.1 *Move a line $[X, Y]$ such that two points X and Y trace a pair of orthogonal lines. Any point P fixed on the line $[X, Y]$ with $P \neq X, Y$ traces an ellipse.*

For the kinematic aspect of this construction we refer to Section 2.3.

Parallel projections of a circle

Let $c = (C; r)$ be a circle in an arbitrary plane ε in Euclidean three-space $\mathbb{E}^3$. Assume that $\pi : \ \mathbb{E}^3 \to \varphi$ is a parallel projection from the Euclidean space $\mathbb{R}^3$ onto some plane φ. The fibers of the projection are parallel to some line f which itself is neither parallel to ε nor to φ (Figure 8.2).

Parallel projections preserve affine ratios, and thus, the π-image $c' = \pi(c)$ of the circle c is centrally symmetric with respect to $C' = \pi(C)$. In order to make sure that c' is an ellipse, we consider the plane σ of symmetry of

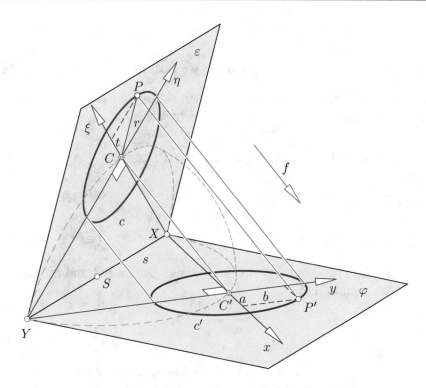

FIGURE 8.2. The image of a circle c under a parallel projection $\pi : \mathbb{E}^3 \to \varphi$ is an ellipse e.

the segment CC'. It meets the line $s := \varepsilon \cap \varphi$ in a point S. (In the unlikely case that $s \subset \sigma$ we are allowed to choose S freely on s.) Obviously, the point S satisfies $\overline{CS} = \overline{C'S}$. There are precisely two points X and Y on s, with $\overline{XS} = \overline{YS} = \overline{CS}$, with $[X, C] \perp [Y, C]$, and $[X, C'] \perp [Y, C']$ since C and C' lie on a sphere centered at S passing through X and Y. (If $\sigma \parallel s$, then the sphere degenerates into a plane and the points S and X become the ideal points of s, while Y becomes the common foot of the normals from C and C' onto s.)

Now, we fix a Cartesian coordinate system centered at C with axes ξ and η through X and Y. Then, an arbitrary point $P \in c$ has coordinates

$$\xi = r \cos t, \quad \eta = r \sin t \tag{8.1}$$

where $t = \sphericalangle \xi C P$. The image C' of C under π can be described in Cartesian coordinates in the image plane φ: The Cartesian system shall be centered at C' and the x- and y-axes shall run through X and Y, respectively. Therefore, we have

$$x = \alpha\xi, \quad y = \beta\eta. \tag{8.2}$$

The factors α and β are those distorsion ratios that are to be multiplied with all ξ- and η-coordinates, respectively. We have

$$\alpha = \frac{\overline{XC'}}{\overline{XC}} \quad \text{and} \quad \beta = \frac{\overline{YC'}}{\overline{YC}}. \tag{8.3}$$

From (8.1) and (8.2) we find

$$x = a\cos t, \quad y = b\sin t \quad \text{where} \quad a = \alpha r, \quad b = \beta r. \tag{8.4}$$

From the first two equations in (8.4) we can eliminate t and arrive at the Cartesian equation of c':

$$\frac{x^2}{a^2} + \frac{y^2}{b^2} = 1.$$

From Example 7.1.1 in Section 7.1 we know that any pair of conjugate diameters of a circle is a pair of orthogonal diameters, and *vice versa*. The tangents at the endpoints of one diameter are parallel to the conjugate diameter. The images of any such pair under a parallel projection will, in general, not be a pair of orthogonal lines any more. Since parallelity is preserved under π, the circumscribed square of tangents at the endpoints of conjugate diameters of c will be mapped to a circumscribed parallelogram of tangents of c'.

Let now an arbitrary pair of orthogonal diameters be the pair of axes of a Cartesian coordinate system. Consequently, c can be parametrized via (8.1). The corresponding frame $[x, y]$ in the image plane φ will not be Cartesian any more. However, it will be a general affine frame wherein the ellipse c' still can be given by (8.3). Now, a and b denote half of the lengths of a pair of conjugate diameters as seen in Figure 8.3. On the other hand, any two segments in φ, say $C'A'$ and $C'B'$, emanating from a point can always be considered as the images of a pair of orthogonal semidiameters MA and MB under a parallel projection. Therefore, in any affine coordinate frame, $x = a\cos t$, $y = b\sin t$ parametrizes an ellipse. So we can say:

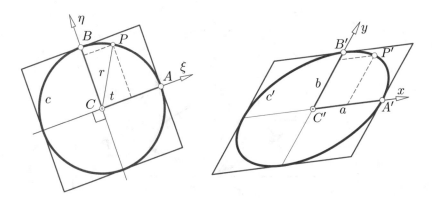

FIGURE 8.3. Left: The undistorted version of a circle shows a pair of conjugate and orthogonal diameters together with the circumscribed square of tangents at the respective endpoints. Right: The image of the circle under a parallel projection still shows a pair of conjugate diameters which in general is not a pair of orthogonal diameters. The square of tangents is mapped to some parallelogram of tangents.

Theorem 8.1.1 *The image of an ellipse in* $\mathbb{E}^3$ *under a parallel projection* $\pi : \mathbb{E}^3 \to \varphi$ *into some plane* φ *is an ellipse, provided that the fibers of* π *are not parallel to the carrier plane of the ellipse.*

It is sufficient to use a pair of conjugate diameters of the given ellipse c as the axes ξ and η of an affine coordinate system. With

$$\xi = a \cos t, \quad \eta = b \sin t \tag{8.5}$$

and (8.2), we obtain a parametrization of the image curve $c' = \pi(c)$.

■ **Example 8.1.2** Prove that for all pairs of conjugate semidiameters CP and CQ of an ellipse the sum $\overline{CP}^2 + \overline{CQ}^2$ of squared lengths is the same.

Ellipse on a pair of conjugate diameters - Rytz's construction

In Constructive Geometry, there are many cases where an ellipse is to be constructed from a pair of conjugate diameters. This means that a pair (CP, CQ) of conjugate semidiameters of an ellipse is given and we have to find the axes as well as the semiaxis lengths of the thus defined ellipse. Very often a construction is used that is due to the Swiss mathematician DAVID RYTZ VON BRUGG (1801–1868).

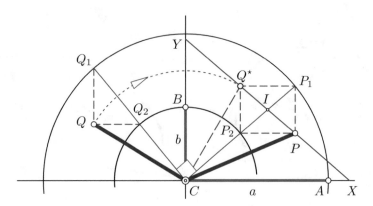

FIGURE 8.4. Points of an ellipse constructed out of the incircle and the circumcircle.

Let us have a look at Figure 8.4: Like in Figure 8.1 we can see that there are two congruent right triangles QQ_1Q_2 and PP_2P_1. The diameters $[C,Q]$ and $[C,P]$ of the ellipse with center C, principal axis $[C,A]$, and auxiliary axis $[C,B]$ form a conjugate pair since $[C,Q_1]\perp[C,P_1]$. If we apply a quarter turn about C in clockwise direction to the triangle QQ_1Q_2, we see $Q_1 \mapsto P_1$, $Q \mapsto Q^\star$, and $Q_2 \mapsto P_2$. Thus, we obtain a rectangle $Q^\star P_1PP_2$ whose sides are parallel to the ellipse's axes. The diagonal $[P,Q^\star]$ meets the axes in the points X and Y and encloses the same angles with the axes as $[P_1,P_2]$ does. Consequently, we have $\overline{YP} = \overline{CP_1} = a$ and $\overline{XP} = \overline{CP_2} = b$ and also for the center I of the rectangle we have $\overline{IX} = \overline{IC} = \overline{IY}$. This justifies the construction given by RYTZ as shown in Figure 8.5.

Note that the result does not depend on the direction in which the point Q is rotated about C through $90°$.

● **Exercise 8.1.1** Conjugate diameters and focal points of an ellipse. Let the center of an ellipse be the origin of Cartesian coordinates (x,y) which are combined to complex numbers $z = x + iy$. Suppose that $p,q \in \mathbb{C}$ are the complex coordinates of the endpoints P and Q of a pair of conjugate semidiameters. Prove that $\pm\sqrt{p^2 + q^2}$ are the complex coordinates of the two real focal points of the given ellipse.

A one-parameter family of triangles of equal area

Affine transformations map hyperbolas to hyperbolas (Figure 8.6). Let $P = (\xi, \eta)$ be a point on a hyperbola h with semimajor axis length a and semiminor axis length b as shown in Figure 8.7. First we observe:

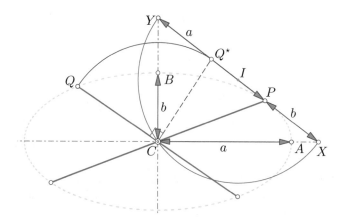

FIGURE 8.5. RYTZ's construction of the axes and vertices of an ellipse on a given pair of conjugate semidiameters CP and CQ.

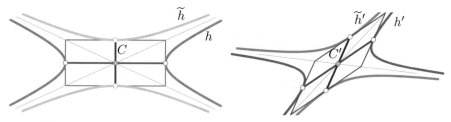

FIGURE 8.6. The image of a pair $(h, \widetilde{h})$ of conjugate hyperbolas under an affine transformation is again a pair $(h', \widetilde{h}')$ of conjugate hyperbolas.

Lemma 8.1.2 *The point P of contact of a hyperbola h and its tangent t at P is the midpoint of the segment $T_1 T_2$ on t between the two asymptotes.*

Proof: The hyperbola's diameter parallel to t meets t in the common ideal point T_u. This diameter is conjugate to the diameter P. Thus, P is the fourth harmonic point to T_u with respect to intersections T_1 and T_2 of t and h's asymptotes (Figure 8.7). ∎

The locus of midpoints of segments on chords parallel to a diameter of a hyperbola (indeed of any conic with center) is the conjugate diameter.

Figure 8.7 (right) shows two triangles, each bounded by a tangent of h and the asymptotes. As can be seen at once, the left (orange) triangle has the area $A = ab$ since the base equals twice the semiminor axis length b and and the altitude through h's center C equals the semimajor axis length a. Surprisingly, we have:

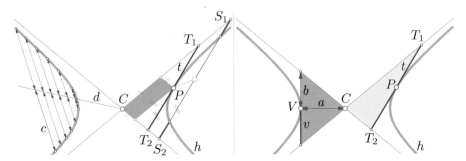

FIGURE 8.7. Left: The point P of contact of the tangent t and the hyperbola h is the midpoint of the segment $T_1 T_2$. Any chord of h determines equally long segments between h and the asymptotes. The midpoints of interior segments on parallel chords gather on a diameter. Right: The area of the triangle built by a tangent t of h and the asymptotes is constant and independent of t.

Theorem 8.1.2 *The area of the triangle built by the asymptotes and a tangent of a hyperbola h at any point $P \in h$ is independent of the choice of P and equals the product of the semiaxis lengths, i.e., $A = ab$.*

Proof: Let P be a generic point on h with the tangent t. Assume further that t meets the asymptotes in the points T_1 and T_2.

From the right-hand side of Figure 8.7 and from Lemma 8.1.2 we can conclude that projections of P parallel to one asymptote onto the other are the midpoints of the segments CT_1 and CT_2 where C is the center. Thus, there is a parallelogram (salmon parallelogram in Figure 8.7, left) defined by P. Its area is half the area of the triangle enclosed by t and the asymptotes.

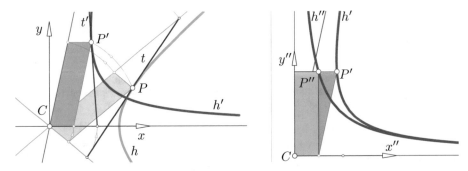

FIGURE 8.8. Rotation (left) and shearing (right) for the proof of Theorem 8.1.2.

Now, we have to find the area of the parallelogram. For that purpose, we rotate the hyperbola h about C through an angle φ such that one asymptote is mapped to the x-axis (see Figure 8.8, left). The angle φ is defined by $\cos\varphi = a/\sqrt{a^2 + b^2}$. The equation $b^2 x^2 - a^2 y^2 = a^2 b^2$ of h changes to $-2abxy - (a^2 - b^2)y^2 = a^2 b^2$. Then, we apply a shear transformation that preserves areas by substituting $x = x' + cy'$ and $y = y'$. The constant c is to be determined such that the coefficient of y^2 vanishes (Figure 8.8, right). This yields $c = (b^2 - a^2)/2ab$ and, once again, the equation of the hyperbola simplifies to $ab + 2xy = 0$ while the area of the parallelogram does not change. Since x and y are the edge lengths of the rectangle (still a parallelogram), the area of the parallelogram equals $\frac{1}{2}ab$, and thus, the area of the triangle between t and the asymptotes is constant and equals ab.　　　　　　　　　　　　　　　　■

Pairs of conjugate hyperbolas

Any hyperbola h with semimajor axis a_h and semiminor axis b_h determines a *conjugate hyperbola* k. The major axis of h equals the minor axis of k; the minor axis of h equals the major axis of k. The same holds true for the semiaxis's lengths: $a_h = b_k$, $b_h = a_k$, see Figure 8.9. The hyperbolas h and k share the asymptotes. From Theorem 8.1.2, we can deduce:

Theorem 8.1.3 *Let h and k be a pair of conjugate hyperbolas with common center C. For $P \in h$ let $[C,Q]$ be the diameter of h conjugate to $[C,P]$ such that $Q \in k$. The tangents of h parallel to the tangent at P together with the tangents of k parallel to the tangent at Q form a parallelogram whith the area $A = a_h b_h = a_k b_k$ independent of the choice of P.*

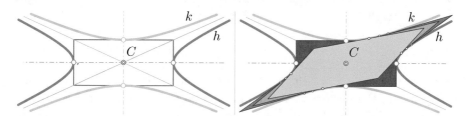

FIGURE 8.9. Left: A pair (h,k) of conjugate hyperbolas. Right: The parallelograms built by tangents of h and k have fixed area.

● **Exercise 8.1.2** Asymptotes and axes of hyperbolas. The two hyperbolas

$$h : \frac{x^2}{a^2} - \frac{y^2}{b^2} = 1 \quad \text{and} \quad \widetilde{h} : \frac{x^2}{a^2} - \frac{y^2}{b^2} = -1$$

share the asymptotes and the involutions of conjugate diameters. Suppose that $P \in h$ and $Q \in \widetilde{h}$ are the given endpoints of conjugate semidiameters. How to find the asymptotes and axes of h and $\widetilde{h}$ (note Figure 8.6)?

● **Exercise 8.1.3** Focal points of a hyperbola. Let $p, q \in \mathbb{C}$ be the complex coordinates of the endpoints P and Q of conjugate semidiameters of a hyperbola h as explained in Exercise 8.1.2 with $P \in h$. Verify that $\pm\sqrt{p^2 + q^2}$ are the complex coordinates of the focal points of h.

■ **Example 8.1.3** Center of a conic. Assume we are given a conic c with the equation

$$c: \ \mathbf{x}^\mathrm{T}\mathbf{A}\mathbf{x} = 0 \tag{8.6}$$

with $\mathbf{A} \in \mathbb{R}^{3\times3}$ being a regular symmetric matrix with entries from the field $\mathbb{R}$ of real numbers and $\mathbf{x} = (x_0, x_1, x_2)$ being homogeneous coordinates in the real projective plane. According to Section 7.1, the dual conic $c^\star$, i.e., the set of tangents of c is given by the equation

$$c^\star: \ \mathbf{u}^\mathrm{T}\mathbf{A}^{-1}\mathbf{u} = 0 \tag{8.7}$$

where $\mathbf{u} = (u_0, u_1, u_2)$ are homogeneous coordinates of c's tangents or of the "points" of $c^\star$. Therefore, the center of c equals $\mathbf{m} = \mathbf{A}^{-1}(1, 0, 0)^\mathrm{T}$. The existence of the inverse $\mathbf{A}^{-1}$ of $\mathbf{A}$ is guaranteed for $\mathbf{A}$ is regular. If $a_{ij} \in \mathbb{R}$ with $i, j \in \{0, 1, 2\}$ are the entries of $\mathbf{A}$, then

$$\mathbf{m} = (a_{11}a_{22} - a_{12}^2, a_{02}a_{12} - a_{01}a_{22}, a_{01}a_{12} - a_{02}a_{11}). \tag{8.8}$$

The point $M = \mathbf{m}\mathbb{R}$ from equation (8.8) is a proper point if, and only if, its first coordinate does not vanish, i.e.,

$$a_{11}a_{22} - a_{12}^2 = \det\begin{pmatrix} a_{11} & a_{12} \\ a_{12} & a_{22} \end{pmatrix} \neq 0.$$

The matrix

$$\overline{\mathbf{A}} := \begin{pmatrix} a_{11} & a_{12} \\ a_{12} & a_{22} \end{pmatrix} \tag{8.9}$$

is the right lower 2×2-submatrix of $\mathbf{A}$ from (8.6).

■ **Example 8.1.4** Axes are conjugate orthogonal diameters of a conic with a center. Given the conic $c: \ \mathbf{x}^\mathrm{T}\mathbf{A}\mathbf{x} = 0$ in homogeneous Cartesian coordinates with $\mathbf{A} = \mathbf{A}^\mathrm{T}$ in $\mathbb{P}^2(\mathbb{F})$. Compute the axes of c as the orthogonal conjugate diameters of c, provided c is no parabola.

Principal axes transform within the affine setting

Now we want to treat conics in the affine plane $\mathbb{A}^2(\mathbb{F})$ in terms of affine coordinates. We start with the homogeneous equation $x: \ \overline{\mathbf{x}}^\mathrm{T}\,\overline{\mathbf{A}}\overline{\mathbf{x}} = 0$ with $\overline{\mathbf{x}} = (x_0, x_1, x_2)$ and $\overline{\mathbf{A}} \in \mathbb{F}^{3\times3}$. We switch to affine coordinates be setting $(x_0 : x_1 : x_2) = (1 : x : y)$. Thus, we obtain

$$c: \ a_{11}x^2 + 2a_{12}xy + a_{22}y^2 + 2a_{01}x + 2a_{02}y + a_{00} = 0 \tag{8.10}$$

with coefficients $a_{ij} \in \mathbb{F}$ and $i, j \in \{0, 1, 2\}$. This time, we allow $\overline{\mathbf{A}} \in \mathbb{F}^{3\times3}$ to be singular. Let the vector $\mathbf{x}$ stand for $\mathbf{x} = (x, y) \in \mathbb{F}^2$.

Now the equation of c can also be written as the sum of a quadratic form in x and y, a linear from in x and y, and a constant, as

$$\mathbf{x}^\mathrm{T}\mathbf{A}\mathbf{x} + 2\mathbf{a}^\mathrm{T}\mathbf{x} + a_{00} = 0 \tag{8.11}$$

with the vector $\mathbf{a} = (a_{01}, a_{02})$ and the matrix $\mathbf{A} = \begin{pmatrix} a_{11} & a_{12} \\ a_{12} & a_{22} \end{pmatrix} \in \mathbb{F}^{2\times2}$.

In a first step, we apply a translation to c such that the linear form in (8.11) becomes the zero-form. For that purpose we let

$$\mathbf{x} = \mathbf{x}' + \mathbf{m} \tag{8.12}$$

with $\mathbf{m} \in \mathbb{R}^2$ to be determined. We insert (8.12) into (8.11) and find

$$\mathbf{x}'^{\mathrm{T}} \mathbf{A} \mathbf{x}' + 2(\mathbf{m}^{\mathrm{T}} \mathbf{A} + \mathbf{a}^{\mathrm{T}}) \mathbf{x}' + \mathbf{m}^{\mathrm{T}} \mathbf{A} \mathbf{m} + 2\mathbf{a}^{\mathrm{T}} \mathbf{m} + a_{00} = 0, \tag{8.13}$$

since $\mathbf{m}^{\mathrm{T}} \mathbf{A} + \mathbf{a}^{\mathrm{T}} = (\mathbf{A}^{\mathrm{T}} \mathbf{m} + \mathbf{a})^{\mathrm{T}}$ and $\mathbf{A}^{\mathrm{T}} = \mathbf{A}$. Only in the cases where $\mathbf{A}$ is regular, the vector $\mathbf{m}$ can be chosen such that the linear form in (8.13) vanishes. At this point we have to distinguish between two cases: $\mathbf{A}$ is either regular or singular:

1. $\mathbf{A}$ is regular. Then, we have

$$\mathbf{m} = -\mathbf{A}^{-1}\mathbf{a} \tag{8.14}$$

which represents the center. (8.13) changes to

$$F(\mathbf{x}) = \mathbf{x}'^{\mathrm{T}} \mathbf{A} \mathbf{x}' + \alpha = 0 \tag{8.15}$$

where $\alpha = \mathbf{a}^{\mathrm{T}} \mathbf{A}^{-1} \mathbf{a} - 2\mathbf{a}^{\mathrm{T}} \mathbf{A}^{-1} \mathbf{a} + a_{00}$. It remains to discuss the cases $\alpha = 0$ or $\alpha = -1$ since otherwise we can divide the equation by $-\alpha$.

Though $\mathbf{A} \in \mathbb{F}^{2 \times 2}$ is symmetric, it will in general not be a diagonal matrix. However, it is easy to diagonalize $\mathbf{A}$ by an appropriate choice of our affine coordinate frame. We only have to pay attention to the fact that $\mathbf{A}$ defines an involution of conjugate diameters as well as that of conjugate ideal points. Let $E_1 = \mathbf{e}_1 = (1,0)$ and $E_2 = \mathbf{e}_2 = (0,1)$ be the unit points. Then, we can express the entries of $\mathbf{A}$ as

$$a_{11} = \mathbf{e}_1^{\mathrm{T}} \mathbf{A} \mathbf{e}_1, \quad a_{12} = \mathbf{e}_1^{\mathrm{T}} \mathbf{A} \mathbf{e}_2, \quad a_{21} = \mathbf{e}_2^{\mathrm{T}} \mathbf{A} \mathbf{e}_1, \quad a_{22} = \mathbf{e}_2^{\mathrm{T}} \mathbf{A} \mathbf{e}_2.$$

This reveals that $\mathbf{A}$ has diagonal form if, and only if, the base vectors point into conjugate directions. So, we can choose one basis vector $\mathbf{e}_1'$ such that $a_{11}' = \mathbf{e}_1'^{\mathrm{T}} \mathbf{A} \mathbf{e}_1' \neq 0$. Then, we have $\mathbf{e}_1'^{\mathrm{T}} \mathbf{A} \mathbf{e}_2' = 0$ for the second basis vector. If $\mathbb{F} = \mathbb{R}$, we can still scale $\mathbf{e}_1'$ and $\mathbf{e}_2'$ in an appropriate way. In the case $a_{11}' > 0$ we let $\mathbf{e}_1'' = \frac{1}{\sqrt{a_{11}'}} \mathbf{e}_1'$, otherwise $\mathbf{e}_1'' = \frac{1}{\sqrt{-a_{11}'}} \mathbf{e}_1'$. Thus, the new entry a_{11}'' can be $+1$ or -1. Similarly, we obtain $a_{22}'' = +1$, or $a_{22}'' = -1$, or $a_{22}'' = 0$. The latter case only arises if $\mathbf{A}$ is singular.

2. In the case $\det \mathbf{A} = 0$, we have no solution for $\mathbf{m}$. But after changing to the new basis $\{\mathbf{e}_1', \mathbf{e}_2'\}$ with $\mathbf{e}_2'^{\mathrm{T}} \mathbf{A} \mathbf{e}_2' = 0$, we get $(x + a_{01})^2 + 2a_{02}y + a_{00} - a_{01}^2 = 0$. We set $x' = x + a_{01}$, and under the assumption $a_{02} \neq 0$, we rescale $\mathbf{e}_2$ in order to obtain $a_{02} = 1$, and eliminate the constant by a translation along the y-axis.

If $a_{02} = 0$, the rescaling of $\mathbf{e}_1$ reduces the constant either to 0, -1, $+1$.

Theorem 8.1.4 *In the real affine plane $\mathbb{A}^2(\mathbb{R})$ there is an affine coordinate frame such that the equation of any regular or singular conic can be reduced to one of the following standard equations:*

$$x^2 + y^2 - 1 = 0, \quad x^2 - y^2 - 1 = 0, \quad x^2 + y^2 + 1 = 0, \quad x^2 + 2y = 0,$$
$$x^2 + y^2 = 0, \quad x^2 - y^2 = 0, \quad x^2 + 1 = 0, \quad x^2 - 1 = 0, \quad x^2 = 0.$$

• **Exercise 8.1.4** Center of a conic and the critical points of quadratic functions.

Assume $f(x, y) = a_{11}x^2 + 2a_{12}xy + a_{22}y^2 + 2a_{01}x + 2a_{02}y + a_{00}$ is a quadratic function over $\mathbb{R}^2$. Show that the maxima/minima of f are found at the zeros of the gradient $\mathrm{grad} f = 2\overline{\mathbf{A}}\mathbf{x} + 2\mathbf{a}$, or equivalently, at $\mathbf{x}_m = -\overline{\mathbf{A}}^{-1}\mathbf{a}$.

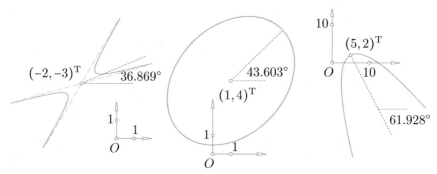

FIGURE 8.10. The conics from Exercise 8.1.5

• **Exercise 8.1.5** The following equations describe conics in the Euclidean plane (cf. Figure 8.10). Find out the type and all metrical invariants, such as centers (if there is one), the lengths of axes, and the angle enclosed by the principal axis and the x-axis of the coordinate system. In case of a parabola, find the vertex.

Hint: The eigenvectors of the coefficient matrix $\mathbf{A}$ of the quadratic part deliver a basis of an orthonormal frame centered either at the center or the vertex of the conic depending on whether it is an ellipse/hyperbola or a parabola.

1. $-38x^2 + 168xy - 87y^2 - 656x + 858y - 1968 = 0$,
2. $10656x^2 - 5880xy + 10369y^2 + 2208x - 77072y + 31936 = 0$,
3. $64x^2 + 240xy + 225y^2 - 2650x - 1284y + 10918 = 0$.

8.2 Conics are rational quadratic Bézier curves

Bézier curves and the algorithm of de Casteljau

We shall describe briefly and without going into too much detail what a Bézier curve is. We restrict ourselves to quadratic Bézier curves since conics allow rational quadratic parametrizations.

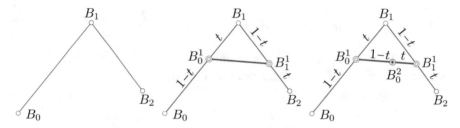

FIGURE 8.11. Repeated affine combinations of points reduces the number of control points in each step of DE CASTELJAU's algorithm. Finally, the algorithm terminates if the last polygon consists of just one point.

Let $B_0 = \mathbf{b}_0$, $B_1 = \mathbf{b}_0$, and $B_2 = \mathbf{b}_2$ be three different non-collinear points in an affine plane $\mathbb{A}^2(\mathbb{F})$, *e.g.*, in the Euclidean plane. We call these points *control points* and the open polygon is called a *control polygon*. Now, by choosing a value $t \in [0,1]$, we construct points B_0^1 and B_1^1 in between the given control points as affine combinations $\mathbf{b}_0^1 = (1-t)\mathbf{b}_0 + t\mathbf{b}_1$ and $\mathbf{b}_1^1 = (1-t)\mathbf{b}_1 + t\mathbf{b}_2$. Once again, we can introduce a new "in between point" B_0^2 as

$$\mathbf{b}_0^2 = (1-t)\mathbf{b}_0^1 + t\mathbf{b}_1^1$$

with the same fixed $t \in [0,1]$.

In the first step, we defined two new points from three given points. In the second step, we gained one new point from two points. This procedure is called *algorithm of de Casteljau* due to the French physicist and mathematician PAUL DE CASTELJAU (born 1930 in Besançon). (It can easily be guessed how it works with an arbitrary number of initial points.) Figure 8.11 shows how to construct the point B_0^2 according to DE CASTELJAU.

It is obvious that the $\mathbf{b}_0^2$ depends on t which plays the role of an affine ratio, and if t runs through the entire interval $[0,1]$, then $\mathbf{b}_0^2$ is the parametrization of a *quadratic Bézier curve*[1]. We obtain this parametrization in full length by inserting the affine combinations for $\mathbf{b}_0^1$ and $\mathbf{b}_1^1$. This leads to

$$\mathbf{b}_0^2(t) = (1-t)^2\mathbf{b}_0 + 2t(1-t)\mathbf{b}_1 + t^2\mathbf{b}_2. \tag{8.16}$$

Equation (8.16) parametrizes a parabola if $t \in \mathbb{R}$. The part of the Bézier curve that is parametrized over the unit interval $[0,1]$ is shown in Figure 8.12 (left). The three quadratic functions

$$f_0 = (1-t)^2, \quad f_1 = 2t(1-t), \quad f_2 = t^2 \tag{8.17}$$

are the well-known *Bernstein polynomials* of degree 2 which constitute a basis in the space of univariate polynomials of degree 2.

Now, we have:

Lemma 8.2.1 *Any quadratic Bézier curve is a parabola. Any parabola can be parametrized as a quadratic Bézier curve.*

Proof: In order to show that (8.16) parametrizes a parabola, we assume (without loss of

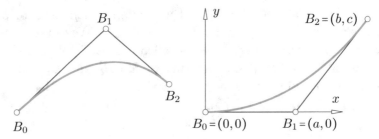

FIGURE 8.12. Left: A parabola as a quadratic Bézier curve with control points B_0, B_1, and B_2. Right: An appropriately chosen coordinate system, cf. proof of Corollary 8.2.1.

generality) that $\mathbf{b}_0 = (0,0)^{\mathrm{T}}$, $\mathbf{b}_1 = (a,0)^{\mathrm{T}}$, and $\mathbf{b}_2 = (b,c)^{\mathrm{T}}$ with $a \neq 0$, $a \neq b$, $b \neq 0$, $c \neq 0$ (see Figure 8.12). Then, we compute an implicit equation of the curve by eliminating the parameter t from $x = 2at(1-t) + bt^2$ and $y = ct^2$ and find $(cx + (2a-b)y)^2 - 4a^2cy = 0$ which is a parabola with axis direction $(2a-b,c)^{\mathrm{T}}$.

For the second part, we refer to Example 6.3.1 in Section 6.3. Here we have shown that any conic allows a rational parametrization. For a parabola we choose the center of the stereographic projection (that gives the rational parametrization) as the one and only ideal point of the parabola. Then the parametrization becomes polynomial.   ∎

---

[1]PIERRE ÉTIENNE BÉZIER (1910–1999) was a French engineer.

One important property of Bézier curves is the affine invariance of their representation. Since $B_0^2(t)$ is a composition of affine combinations, the affine image of any Bézier curve is the Bézier curve defined by the transformed control polygon.

Conics with center

The quadratic Bézier representation can only be used for the parametrization of parabolas (cf. Lemma 8.2.1). What about the other conics? From Section 6.3 we already know that conics admit rational parametrizations. Thus, we can use the concept of *rational Bézier curves*.

A rational quadratic Bézier curve can be written in the form

$$\mathbf{c}(t) = \left(\sum_{i=0}^{2} w_i f_i(t)\right)^{-1} \cdot \sum_{i=0}^{2} w_i \mathbf{b}_i f_i(t) \tag{8.18}$$

where $\mathbf{b}_i$ are the coordinate vectors of the control points, $f_i(t)$ are the Bernstein polynomials (8.17) of degree 2, and $w_i \in \mathbb{R}$ are *weights*. Increasing weights of any control point, pulls the curve towards this point. With appropriately chosen weights, we change not only the shape of the rational Bézier curve, we can even change the type of conic. Thus, we can parametrize ellipses, parabolas, and hyperbolas in this way.

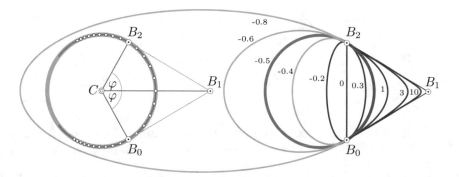

FIGURE 8.13. Left: A circle as a rational Bézier curve and the geometric meaning of the weight w_1. Right: Some rational quadratic Bézier curves with common control polygon and different weights of B_1.

With the help of the unit circle centered at $(0,0)^{\mathrm{T}}$, we demonstrate how to find weights and control points in order to make the rational Bézier curve a circle. We are interested in a segment of the circle. The segment shall have the central angle $0 < 2\varphi < \pi$. Thus, we can fix the control points B_0 and B_2 and the point B_1 lies on the tangents at both endpoints of the curve segment. We obtain

$$\mathbf{b}_0 = (\cos\varphi, -\sin\varphi)^{\mathrm{T}}, \quad \mathbf{b}_1 = (1/\cos\varphi, 0)^{\mathrm{T}}, \quad \mathbf{b}_2 = (\cos\varphi, \sin\varphi)^{\mathrm{T}}.$$

Note that the point B_1 is inverse to the point $[C, B_1] \cap [B_0, B_2]$ with respect to the circle. We choose $w_0 = w_2 = 1$ and insert the latter control points into (8.18). With the unknown weight w_1, we find the parametrization

$$\mathbf{c}(t) = \left(\frac{(1 - 2t + 2t^2)\cos\varphi + 2w_1(1-t)t\cos^{-1}\varphi}{(1-t)^2 + 2w_1(1-t)t + t^2}, \frac{(2t-1)\sin\varphi}{(1-t)^2 + 2w_1(1-t)t + t^2} \right)$$

which obviously differs from that given in (6.6) (cf. Example 6.3.1 in Section 6.3) since it parametrizes a certain part of the curve over the unit interval. The parametrization fulfills the equation $x^2 + y^2 = 1$ of the unit circle if, and only if, $w_1 = \pm\cos\varphi$. Therefore, the weight of the control point B_1 has to be $\pm\cos\varphi$ if the radii at the two endpoints B_0 and B_2 enclose an angle of 2φ.

Figure 8.13 (left, enclosed in the large ellipse) shows how the control points of a rational quadratic Bézier curve have to be chosen such that it becomes a circle.

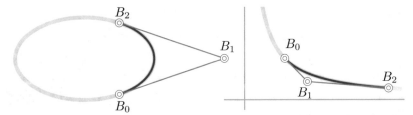

FIGURE 8.14. Control points for pieces of conics: ellipse (left), hyperbola (right).

● **Exercise 8.2.1** An equilateral hyperbola and an ellipse. Compute a rational parametrization of the part of the equilateral hyperbola starting at $\mathbf{b}_0 = (1,1)^{\mathrm{T}}$ ending at $\mathbf{b}_2 = (e, 0)^{\mathrm{T}}$ for arbitrary $e \in \mathbb{R}$. What happens if $e \to \infty$? Where is the control point $\mathbf{b}_1$ and what are the weights (see Figure 8.14, right)?

● **Exercise 8.2.2** Symmetric arc on an ellipse. Give a rational parametrization of the symmetric arc of an ellipse with $\mathbf{b}_0 = (a\cos\varphi, -b\sin\varphi)^{\mathrm{T}}$ and $\mathbf{b}_2 = (a\cos\varphi, b\sin\varphi)^{\mathrm{T}}$ with $0 < \varphi < \pi/2$ and $a, b \in \mathbb{R} \setminus \{0, \}$. Show that $\mathbf{b}_1 = (\frac{a}{\cos\varphi}, 0)^{\mathrm{T}}$ and $w_1 = \pm\cos\varphi$ (cf. Figure 8.14, left).

8.3 Conics and number theory

Diophantine quadratic equations

The solutions (x, y) of a quadratic equation in two variables x and y

$$a_{00} + 2a_{01}x + 2a_{02}y + a_{11}x^2 + 2a_{12}xy + a_{22}y^2 = (1\ x\ y)\mathbf{A}\begin{pmatrix} 1 \\ x \\ y \end{pmatrix} = 0 \quad (8.19)$$

with coefficients $a_{00}, \ldots, a_{22} \in \mathbb{F}$ are the points of a conic in the affine plane $\mathbb{A}^2(\mathbb{F})$. Special interest was given to the case where $\mathbb{F}$ is replaced with the ring $\mathbb{Z}$ of integers. This was studied by CARL FRIEDRICH GAUSS (1777–1855) and many others before. The case $a_{01} = a_{02} = a_{12} = 0$ and $a_{11} = a_{22} = -a_{00} = 1$ leads to the unit circle with four integer solutions $(\pm 1, 0)$ and $(0, \pm 1)$.

We have derived a rational parametrization of the unit circle in Section 6.3 (see page 231):

$$\left(\frac{1 - t^2}{1 + t^2}, \frac{2t}{1 + t^2} \right).$$

We recall that the stereographic projection was used in order to elaborate this. Now we replace the affine parameter t with homogeneous parameters (u, v) by $t = \frac{v}{u}$ and then we switch to the u homogeneous representation

$$(u^2 - v^2, 2uv, u^2 + v^2) \quad \text{with} \quad (u, v) \in \mathbb{F}^2 \setminus \{(0,0)\}. \quad (8.20)$$

Thus, we have parametrized the *Pythagorean triplets* (x, y, z) satisfying $x^2 + y^2 = z^2$ over any field $\mathbb{F}$ with $\mathrm{char}\,\mathbb{F} \neq 2$ and even over $\mathbb{Z}$.

If the integer square z^2 equals the sum $x^2 + y^2$ of two integer squares, then the circle with radius z carries at least twelve points with integer coordinates: $(0, \pm z)$, $(\pm z, 0)$, $(\pm x, \pm y)$, and $(\pm y, \pm x)$. Note that in any case $x \neq y$, for $2x^2$ cannot be an integer square. The decomposition of integer squares into sums of integer squares is not unique (Figure 8.15).

● **Exercise 8.3.1** Rational parametrization of the pseudo-Euclidean unit circle.

Apply the stereographic projection to an equilateral hyperbola $x^2 - y^2 = 1$, in order to obtain a rational parametrization which (after homogenizing) yields

$$(u^2 + v^2, 2uv, u^2 - v^2) \quad \text{with} \quad (u, v) \in \mathbb{F}^2 \setminus \{(0,0)\}. \quad (8.21)$$

This is a description of the rational points on the pseudo-Euclidean unit circle.

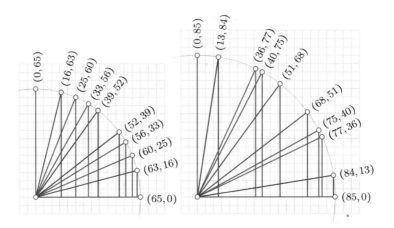

FIGURE 8.15. The decomposition of integer squares need not be unique: The circle with radius 65 carries 36 lattice points since $65^2 = 16^2 + 63^2 = 25^2 + 60^2 = 33^2 + 56^2 = 39^2 + 52^2$ (left). The same works with $85^2 = 13^2 + 82^2 = 36^2 + 77^2 = 40^2 + 75^2 + 51^2 + 68^2$ (right).

Equation (8.20) gives a rational representation of the solutions of the quadratic diophantine equation (8.19) in two variables in a normal form (diagonalized). However, these solutions are in general not integer solutions.

For any $n \in \mathbb{N}$ at least one ellipse can be given such that it contains exactly n lattice points (points with integer coordinates). However, there are only finitely many lattice point on each such ellipse. For the hyperbola it is feasible that there are infinitely many integer points on it. This is also the case for the parabola: If a conic c has the normal form $x^2 - 2ay = 0$ with $a \in \mathbb{F} \backslash \{0\}$ and $\mathrm{char}\mathbb{F} \neq 2$, then it is easy to find all points with coordinates in $\mathbb{F}^2$ on c.

Problems of this kind are, indeed, number theoretic in nature. Results including bounds on the number of integer lattice points on arcs contained in ellipses can be found in [14]. The case of hyperbolas is discussed in [15].

In Table 8.1, we have collected the smallest ellipses passing through $n = 3, 4, \ldots, 20$ integer lattice points. Here, an ellipse is called small if the semimajor axis is as small as possible. The attached figure illustrates some of the ellipses from the left-hand side of the Table 8.1. To be more precise: The figure shows some translata of some of the ellipses from the table in order to make it easier to illustrate the curves.

n	equation of the ellipse
3	$x^2+xy+y^2-x-y=0$
4	$x^2+y^2-x-y=0$
5	$2x^2-xy+2y^2-x+y-3=0$
6	$x^2-xy+y^2-1=0$
7	$2x^2-xy+2y^2-5x-4y-6=0$
8	$x^2+y^2-x-y-2=0$
9	$x^2-xy+3y^2-7x-6y=0$
10	$x^2-xy+4y^2-5x-5y-6=0$
11	$3x^2-3xy+4y2-21x-21y-10=0$
12	$x^2+y^2-5x-5y=0$
13	$2x^2-xy+3y^2-31x-26y-25=0$
14	$x^2-xy+4y^2-12x-9y-13=0$
15	$2x^2-xy+5y^2-39x-36y-41=0$
16	$x^2+y^2-7x-7y-8=0$
17	$2x^2-xy+3y^2-51x-51y-54=0$
18	$x^2-xy+y^2-7x-7y=0$
19	$2x^2-xy+3y^2-61x-61y-6=0$
20	$2x^2-xy+2y^2-27x-27y-29=0$

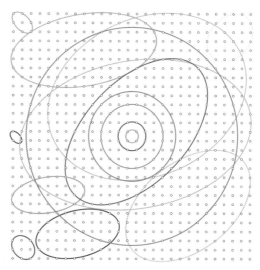

TABLE 8.1. Smallest ellipses passing through $n = 3, \ldots, 20$ integer lattice points. A bold-faced number indicates that this particular ellipse is a circle.

For the general form of quadratic diophantine equation as given in (8.19) solutions can easily be parametrized if at least one integer solution (ξ, η) is known. The pencil of lines with vertex (ξ, η) is now parametrized by $p = (\xi + t, \eta + kt)$ with the parameter t on the lines and the parameter k in the pencil. Both parameters can be replaced by homogeneous ones if necessary. Inserting p in (8.19) yields a linear equation in t because t splits off from the quadratic polynomial for p is a solution of (8.19). From the remaining linear equation we find t as rational expression in a_{ij} and k:

$$t = -2\frac{\mathbf{k}^{\mathrm{T}}\mathbf{A}\mathbf{x}}{\mathbf{x}^{\mathrm{T}}\mathbf{A}\mathbf{x}}$$

where $\mathbf{x} = (1, \xi, \eta)$ and $\mathbf{k} = (0, 1, k)$.

Pell equation

The diophantine quadratic equation

$$x^2 - dy^2 = \pm 1 \tag{8.22}$$

with $d \in \mathbb{N} \setminus \{0\}$ is usually called *Pell equation* after the English mathematician JOHN PELL (1611–1685). The solutions $(\pm 1, 0)$ are called trivial and are obviously independent of d. The case $d = 1$ has no integer solution

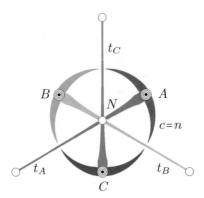

FIGURE 8.16. Just a rough idea what a conic in $\mathbb{P}^2(\mathbb{Z}_2)$ may look like: N is c's nucleus where the three tangents of c meet each other.

(besides the trivial ones) since the difference of two integer squares cannot be 1.

This special type of diophantine equation can be interpreted as the equation of a hyperbola. The task is to find integer points on the hyperbola for any $d \in \mathbb{N}$. Once a solution (ξ_0, η_0) is found, many others can be generated by the linear recurrence

$$\begin{pmatrix} \xi_{k+1} \\ \eta_{k+1} \end{pmatrix} = \begin{pmatrix} \xi_0 & d\eta_0 \\ \eta_0 & \xi_0 \end{pmatrix} \begin{pmatrix} \xi_k \\ \eta_k \end{pmatrix}.$$

However, the linear recurrence does not necessarily produce all solutions. Solving ARCHIMEDES's cattle problem requires to solve a Pell equation[2].

Conics over finite fields

In a projective plane over a finite field $\mathbb{F}$ there are only finitely many points. Let $N = \#\mathbb{F}$ be the number of elements in $\mathbb{F}$. Then, any line has $N + 1$ points and any conic contains that many points (see Example 6.4.8 in Section 6.4). The number of points in $\mathbb{P}(\mathbb{F}^3)$ equals $N^2 + N + 1$.

The very special case $N = 2$, $i.e.$, the projective minimal plane, shows conics with a completely different behavior than anywhere else: Assume that a conic c in the projective plane is given by the homogeneous equation

$$c: \ x_0^2 + x_1^2 + x_2^2 = 0.$$

---

[2]H.W. LENSTRA: *Solving the Pell equation*. Notices of the AMS **29**/2 (2002), 182–192.

Clearly, there are only three points on c because it contains as much points as a line does. These points are

$$A = (0:1:1), \quad B = (1:0:1), \quad C = (1:1:0).$$

These three points lie on one line n: $x_0 + x_1 + x_2 = 0$ which never happens in any other projective plane! The tangents at A, B, and C are the lines

$$t_A : \; x_1 + x_2 = 0, \quad t_B := x_0 + x_2 = 0, \quad t_C : \; x_0 + x_1 = 0$$

which are *concurrent in one point*, namely $N = (1:1:1)$. The point N is called the *nucleus* of c, and it is the pole of n with regard to c. Note that the equations of c and n describe the same set of points. Therefore, $c = n$.

What about an affine classification? Assume that n is the line at infinity. Then, the affine part of c is empty. Only the center of c is present. It is the point N since it is the pole of n with regard to c. Figure 8.16 shall illustrate the incidence relations of c's tangents.

Note that polarities in $\mathbb{P}^2(\mathbb{Z}_2)$ are nullpolarities at the same time, *i.e.*, each polar line contains the corresponding pole.

For more details on conics in finite geometry we refer to [36].

● **Exercise 8.3.2** Study the normal forms of conics in $\mathbb{P}^2(\mathbb{Z}_3)$ and give an affine classification of the conics in this plane.

The One-Seventh-conic

The rational number $1/7$ can be written in decimal expansion as

$$\frac{1}{7} = 0,142857\ldots$$

Now we take overlapping pairs of subsequent digits of the expansion and interpret them as affine coordinates of points in a plane. Thus, we have the six points

$$(1,4), \; (4,2), \; (2,8), \; (8,5), \; (5,7), \; (7,1).$$

Surprisingly, these six points lie on an ellipse with the equation

$$s: \; 19x^2 + 36xy + 41y^2 - 333x - 531y + 1638 = 0.$$

The ellipse s is called the *one-seventh-conic* (see Figure 8.17).

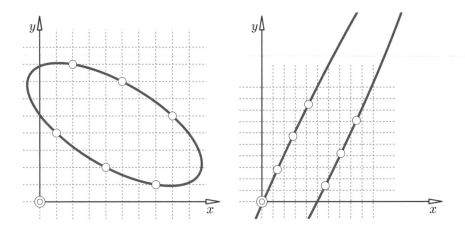

FIGURE 8.17. The one-seventh-ellipses carry six points with integer coordinates derived from the decimal expansion of $\frac{1}{7}$.

● Exercise 8.3.3 More one-seventh-conics

From the decimal expansion of $\frac{1}{7}$ we can create more sixtuples of overlapping pairs of integers:

$(1,4)$,	$(4,2)$,	$(2,8)$,	$(8,5)$,	$(5,7)$,	$(7,1)$,
$(14,42)$,	$(42,28)$,	$(28,85)$,	$(85,57)$,	$(57,71)$,	$(71,14)$,
$(142,428)$,	$(428,285)$,	$(285,857)$,	$(857,571)$,	$(571,714)$,	$(714,142)$,
$(1428,4285)$,	$(4285,2857)$,	$(2857,8571)$,	$(8571,5714)$,	$(5714,7142)$,	$(7142,1428)$,

$\vdots$

Show that these sixtuples of points are located on an ellipse if the number of digits in each coordinate is either $3k+1$ or $3k+2$ (with $k \in \mathbb{N}$). The ellipses have the centers

$$\frac{9}{2}(1,1), \quad \frac{99}{2}(1,1), \quad \frac{9999}{2}(1,1), \quad \frac{99999}{2}(1,1), \quad \dots.$$

If the number of digits in each place of the coordinate vectors equals $3k$ (with $k \in \mathbb{N} \setminus \{0\}$), then the one-seventh-conics degenerate into pairs of parallel lines with the direction $(2,-1)$.

There also exists a family of *one-thirteenth-conics*: Write down the decimal expansion of $1/13$ and define points in the same way as above. Show that these points define either ellipses or pairs of parallel lines. The centers of the ellipses defined from $1/13$ are identical with the ones given above. The parallels have the direction $(1,-3)$.

9 Special problems

A one-sheeted hyperboloid is a ruled quadric and carries two one-parameter
families of lines. A top view shows the pattern of rulings as a part of a Poncelet
grid. According to theoretical kinematics, the framework of crossing rods is
flexible if all crossings of generators are realized as hinges.

9.1 Conics in triangle geometry

There are many ways to define conics related to a triangle Δ with vertices A, B, C. We are going to pick out very few examples. All triangles have a circumcircle, *i.e.*, the uniquely defined conic passing through the triangles vertices and the absolute points[1] of Euclidean geometry. In Section 7.3, we have seen that each triangle in the Euclidean plane defines a pencil of equilateral hyperbolas including the Kiepert hyperbola, cf. Example 7.3.4. The base points of this pencil are the three vertices of the triangle together with the orthocenter. By the same token, the set of the vertices together with the orthocenter H is called an *orthocentric quadrangle* and has some symmetry: The orthocenter of any triangle built from three of these points is the fourth point.

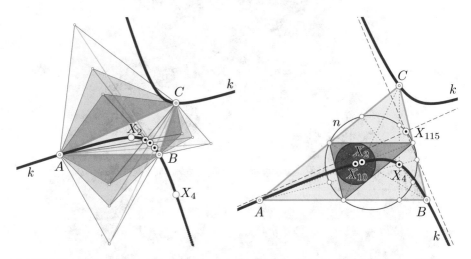

FIGURE 9.1. Left: The Kiepert hyperbola k is the locus of perspectors of isosceles triangles attached to Δ's sides and contains the centroid[2] X_2 and the orthocenter X_4. The three unlabeled points are the first Fermat point X_{13}, the outer Vecten point X_{485}, and the first Napoleon point X_{17}. Right: The Kiepert center X_{115} is located on the nine point circle n. The Kiepert hyperbola also contains the Spieker center X_{10}.

---

[1] The absolute points of Euclidean geometry are explained in Section 6.4. Every Euclidean circle passes through this pair of complex conjugate points on the line at infinity. Conversely, any non-degenerate conic that passes through the absolute points of Euclidean geometry is a Euclidean circle.

Figure 9.1 shows an elementary approach to Kiepert's hyperbola k: We attach three similar isosceles triangles to the sides of Δ either to the outside or to the inside. The additional points of the attached triangles form triangles which are perspective to the base triangle Δ. This means that the three connecting lines with the opposite vertices have a common point, the *perspector*. It turns out that all perspectors lie on k independent on whether the triangles are erected inside or outside Δ.

Each triangle has four tritangent circles, the incircle and the three excircles (see Figure 9.2, left). These are the conics tangent to three lines passing through the absolute points. From Example 7.2.2, we learn that, generally speaking, there are four solutions to this problem.

An exhaustive list of circles and conics related to triangles is available at [41] or [69]. In triangle geometry, special circles and conics are of interest: those that carry a huge bunch of centers or so-called *central conics*, *i.e.*, conics whose equations in terms of homogeneous trilinear or homogeneous barycentric coordinates show a cyclic symmetry, cf. [40].

In the following, we list and illustrate some of the central conics, not just circles, that frequently appear in triangle geometry.

Homogeneous trilinear coordinates, barycentric coordinates

In Section 5.3, we defined homogeneous coordinates in a projective plane. Here we need a special kind of homogeneous coordinates. The plane we are dealing with is the Euclidean plane $\mathbb{E}^2$. Whenever necessary, we use the projective and the complex extension. The vertices A, B, C shall serve as the base points which then have homogeneous coordinates $A = (1 : 0 : 0)$, $B = (0 : 1 : 0)$, $C = (0 : 0 : 1)$. Any point in the plane of the triangle $\Delta = ABC$ (except the ones on $[A, B] \cup [B, C] \cup [C, A]$) can be used for a unit point. The choice of the unit point specializes the coordinates. Two choices are frequently used: The choice of the triangle's incenter X_1 as the unit point leads to the *(homogeneous) trilinear coordinates*, sometimes called *(homogeneous) trilinear distances*. The choice of the centroid X_2 for the unit point yields the *(homogeneous) barycentric coordinates*.

---

[2]When we deal with triangle geometry, we shall follow the usual notation: X_1 is the *incenter* of the triangle Δ, X_2 is the *centroid* or *barycenter*, X_3 is the *circumcenter*, X_4 is the *orthocenter*. For a complete list of named triangle centers we refer to [41].

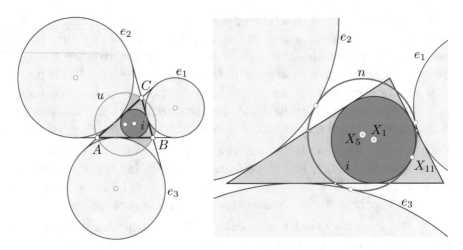

FIGURE 9.2. The prominent representatives of circles related to a triangle: Left: The incircle i, the three excircles e_1, e_2, e_3, and the circumcircle u. Right: The nine point circle n touches the four tritangent circles of a triangle. The point of contact of n and i is the so called Feuerbach point X_{11}. Further, n passes through the midpoints of Δ's edges and the midpoints of certain segments of the altitudes.

The term homogeneous should be used with care when dealing with trilinear distances. The geometric meaning of the coordinates based on (A, B, C, X_1) is the following: The incenter X_1 has coordinates $(1 : 1 : 1)$ for it is the chosen unit point. The incenter has equal distances to the three side lines of Δ, it is the radius ϱ of the incircle, thus $(\varrho, \varrho, \varrho)$ is also a representation of X_1, expressing that X_1 is actually at distance ϱ to any side of Δ. The vector $(\varrho, \varrho, \varrho)$ is, therefore, called the *actual trilinear coordinates* of X_1. These coordinates are no longer homogeneous.

Any point P in the plane of Δ can be described by trilinear coordinates which are simply the oriented distances of P to the side lines $[B, C]$, $[C, A]$, $[A, B]$ of Δ in that particular ordering. These coordinates show signs in the different areas defined by Δ's sides. Figure 9.3 shows the distribution of signs of the trilinear coordinates.

We denote the side lengths of Δ by $c = \overline{AB}$, $a = \overline{BC}$, $b = \overline{CA}$. Let further F denote the area of Δ. Then, homogeneous trilinear coordinates $(p_0 : p_1 : p_2)$ can be transformed into actual trilinear coordinates $(d_0 : d_1 : d_2)$

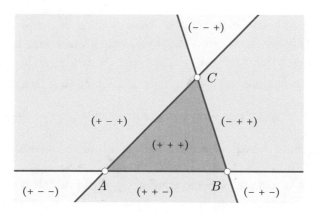

FIGURE 9.3. The distribution of signs of trilinear coordinates.

via

$$(d_0, d_1, d_2) = \frac{2F}{ap_0 + bp_1 + cp_2}(p_0, p_1, p_2)$$

which can easily be seen by computing the area F of Δ as the sum of the areas of the three subtriangles ABP, BCP, CAP. The scaling of coordinates fails if

$$ap_0 + bp_1 + cp_2 = 0$$

which characterizes points on the line $\mathcal{L}_\infty$ at infinity. Thus, $\mathcal{L}_\infty$ has (homogeneous) trilinear coordinates $(a : b : c)$. A point P is called a *triangle center* or simply *center* of Δ if its trilinear coordinates are cyclic symmetric functions in the side lengths of Δ. The first coordinate is usually called a *center function*. For example $X_1 = (1, 1, 1)$ (incenter) is a center. The centroid $X_2 = (bc : ca : ab)$ is also a center, since $a \to b \to c \to a$ changes bc to ca, ca to ab, and ab to bc. A *central line* is a line whose homogeneous trilinear coordinates are center functions.

Central circles

In terms of trilinear coordinates $(x_0 : x_1 : x_2)$, conics are given by homogeneous quadratic equations

$$(x_0, x_1, x_2) \begin{pmatrix} q_{00} & q_{01} & q_{02} \\ q_{01} & q_{11} & q_{12} \\ q_{02} & q_{12} & q_{22} \end{pmatrix} \begin{pmatrix} x_0 \\ x_1 \\ x_2 \end{pmatrix} = 0 \qquad (9.1)$$

as is the case in terms of any kind of homogeneous coordinates (cf. Section 6.3).

In triangle geometry, especially to those conics is payed attention to, whose equation can be written as a cyclic sum. The terms in the sum are cyclic symmetric functions of the side lengths a, b, c of Δ.

In [40] it is shown that all circles have a trilinear equation of the form

$$(\lambda_0 x_0 + \lambda_1 x_1 + \lambda_2 x_2)(a x_0 + b x_1 + c x_2) + \kappa(a x_1 x_2 + b x_2 x_0 + c x_0 x_1) = 0 \quad (9.2)$$

with appropriate $\lambda_0, \lambda_1, \lambda_2, \kappa \in \mathbb{R}$. This is an affine parametrization of a pencil of conics (see Section 7.3) with the affine parameter $\kappa \in \mathbb{R}$. The first term of (9.2) is a singular conic consisting of the common radical line of all circles in the pencil and the line at infinity. The representation of the pencil can be homogenized by replacing κ with $\kappa_1 : \kappa_0$. Then, $\kappa_0 = 0$, or equivalently $\kappa = \infty$ yields the interpretation of the second term in (9.2): it is the circumcircle of Δ with the equation

$$a x_1 x_2 + b x_2 x_0 + c x_0 x_1 = 0.$$

● Exercise 9.1.1 Center of a circle. Show that the center of a circle given by (9.2) has the following trilinear coordinates

$$(-\lambda_0 + \lambda_1 \cos C + \lambda_2 \cos B - \kappa \cos A :$$
$$\lambda_0 \cos C - \lambda_1 + \lambda_2 \cos A - \kappa \cos B :$$
$$\lambda_0 \cos B + \lambda_1 \cos A - \lambda_2 - \kappa \cos C)$$

where $\cos A = (b^2 + c^2 - a^2)/(2bc)$ is the cosine of Δ's interior angle at A.[3] The values $\cos B$ and $\cos C$ are the cosines of the interior angles at B and C. Make use of the fact that the center of a conic is the conic's pole of the ideal line (cf. page 266).

● Exercise 9.1.2 Center of an arbitrary conic. Show that the centers of the circumconics given by (9.4) have the trilinear coordinates

$$(q_{01}(-a q_{01} + b q_{02} + c q_{12}) : q_{02}(a q_{01} - b q_{02} + c q_{12}) : q_{12}(a q_{01} + b q_{02} - c q_{12})). \quad (9.3)$$

Simson line

A well-known theorem which is erroneously ascribed to the English mathematican ROBERT SIMSON (1687–1768) is actually due to the Scottish mathematician WILLIAM WALLACE (1768–1843). It characterizes the points on the circumcircle:

---

[3]This notation is standard in triangle geometry and shall not lead to any confusion.

Theorem 9.1.1 *The pedal points of the normals from a point P to the side lines of a triangle are collinear if, and only if, P is a point on the circumcircle u.*

The locus of collinearity is called the *Simson line*. A proof of this result can be found in the textbook [19]. Figure 9.4 illustrates the contents of Theorem 9.1.1 and some more results that relate to Theorem 9.1.1.

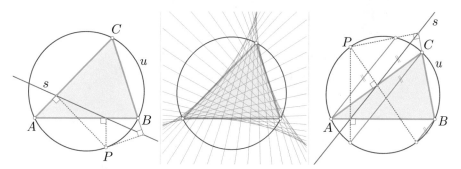

FIGURE 9.4. Some results related to Theorem 9.1.1: Left: the SIMSON line. Middle: If the point P traverses u, then the Simson line envelopes a Steiner hypocycloid. Right: The remaining intersections of the normals through P and the circumcircle u give rise to three lines parallel to s.

SIMSON's theorem appears in connection with parabolas:

Theorem 9.1.2 *Let p be a parabola with focus F and assume that u, v, w are three (mutually distinct and finite) tangents of p. The circumcircle of the triangle built by u, v, w passes through F.*

Proof: In Figure 2.13, we can see that the normals of a parabola's tangent from the focus F meet the tangents at the tangent line of the vertex. Thus, the pedal points of the normals to the sides of the tangent triangle are collinear and F must be a point on the circumcircle of the tangent triangle according to Theorem 9.1.1. ∎

Figure 9.5 illustrates the contents of Theorem 9.1.2.

Central conics

For conics that pass through the vertices, *i.e.*, conics circumscribed to Δ, (9.1) is characterized by

$$q_{00} = q_{11} = q_{22} = 0.$$

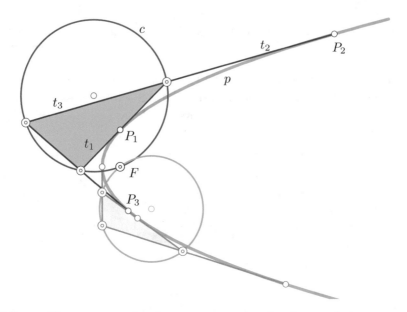

FIGURE 9.5. The circumcircle of any tangent triangle of a parabola p passes through p's focus (cf. theorem 9.1.2).

Therefore, the equations of the circumconics can be written in the form

$$\sum_{\text{cyclic}} \frac{q_{01}}{x_2} = 0 \tag{9.4}$$

where the cyclic sum means that all terms are obtained from the given one by a cyclic shift in all indices in $\{1,2,3\}$ and components. For example $\sum_{\text{cyclic}} \frac{q_{01}}{x_2} = \frac{q_{01}}{x_2} + \frac{q_{12}}{x_0} + \frac{q_{20}}{x_1}$.

Obviously, it is sufficient to prescribe one coefficient (center function) in order to determine the conic. In Section 7.5, we have learned that the isogonal conjugation and the isotomic conjugation are quadratic Cremona transformations. Comparing the normal form (7.15) of a Cremona transformation with three base points and (9.4), we can see that the circumconics are the isogonal transforms of certain central lines.

● Exercise 9.1.3 Show that (9.1) is the equation of a parabola circumscribed to Δ if, and only if, the coefficients q_{ij} satisfy

$$2abq_{01}q_{02} - a^2q_{01}^2 + 2bcq_{02}q_{12} - b^2q_{02}^2 + 2caq_{01}q_{12} - c^2q_{12}^2 = 0$$

together with the regularity condition $\det(q_{ij}) \neq 0$.

Steiner ellipses

Among the conics circumscribed to Δ there is a special well-defined ellipse s called *Steiner circumellipse*. The tangents of s at Δ's vertices are parallel to the opposite side lines. It is clear that these three line elements fit to a conic, since the tangents form the *anticomplementary triangle* of Δ which is perspective to Δ with the centroid S for the perspector. The perspector S also serves as the Brianchon point in a special version of Theorem 6.2.3 of these three line elements (see Figure 6.7).

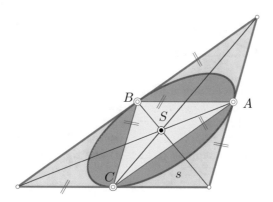

FIGURE 9.6. The common centroid S of $\Delta = ABC$ and its anticomplementary triangle is the Brianchon point of Δ's Steiner ellipse s.

We shall compare the area A_Δ of an equilateral triangle Δ with the area A_u of its circumcircle u (Figure 9.6): Provided that the circumradius of Δ equals R, we find $A_\Delta = \frac{3}{2}R^2 \sin \frac{2\pi}{3} = \frac{3\sqrt{3}}{4}R^2$ and $A_u = R^2\pi$. Note that for an equilateral triangle the circumcircle is its Steiner circumellipse. Surprisingly, the ratio of these two areas is constant:

Theorem 9.1.3 *The area A_s of the Steiner circumellipse s of a triangle Δ and the area A_Δ of Δ are related via*

$$A_\Delta : A_s = 3\sqrt{3} : 4\pi \tag{9.5}$$

independently of the choice of the triangle.

Proof: There exist affine transformations α which map the given triangle $\Delta = ABC$ onto an equilateral triangle $A_1B_1C_1$ and, hence, the Steiner circumellipse s of Δ onto the circumcircle s_1 of $A_1B_1C_1$. Choose, e.g., $A_1 = A$, $B_1 = B$. Then, the affine transformations multiply all (signed) areas with a constant factor $f = \det \mathbf{A}$ if $\mathbf{A} \in \mathbb{R}^{2\times2}$ represents the associated linear mapping in a Cartesian or an affine frame. Thus, the ratio $A_\Delta : A_s$ is affinely invariant. ∎

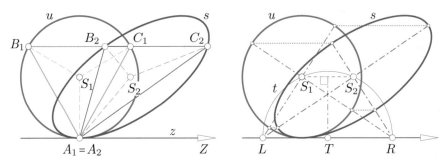

FIGURE 9.7. Left: The circumcircle u of an equilateral triangle can be mapped via an area preserving affine transformation (affine shearing, see page 186) to the Steiner circumellipse s of an arbitray triangle. Right: The construction of the principal axis of the Steiner ellipse s uses the Thales circle t centered at T on the axis of the affine mapping (shearing) and the bisector of the two centers S_1 and S_2.

Figure 9.7 shows an elementary way to find the principal axis of the Steiner ellipse s being the affine image of the circumcircle u. In fact, this construction can be used for arbitrary affine mappings applied to a circle, provided that there is an axis, $i.e.$, the affine mapping is a perspective affine mapping. Note that α is center preserving, for it preserves affine ratios.

The construction in Figure 9.7 (right-hand side) uses the fact that conjugate diameters of a circle are orthogonal. Furthermore, the axes of the image s of u are conjugate and orthogonal. Thus, the fixed points L and R on the axis z (of α) have to lie on the Thales circle t through the centers of u and s.

The Steiner circumellipse also appears in combination with three concurrent osculating circles of an ellipse (see page 97).

The following theorem is usually ascribed to MORRIS MARDEN (US mathematician, 1905–1991),[4] though it was first published and proved by

---

[4]MARDEN published this result in 1945, never claiming that it was his result, see: M. MARDEN: *A note on the zeroes of the sections of a partial fraction.* Bull. AMS **51**/12 (1945), 935–940.

JÖRG SIEBECK (German mathematician).[5] It relates the Steiner inellipse of a triangle to the roots of a complex polynomial:

Theorem 9.1.4 *The zeros φ_1 and φ_2 of the derivative $p'(z)$ of the complex cubic polynomial $p(z) = (z - \alpha)(z - \beta)(z - \gamma)$ are the focal points of the Steiner inellipse of the triangle with vertices α, β, γ, provided that the complex numbers are identified with points in the Euclidean plane.*

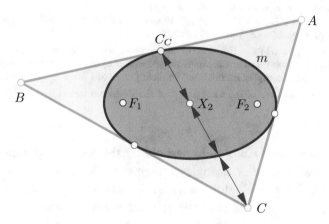

FIGURE 9.8. The Steiner inellipse m of the triangle ABC with focal points F_1 and F_2 is centered at X_2.

Proof: We assume that the Steiner inellipse of Δ is given by the equation

$$m: \frac{x^2}{a^2} + \frac{y^2}{b^2} = 1.$$

The Steiner inellipse of a triangle Δ touches the sides at the midpoints. Therefore, we assume that one point C_1 of contact has the complex coordinate

$$\gamma_1 = a\frac{1 - t^2}{1 + t^2} + ib\frac{2t}{1 + t^2}.$$

Then, Δ's vertex C opposite to C_1 is given by $\gamma = -2\gamma_1$, because the center of the inellipse m is the centroid X_2 of Δ and divides the Cevian at the ratio $1 : 2$ (cf. Figure 9.8). The tangents from C to m meet the tangent $[A, B]$ at the triangle vertices A and B with coordinate vectors

$$\alpha = \tfrac{1}{1+t^2}\left(a(1 - t^2 - 2\sqrt{3}t) + ib(2t + \sqrt{3}(1 - t^2))\right),$$

$$\beta = \tfrac{1}{1+t^2}\left(a(1 - t^2 + 2\sqrt{3}t) + ib(2t - \sqrt{3}(1 - t^2))\right).$$

Now, we build

$$p(z) = (z - \alpha)(z - \beta)(z - \gamma).$$

---

[5] JÖRG SIEBECK: *Über eine neue analytische Behandlungsweise der Brennpunkte.* J. reine u. angew. Mathematik **64** (1864), 175–182.

Differentiating once with respect to z, we find $p'(z) = 3(z^2 - a^2 + b^2)$ whose zeros in $\mathbb{C}$ are

$$\varphi_1 = \sqrt{a^2 - b^2} \quad \text{and} \quad \varphi_2 = -\sqrt{a^2 - b^2}$$

which are the complex coordinates of the focal points of m.   ■

Remark 9.1.1 Another proof of Theorem 9.1.4 can be found in: D. KALMAN: *An Elementary Proof of Marden's Theorem*. The American Mathematical Monthly **115** (2008), 330–338.

Marden's theorem can be generalized: Assume that α, β, γ are roots of a polynomial $p(z)$ and have multiplicities r, s, t, *i.e.*, they are the roots of $p(z) = (z - \alpha)^r (z - \beta)^s (z - \gamma)^t$. The two non-trivial zeros of $p'(z)$ are the focal points of the inellipse of Δ that touches the sides at points dividing the edges of Δ at ratios $r : s$, $s : t$, and $t : r$.[6]

If V_i (with $i \in \{1, \ldots, n\}$) are the vertices of an n-gon (not necessarily regular) that has an inscribed ellipse i, then there is a complex polynomial $p(z)$ of degree n that vanishes at all V_i (or at the respective complex numbers). The focal points of i are among the roots of $p'(z)$.[7]

A characterizing property of the focal points of the Steiner circumellipse is given in [1].

We have found the Kiepert hyperbola as the locus of perspectors of Δ and the "new" points of triangles attached to Δ's sides. What about the perspectrices? Surprisingly, the perspectrices turn out to be the set of tangents of a conic p, namely the *Kiepert parabola* (see Figure 9.9).

Like in the case with the Kiepert hyperbola, we attach isosceles and similar triangles to Δ's sides. The outermost points of the attached triangles again form triangles which are perspective to Δ. Let us perform the projective closure of the Euclidean plane. This allows us to study extremal attached triangles. If the base angles of the isosceles triangles attached to Δ reach $90°$, then the triangle of outermost points degenerates into the line at infinity. Thus, the line at infinity is the perspectrix in this case. Therefore, one of the tangents of the envelope is the line at infinity.

The Brianchon point of p equals the Steiner point X_{99} of Δ which is located on the Steiner circumellipse s.

The famous nine point circle n (Figure 9.2) is centered at X_5 (the nine point center) and is half the size of the circumcircle. The circle n is the circumcircle of the medial triangle, *i.e.*, the triangle of the midpoints of

---

[6] B.Z. LINFIELD: *On the relation of the roots and poles of a rational function to the roots of its derivative*. Bull. AMS **27** (1920), 17–21.

[7] J.L. PARISH: *On the derivative of a vertex polynomial*. Forum Geom. **6** (2006), 285–288.

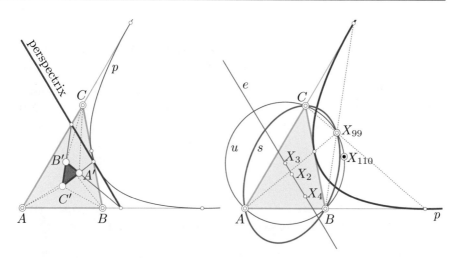

FIGURE 9.9. Left: The Kiepert parabola p is the envelope of all perspectrices x of Δ and the triangles Δ_x built by the "new" points A', B', C' of the attached triangles. Right: The Euler line e of Δ is the directrix of the Kiepert parabola p. The focal point of the Kiepert parabola is the triangle center X_{110}.

Δ's sides. It contains six more points: The three feet of the altitudes of Δ and the three midpoints of the altitudes' segments between the orthocenter and the vertices.

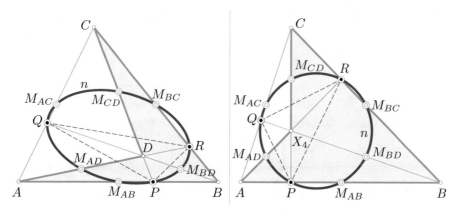

FIGURE 9.10. Left: The nine point conic n is a circumconic of the diagonal triangle of a complete quadrangle and carries the midpoints of all six sides. Right: The nine point circle of a triangle ABC is the nine point conic of an orthocentric quadrangle which means of the quadrangle $ABCX_4$ with X_4 being the orthocenter of ABC.

However, the nine point circle is a special appearance of a more general conic which is associated with any *complete quadrangle*, *i.e.*, the six lines joining the four vertices of a quadrangle.

Assume that $Q = ABCD$ is a quadrangle. Each side has a midpoint, for example, M_{AB} is the midpoint of the edge AB. Thus, there are six midpoints of edges. Further, there are three diagonal points $P = [A, B] \cap [C, D]$, $Q = [A, C] \cap [B, D]$, and $R = [A, D] \cap [B, C]$. Now, we have constructed a total of nine points out of the four vertices of Q and the following holds:

Corollary 9.1.1 *There exists a unique conic passing through the six midpoints of the edges and the three diagonal points of a quadrangle.*

The left-hand side of Figure 9.10 shows the nine point conic of a non-convex quadrangle. It doesn't matter if the quadrangle is convex or not. The nine point conics always exists, but may sometimes degenerate as is the case with a rectangle. In Section 7.5, we use a quadratic birational transformation to construct the nine point conic as the image of a line (see page 339).

• Exercise 9.1.4 Proof of Corollary 9.1.1. Proof the existence of the nine-point conic by assuming that $A = (1 : 0 : 0)$, $B = (0 : 1 : 0)$, $C = (0 : 0 : 1)$, and $D = (1 : 1 : 1)$. Assume further that $(a : b : c)$ are the homogeneous coordinates of the line at infinity $\mathcal{L}_\infty$ which helps to define midpoints of the edges as fourth harmonic points of the ideal points of the edges with respect to the edges.

Show that the diagonal points are $P = (1 : 1 : 0)$, $Q = (1 : 0 : 1)$, $R = (0 : 1 : 1)$ and the midpoints are

$$M_{AB} = (b : a : 0), \quad M_{AC} = (c : 0 : a), \quad M_{AD} = (2a + b + c : a : a),$$
$$M_{BC} = (0 : c : b), \quad M_{BD} = (b, a + 2b + c, b), \quad M_{CD} = (c : c : a + b + 2c).$$

The coefficient matrix of the equation of the nine point conic equals

$$\begin{pmatrix} 2a & a+b & a+c \\ a+b & 2b & b+c \\ a+c & b+c & 2c \end{pmatrix}$$

and the center equals $X_{1125} = (bc(2a + b + c) : ca(a + 2b + c) : ab(a + b + 2c))$. If $D = X_4$ (the orthocenter), then n is the nine-point circle.

Two famous quadratic Cremona transformations

Isogonal conjugation - a Cremona transform

In triangle geometry, the *isogonal conjugation* plays an important role. Besides the fact that many triangle centers are related via this particular mapping, it relates lines and circumconics of the base triangle $\Delta = ABC$.

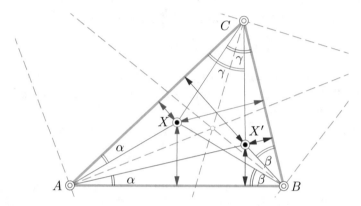

FIGURE 9.11. Corresponding points of the isogonal conjugation are projected from the vertices by lines symmetric w.r.t. the angle bisectors.

The isogonal conjugation κ is defined for all points in the Euclidean plane off the lines $[A,B]$, $[B,C]$, $[C,A]$, *i.e.*, the side lines of Δ. The mapping κ is defined as follows (see Figure 9.11): Any point $X \notin [A,B] \cup [B,C] \cup [C,A]$ can be joined with Δ's vertices. Then, we have

$$\alpha = \sphericalangle CAX, \quad \beta = \sphericalangle ABX, \quad \gamma = \sphericalangle BCX.$$

We reflect $[A,X]$ in the interior (or exterior) angle bisector at A; and we reflect $[B,X]$ in the angle bisector at B. The κ-image X' of X is the intersection of these reflected lines. Thus, $\sphericalangle X'AB = \alpha$ and $\sphericalangle X'BC = \beta$. It is elementary to verify that the third reflected line also passes through X'. This can be done using CEVA's theorem.

It is clear from the definition of κ that the interior and exterior angle bisectors are fixed as a whole but not pointwise.

A coordinate representation can be derived in the following way: Assume that (ξ, η, ζ) are the actual trilinear coordinates of a point X in an admissible position. We are looking for the trilinear coordinates $(\xi' : \eta' : \zeta')$

of X'. The latter may in general not be actual trilinear, so we aim at homogeneous coordinates. Then, note that

$$\overline{XC}\sin(C-\gamma) = \overline{XB}\sin\beta = \xi,$$

where the C in the sine denotes the interior angle at the vertex C. Furthermore, we obtain four more equations of the same type by cyclically shifting $a \to b \to c \to a$, $\xi \to \eta \to \zeta \to \xi$, and $A \to B \to C \to A$. For the image point $X' = \kappa(X)$, we find six similar equations:

$$\overline{X'C}\sin\gamma = \overline{X'B}\sin(B-\beta) = \xi'$$

plus four equations obtained by cyclically shifting all involved quantities. Now, we eliminate $\sin(A-\alpha)$, $\sin(B-\beta)$, $\sin(C-\gamma)$ in a first step. In the second step, we do the same with the remaining sines. In the third step, the ratios $\overline{X'A}:\overline{XA}$, $\overline{X'B}:\overline{XB}$, $\overline{X'C}:\overline{XC}$ are eliminated. This yields

$$(\xi':\eta':\zeta') = (\eta\zeta:\zeta\xi:\xi\eta). \tag{9.6}$$

Obviously, this agrees with the normal form of a quadratic Cremona transform of the first type (cf. equation (7.15) in Section 7.5) with three distinct base points. These exceptional points are the vertices of Δ. This can easily be seen by inserting their homogeneous coordinates into (9.6) which produces $(0:0:0)$, i.e., no point.

Though the coordinate representation (9.6) of the isogonal conjugation equals that of the harmonic conjugation from Section 5.4, it differs in a crucial point: Here, the coordinates are that of a point, in the other case the coordinate vector represents a line.

It is not only the algebraic representation (9.6) that makes clear that algebraic curves of degree n are mapped to curves of degree $2n$, generally speaking. (Exceptions occur if a curve passes through at least one exceptional point.) If we look at the pencils about Δ's vertices, we see that, e.g., the lines $[A, X]$ and $[A', X]$ are corresponding lines in the symmetric involution about A. This holds true in B and C as well. If now a point X traces a line l, X and X' are projected by pencils which are related by a projective mapping that is not perspective. Thus, $l' = \kappa(l)$ is a conic as defined in Definition 6.1.1, provided l does not contain a single exceptional point.

Figure 9.12 shows the action of κ applied to some special curves. The circumcircle u is the isogonal transform of the line $\mathcal{L}_\infty$ (the line at infinity).

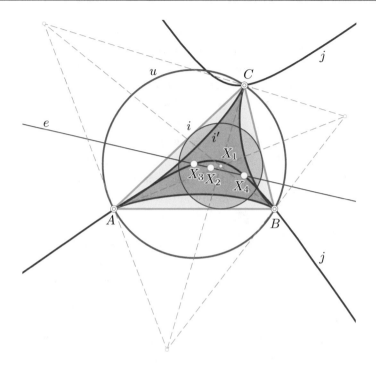

FIGURE 9.12. The circumcircle u is the image of the line $\mathcal{L}_\infty$, the Euler line maps to the Jeřabek hyperbola j, and the incircle i is mapped to a quartic i' with three cusps at the exceptional points A, B, C.

● Exercise 9.1.5 The circumcircle as isogonal transform of the line at infinity. Use the theorem of the angle of circumference and Definition 6.1.1 and show that the circumcircle of Δ is the isogonal transform of the ideal line. It turns out that the isogonal transform of $\mathcal{L}_\infty$ is generated with the help of congruent pencils of lines, cf. Example 6.2.2.

From $\kappa(\kappa(x_0, x_1, x_2)) = \kappa(x_1 x_2, x_2 x_0, x_0 x_1) = x_0 x_1 x_2 (x_0, x_1, x_2)$ and the definition of κ by means of involutive projectivities, we can immediately see that κ is involutive. Thus, the κ-image of u is $\mathcal{L}_\infty$. The Euler line $e = [X_2, X_3]$ (or $\mathcal{L}_{2,3}$ in Kimberling's notation) is mapped to the Jeřabek hyperbola h, see Figure 9.12.

In fact, any line l with homogeneous trilinear coordinates $(l_0 : l_2 : l_2)$ is mapped to a conic with the equation

$$c: \; l_0 x_1' x_2' + l_1 x_2' x_0' + l_2 x_0' x_1' = 0$$

which is circumscribed to Δ. If $(l_0 : l_1 : l_2)$ are the coordinates of a triangle center, then they can be identified with the trilinear coordinates of a central line which has a corresponding central conic circumscribed conic to Δ. So there are three one-to-one correspondences: 1. between triangle centers and central lines, 2. between central lines and central conics, and 3. between central conics and triangle centers.

In Figure 9.12 we can also see what happens with the incircle i if we apply the isogonal transformation κ. The image curve i' is a quartic curve with three cusps since i touches the three base lines (*i.e.*, the joins of the three exceptional points). In case of an equilateral triangle Δ, the quartic i' has a three-fold symmetry and is frequently called *deltoid*, but it is a *hypocycloid*, namely STEINER's hypocycloid.

Isotomic conjugation

Another transformation that appears frequently in triangle geometry is called the *isotomic conjugation*. Like the isogonal conjugation, this mapping is also an involutive quadratic Cremona transformation.

Figure 9.13 shows how to apply the isotomic conjugation to a point X in the plane of the triangle Δ: The vertices X_A, X_B, X_C of the Cevian-triangle of a point X are reflected in the midpoints of Δ's sides. For example, $X_C \in [B, C]$ is mapped to X_C' via the reflection in the midpoint M_C of AB. The points X_A', X_B', and X_C' are the vertices of the Cevian triangle of the isotomic conjugate X' of X.

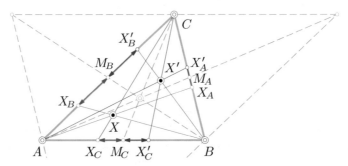

FIGURE 9.13. The isotomic transformation is somehow dual to the isogonal transformation: The vertices X_A, X_B, X_C of the Cevian-triangle of X are reflected in the midpoints of the sides of Δ which results in the Cevian-triangle of the isotomic conjugate X' of X.

● **Exercise 9.1.6** Isotomic conjugation. Show that the isotomic conjugation as explained above is a quadratic Cremona transformation of type 1 (cf. the normal form given in (7.15)).

Show that a point X with homogeneous trilinear coordinates $(\xi : \eta : \zeta)$ is mapped to

$$(b^2 c^2 \eta \zeta : c^2 a^2 \zeta \xi : a^2 b^2 \xi \eta).$$

• **Exercise 9.1.7** The Steiner ellipse as the isotomic image of the line at infinity. Show that the image of the ideal line $\mathcal{L}_\infty$ under the isotomic conjugation is the Steiner circumellipse s of Δ with the equation

$$s : \quad bc x_1 x_2 + ca x_2 x_0 + ab x_0 x_1 = 0.$$

Figure 9.14 shows the action of the isotomic mapping on the grid of lines parallel to the sides of the base triangle Δ.

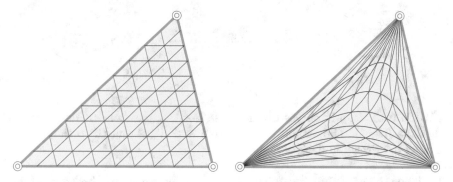

FIGURE 9.14. Left: grid of lines parallel to Δ's sides, Right: the isotomic transform of the grid.

9.2 Isoptic curves of conics

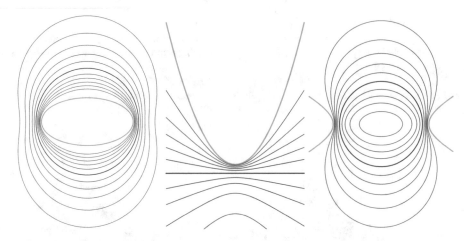

FIGURE 9.15. Isoptic curves of an ellipse (left), a parabola (middle), and a hyperbola (right). The orthoptic circles (left and right) and the orthoptic line (middle) are drawn in dark blue.

The locus i_φ of points from which a given curve c can be seen under a constant angle is called *isoptic curve*, or simply, *isoptic*. Especially, if $\varphi = \frac{\pi}{2}$, we call i_φ the *orthoptic curve* of c. The orthopic curves of ellipses are circles (cf. 2.2.6). However, it is clear that the isoptic curves of circles are concentric circles. The radius r_φ of the isoptic circle c_φ of a circle c with radius r equals

$$r_\varphi = \frac{r}{\cos \frac{\varphi}{2}}.$$

In the following we shall derive the isoptic curve of conics. We start with an ellipse and pay attention to the isoptics of parabolas. The computations of isoptics of a hyperbola shall be left to the reader as an exercise.

Isoptic curves of ellipses

The ellipse e is given by the equation

$$e : \frac{X^2}{a^2} + \frac{Y^2}{b^2} = 1$$

where $a > b$ is an admissible assumption. If the point with Cartesian coordinates (x, y) is a point on the isoptic curve i_φ of e, then there are two tangents t and τ of e passing through (x, y). The tangents t and τ touch e at the points (u, v) and (ξ, η) which yields

$$t : \frac{ux}{a^2} + \frac{vy}{b^2} = 1, \quad \tau : \frac{x\xi}{a^2} + \frac{y\eta}{b^2} = 1 \tag{9.7}$$

and

$$\frac{u^2}{a^2} + \frac{v^2}{b^2} = 1, \quad \frac{\xi^2}{a^2} + \frac{\eta^2}{b^2} = 1. \tag{9.8}$$

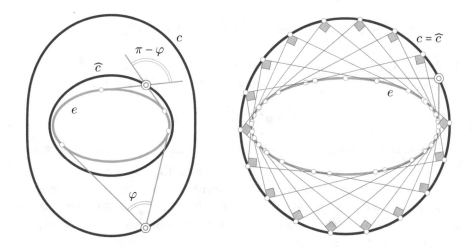

FIGURE 9.16. Left: The two branches c and $\widehat{c}$ of the isoptic curve of an ellipse e appear as the zeros of one algebraic equation. From points on c, the ellipse e is seen under the angle φ; from points on $\widehat{c}$ the ellipse is seen under the angle $\pi - \varphi$. Right: If $\varphi = \pi$, then the two branches c and $\widehat{c}$ become one branch of multiplicity two. It is the director circle with center $(0,0)$ and radius $\sqrt{a^2 + b^2}$.

The measure of the angle enclosed by the tangents t and τ shall be φ. From the equations (9.7) of t and τ, we can determine the slope angles against the x-axis:

$$\tan \sphericalangle (t, x\text{-axis}) = \frac{ub^2}{va^2} = \tan \varphi_1, \quad \tan \sphericalangle (\tau, x\text{-axis}) = \frac{\xi b^2}{\eta a^2} = \tan \varphi_2.$$

Since $\varphi = \varphi_2 - \varphi_1$, we use the formula

$$\tan(\varphi_2 - \varphi_1) = \frac{\tan \varphi_2 - \tan \varphi_1}{1 + \tan \varphi_1 \tan \varphi_2}$$

which results in

$$\tan \varphi = a^2 b^2 \frac{\xi v - \eta u}{a^4 \eta v + b^4 \xi u}. \tag{9.9}$$

The isoptic curve i_φ of e is now determined by the equations (9.7), (9.8), and (9.9). These are five equations in the six unknowns u, v, ξ, η, x, and y. We eliminate all variables but x and y and arrive at an implicit equation of i_φ. In a first elimination step we use (9.7) and eliminate for example u and ξ from (9.8) and (9.9). In a final step, v and η are to be eliminated from the modified equation in (9.9) and we arrive at the quartic curve

$$i_\varphi: \ (x^2 + y^2 - a^2 - b^2)^2 + 4 \cot^2 \varphi (a^2 b^2 - b^2 x^2 - a^2 y^2) = 0. \tag{9.10}$$

From (9.10), we can see that the orthoptic curve $i_{\pi/2}$ is a circle of multiplicity two: With $\varphi = \frac{\pi}{2}$, or equivalently, $\cot \varphi = 0$, (9.10) simplifies to

$$(x^2 + y^2 - a^2 - b^2)^2 = 0,$$

and thus, the isoptic circle $i_{\pi/2}$ of e is centered at $(0,0)$, *i.e.*, it is concentric with e and has the radius $\sqrt{a^2 + b^2}$ (compare with Theorem 2.2.6). Figure 9.15 shows some isoptic curves of a hyperbola, a parabola and an ellipse. The orthoptic curves are shown in dark blue. In Figure 9.16, the isoptic curve of an ellipse with $\varphi = \frac{\pi}{3}$ is shown. Here, we observe that the isoptic i_φ consists of two separated branches c and $\widehat{c}$ provided that $\varphi \neq 0, \frac{\pi}{2}, \pi$. This is a consequence of the fact that $\cot(\pi - \varphi) = -\cot \varphi$ and $\cot^2(\pi - \varphi) = \cot^2 \varphi$. However, the two branches c and $\widehat{c}$ are two disjoint components of one irreducible algebraic variety given by (9.10).

We introduce homogeneous coordinates by letting $x = x_1 x_0^{-1}$ and $y = x_2 x_0^{-1}$ and multiply (9.10) by $x_0^4 \tan^2 \varphi$ which gives

$$\tan^2 \varphi (x_1^2 + x_2^2 - (a^2 + b^2) x_0^2)^2 + 4 x_0^2 (a^2 b^2 x_0^2 - b^2 x_1^2 - a^2 x_2^2) = 0.$$

If $\varphi = 0$ or $\varphi = \pi$, we have

$$x_0^2 (a^2 b^2 x_0^2 - b^2 x_1^2 - a^2 x_2^2) = 0$$

as the equation of i_0 or i_π. Thus, i_0 and i_π consist of e and the ideal line of the real projective plane. The ideal line appears with multiplicity two since the tangents drawn from ideal points to e enclose an angle of $0°$ as well as $180°$.

The absolute points $(0 : 1 : i)$ and $(0 : 1 : -i)$ are double points on i_φ $(\varphi \neq 0, \frac{\pi}{2}, \pi)$: This can easily be seen if i_φ is intersected with the ideal line $\omega : x_0 = 0$. This gives $(x_1^2 + x_2^2)^2 = (x_1 + ix_2)^2(x_1 - ix_2)^2 = 0$.

We summarize:

Theorem 9.2.1 *The isoptic curves i_φ of ellipses are quartic curves with double points at the absolute points of Euclidean geometry if $\varphi \neq 0, \frac{\pi}{2}, \pi$. The orthoptic of an ellipse is a circle of multiplicity two concentric with the ellipse and with radius $\sqrt{a^2 + b^2}$ if a and b are the ellipse's semimajor and semiminor axes.*

This result holds similarly for hyperbolas, see the exercise on page 399.

Isoptic curves of a parabola

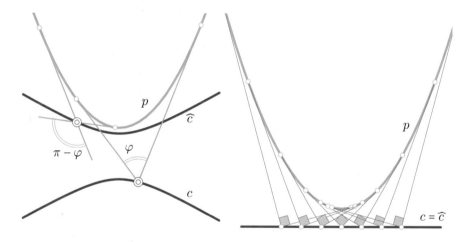

FIGURE 9.17. Left: The two branches c and $\widehat{c}$ of the isoptic curve of a parabola p are part of one conic. From points on c, the parabola p is seen under the angle φ; from points on $\widehat{c}$ the ellipse is seen under the angle $\pi - \varphi$. Right: If $\varphi = \pi$, then the two branches c and $\widehat{c}$ become one line of multiplicity two. It is p's directrix.

The equation of a parabola p in Cartesian coordinates (X, Y) is

$$p : X^2 - 2qY = 0$$

where $q \in \mathbb{R} \setminus \{0\}$. The point (x, y) is a point of the isoptic curve i_φ of p if it is the intersection of two tangents

$$t : xu - q(v + y) = 0, \quad \tau : \xi x - q(\eta + y) = 0 \qquad (9.11)$$

which touch p at the points (u,v) and (ξ,η), respectively. The contact points satisfy

$$u^2 - 2qv = 0, \quad \xi^2 - 2q\eta = 0. \tag{9.12}$$

Again, we use the equations of the tangents (9.11) and read off the tangent of the angles enclosed with the x-axis of the underlying Cartesian coordinate system. We denote these angles by the same Greek letters as done in the case of the ellipse and arrive at

$$\tan\varphi_1 = \frac{u}{q} \quad \text{and} \quad \tan\varphi_2 = \frac{\xi}{q}.$$

With the formula for $\tan(\varphi_2 - \varphi_1)$ we have

$$\tan\varphi = \tan(\varphi_2 - \varphi_1) = q\frac{\xi + u}{q^2 + \xi u} \tag{9.13}$$

for the tangent of $\varphi = \mathord{\}}(t,\tau)$. From (9.11), (9.12), and (9.13) we eliminate u, v, ξ, and η in order to find the implicit equation of the isoptic curves of the parabola p. This yields

$$i_\varphi : \ (q + 2y)^2 - 4\cot^2\varphi(x^2 - 2qy) = 0. \tag{9.14}$$

Surprisingly, the equation of the isoptics of parabolas are quadratic. Thus, these curves are conics. Moreover, the quadratic part of (9.14) equals

$$y^2 - x^2\cot^2\varphi$$

from which we infer that i_φ is a hyperbola as long as $\varphi \neq 0, \frac{\pi}{2}, \pi$. For an elementary approach to this result, we refer to Exercise 2.2.7. Figure 9.15 (middle) shows some isoptic hyperbolas of a parabola.

The unexpected reduction of the degree of the isoptics in the present case is caused by the following fact: In the last elimination step, the factor

$$4x^2 + (q - 2y)^2$$

splits off from the final resultant. This quadratic polynomial does not involve φ and can, therefore, not describe a part of the isoptic curve. If it is set equal to zero, this is the equation of the pair of isotropic lines through the focus of the parabola p.

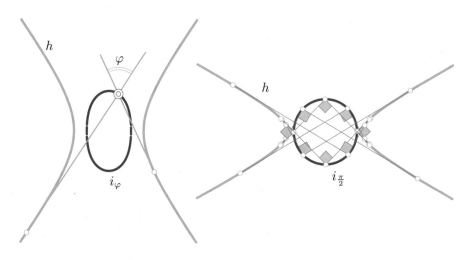

FIGURE 9.18. Left: Isoptic curve of a hyperbola h. Right: Orthoptic curve of a hyperbola h.

Since (9.14) can be rearranged and homogenized as done above for the isoptics of the ellipse, we can write

$$\tan^2 \varphi (px_0 + 2x_2)^2 - 4(x_1^2 - 2px_0x_2) = 0.$$

Now, the equation of the orthoptic is found by inserting $\varphi = \frac{\pi}{2}$ which gives

$$(px_0 + 2x_2)^2 = 0.$$

Thus, the orthoptic of a parabola is its directrix with multiplicity two. In other words: Tangents of a parabola that intersect in points of the directrix are orthogonal. Figure 9.15 as well as Figure 9.17 shows some isoptic curves of a parabola and the directrix as the orthoptic.

Now we can summarize the results:

Theorem 9.2.2 *The isoptic curves i_φ of parabolas are hyperbolas if $\varphi \neq 0, \frac{\pi}{2}, \pi$. The directrix of a parabola is the parabola's orthoptic curve.*

● **Exercise 9.2.1** Isoptic curves of a hyperbola.

Show that the equation of the isoptic curve i_φ of the hyperbola

$$\frac{x^2}{a^2} - \frac{y^2}{b^2} = 1$$

is

$$(x^2 + y^2 - a^2 + b^2)^2 + 4\cot^2 \varphi (b^2 x^2 - a^2 y^2 - a^2 b^2) = 0.$$

The orthoptic curves of the given hyperbola is the circle centered at $(0,0)$ with radius $\sqrt{a^2 - b^2}$ provided that $a > b$. What happens in the cases $a = b$ or $a < b$ (cf. Figure 9.18)? What is the difference between i_φ and $i_{\pi-\varphi}$?

The double role of an isoptic of a conic with center

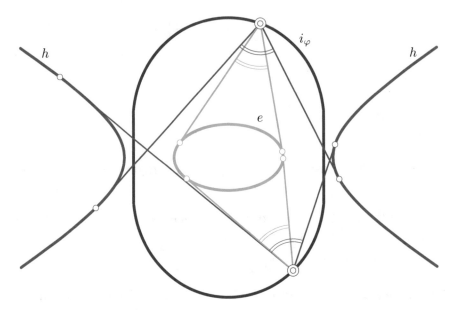

FIGURE 9.19. Any isoptic curve i_φ of a conic e is also an isoptic curve to another conic h, but for a different angle.

The only coefficients in the equation of the isoptic of an ellipse as given in (9.10) which depend on a, b, and φ are those of the monomials x^2, y^2, and the of the term free of x and y:

$$
\begin{aligned}
C_x &:= -2(a^2 + b^2 - 2b^2 \cot^2 \varphi), \\
C_y &:= -2(a^2 + b^2 - 2a^2 \cot^2 \varphi), \\
C_0 &:= (a^2 + b^2)^2 + 4a^2 b^2 \cot^2 \varphi.
\end{aligned}
\tag{9.15}
$$

Assume that we are given an isoptic curve i_φ of an ellipse e for some $\varphi \neq 0, \frac{\pi}{2}, \pi$. Then, the coefficients C_x, C_y, and C_0 are uniquely determined. On the other hand, we can consider only C_x, C_y, and C_0 as given values and assume that we have forgotten a, b, and φ. So, the equations (9.15) are a system in the unknowns a^2, b^2, and $\cot^2 \varphi$. Since the first two equations are linear and the second equation is quadratic, we find up to two values for a^2, b^2, $\cot^2 \varphi$. Therefore, we can say:

Theorem 9.2.3 *Any isoptic curve i_φ of a conic e with center is an isoptic curve $i_{\tilde{\varphi}}$ for another concentric conic h. The squares of the semiaxes lengths and the angle are the solutions of the system given in (9.15).*

Figure 9.19 shows two conics e and h with the same curve i_φ for an isoptic of both conics.

The isoptic curves of conics are spiric curves. Some of these curves can be seen in Figure 3.37 in Section 3.2 where spiric curves appeared as generalizations of conics.

9.3 Pedal points on conics, Apollonian hyperbola

Pedal points on ellipses and hyperbolas

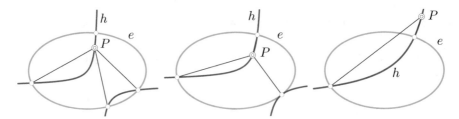

FIGURE 9.20. Different appearances of Apollonian hyperbolas correspond to different numbers of pedal points: from left to right: four, three, and two normals from P to e.

Let us now face a problem that may be important for various problems in computational geometry, $e.g.$, for collision detection. Assume we are given an ellipse e with the Cartesian equation

$$e : \frac{x^2}{a^2} + \frac{y^2}{b^2} = 1$$

with $a > b$ and a point $P = (u, v)$ in the plane of the ellipse. We ask for a point $N \in e$ that is closest to the given point P.

For an extremal distance between a parametrized curve $\mathbf{c}(t)$ and the point P with coordinate vector $\mathbf{p}$ it is necessary that

$$\frac{\mathrm{d}}{\mathrm{d}t}\langle \mathbf{c} - \mathbf{p}, \mathbf{c} - \mathbf{p}\rangle = \langle \dot{\mathbf{c}}, \mathbf{c} - \mathbf{p}\rangle = 0.$$

This means, at points in extremal distance to P the normal of the curve has to pass through P. Therefore, we measure the distance of P to the ellipse e on a normal n of the ellipse that is passing through P. The point $F \in e$ where n is the ellipse's normal is called *pedal point* of the normal n drawn from P to e. Thus, the distance of P to the ellipses satisfies $\overline{eP} = \overline{FP}$.

The normal n at $N = (x, y) \in e$ is parallel to the gradient

$$\frac{1}{2}\mathrm{grad}_c(N) = \left(\frac{x}{a^2}, \frac{y}{b^2}\right)$$

of the left-hand side of e at N, and thus, we have the condition

$$\left\langle \begin{pmatrix} u - x \\ v - y \end{pmatrix}, \begin{pmatrix} -\dfrac{y}{b^2} \\ \dfrac{x}{a^2} \end{pmatrix} \right\rangle = 0.$$

The latter equation can be written as

$$h : \ (a^2 - b^2)xy + vb^2 x + ua^2 y = 0. \tag{9.16}$$

Equation (9.16) is the equation of an equilateral hyperbola h passing through the center M of the ellipse since $(0,0)$ satisfying (9.16). The hyperbola is called *Apollonian hyperbola* and does not change if (a,b) is replaced with $(\lambda a, \lambda b)$ while $\lambda \neq 0$. Its asymptotes are parallel to the axes of the ellipse. Obviously, we can find the candidates of closest points as the at most four points of intersection of h and the given ellipse e. Thus, we have:

Theorem 9.3.1 *The pedal points of the normals drawn from a point P to an ellipse e are located on the Apollonian hyperbola h. The hyperbola h contains the point P and the center M of the ellipse. Furthermore, it is equilateral since its asymptotes are parallel to the axis of the e. The hyperbola h remains the same when the ellipse is transformed by a central similarity with center M and scaling factor $\lambda \neq 0$.*

The number of solutions to the initial problem is at most four since there are at most four common points of e and h. We can find at least two real solutions, *i.e.*, there are at least two real normals of e passing through a point P in e's plane because the hyperbola h contains the center M of the ellipse e. Figure 9.20 shows three different choices of P such that there are four, three, and two real normals form P to e.

• **Exercise 9.3.1** Normals from a point P to a hyperbola.

Show that Theorem 9.3.1 holds in similar way if we replace the ellipse e by the hyperbola l with the equation

$$l : \ \frac{x^2}{a^2} - \frac{y^2}{b^2} = 1.$$

In this case, the Apollonian hyperbola has the equation

$$(a^2 + b^2)xy - vb^2 x - ua^2 y = 0. \tag{9.17}$$

Figure 9.21 shows some appearances of Apollonian hyperbolas depending on P.

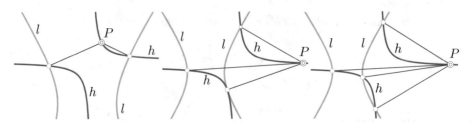

FIGURE 9.21. Different appearances of Apollonian hyperbolas h correspond to different numbers of pedal points: from left to right: two, three, and four normals from P to the hyperbola l.

The Apollonian hyperbola for an ellipse or a hyperbola degenerates and splits into a pair of different lines if $u = 0$ or $v = 0$, *i.e.*, if the point P lies on one of the axes of the ellipse or hyperbola, cf. (9.16) and (9.17). If we insert $a = b$ into (9.16), then we see that the Apollonian hyperbola of a circle is the circle's diameter through P.

Pedal points on parabolas

● Exercise 9.3.2 Pedal points on a parabola. Now we treat the normals from a point $P = (u, v)$ to the parabola p with the equation

$$p:\ x^2 - 2qy = 0$$

with $q \neq 0$. The techniques are almost the same. We treat this case separately since the number of solutions differs from that in the case of ellipses and hyperbolas. Again, we find an Apollonian hyperbola h. Here, it has the equation

$$h:\ xy + (q - v)x - uq = 0.$$

We cannot expect the same behavior of h for there is no center to pass through. If $u = 0$, then P is a point on the axis of the parabola. Exactly in this case, the Apollonian hyperbola of p splits into p's axis (with equation $x = 0$) and a further line (with equation $y = v - q$).

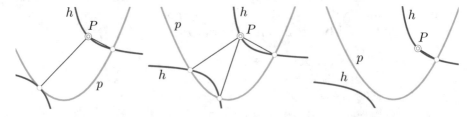

FIGURE 9.22. The normals from a point P to a parabola p (from left to right): two solutions, one with multiplicity two, the generic case with three solutions, one solution.

The two conics c and h share one ideal point $(0 : 0 : 1)$, and thus, we obtain at most three common points and at least one real common point (see Figure 9.22). So, we have

Theorem 9.3.2 *The number of real normals which can be drawn from a generic point P to a parabola p ranges between one and three. The pedal points lie on an equilateral hyperbola h which also contains P. The asymptotes of h are parallel to the axis and the tangent at the vertex of the parabola p.*

9.4 Pedal curves of conics

Pedal curve of a circle

Assume we are given a conic c and a point F in c's plane. We define the *pedal curve p of c with respect to F* as the set of pedal points of normals from F to c's tangents. In the simple case of a circle c and the choice of F as the center of c the pedal curve p equals the initial circle. Theorem 2.2.2 was a first non-trivial result on pedal curves of conics and showed that circles may appear as pedal curves.

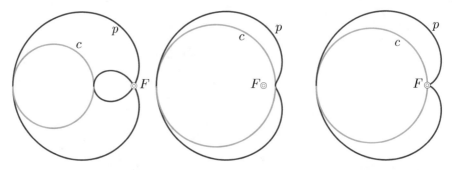

FIGURE 9.23. Pascal's limaçons as pedal curves p of a circle c for three different choices of F. Left: F is an ordinary double point on p if c sends two tangents through F. Middle: F is an isolated double point of p if F is an interior point of c. Right: The pedal curve has a cusp at F if $F \in c$; p is a cardioid.

Some basic properties of pedal curves (as we can observe them in Figure 9.23) can be seen from the definition and hold if c is an arbitrary conic (ellipse, parabola, hyperbola). There are two, one, or no real tangents of c passing through F. Thus, there are two, one, or zero real normals and the same number of pedal points coinciding with F. Therefore, the point F is an ordinary double point on p in the first case, a cusp of the first kind on p (like $(0,0)$ on $x^3 = y^2$), or an isolated double point on p.

If c is a circle and F does not coincide with the center C, then let $d = \overline{CF}$. The equations of the pedal curves of c now depend on d. Without loss of generality we may assume that c is the circle with equation $(x-d)^2 + y^2 = r^2$ with $r > 0$ and $d \in \mathbb{R}$ and $F = (0,0)$. The tangents of c are given by

$$t : (x - d) \cos \varphi + y \sin \varphi = r$$

and the corresponding normals through F have the equation

$$n: \quad -x\sin\varphi + y\cos\varphi = 0$$

with $\varphi \in [0, 2\pi[$ defining the point of contact. The intersection of t and n results in a parametrization of the pedal curve p and reads

$$p(\varphi) = (\cos\varphi(r + d\cos\varphi), \ \sin\varphi(r + d\cos\varphi)). \tag{9.18}$$

We eliminate the parameter φ from 9.18 and find the implicit equation of the circles' pedal curves as

$$p: \ (x^2 + y^2)^2 - 2dx(x^2 + y^2) + (d^2 - r^2)x^2 - r^2y^2 = 0. \tag{9.19}$$

Since the terms of lowest degree in x and y are quadratic, the curve p has a double point at $F = (0,0)$. The linear factors

$$\left(x\sqrt{d^2 - r^2} + ry\right)\left(x\sqrt{d^2 - r^2} - ry\right) = 0$$

of the quadratic term of (9.19) give the equations of the tangents of c at F. This again shows that the point F is an ordinary double point if $d > r$, that F is a cusp if $d = r$, that F is isolated if $d < r$.

The curve p given in (9.19) has cusps at the absolute points $(0 : 1 : \pm i)$. This can be concluded from properties of the pedal transformation (cf. Theorem 7.5.3 on page 335).

Conchoids of a circle

Let $F = (0,0)$ be a point in the plane of the circle c with center $C = (d,0)$ and radius $r > 0$. Now we construct a curve k from F and the points of c by adding a segment of constant length $l \in \mathbb{R}$ to the point $X \in c$ on the lines $[F, X]$. The curve k is called the *conchoid of the circle c at distance l*, cf. Figure 9.24. Note that k has two branches k_1 and k_2, one branch corresponding to l and the other one corresponding to $-l$. If K is the endpoint of such segment, then its trace can be parametrized by first parametrizing the set of points $X \in c$ as $(d + r\cos t, r\sin t)$ with $t \in [0, 2\pi[$. Then, we find a parametrization of the conchoid k

$$\mathbf{k}(t) = \left((d + r\cos t)\left(1 + \frac{l}{W}\right), \ r\sin t\left(1 + \frac{l}{W}\right)\right) \tag{9.20}$$

where $W := \sqrt{d^2 + r^2 + 2dr\cos t}$.

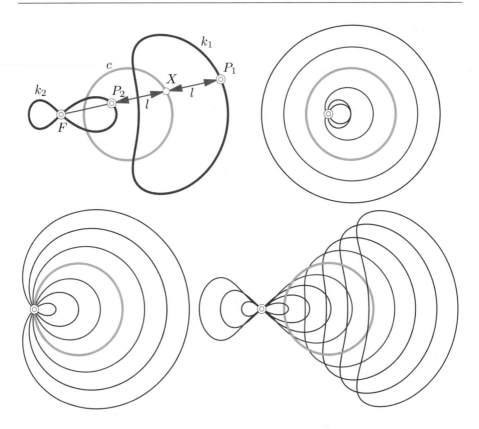

FIGURE 9.24. Construction of conchoids of a circle c: Top row: Determine points P_1, P_2 on $[F, X]$ such that $\overline{XP_1} = d = \overline{XP_2} = d$ (left). Some conchoids of a circle with F as interior point (right). Bottom row: Some conchoids of a circle with F on the circle (left), and an outer point F (right).

By eliminating the parameter t we obtain an implicit equation of the conchoid k which reads

$$
\begin{aligned}
(x^2 + y^2)^3 &- 2(l^2 + r^2 - 3d^2 + 2dx)(x^2 + y^2)^2 \\
&- 4d^2 x^2 y^2 + 4dx(l^2 + r^2 - d^2)(x^2 + y^2) \\
&+ ((l^2 - d^2 - r^2)^2 - 4d^2 r^2)(x^2 + y^2) = 0
\end{aligned}
\tag{9.21}
$$

Obviously, this is a sextic which has triple points at the absolute points since $(x^2 + y^2)^3 = 0$ describes the directions to the points of intersection with the ideal line.

If now $F \in c$, i.e., $d = \pm r$, the sextic splits into a circle about F with radius l and a quartic:

$$(x^2 + y^2 - l^2)((x^2 + y^2)^2 - 4rx(x^2 + y^2) + (4r^2 - l^2)x^2 - l^2y^2) = 0$$

A comparison of the quartic factor with (9.19) shows that, in the special case $F \in c$, the pedal curve p equals the conchoid k if, and only if, $l = \pm r$ and $r = \frac{d}{2}$. In this special case, k is known as *Pascal's limaçon*, named after ÉTIENNE PASCAL (1588–1651), the father of BLAISE PASCAL. So, we can state:

Theorem 9.4.1 *Pascal's limaçon is a pedal curve and a conchoid of a circle at the same time.*

Figure 9.24 shows some conchoids of a circle for some choices of F and l.

Pedal curves of parabolas

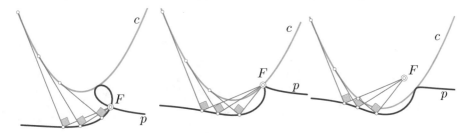

FIGURE 9.25. Pedal curves of a parabola c with respect to F: an outer point F (left), $F \in c$ (middle), and an interior point F (right).

Among the pedal curves of a parabola we find a lot of well-known curves. We shall give some examples here. Let us assume that a parabola c is given by the equation

$$c: \ x^2 - 4ay = 0,$$

or equivalently, by the parametrization

$$c(t) = \left(4at, 4at^2\right)$$

where $a \in \mathbb{R} \setminus \{0\}$ and $t \in \mathbb{R}$. The pedal curve will be constructed with respect to the point $F = (u, v)$. The tangents of c and the normals n passing through F are now given by their Cartesian equations

$$4at^2 - 2tx + y = 0 \quad \text{and} \quad u + 2tv - x - 2ty = 0. \tag{9.22}$$

We eliminate the parameter t from these two equations and find an implicit equation of c's pedal curve p with respect to F as

$$p: (y^2 + x^2)y + (a - v)x^2 - uxy - 2vy^2 + u(v - 2a)x + v^2 y + au^2 = 0. \quad (9.23)$$

On the other hand, a rational parametrization of p can be found by simply solving the system of linear equations given in (9.22) with respect to x and y. Thus, we can say:

Theorem 9.4.2 *The pedal curve p of a parabola c with respect to any point F is a rational cubic curve with its singularity at F. The cubic p is circular, i.e., it contains the absolute points $(0:1:\pm i)$.*

The point F is singular on p, to be more precise, it is a double point by the same reasons as in the case of a circle. Figure 9.25 shows the pedal curves p of a parabola c for three different choices of F.

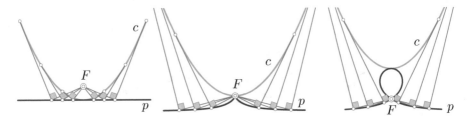

FIGURE 9.26. Special choices of F result in special pedal curves.

Notable pedal curves appear for some special choices of F. If we choose F equal to the focus of the parabola c, *i.e.*, $u = 0$ and $v = a$, then (9.23) simplifies and factors into

$$y(x^2 + (y - a)^2) = 0$$

which is the equation of a reducible cubic curve being the union of c's tangent at the vertex (with the equation $y = 0$) and the two isotropic lines $x \pm i(y - a) = 0$ through c's focus F (see Figure 9.26, left). This means that the lines through the focus of a parabola which are orthogonal to the tangents of the parabola meet the tangents in points on the tangent at the parabola's vertex (compare with Theorem 2.2.2 and Figure 2.13). This gives the following kinematic generation of a parabola:

Theorem 9.4.3 *Let the vertex of a rigid right angle glide along a line l such that one leg m is incident with a fixed point $F \notin l$. Then, the envelope of the other leg n is a parabola with focus F and the tangent l at the vertex.*

If $F = (0,0)$, *i.e.*, F is the vertex of c, then (9.23) simplifies to

$$y(x^2 + y^2) + ax^2 = 0.$$

This curve is called *right cissoide* (see Figure 9.26 in the middle) and is a special case of cissoides. The more general forms of cissoides appear for any other choice of F on the parabola c.

We obtain MacLaurin's *trisectrix* if $F = (0, -3a)$. Then, the tangents at p's double point F enclose an angle of $120°$.

Some prominent examples of pedal curves of conics with center

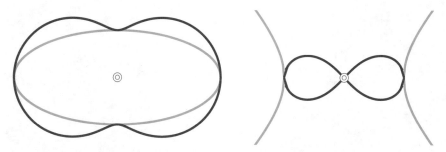

FIGURE 9.27. Some lemniscates as pedal curves of conics with center: Left: The pedal curve of an ellipse with respect to the ellipse's center is BOOTH's lemniscate. Right: The pedal curve of an equilateral hyperbola with respect to its center is BERNOULLI's lemniscate.

• Exercise 9.4.1 Two lemniscates as pedal curves.
Assume $a, b \in \mathbb{R}$ and $a > 0$, $b > 0$. Then,

$$c: \frac{x^2}{a^2} + \varepsilon\frac{y^2}{b^2} = 1$$

is the equation of a conic. The curve c is an ellipse/hyperbola if $\varepsilon = +1/\varepsilon = -1$. Show that

$$p: \ \varepsilon^2(x^2 + y^2)^2 - 2\varepsilon^2(ux + vy)(x^2 + y^2) - \varepsilon^2(a^2 - u^2)x^2) - (\varepsilon^2 v^2 - b^2)y^2$$
$$-2\varepsilon^2 uvxy - 2\varepsilon^2 a^2 ux - 2b^2 vy + a^2\varepsilon^2 u^2 + b^2 v^2$$

is the equation of the pedal curve of c with respect to $F = (u, v)$. For the special choice $F = (0,0)$ the pedal curve p is BOOTH *'s lemniscate* if $\epsilon = +1$ and it is BERNOULLI *'s lemniscate* if $\epsilon = -1$ (see Figure 9.27).

9.5 Poncelet porisms

FIGURE 9.28. A Poncelet polygon defines a quadrangular Poncelet grid. There are two families of conics supplying the vertices of the grid.

A porism is a theorem about a closure property of a geometric figure or construction. It is a theorem of the form: If some problem has a solution for a specific choice of initial elements, then it can be solved infinitely many times, *i.e.*, for any admissible choice of initial elements.

At first we describe the *Braikenridge-MacLaurin problem* of closed polygons which are inscribed in one polygon and circumscribed to another

polygon at the same time. Then, we pay attention to the *porism of* PON-
CELET which deals with polygons which are inscribed in one conic and
circumscribed to another conic. The conics appearing in the Poncelet porism can be seen as a generalization of the polygons which occur in the
Braikenridge-MacLaurin problem. In this section, conics are alway assumed to be regular unless otherwise stated.

Closure problems in polygons

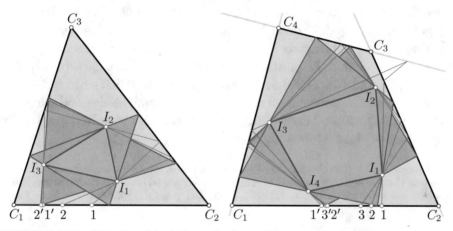

FIGURE 9.29. Closed polygons inscribed in and circumscribed to triangles or
quadrangles always correspond to the solutions of quadratic equations.

For given $n \geq 3$ we want to construct an n-gon that is circumscribed to a
given n-gon and inscribed in a given n-lateral in $\mathbb{P}^2(\mathbb{R})$. Literally, we shall
try out the problem on a triangle $I_1I_2I_3$ and a trilateral consisting of the
sides of the triangle $C_1C_2C_3$. We start with a point labeled by 1 on the
side $[C_1, C_2]$. We join 1 with I_1 and extend the segment to its intersection
with the side $[C_2, C_3]$. Then, we extend the line from this point through
I_2 and intersect with the line $[C_3, C_1]$, and finally extend the line from
this point through I_3 to intersect $[C_1, C_2]$ again. If we have the good
fortune to discover that the new point $1'$ coincides with 1, then we have
found a solution. Figure 9.29, left, illustrates our attempts. However, this
event will not be the case in general, and so we try again a second and
third time. The sequences of points 1, 2, 3, ... and $1'$, $2'$, $3'$, ... are related
via the projectivity

$$\alpha\colon\ [C_1, C_2] \overset{I_1}{\barwedge} [C_2, C_3] \overset{I_2}{\barwedge} [C_3, C_1] \overset{I_3}{\barwedge} [C_1, C_2].$$

Now the question arises: Is there a fixed point, *i.e.*, a point 1 which coincides with the corresponding point $1' = \alpha(1)$? We learned in Section 5.2 that $\alpha\colon \mathbf{x}\mathbb{R} \to \mathbf{x}'\mathbb{R}$ is represented by a linear map $\mathbf{x}' = \mathbf{A}\mathbf{x}$ with a 2×2 matrix $\mathbf{A}$ when on the line $[C_1, C_2]$ homogeneous coordinates are used. $\mathbf{x}\mathbb{R}$ is fixed if, and only if, there is an *eigenvalue* $\lambda \in \mathbb{R}$ with $\mathbf{A}\mathbf{x} = \lambda\mathbf{x}$. The eigenvalues of $\mathbf{A}$ are the zeros of the quadratic characteristic polynomial of $\mathbf{A}$, and each eigenvalue λ gives at least one fixed point as nontrivial solution of the homogeneous linear system $(\mathbf{A} - \lambda\mathbf{I}_2)\mathbf{x} = \mathbf{0}$. Hence, either two fixed points (two distinct real, or one real of multiplicity 2, or a conjugate complex pair) are to be expected, or, when $\mathbf{A}$ happens to be a multiple of the unit matrix $\mathbf{I}_2$, then α is the identity, and the triangle closes for each choice. The latter occurs when the point triples $\{C_1, I_1, I_2\}$, $\{C_2, I_2, I_3\}$ and $\{C_3, I_3, I_1\}$ are collinear (Figure 9.30, left). Since the solutions of a quadratic equation can be found by straightedge and compass, the BRAIKENRIDGE-MACLAURIN problem can also be solved graphically by means of classical drawing tools.

● Exercise 9.5.1 The Braikenridge-MacLaurin problem asks for triangles inscribed in the triangle $C_1 C_2 C_3$ and circumscribed to $I_1 I_2 I_3$. Why is there an infinity of such triangles if, and only if, the point triples $\{C_1, I_1, I_2\}$, $\{C_2, I_2, I_3\}$ and $\{C_3, I_3, I_1\}$ are collinear (note Figure 9.30, left)?

The number n of vertices (or equivalently, the number of edges) of the involved polygons $C_1 C_2 \ldots C_n$ and $I_1 I_2 \ldots I_n$ (see the case $n = 4$ in Figure 9.29) plays no role at this problem; the presented procedure works in the same way for $n > 3$. There is always a quadratic equation to be solved, *i.e.*, there are either two real solutions, or one real solution of multiplicity two, or two conjugate complex solutions, or even infinitely many solutions. However, the BRAIKENRIDGE-MACLAURIN problem is no example of a porism.

There is no big difference between this problem and the modification in which the polygon $C_1 C_2 \ldots C_n$ is replaced by a conic c. Under the assumption that none of the vertices $I_1, I_2, \ldots I_n$ is a point of c we can again start with points $1, 2, 3 \in c$ like before. Thus, we obtain on c a projectivity α which is the product of involutions with the centers I_1, I_2, and so on. Again, the fixed points of α are the starting points for closing n-gons inscribed in c and circumscribed to the n-gon $I_1 I_2 \ldots I_n$.

In the case $n = 3$ the projectivity α is the identity if, and only if, $I_1 I_2 I_3$ is auto-polar with respect to c. But this implies that two vertices of this

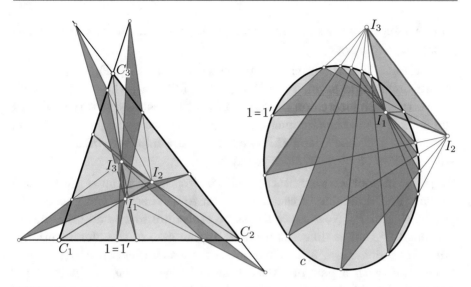

FIGURE 9.30. These are particular cases where each triangle closes which is inscribed in the triangle C_1C_2C (left) or to the conic c (right) and which has sides passing in turn through I_1, I_2 and I_3.

auto-polar triangle are outside of c, *i.e.*, in the area of points where two real tangent lines of c are meeting (note Figure 9.30, right).

Special porisms

What about the following? We replace the outer and inner polygon by conics c and i and ask for n-gons which are inscribed in c and circumscribed

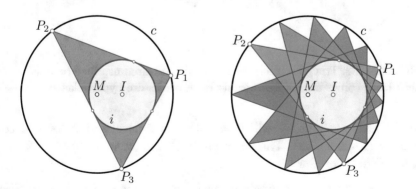

FIGURE 9.31. Left: triangle with incircle i and circumcircle c. Right: poristic family of triangles.

to i. A famous and well-known example comes from the elementary geometry of the triangle.

Any triangle $\mathcal{T}$ in the Euclidean plane with vertices P_1, P_2, and P_3 has a circumcircle c and an incircle i. The sides of $\mathcal{T}$ touch the incircle i while the vertices are located on c (Figure 9.31). Surprisingly, it turns out that the incircle i and the circumcircle c of $\mathcal{T}$ are incircle and circumcircle for more than just one triangle. Indeed, there are infinitely many triangles with vertices on c whose sides touch i.

To put it in another way: Start with two circles, say c and i. Pick a point P_0 on c and draw one of the tangents t_0 to i. Then, t_0 and c share two points P_0 and P_1. Now apply the same procedure to P_1 and let t_1 be the tangent to i that is different from t_0. In the case of the aforementioned triangle we find $P_3 = P_0$, i.e., the polygon $P_0 P_1 P_2$ is closed. Furthermore, if we start with a pair of circles that is known to be the incircle and circumcircle of a triangle $\mathcal{T}$, then the polygon closes independently of the choice of P_0. There is a one-parameter family of triangles sharing the incircle and the circumcircle.

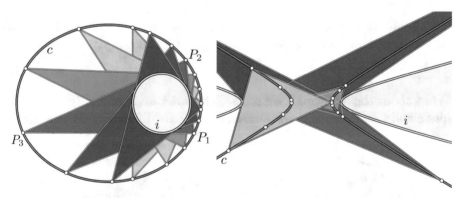

FIGURE 9.32. Triangles inscribed in a conic and circumscribed to another conic. Left: Both conics are ellipses. Right: Both conics are hyperbolas.

In the next sections we will deal with the generalization for two conics c and i (Figure 9.32). The theorem of COLIN MACLAURIN (1698–1746) states (note Corollary 9.5.5 on page 425)

> *If there exists one triangle inscribed in a conic c and circumscribed to another conic i, then there are infinitely many triangles inscribed in and circumscribed to these conics c and i.*

This result is remarkable for the following reason: We learned already in Section 6.4 that all pairs consisting of a conic and an inscribed triangle are projectively equivalent. However, for any triangle $P_1P_2P_3$ inscribed in c there is still a two-parametric family of inscribed conics i. Nevertheless, each of them constitutes a poristic family together with c. It will turn out that by Theorem 9.5.2 for any given conic c there is a four-parametric family of conics i admitting a poristic family of triangles.

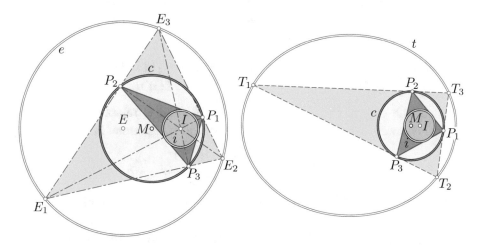

FIGURE 9.33. Further porisms. Left: While a triangle $P_1P_2P_3$ is traversing its poristic family, the triangle's excenters E_1, E_2, E_3 trace a circle e. Thus, there are three nested porisms. Right: The vertices T_1, T_2, T_3 of the tangent triangles of all triangles in a poristic family trace an ellipse t.

There are some more porisms involving one-parameter families of triangles. Figure 9.33 shows a few examples: If a triangle moves through its poristic family, then the vertices of its *excentral triangle, i.e.,* the triangle built by the centers of the excircles, move on a circle e. The radius of e is twice the circumradius of the triangles in the poristic family. Since e's center E is the reflection of the incenter I in the circumcenter M, e can be obtained from the circumcircle c by applying a central dilation with center I and scale factor 2. On the right-hand side of Figure 9.33 we see a kind of porism as described in Theorem 9.5: The vertices of the so-called *tangent triangle* move on an ellipse while the base triangle is traversing its poristic family.

Inpolar conics

There are many porisms connected with pairs of conics. This is the reason why in the coming sections we study relations between two conics which are projectively invariant. Most important in this respect is the (generalized) *characteristic polynomial* of two 3×3 matrices[8] $\mathbf{C}, \mathbf{D}$

$$F_{\mathbf{C},\mathbf{D}}(\sigma,\tau) := \det(\sigma\mathbf{C} + \tau\mathbf{D}) = k_0\,\sigma^3 + k_1\,\sigma^2\tau + k_2\,\sigma\tau^2 + k_3\,\tau^3. \quad (9.24)$$

The substitutions $\tau = 0$ or $\sigma = 0$ reveal $k_0 = \det\mathbf{C}$ and $k_3 = \det\mathbf{D}$. But there are also formulas for the remaining coefficients. For regular $\mathbf{C}$ and $\mathbf{D}$ we can rewrite $F_{\mathbf{C},\mathbf{D}}(\sigma,\tau)$ as

$$\det(\sigma\mathbf{C} + \tau\mathbf{D}) = \det\left[\mathbf{C}(\sigma\mathbf{I}_3 + \tau\mathbf{C}^{-1}\mathbf{D})\right] = \det\left[(\sigma\mathbf{I}_3 + \tau\mathbf{D}\mathbf{C}^{-1})\mathbf{C}\right] =$$
$$= \det\left[\mathbf{D}(\sigma\mathbf{D}^{-1}\mathbf{C} + \tau\mathbf{I}_3)\right] = \det\left[(\sigma\mathbf{C}\mathbf{D}^{-1} + \tau\mathbf{I}_3)\mathbf{D}\right].$$

Hence, the generalized characteristic polynomial is related to the (ordinary) characteristic polynomial of $\mathbf{C}\mathbf{D}^{-1}$ and its inverse $\mathbf{D}\mathbf{C}^{-1}$ by

$$\begin{aligned} F_{\mathbf{C},\mathbf{D}}(1,-X) &= \det\mathbf{D} \cdot \det(\mathbf{C}\mathbf{D}^{-1} - X\mathbf{I}_3), \\ F_{\mathbf{C},\mathbf{D}}(-X,1) &= \det\mathbf{C} \cdot \det(\mathbf{D}\mathbf{C}^{-1} - X\mathbf{I}_3). \end{aligned} \quad (9.25)$$

The coefficient of X^2 in the (ordinary) characteristic polynomial $\det(\mathbf{M} - X\mathbf{I}_3)$ equals the *trace* $\operatorname{tr}\mathbf{M}$. This implies for the coefficients in the generalized characteristic polynomial $F_{\mathbf{C},\mathbf{D}}(\sigma,\tau)$ introduced in (9.24)

$$\begin{aligned} k_0 &= \det\mathbf{C}, & k_1 &= \det\mathbf{C} \cdot \operatorname{tr}(\mathbf{C}^{-1}\mathbf{D}) = \det\mathbf{C} \cdot \operatorname{tr}(\mathbf{D}\mathbf{C}^{-1}), \\ k_3 &= \det\mathbf{D}, & k_2 &= \det\mathbf{D} \cdot \operatorname{tr}(\mathbf{D}^{-1}\mathbf{C}) = \det\mathbf{D} \cdot \operatorname{tr}(\mathbf{C}\mathbf{D}^{-1}). \end{aligned} \quad (9.26)$$

Let $\mathbf{C}, \mathbf{D}$ be symmetric coefficient matrices of two conics in $\mathbb{P}^2(\mathbb{R})$. Changes of the underlying coordinate frame leave the ratio of coefficients $k_0 : k_1 : k_2 : k_3$ in $F_{\mathbf{C},\mathbf{D}}(\sigma,\tau)$ invariant since for $\mathbf{C}' = \mathbf{T}^{\mathsf{T}}\mathbf{C}\mathbf{T}$ and $\mathbf{D}' = \mathbf{T}^{\mathsf{T}}\mathbf{D}\mathbf{T}$ we get

$$F_{\mathbf{C}',\mathbf{D}'}(\sigma,\tau) = \det(\sigma\mathbf{D}'+\tau\mathbf{C}') = (\det\mathbf{T})^2 \det(\sigma\mathbf{D}+\tau\mathbf{C}) = (\det\mathbf{T})^2 F_{\mathbf{C},\mathbf{D}}(\sigma,\tau).$$

Nevertheless, the ratios of any two coefficients are still no invariants of the pair of conics since the replacement of $\mathbf{C}$ by a scalar multiple $\lambda\mathbf{C}$ yields

$$F_{\lambda\mathbf{C},\mathbf{D}}(\sigma,\tau) = (\lambda^3 k_0)\sigma^3 + (\lambda^2 k_1)\sigma^2\tau + (\lambda k_2)\sigma\tau^2 + k_3\,\tau^3. \quad (9.27)$$

The first binary relation on the set of conics is as follows:

---

[8]The characteristic polynomial is of course defined more generally for $n \times n$ matrices, and it is easy to rephrase the following particular results for general matrices.

Definition 9.5.1 In $\mathbb{P}^2(\mathbb{R})$, a conic p is called *inpolar* to the conic c if there exists a triangle ABC auto-polar with respect to p and inscribed in c (Figure 9.34).[9]

Since the conic p itself does not play a role here but only the polarity with respect to p, we want to extend this definition to the case that p has no real points. In any case, the conic c must have real points.

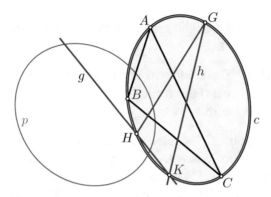

FIGURE 9.34. The conic p is inpolar to c

Below is a standard result of Projective Geometry.

Lemma 9.5.1 *Let a conic p be inpolar to a conic c. Then, there is a one-parameter set of triangles ABC which are auto-polar with respect to p and inscribed in c.*

Note that this example shows again a porism: If there is one auto-polar triangle inscribed in c, then there exist infinitely many.

Proof: Let ABC and GHK be two triangles which both are auto-polar with respect to p. Then, due to a result of VON STAUDT, the six vertices are located either on two lines or on a conic. This holds in each Pappian projective plane. A proof can, *e.g.*, be found in [18, p. 87].

Let the defining triangle ABC be given. Then, specify another point $G \in c$ such that its p-polar line g intersects c at a point $H \neq G$ (Figure 9.34). Continuity arguments guarantee the existence of G sufficiently close to A, B, or C. The line g and the polar h of H meet at a point K which completes a second auto-polar triangle GHK. There must be a conic on A, B, C, G, H, K. Since this conic is uniquely defined by the first five points, it coincides with c. ∎

---

[9]Compare [3, p. 33]. Sometimes in the literature the pair (p, c) is called *apolar*, but this notation does not reveal clearly that this relation is unsymmetric.

The intersection points H, K of g with c are conjugate with respect to p, *i.e.*, they are harmonic with respect to the intersection points of g with p. Hence, the polars g of points $G \in c$ intersect the two conics p and c in harmonic quadruples of points. At most two of them can be conjugate complex since two pairs of conjugate complex points can never be harmonic. The desired harmonic position implies that on g the two involutions of conjugate points with respect to p and to c commute (note Example 5.4.4).

This is the reason for a slight generalization of inpolarity which is needed in the following lemma.

Definition 9.5.1' The conic p in $\mathbb{P}^2(\mathbb{R})$ is called *inpolar* to c if there is a point $G \in c$, $G \notin p$, such that on the p-polar g of G the induced involutions of conjugate points with respect to p and c commute.

In this case, the point G together with the two real or complex conjugate points of intersection between g and c forms a triangle which is autopolar with respect to p. Then, Lemma 9.5.1 again implies that this holds for each choice of $G \in (c \setminus p)$. In fact, Figure 9.37 on page 428 shows an example where p is inpolar to c though for all $G \in c$ the p-polar g has no real intersection with c.

Lemma 9.5.2 *Let the conics p and c be given by the equations*

$$p: \ \mathbf{x}^{\mathrm{T}}\mathbf{P}\,\mathbf{x} = 0, \quad c: \ \mathbf{x}^{\mathrm{T}}\mathbf{C}\,\mathbf{x} = 0$$

with symmetric matrices $\mathbf{P}$ and $\mathbf{C}$. We suppose that c contains real points. Then, p is inpolar to c if, and only if, in the characteristic polynomial

$$F_{\mathbf{P},\mathbf{C}}(\sigma, \tau) = \det(\sigma\mathbf{P} + \tau\mathbf{C}) = j_0\,\sigma^3 + j_1\,\sigma^2\tau + j_2\,\sigma\tau^2 + j_3\,\tau^3$$

the coefficient $j_1 = \det\mathbf{P}\cdot\mathrm{tr}(\mathbf{CP}^{-1})$ is zero.

Proof: We learned that the ratio $j_0 : j_1 : j_2 : j_3$ of coefficients in the characteristic polynomial $F_{\mathbf{P},\mathbf{C}}(\sigma, \tau)$ does not depend on the choice of the coordinate frame. For the conics p inpolar to c, we use a coordinate frame defined by the fundamental triangle ABC. Then, $\mathbf{P}$ is a diagonal matrix while $\mathbf{C}$ has vanishing diagonal entries, hence

$$\sigma\mathbf{P} + \tau\mathbf{C} = \begin{pmatrix} \sigma p_{00} & \tau c_{01} & \tau c_{02} \\ \tau c_{01} & \sigma p_{11} & \tau c_{12} \\ \tau c_{02} & \tau c_{12} & \sigma p_{22} \end{pmatrix}$$

and

$$F_{\mathbf{P},\mathbf{C}}(\sigma, \tau) = \det(\sigma\mathbf{P} + \tau\mathbf{C}) = p_{00}p_{11}p_{22}\sigma^3 + \tau^2(e\sigma + f\tau)$$

with certain coefficients e, f. Obviously, the coefficient j_1 of $\sigma^2\tau$ is zero, and by (9.26) j_1 equals the product $\det \mathbf{P} \cdot \operatorname{tr}(\mathbf{CP}^{-1})$.

In order to prove the converse, we specify a coordinate frame which diagonalizes $\mathbf{P}$ and where the fundamental point $A = (1 : 0 : 0)$ is located on c. This implies $p_{00} \neq 0$ and $c_{00} = 0$.[10] Suppose that in the polynomial

$$F_{\mathbf{P},\mathbf{C}}(\sigma, \tau) = \det(\sigma\mathbf{P} + \tau\mathbf{C}) = \det \begin{pmatrix} \sigma p_{00} & \tau c_{01} & \tau c_{02} \\ \tau c_{01} & \sigma p_{11} + \tau c_{11} & \tau c_{12} \\ \tau c_{02} & \tau c_{12} & \sigma p_{22} + \tau c_{22} \end{pmatrix}$$

the coefficient of $\sigma^2\tau$ is zero, i.e., $j_1 = p_{00}(p_{11}c_{22}+p_{22}c_{11}) = 0$. On the line a: $x_0 = 0$ which is p-polar to A, the conics p and c induce (regular or singular) involutions $(0 : x_1 : x_2) \mapsto (0 : x_1' : x_2')$ of conjugate points, satisfying respectively

$$p_{11}x_1x_1' + p_{22}x_2x_2' = 0 \quad \text{or} \quad c_{11}x_1x_1' + c_{12}(x_1x_2' + x_2x_1') + c_{22}x_2x_2' = 0.$$

According to Example 5.4.4, $j_1 = 0$, i.e., $p_{11}c_{22} + p_{22}c_{11} = 0$, is equivalent to the condition that these two involutions commute. This proves (in the complex extension) the existence of a triangle ABC inscribed in c and auto-polar with respect to p. ∎

When we dualize p and c and reverse their order, i.e., when we replace the ordered pair $(\mathbf{P}, \mathbf{C})$ by $(\mathbf{C}^{-1}, \mathbf{P}^{-1})$, then the characterization of 'in-polarity' given in Lemma 9.5.2 remains valid since

$$\det(\sigma\mathbf{C}^{-1} + \tau\mathbf{P}^{-1}) = \det\left[\mathbf{C}^{-1}(\sigma\mathbf{P} + \tau\mathbf{C})\mathbf{P}^{-1}\right].$$

This means that there are infinitely many triangles auto-polar with respect to c and circumscribed to p, provided that also p has real points (and tangents). Thus, we obtain[11]

Corollary 9.5.3 *Let p and c be conics with real points. Then, p is inpolar to c if, and only if, c is "outpolar" to p, i.e., there are triangles auto-polar with respect to c and circumscribed to p.*

For given p there is a two-parametric family $\mathcal{C}$ of conics c such that p is inpolar to c. The set $\mathcal{C}$ is linear, i.e., with $c_1, c_2 \in \mathcal{C}$ all conics of the pencil spanned by c_1 and c_2 are in $\mathcal{C}$ since

$$\operatorname{tr}\left[\mathbf{P}^{-1}(\lambda_1\mathbf{C}_1 + \lambda_2\mathbf{C}_2)\right] = \lambda_1 \operatorname{tr}(\mathbf{P}^{-1}\mathbf{C}_1) + \lambda_2 \operatorname{tr}(\mathbf{P}^{-1}\mathbf{C}_2).$$

Conversely, for a given conic c the family $\mathcal{P}$ of conics p inpolar to c is two-parametric, and the duals of p constitute a linear family.

---

[10] Only for $p = c$ this choice would be impossible, but under $\mathbf{P} = \mathbf{C}$ we get $j_1 = 3\det\mathbf{C} \neq 0$.

[11] Alternative proofs can be found in [62, p. 213], [3, p. 33–34], or [8, p. 85–86]. The three-dimensional versions of Lemmas 9.5.1 and 9.5.1 can be found in [62, p. 213], the n-dimensional versions in [55, p. 862, footnote 287].

Triangles inscribed in and circumscribed to conics

Let p be inpolar to c. The polarity in p transforms points of c into tangents of a second conic i. The auto-polar triangles inscribed in c are at the same time circumscribed to i. So, there are infinitely many triangles inscribed in c and circumscribed i (see Figure 9.35).

We are going to show that this is the only way to obtain such a pair of conics, and start with the following

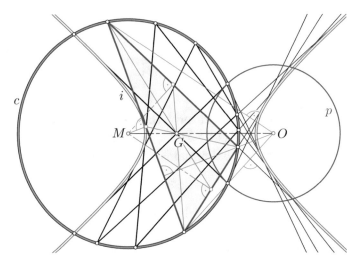

FIGURE 9.35. There is an infinite set of triangles inscribed in the circle c and circumscribed to the hyperbola i. Two triangles are degenerate; the others are auto-polar with respect to the circle p. They share the orthocenter O and the centroid G.

Lemma 9.5.4 *Let c and i be two different conics with real points such that there exists a triangle inscribed in c and circumscribed to i. Then, the tangent lines of i intersect c in points which are conjugate with respect to a regular or singular conic p.*

When $\mathbf{C}$, $\mathbf{P}$ and $\mathbf{D}^{-1}$ are symmetric coefficient matrices of the conics c and p and of the dual of i, resp., then up to a factor $\varrho \in \mathbb{R} \setminus \{0\}$

$$\varrho \mathbf{P} = -\operatorname{tr}(\mathbf{C}\mathbf{D}^{-1})\,\mathbf{C} + 2\,\mathbf{C}\mathbf{D}^{-1}\mathbf{C}.$$

Proof: It is easy to show that the stated representation of $\mathbf{P}$ is invariant against changes of the coordinate frame. We can proceed like in the proof for the influence of coordinate transformations on the characteristic polynomial on page 418. Hence, we can specify

$$c: \; 2(x_0 x_2 - x_1^2) = \mathbf{x}^{\mathsf{T}} \mathbf{C} \mathbf{x} \quad \text{with} \quad \mathbf{C} = \begin{pmatrix} 0 & 0 & 1 \\ 0 & -2 & 0 \\ 1 & 0 & 0 \end{pmatrix}.$$

Any two points of c can be given in parameter form as

$$\mathbf{p}\mathbb{R} = (u^2 : uv : v^2) \quad \text{and} \quad \mathbf{p}'\mathbb{R} = (u'^2 : u'v' : v'^2).$$

The connecting line $\mathbf{p}\mathbb{R} \vee \mathbf{p}'\mathbb{R}$ is represented by the vector

$$\mathbf{u} = \mathbf{p} \times \mathbf{p}' = (uv' - vu') \begin{pmatrix} vv' \\ -uv' - vu' \\ uu' \end{pmatrix},$$

which we divide by $(uv' - vu')$. This factor does not vanish under the assumption $\mathbf{p}\mathbb{R} \neq \mathbf{p}'\mathbb{R}$. Let $\mathbf{D}^{-1} = (\widehat{d}_{ij})$ with $\widehat{d}_{ji} = \widehat{d}_{ij}$ be the coefficient matrix of the dual $\widehat{i}$ of the conic i. Then, our chord $\mathbf{u}\mathbb{R}$ of c is tangent to i if, and only if, $\mathbf{u}^{\mathsf{T}} \mathbf{D}^{-1} \mathbf{u} = 0$ (note the points P and P' in Figure 9.36). This is equivalent to

$$\widehat{d}_{00} v^2 v'^2 - 2\widehat{d}_{01}(uvv'^2 + v^2 u'v') + 2\widehat{d}_{02} uvu'v' + \widehat{d}_{11}(u^2 v'^2 + 2uvu'v' + v^2 u'^2)$$
$$-2\widehat{d}_{12}(u^2 u'v' + uvu'^2) + \widehat{d}_{22} u^2 u'^2 = 0.$$

We can re-substitute $\mathbf{p}$ and $\mathbf{p}'$ and obtain the vanishing symmetric bilinear form

$$\mathbf{p}^{\mathsf{T}} \mathbf{P} \mathbf{p}' = 0 \quad \text{with} \quad \mathbf{P} = \begin{pmatrix} \widehat{d}_{22} & -2\widehat{d}_{12} & \widehat{d}_{11} \\ -2\widehat{d}_{12} & 2(\widehat{d}_{02} + \widehat{d}_{11}) & -2\widehat{d}_{01} \\ \widehat{d}_{11} & -2\widehat{d}_{01} & \widehat{d}_{00} \end{pmatrix}.$$

This means that the points $\mathbf{p}\mathbb{R}, \mathbf{p}'\mathbb{R} \in c$ are conjugate with respect to a regular or singular polarity with the symmetric matrix $\mathbf{P}$. By straightforward computation, we obtain due to the symmetry of $\mathbf{D}^{-1}$

$$\mathbf{C}\mathbf{D}^{-1} = \begin{pmatrix} \widehat{d}_{20} & \widehat{d}_{21} & \widehat{d}_{22} \\ -2\widehat{d}_{10} & -2\widehat{d}_{11} & -2\widehat{d}_{12} \\ \widehat{d}_{00} & \widehat{d}_{01} & \widehat{d}_{02} \end{pmatrix}, \quad \mathbf{C}\mathbf{D}^{-1}\mathbf{C} = \begin{pmatrix} \widehat{d}_{22} & -2\widehat{d}_{21} & \widehat{d}_{20} \\ -2\widehat{d}_{12} & 4\widehat{d}_{11} & -2\widehat{d}_{10} \\ \widehat{d}_{02} & -2\widehat{d}_{01} & \widehat{d}_{00} \end{pmatrix},$$

and we verify $(\widehat{d}_{11} - \widehat{d}_{02})\mathbf{C} + \mathbf{C}\mathbf{D}^{-1}\mathbf{C} = \frac{1}{2}\mathbf{P}$. Of course, the coefficient matrix of any polarity is unique only up to a non-vanishing factor ϱ. ∎

The matrix $\mathbf{P}$ in Lemma 9.5.4 is a linear combination of $\mathbf{C}$ and $\mathbf{C}\mathbf{D}^{-1}\mathbf{C}$. Therefore, the conic p belongs to the pencil spanned by c and by the polar of i with respect to c.

If any point A has the same polar line a w.r.t. c and i, then a is also the polar line of A w.r.t. to p or A is a singular point of p. This can be proved as follows: $\mathbf{C}\mathbf{a} = \lambda \mathbf{D}\mathbf{a} = \mathbf{u}$ for $\mathbf{a} \neq \mathbf{0}$ implies $\mathbf{D}^{-1}\mathbf{C}\mathbf{a} = \lambda \mathbf{a}$, hence

$$\mathbf{P}\mathbf{a} = \left[-\mathrm{tr}(\mathbf{C}\mathbf{D}^{-1})\,\mathbf{C} + 2\,\mathbf{C}\mathbf{D}^{-1}\mathbf{C} \right]\mathbf{a} = \left[-\mathrm{tr}(\mathbf{C}\mathbf{D}^{-1}) + 2\lambda \right]\mathbf{u} \in \mathbf{u}\mathbb{R}.$$

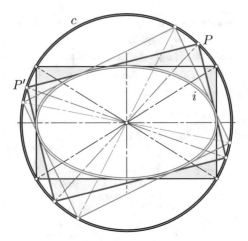

FIGURE 9.36. There is an infinite set of rectangles circumscribed to the ellipse i and inscribed in its director circle c. Their diagonals are conjugate w.r.t. i.

■ **Example 9.5.1** Rectangles inscribed in the director circle. Figure 9.36 illustrates an example where the matrix $\mathbf{P}$ has rank 2: Here, c is the director circle of the ellipse i, *i.e.*,

$$\mathbf{C} = \begin{pmatrix} -a^2 - b^2 & 0 & 0 \\ 0 & 1 & 0 \\ 0 & 0 & 1 \end{pmatrix}, \quad \mathbf{D}^{-1} = \begin{pmatrix} -1 & 0 & 0 \\ 0 & a^2 & 0 \\ 0 & 0 & b^2 \end{pmatrix}, \quad \mathbf{P} = \begin{pmatrix} 0 & 0 & 0 \\ 0 & -2b^2 & 0 \\ 0 & 0 & -2a^2 \end{pmatrix}.$$

The tangents of i intersect c at points $\mathbf{p}\mathbb{R}, \mathbf{p}'\mathbb{R}$ which are placed on conjugate diameters of i. This means exactly that for of each rectangle which is inscribed in the director circle c and circumscribed to the ellipse i, adjacent vertices are conjugate w.r.t. the singular polarity in p.

Theorem 9.5.1 *Let c and i be two different conics with real points, and let the tangents of i intersect c at points conjugate with respect to p. Then, the following three statements are equivalent:*

    (i) p is inpolar to c in the generalized sense (Definition 9.5.1', page 420).

    (ii) There exist infinitely many triangles (including such with a pair of complex conjugate vertices) inscribed in c and circumscribed to i.

    (iii) c and i are polar with respect to p.

Proof: (i) ⟺ (ii): By Lemma 9.5.1, there is an infinite set of triangles ABC inscribed in c and auto-polar with respect to p. The vertices B and C are the only points of c which are conjugate to A. Since the tangents of i which pass through A must intersect c in points conjugate to A, the sides AB and AC are tangent to i. The same holds for vertex B. Hence, each triangle ABC is circumscribed to i. When the two tangents from A to i are complex conjugate, then also B and C are complex conjugate, but their connecting line $[B, C]$ is real.

Conversely, according to the definition of p each triangle inscribed in c and circumscribed to i has pairwise conjugate vertices with respect to p. Therefore, it is auto-polar.

(ii) ⇔ (iii): The polarity with respect to p maps the vertices $A, B, C \in c$ of the auto-polar triangles onto the corresponding opposite sides. Hence, the envelope i is polar to c.

Conversely, for each point $A \in c$ the p-polar a is a tangent of i. Let a intersect c in two real or complex conjugate points B and C. Then, by the definition of p, the lines $[A, B]$ and $[A, C]$ are tangents of i, too. Hence, ABC is circumscribed to i and inscribed in c. ∎

Now, the announced theorem of COLIN MACLAURIN (see page 416) is just a corollary of Theorem 9.5.1.

Corollary 9.5.5 *If for two conics c, i there exists one triangle inscribed in c and circumscribed to i, then there are infinitely many.*

Proof: Suppose, ABC is such a triangle. Then, the vertices A, B, C are pairwise conjugate with respect to the uniquely defined polarity p (Lemma 9.5.4). Hence, ABC is auto-polar with respect to p, and thus, p inpolar to c. The rest follows from Theorem 9.5.1. ∎

■ **Example 9.5.2** Triangles sharing the circumcircle and the centroid. For the conics displayed in Figures 9.35 and 10.19 (page 463), we use a coordinate frame with the center M of the circle c as origin and the centroid G with cartesian coordinates $(g, 0)$. In Figure 9.35, we have $3g > r$, in Figure 10.19, we have $3g < r$. The equations of the involved conics are

$$c: \ x^2 + y^2 = r^2, \qquad i: \ x^2 - 3gx + \frac{r^2}{r^2 - 9g^2} y^2 = \frac{r^2 - 9g^2}{4},$$

$$p: \ x^2 - 6gx + y^2 = -\frac{9g^2 + r^2}{2}.$$

In homogeneous coordinates the symmetric coefficient matrices of c, i, and p are

$$\mathbf{C} = \begin{pmatrix} -r^2 & 0 & 0 \\ 0 & 1 & 0 \\ 0 & 0 & 1 \end{pmatrix}, \quad \mathbf{D} = \begin{pmatrix} \frac{9g^2 - r^2}{4} & -\frac{3g}{2} & 0 \\ -\frac{3g}{2} & 1 & 0 \\ 0 & 0 & \frac{r^2}{r^2 - 9g^2} \end{pmatrix},$$

$$\mathbf{P} = \begin{pmatrix} \frac{9g^2 + r^2}{2} & -3g & 0 \\ -3g & 1 & 0 \\ 0 & 0 & 1 \end{pmatrix}, \quad \mathbf{P}\mathbf{C}^{-1}\mathbf{P} = \frac{r^2 - 9g^2}{r^2}\mathbf{D}.$$

The characteristic polynomials read

$$F_{\mathbf{C},\mathbf{D}}(\sigma, \tau) = -r^2\sigma^3 + \frac{9(3g^2 - r^2)^2}{4}\sigma^2\tau + \frac{3r^2(r^2 - 3g^2)}{2(9g^2 - r^2)}\sigma\tau^2 + \frac{r^4}{4(9g^2 - r^2)}\tau^3,$$

$$F_{\mathbf{P},\mathbf{C}}(\sigma, \tau) = \det(\sigma\mathbf{P} + \tau\mathbf{C}) = \frac{r^2 - 9g^2}{2}\sigma^3 + \frac{9g^2 - 3r^2}{2}\sigma\tau^2 - r^2\tau^3.$$

We note that in the last polynomial the coefficients j_1 of $\sigma^2\tau$ is zero (cf. Lemma 9.5.2). In $F_{\mathbf{C},\mathbf{D}}(\sigma, \tau)$ the coefficients $k_0, \ldots, k_3$ satisfy $k_2^2 = 4k_1 k_3$, and this is in accordance with the Theorem 9.5.2 below.

We conclude this section with the analytic characterization of two conics c and i with the property stated in Theorem 9.5.1,(ii). As a preparation, we pick out the statement (i) $\Rightarrow$ (iii) from the aforementioned theorem and present additionally an analytical proof:

By virtue of Lemma 9.5.4, there is a polarity in a regular or singular conic p with the coefficient matrix $\mathbf{P} = -\mathrm{tr}(\mathbf{CD}^{-1})\,\mathbf{C} + 2\,\mathbf{CD}^{-1}\mathbf{C}$. This implies

$$\mathbf{PC}^{-1} = -\mathrm{tr}(\mathbf{CD}^{-1})\,\mathbf{I}_3 + 2\,\mathbf{CD}^{-1} \qquad (9.28)$$

and

$$\mathrm{tr}(\mathbf{PC}^{-1}) = -3\,\mathrm{tr}(\mathbf{CD}^{-1}) + 2\,\mathrm{tr}(\mathbf{CD}^{-1}) = -\mathrm{tr}(\mathbf{CD}^{-1}). \qquad (9.29)$$

Let p be inpolar to c. Then, due to Lemma 9.5.2, the coefficient j_1 in $F_{\mathbf{P},\mathbf{C}}(\sigma,\tau) = \det(\sigma\mathbf{P} + \tau\mathbf{C})$ vanishes. By (9.25), we obtain

$$F_{\mathbf{P},\mathbf{C}}(1,-X) = \det\mathbf{C}\,\det(\mathbf{PC}^{-1} - X\mathbf{I}_3) = j_0 X^0 + j_2 X^2 - j_3 X^3.$$

Now, we use the theorem of Cayley-Hamilton. When in the characteristic polynomial $\det(\mathbf{PC}^{-1} - X\mathbf{I}_3)$, the indeterminate X is replaced by the matrix $\mathbf{PC}^{-1}$, then we obtain the zero matrix $\mathbf{O}_3$. This yields

$$j_0\mathbf{I}_3 + (\mathbf{PC}^{-1})^2\left[j_2\mathbf{I}_3 + j_3\mathbf{PC}^{-1}\right] = \mathbf{O}_3 \qquad (9.30)$$

with

$$j_0 = \det\mathbf{P}, \quad j_2 = \det\mathbf{C}\,\mathrm{tr}(\mathbf{PC}^{-1}), \quad j_3 = \det\mathbf{C}.$$

By (9.28) and (9.29), we obtain from (9.30)

$$\det\mathbf{P}\,\mathbf{I}_3 - (\mathbf{PC}^{-1})^2\left[\det\mathbf{C}\,\mathrm{tr}(\mathbf{CD}^{-1})\mathbf{I}_3 - \mathrm{tr}(\mathbf{CD}^{-1})\det\mathbf{C}\mathbf{I}_3 - 2\mathbf{CD}^{-1}\right] = \mathbf{O}_3,$$

$$\det\mathbf{P}\,\mathbf{I}_3 - 2(\det\mathbf{C})\,\mathbf{PC}^{-1}\mathbf{PC}^{-1}\mathbf{CD}^{-1} = \mathbf{O}_3,$$

hence, $(\det\mathbf{P})\mathbf{DC}^{-1} = 2(\det\mathbf{C})\mathbf{PC}^{-1}\mathbf{PC}^{-1}$ or

$$\mathbf{D} = 2\det\mathbf{C}(\det\mathbf{P})^{-1}\,\mathbf{PC}^{-1}\mathbf{P}, \qquad (9.31)$$

and this shows that i is p-polar to c.

Some of these formulas will also be needed in the proof of

Theorem 9.5.2 *Let c and i be two conics with symmetric coefficient matrices $\mathbf{C}$, $\mathbf{D}$, respectively. Then, there exist infinitely many triangles (with possibly two complex conjugate vertices) inscribed in c and circumscribed to i if, and only if, in the corresponding characteristic polynomial $F_{\mathbf{C},\mathbf{D}}(\sigma,\tau)$ shown in (9.24) the coefficients k_1, k_2, k_3 satisfy $4k_1 k_3 - k_2^2 = 0$.*

Proof: We recall that by (9.26)

$$k_1 k_3 = \det \mathbf{C} \det \mathbf{D} \operatorname{tr}(\mathbf{DC}^{-1}), \quad k_2 = \det \mathbf{D} \operatorname{tr}(\mathbf{CD}^{-1}).$$

Note that the condition $4k_1 k_3 - k_2^2 = 0$ does not change when $\mathbf{C}$ or $\mathbf{D}$ is replaced by a scalar multiple.

Suppose there are infinitely many triangles inscribed in c and circumscribed to i or — equivalently — p is inpolar c. Then, Lemma 9.5.2 implies $\operatorname{tr}(\mathbf{CP}^{-1}) = 0$, and with (9.31) follows

$$\det \mathbf{D} = \frac{8(\det \mathbf{C})^2}{\det \mathbf{P}} \quad \text{and} \quad \mathbf{DC}^{-1} = \frac{2\det \mathbf{C}}{\det \mathbf{P}}(\mathbf{PC}^{-1})^2. \tag{9.32}$$

For any 3×3 matrix $\mathbf{M} = (m_{jk})$, we have

$$[\operatorname{tr}(\mathbf{M})]^2 - \operatorname{tr}(\mathbf{M}^2) = \sum_{0 \le j < k \le 2} 2(m_{jj}m_{kk} - m_{jk}m_{kj}) = 2\det \mathbf{M} \operatorname{tr}(\mathbf{M}^{-1}).$$

This gives for $\mathbf{M} = \mathbf{PC}^{-1}$

$$[\operatorname{tr}(\mathbf{PC}^{-1})]^2 - \operatorname{tr}(\mathbf{PC}^{-1})^2 = -2\det(\mathbf{PC}^{-1})\operatorname{tr}(\mathbf{CP}^{-1}) = 0,$$

hence, by (9.32) and (9.29),

$$\begin{aligned} k_1 k_3 &= \det \mathbf{C} \det \mathbf{D} \operatorname{tr}(\mathbf{DC}^{-1}) = \frac{2(\det \mathbf{C})^2 \det \mathbf{D}}{\det \mathbf{P}}[\operatorname{tr}(\mathbf{PC}^{-1})]^2 \\ &= \frac{(\det \mathbf{D})^2}{4}[\operatorname{tr}(\mathbf{PC}^{-1})]^2 = \frac{1}{4}k_2^2. \end{aligned}$$

Conversely, under $4k_1 k_3 = k_2^2$ we have by (9.24)

$$F_{\mathbf{C},\mathbf{D}}(\sigma, \tau) := k_0 \sigma^3 + k_1 \sigma^2 \tau + k_2 \sigma \tau^2 + k_3 \tau^3 = k_0 \sigma^3 + \frac{\tau}{4k_3}(k_2 \sigma + 2k_3 \tau)^2.$$

Due to (9.25), we obtain

$$F_{\mathbf{C},\mathbf{D}}(1, -X) = \det \mathbf{D} \det(\mathbf{CD}^{-1} - X\mathbf{I}_3) = k_0 - \frac{X}{4k_3}(k_2 - 2k_3 X)^2.$$

Now, we substitute the coefficients k_i from (9.26) and apply the theorem of Cayley-Hamilton:

$$4(\det \mathbf{C} \det \mathbf{D})\mathbf{I}_3 = (\mathbf{CD}^{-1})(\det \mathbf{D})^2 [\operatorname{tr}(\mathbf{CD}^{-1})\mathbf{I}_3 - 2\mathbf{CD}^{-1}]^2.$$

By (9.28) the term in brackets equals $-\mathbf{PC}^{-1}$, and we conclude with

$$4\frac{\det \mathbf{C}}{\det \mathbf{D}}\mathbf{DC}^{-1} = (\mathbf{PC}^{-1})^2, \quad \mathbf{D} = \frac{\det \mathbf{D}}{4\det \mathbf{C}}\mathbf{PC}^{-1}\mathbf{P}.$$

This shows that i is p-polar to c which by Theorem 9.5.1 is equivalent to the existence of triangles inscribed in c and circumscribed to i. ∎

■ **Example 9.5.3** Theorem 9.5.2 is only valid in $\mathbb{P}^2(\mathbb{C})$. We show that the condition $4k_1 k_3 - k_2^2 = 0$ is not sufficient for the existence of *real* triangles inscribed in c and circumscribed to i. The following symmetric coefficient matrices of the conics c, i and p displayed in Figure 9.37 are all diagonalized:

$$\mathbf{C} = \operatorname{diag}\left(-1, \frac{1}{8}, \frac{1}{5}\right), \quad \mathbf{D} = \operatorname{diag}\left(-1, \frac{1}{128}, \frac{1}{125}\right), \quad \mathbf{P} = \operatorname{diag}\left(-1, \frac{1}{32}, -\frac{1}{25}\right).$$

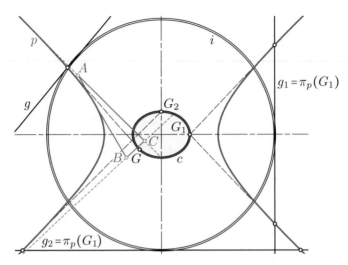

FIGURE 9.37. The conics c, i, and p satisfy the conditions of the Theorems 9.5.1 and 9.5.2, but all the infinitely many triangles inscribed in c, circumscribed to i, and auto-polar with respect to p have two complex conjugate vertices. The displayed triangle ABC is circumscribed to p and auto-polar with respect to c; it confirms by Corollary 9.5.3 that p is inpolar to c.

It is easy to verify that p corresponds to c and i in the sense of Lemma 9.5.4. Furthermore, i is polar to c with respect to p as $\mathbf{C} = \mathbf{P}\mathbf{D}^{-1}\mathbf{P}$, and in accordance with Theorem 9.5.1 the conic p is inpolar to c in the generalized sense (see page 420). Hence, the matrices satisfy the conditions presented in Lemma 9.5.2 and in Theorem 9.5.2. But all triangles inscribed in c and circumscribed to i and consequently auto-polar w.r.t. p have two conjugate complex vertices. Hence, there is nothing to see of the porism in Figure 9.37, except a real triangle ABC auto-polar with respect to c and with sides tangent to p (note Corollary 9.5.3 on page 421).

Also the fact that the tangents of i intersect p and c in harmonic quadruples cannot be visualized. We can only check: For example, the p-polar g_1 of the vertex $G_1 = (\sqrt{8}, 0)$ of c has the equation $x = 8\sqrt{2}$. Its intersection points with c are $(8\sqrt{2}, \pm 5i\sqrt{3})$, those with p have the coordinates $(8\sqrt{2}, \pm 5\sqrt{3})$, and these two pairs separate harmonically (see Example 5.4.1). Analogously, the p-polar g_2 of $G_2 = (0, \sqrt{5}) \in c$ has the equation $y = -5\sqrt{5}$. It intersects c at $(\pm 8i\sqrt{3}, -5\sqrt{5})$ and p at $(\pm 8\sqrt{3}, -5\sqrt{5})$, and again this is a harmonic quadrupel of points.

Instead of triangles inscribed in c and circumscribed to i we can look for *quadrangles* or generally for *n-gons*, $n > 3$. In view of this, note the Figures 9.36, 9.38, and 9.28 for $n = 4$, $n = 7$, and $n = 36$, respectively.

Suppose $P_1 P_2 \ldots P_n$ is an n-gon inscribed in c with pairwise different vertices and with sides $[P_1, P_2]$, $[P_2, P_3], \ldots$ tangent to i. Then, by virtue of Lemma 9.5.4, P_1 and P_3 are conjugate to P_2 with respect to p, *i.e.*, the diagonals $[P_1, P_3]$, $[P_2, P_4], \ldots, [P_{n-2}, P_n]$ are tangent to the polar i' of c

with respect to p. Hence, for $n = 2m$ the polygon $P_1 P_2 \ldots P_{2m}$ closes, *i.e.*, the last side $[P_{2m}, P_1]$ is tangent to i if, and only if, the sides of the closed polygon $P_1 P_3 \ldots P_{2m-1}$ are tangent to i'. By iteration we would be able to find characterizations for pairs of conics (c, i) with the property that there are closing n-gons inscribed in c and circumscribed to i. However, in the next section we follow the traces of PONCELET and CAYLEY on a direct approach to the remarkable Theorem 9.5.4.

● Exercise 9.5.2 Formulate the dual of Lemma 9.5.4 on page 422.
Suppose there is an n-gon $P_0, \ldots, P_{n-1}$ inscribed in c and circumscribed to i with $n > 4$. Where are the intersection points $[P_i, P_{i+1}] \cap [P_{i+2}, P_{i+3}]$ (indices modulo n) located?

● Exercise 9.5.3 Prove the following converse of Lemma 9.5.4:
For any two conics c and p there is a regular or singular conic i contacting all lines l which intersect c in points conjugate with respect to p. The conic i is called *von Staudt's conic* of the pair (c, p) (see [8, p. 91]). The symmetric coefficient matrix $\widehat{\mathbf{D}}$ of the tangent equation of i can be expressed in terms of the symmetric coefficient matrices $\mathbf{C}$ and $\mathbf{P}$ by

$$\varrho\, \widehat{\mathbf{D}} = -\mathrm{tr}(\mathbf{P}\mathbf{C}^{-1})\mathbf{C}^{-1} + \mathbf{C}^{-1}\mathbf{P}\mathbf{C}^{-1} \quad \text{or}$$
$$\sigma\, \widehat{\mathbf{D}} = -\mathrm{tr}(\mathbf{C}\mathbf{P}^{-1})\mathbf{P}^{-1} + \mathbf{P}^{-1}\mathbf{C}\mathbf{P}^{-1} \quad \text{for any } \varrho, \sigma \in \mathbb{R}.$$

● Exercise 9.5.4 Suppose there is a quadrangle $P_1 \ldots P_4$ inscribed in c and circumscribed to i. Then, the coefficients in the characteristic polynomial $F_{\mathbf{C}, \mathbf{D}}(\sigma, \tau)$ of (9.24) satisfy $8k_0 k_3^2 - 4k_1 k_2 k_3 + k_2^3 = 0$. Prove this statement by the following steps:

1. The matrix $\mathbf{P}$ of the polarity stated in Lemma 9.5.4 has to be of rank 1.

2. From $\det \mathbf{P} = 0$ one can conclude that $\frac{1}{2}\mathrm{tr}(\mathbf{C}\mathbf{D}^{-1})$ is an eigenvalue of $\mathbf{C}\mathbf{D}^{-1}$.

3. Express this eigenvalue in terms of the coefficients $k_1, \ldots, k_3$ and plug this into the characteristic polynomial.

Poncelet porisms

JEAN-VICTOR PONCELET (1788–1867) studied "porisms" and proved that MACLAURIN's theorem (see page 416) holds for general n-gons. This particular example of a porism involving two conics is called *Poncelet porism*.

Theorem 9.5.3 *Let c and i be two conics in a projective plane. If there exists a closed polygon $P_0 \ldots P_{n-1}$ such that the vertices P_i are on c and the side lines $[P_k, P_{k+1}]$ are tangent to i for all $k \in \{0, 1, \ldots, n-1\}$ with indices taken modulo n, then there are infinitely many such closed polygons.*

Theorem 9.5.3 gives a necessary and sufficient condition for the existence of infinitely many inscribed and circumscribed closed polygons. But it

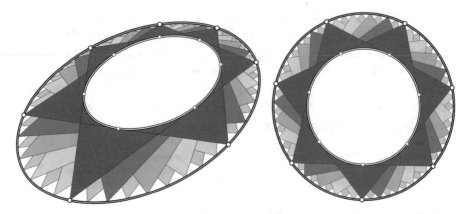

FIGURE 9.38. Poristic figures may be regular, or unsymmetric, or symmetric. The inscribed polygons can be regular, or convex, or star-shaped.

says nothing about how to specify two conics c and i such that they admit such polygons.

Figure 9.38 shows some porisms of Poncelet type. The theorem as well as its first proof is due to PONCELET. He was a French mathematician and engineer who joined the French army and participated in NAPOLEON's invasion of Russia from 1812 to 1814. However, he became a prisoner of war and, while imprisoned at Saratow, he found the time to write his famous work on the figures in geometry, a work that could be considered one of the fundamentals of Projective Geometry, containing also Theorem 9.5.3 and its proof. It should be noted that none of the techniques that are used nowadays for the proof were developed at that time.

The proof of Theorem 9.5.3 involves complex functions, differential geometry, and elliptic curves. Because of its relations to many subdisciplines of mathematics, this theorem is often called *a pearl of mathematics*. We shall only give a sketch of the proof, since it is far beyond the scope of our book. Details can be found in [26, 24, 25].

Proof: (Sketch) Let P and t be a point and a line in a projective plane $\mathbb{P}^2$. We describe P as well as t by homogeneous coordinates, *i.e.*, $P = (p_0 : p_1 : p_2)$ and $t = (t_0 : t_1 : t_2)$. Since $P \in c$, the point P can be represented by quadratic functions $p_i = p_i(u)$ in the parameter u for $i \in \{0, 1, 2\}$. The same holds true for the tangents t of c; this one-parameter family of tangents allows an analogous parametrization $t_j = t_j(v)$ with $v \in \mathbb{R}$ for $j \in \{0, 1, 2\}$. The incidence of P with t can be written in terms of coordinates as

$$p_0 t_0 + p_1 t_1 + p_2 t_2 = 0. \tag{9.33}$$

This is a biquadratic equation in u and v. It can be interpreted as the equation of a curve m in the (u,v)-plane. The points of the curve correspond to incident pairs (P,t) where P is on the conic c and t is tangent to i.

The algebraic curve m of degree 4 is elliptic if the conics c and i are in general position, *i.e.*, if they span a pencil of conics of the first kind. In this case, the conics c and i have four mutually different points and four mutually different tangents in common if we extend into the complex (u,v)-plane. The curve m is called the *Poncelet correspondence*.

There acts an involution σ on m mapping the pair (P,t) to the pair (P',t) where P and P' are the two points of intersection of t with c; points of m corresponding under σ share the v-coordinate. Obviously, by virtue of Lemma 9.5.4, the involution σ is the restriction of the conjugacy with respect to p onto the conic c. A further involution τ sends the pair (P,t) to the pair (P,t') where t and t' are the tangents of i which pass through P. The composition $\mu := \tau \circ \sigma$ acts like a translation on m; it is called *Poncelet map*.

Let $P_1 \ldots P_n$ be a polygon inscribed in c and circumscribed to i with $t_1 = [P_1, P_2]$, $t_2 = [P_2, P_3]$, ... Then, we have

$$\mu \colon\ (P_1, t_1) \mapsto (P_2, t_2),\ (P_2, t_2) \mapsto (P_3, t_3),\ \ldots.$$

The polygon closes if μ^n but not μ keeps (P_1, t_1) fixed.

It can be proved that μ^n equals the identity on m as soon as it has at least five fixed points. However, four fixed points (P,t) of μ and μ^n are already known from the beginning: t is one of the four common tangents of c and i, and P is its point of contact with c. Hence, the existence of one closing in- and circumscribed n-gon, *i.e.*, of one additional fixed point, implies that all these n-gons must close.

The proof, sketched so far, was given for two conics in general position. It works in a similar way if the conics c and i are in a special position, *i.e.*, if they span a pencil of conics of the second, third, fourth, or fifth kind. ∎

There is another surprising result which gives a criterion on two conics for the existence of in- and circumscribed n-gons, $n \geq 3$. It includes Theorem 9.5.2 and the statement presented in Exercise 9.5.4.

Let $\mathbf{C}$ and $\mathbf{D}$ be the symmetric coefficient matrices of the conics c and i in any coordinate frame. Now, we recall from (9.24) on page 418 the characteristic polynomial

$$F_{\mathbf{C},\mathbf{D}}(\sigma, \tau) = \det(\sigma \mathbf{C} + \tau \mathbf{D})$$

and expand the following function in a power series:

$$\sqrt{F_{\mathbf{C},\mathbf{D}}(t,1)} = \sqrt{\det(t\,\mathbf{C} + \mathbf{D})} = a_0 + a_1 t + a_2 t^2 + a_3 t^3 + \cdots. \qquad (9.34)$$

The following theorem dates back to the British mathematician ARTHUR CAYLEY (1821–1895).

Theorem 9.5.4 *There exists an n-sided polygon inscribed in c and cir-cumscribed to i if, and only if, the coefficients of the power series given in (9.34) satisfy*

$$\det\begin{pmatrix} a_2 & \cdots & a_{m+1} \\ \vdots & & \vdots \\ a_{m+1} & \cdots & a_{2m} \end{pmatrix} = 0, \quad \text{if } n = 2m+1, \quad m \geq 1,$$

$$\det\begin{pmatrix} a_3 & \cdots & a_{m+1} \\ \vdots & & \vdots \\ a_{m+1} & \cdots & a_{2m+1} \end{pmatrix} = 0, \quad \text{if } n = 2m, \quad m \geq 2.$$

(9.35)

The proof of this result uses the same ideas as the proof of Theorem 9.5.3 (see, *e.g.*, [26]). The determinant $\det(t\,\mathbf{C} + \mathbf{D})$ is a cubic polynomial in t. It has no multiple roots if c and i span a pencil of conics of the first kind. Otherwise there would be less than three singular conics in the pencil (cf. Section 7.3).

CAYLEY's Theorem 9.5.4 is more than just a projectively invariant criteri-on on the equations of two conics. As a byproduct, even metric properties of poristic figures result from the formula presented in (9.35). Consider the following

■ **Example 9.5.4** Criteria for bicentric n-gons. An n-gon is called *bicentric* if it has a circum-circle c and an incircle i. Theorem 9.5.4 delivers conditions on the radii and the distance of the centers of c and i to make them the circumcircle and the incircle of a bicentric n-gon.

Let c and i be two circles with equations in terms of Cartesian coordinates that read

$$c:\ x^2 + y^2 = R^2 \quad \text{and} \quad i:\ (x-d)^2 + y^2 = r^2, \tag{9.36}$$

where R, r, d are real or complex numbers and, without loss of generality, $R > r$ (in the real case). According to Theorem 9.5.4, we have to expand the function

$$\sqrt{-(t+1)\,(t^2 R^2 + t(R^2 + r^2 - d^2) + r^2)}$$

in power series as given in (9.34). We obtain

$$ri + \frac{i}{2r}(R^2 + 2r^2 - d^2)t + \frac{i}{8r^3}(R^2 - 2Rr - d^2)(R^2 + 2Rr - d^2)t^2 + \cdots.$$

Now, we use the condition from Theorem 9.5.4: When we look for triangles circumscribed to i and inscribed in c, the determinant given in (9.35) is that of the 1×1-matrix (a_2) (compare Theorem 9.5.2). Therefore, the radii R and r and the distance d of the incenter and circumcenter of a triangle satisfy

$$R^2 - 2Rr = d^2. \tag{9.37}$$

The second factor of the coefficient a_2 of t^2 only differs by the sign of r. The formula given in (9.37) was at least known to LEONHARD EULER (1707–1783).

Many other formulas relating the radii of the incircle and the circumcircle of an n-gon can be derived from CAYLEY's formula. The case of two circles like in Example 9.5.4 can also be treated by elementary methods, and a huge amount of formulas for arbitrary n-gons was elaborated in the past.

Figures displaying Poncelet porisms as well as such of MacLaurin type are invariant under projective transformations. That means: Once you have found a poristic figure, you can apply any collinear transformation, and you find a new poristic figure. This is also the case for the so-called *Poncelet grid* displayed in Figure 9.28.

The Poncelet grid arises in a natural way. The edges of one Poncelet polygon create a pattern of quadrilaterals. Surprisingly, each vertex of the quadrilateral grid lies on two conics and the grid built by the edges can be replaced by the grid formed by two families of conics [53].

■ **Example 9.5.5** Billiards in ellipses. *Suppose a small ball can move within an ellipse c and its collisions with the boundary are perfectly elastic. Then, the trace of the ball is called a billiard, i.e., a polygon inscribed in c such that any two adjacent sides meet c under equal angles (see Figure 9.39).*

When at a point $P \in c$ any line l is reflected at the conic, then l together with its mirror l' are tangents of the same conic i confocal to c, and i does not change after iterated reflections in c. Hence, a billiard is a polygon inscribed in c and circumscribed to i so that we can claim that if c and i admit one billiard which closes after n reflections, then there are infinitely many with this property.

Suppose, c is an ellipse with axes a, b. Then, we can set up the confocal ellipse i by the equation

$$i: \quad \frac{x^2}{a^2 + k} + \frac{y^2}{b^2 + k} = 1, \quad 0 > k > -b^2.$$

After some computation we obtain closing billiards with $n = 3$ reflections if, and only if, by virtue of Theorem 9.5.2,

$$k = \frac{a^2 b^2}{(a^2 - b^2)^2} \left(a^2 + b^2 - 2\sqrt{a^4 - a^2 b^2 + b^4} \right).$$

■ **Example 9.5.6** Spatial ball-bearings. We end this section with a particular porism for two circles c_1, c_2 in 3-space: If there exists one closed n-gon in form of a zig-zag between c_1 and c_2 such that all sides are of equal length l, then there are infinitely many. The proof (see [38, 67]) is again based on PONCELET's closure theorem since the orthogonal projections of the sides into the plane of one circle c_i envelope a conic [64].

There is an application in mechanical engineering: Suppose the vertices of the n-gon are movable along the respective circles c_1, c_2 and at the same time the centers of balls with radius $l/2$. Then, we obtain the spatial version of a ball-bearing [4], because each ball remains in permanent contact with the two adjacent balls which are centered on the other circle.

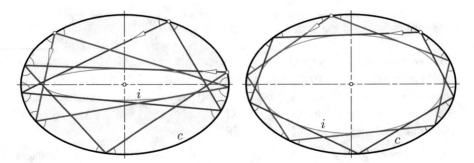

FIGURE 9.39. Billiards in an ellipse c are always circumscribed to a confocal
ellipse i. Here closed billiards with either three or five reflections are displayed.

10 Other geometries

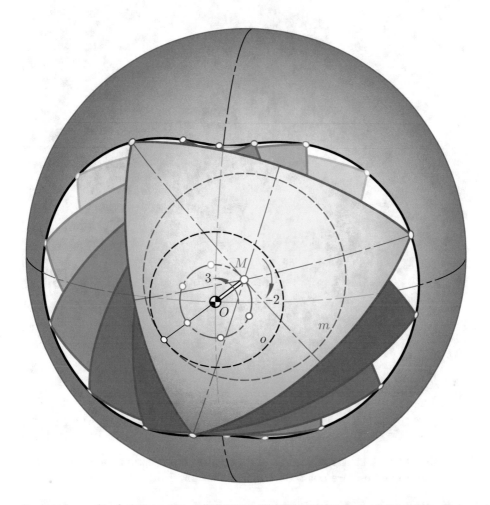

At the *'spherical Wankel engine'* the common path t of the three vertices of the rotor is a trochoid. It can be generated when the circle m which is attached to the rotor, is rolling along the fixed circle o. The angular velocities of the rotor's rotations about M and additionally about the fixed center O build the constant ratio $-3 : 2$.

10.1 Spherical conics

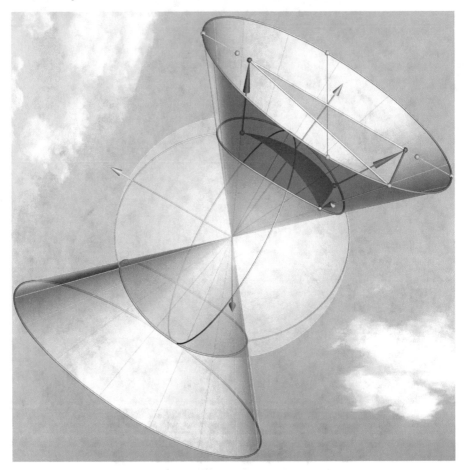

FIGURE 10.1. A spherical conic and its quadratic projection cone. The foci of the projected ellipse are not the images of the spherical foci.

This chapter deals with a generalization of conics on the sphere. We call curves on a sphere "spherical conics" when they fulfill the classical condition that we know from ordinary conics: When the sum or the difference of the distances from two fixed points on the sphere — measured on shortest paths on the sphere — is constant, we speak of spherical ellipses and hyperbolas.

It turns out that any spherical ellipse is at the same time a spherical hyperbola, and *vice versa*. The borderline case of a spherical parabola, where the spherical distance from a point is equal to the distance from a great circle, can also be interpreted as such a spherical conic.

Spherical conics are the intersections of a sphere with quadratic cones with the apex in the center of the sphere (Figure 10.1). Thus, they are algebraic curves of degree four on the sphere. Their images under orthogonal projections into their planes of symmetry are portions of planar conics.

Many theorems about ordinary conics can be transferred to the sphere, *e.g.*, the property that the tangent and the normal of a spherical conic are the bisectors of the focal rays through the spherical foci. PROCLUS's (DE LA HIRE's) construction of a conic and IVORY's theorem about confocal conics work on the sphere as well. Theorems of planar projective geometry, *e.g.*, PASCAL's and BRIANCHON's theorem can be transferred without difficulties onto the sphere.

Geometry on the sphere

In the introduction we reported that there are many analogies in the theory of conics between the spherical and the planar case.

In order to be able to prove these analogies, we first have to provide the reader with some particularities of the geometry on the sphere. The reader may skip this section and refer to it later on when necessary.

It is important that on the sphere we measure only angles: Since we have no straight lines on the sphere, the role of straight lines is taken over by *great circles*.

- Each point P on the sphere has an *antipodal point* or *antipode* P^*, and each great circle passing through P contains also P^*.

- For any two different points P and $Q \neq P^*$ there exists a unique connecting great circle denoted by $[P, Q]$. In the sequel, when we speak of the *segment*, the *arc*, or the *side* PQ for any two non-antipodal points P, Q, we always mean on the great circle $[P, Q]$ the shorter of the two arcs terminated by P and Q.

- The *(spherical) distance* $\widehat{PQ}$ is defined as length of the side PQ, measured on the *unit sphere* $\mathcal{S}^2$ with center O. Thus, we have $0 \leq \widehat{PQ} \leq \pi$, when we include also the limiting cases $Q = P$ and $Q = P^*$. The distance $\widehat{PQ}$ equals the central angle $\sphericalangle POQ$ of the segment. When $\mathbf{p}$, $\mathbf{q}$ with $\|\mathbf{p}\| = \|\mathbf{q}\| = 1$ are the position vectors of the two points, then we use the scalar product to compute the distance by $\cos \widehat{PQ} = \langle \mathbf{p}, \mathbf{q} \rangle$.

Theorem 10.1.1 Triangle inequality: *For any three points A, B, C on the sphere the inequality*

$$\widehat{AB} \le \widehat{AC} + \widehat{BC}$$

holds true. In the case of an equality, the three points must be located on a great circle.

Proof: Firstly, we treat the particular case $\widehat{AB} = \pi$. There exists a great circle passing through A, $B = A^*$ and C, and $\widehat{AB} = \widehat{AC} + \widehat{BC}$.

For the remaining case $\widehat{AB} < \pi$ we can prove that there is no point $X \in \mathcal{S}^2$ satisfying $\widehat{AX} + \widehat{BX} < \widehat{AB}$. This statement is equivalent to the triangle inequality:

The set of points $X \in \mathcal{S}^2$ with a given positive spherical distance $r < \pi$ from a point $M \in \mathcal{S}^2$ is a circle denoted by $(M; r)$ or $(M^*; \pi - r)$. The points M and M^* are the *spherical centers* of this circle. The plane spanned by the circle is orthogonal to the diameter connecting M with M^*.

Let two positive reals r_A, r_B with $r_A + r_B < \widehat{AB}$ be given. Then the intersection of the circles $(A; r_A)$ and $(B; r_B)$ is empty. We see this by inspecting the orthogonal projection onto the plane of the great circle $[A, B]$ (Figure 10.2, left). The two circles are depicted as two disjoint line segments.

In the limiting case $r_A + r_B = \widehat{AB}$ (Figure 10.2, right) the two circles share exactly one point C on the side AB.   ∎

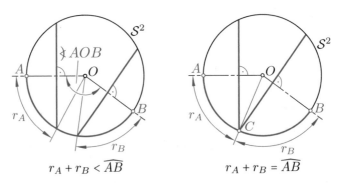

FIGURE 10.2. Proving the spherical triangle inequality by orthogonal projection onto the plane of the great circle $[A, B]$.

Corollary 10.1.1 *The shortest path between any two points P and Q on the sphere $\mathcal{S}^2$ is their connecting arc PQ.*

- The *(spherical) bisector* of any two different points $P, Q \in \mathcal{S}^2$ is defined as the set of points $X \in \mathcal{S}^2$ obeying $\widehat{XP} = \widehat{XQ}$. Since equal spherical distances imply also equal 3D-distances along the chords, the bisector is the great circle in the plane of symmetry of P and Q.

- The *angle* between two great circles g_1, g_2 is defined as the angle of the respective tangent lines t_1, t_2 at their intersection points. It is equal to the angle between the planes spanned by the circles. We may restrict it by $\sphericalangle g_1 g_2 = \sphericalangle t_1 t_2 \leq \frac{\pi}{2}$. The angle between two arcs PQ and PR is bounded by $0 \leq \sphericalangle QPR \leq \pi$.

- Below, when we speak of a *triangle ABC* on $\mathcal{S}^2$, then the three points are never collinear, *i.e.*, they are not located on the same great circle. Each side is shorter than π, and also each (interior) angle is smaller than π.

In the following theorem we summarize some formulas from spherical trigonometry.

Theorem 10.1.2 *Let A, B, and C be a spherical triangle on $\mathcal{S}^2$ with side lengths $a = \widehat{BC}$, $b = \widehat{CA}$, and $c = \widehat{AB}$ and with interior angles α, β, and γ. Then the following formulas hold:*

1. **Spherical Law of Sines:**

$$\frac{\sin a}{\sin \alpha} = \frac{\sin b}{\sin \beta} = \frac{\sin c}{\sin \gamma}.$$

2. **Spherical Laws of Cosines:**

$$\begin{aligned}
\cos a &= \cos b \cos c + \sin b \sin c \cos \alpha, \\
\cos \alpha &= -\cos \beta \cos \gamma + \sin \beta \sin \gamma \cos a.
\end{aligned}$$

3. **Spherical Pythagoras theorem** *for $\alpha = \frac{\pi}{2}$:*

$$\cos a = \cos b \cos c.$$

4. **Area of the triangle:**

$$A = \alpha + \beta + \gamma - \pi.$$

A planar triangle ABC can be defined (up to rotation, translation or reflection) by the lengths a, b, c of its sides. This allows us to calculate the interior angles α, β, γ, since we have many formulas connecting a, b, c and α, β, γ. Note that on the sphere a triangle is even uniquely defined by its three angles α, β, and γ.

Ellipses, hyperbolas and parabolas on a sphere?

Any planar section of a sphere is a circle, a point, or the empty set. Since conics lie in a plane, there are no ordinary conics on a sphere. However, we can carry over the gardener's construction of an ellipse to the sphere.

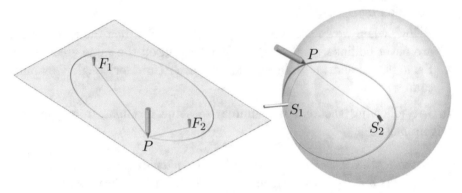

FIGURE 10.3. Gardener's constructions in the plane and on the sphere.

Let us consider the well-known construction of an ellipse in the plane as the locus of points whose distances to two fixed points – the foci F_1 and F_2 – sum up to a constant (Figure 10.3, left). A rope of constant length $2a$ is stretched tight in every position P of the pencil so that we have

$$\overline{F_1 P} + \overline{F_2 P} = 2a.$$

This construction also works on the sphere. Again, we consider two fixed points S_1 and $S_2 \neq S_1, S_1^*$ on the unit sphere (Figure 10.3, right) and choose a rope of constant length $2a$ which we preliminarily restrict by requiring $2a < \pi$. Now, we move the pencil such that the rope is stretched tight all the time. This means that in every position the two portions of the rope run along geodesic curves on the sphere, $i.e.$, great circles. According to Corollary 10.1.1, they form two segments $S_1 P$ and $P S_2$, and the distances again satisfy

$$\widehat{S_1 P} + \widehat{S_2 P} = 2a.$$

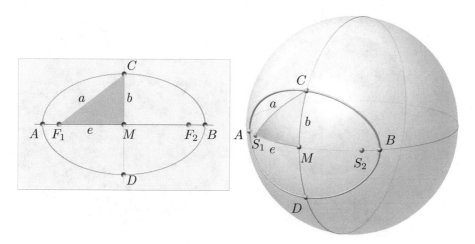

FIGURE 10.4. The lengths a, b, and e at planar and spherical conics.

Obviously, the point P traces a spherical curve with two-fold symmetry which shall be called a *spherical ellipse*. The points S_1 and S_2 are called the *spherical foci* or *spherical focal points*.

Now, we recall some basic facts (Figure 10.4): A planar ellipse has a major axis and a minor axis on its only two perpendicular and conjugate diameters which intersect at the ellipse's center M. On the major axis we find the vertices A and B. The intercepts of the ellipse with the minor axis are the *auxiliary vertices* C and D. The ellipse is symmetric with respect to both the major and the minor axes. Therefore, it is also symmetric w.r.t. the center M. By definition, we have $\overline{AB} = 2a$. The two given foci are at a distance $\overline{F_1 F_2} = 2e$. According to PYTHAGORAS's theorem, the semimajor axis a, the semiminor axis b, and the linear eccentricity e are related via $\overline{CD} = 2b = 2\sqrt{a^2 - e^2}$.

Let us do the rope construction again on the unit sphere $\mathcal{S}^2$ (Figure 10.5, left): We choose two fixed points S_1 and S_2 and a tightened rope with given length $2a$ through a point P. The locus of P is what we called a spherical ellipse, and we have $\widehat{S_1 P} + \widehat{S_2 P} = 2a$. Now consider the antipode S_1^* of S_1. Since we have $\widehat{S_1 P} + \widehat{S_1^* P} = \pi$ (half the circumference of $\mathcal{S}^2$), we get $(\pi - \widehat{S_1^* P}) + \widehat{S_2 P} = 2a$ or, equivalently,

$$\widehat{S_1^* P} - \widehat{S_2 P} = \pi - 2a.$$

It becomes an equivalent if we interchange of S_1 with S_2 and S_1^* with S_2^*.

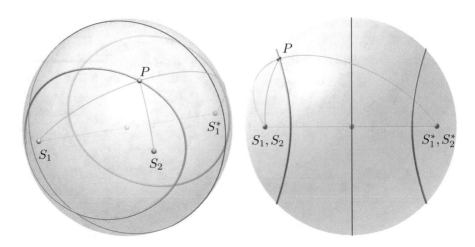

FIGURE 10.5. A spherical ellipse is at the same time a spherical hyperbola

Therefore, the points on the spherical ellipse satisfy a *sum relation* and a *difference relation on their distances* to the foci *at the same time*. The constancy of the absolute value of differences was the defining property of hyperbolas in the plane. When in the plane the order of the two focal points is prescribed and the difference is signed, then the locus of points X is one branch of a hyperbola.

Theorem 10.1.3 *Any spherical ellipse with focal points S_1 and S_2 and major axis $2a$ is a branch of a spherical hyperbola with ordered foci S_2 and S_1^* and major axis $\pi - 2a$, where S_1^* is the antipode of S_1.*

Now, we consider the symmetric curve on the sphere with respect to the bisector plane of S_2 and S_1^* (Figure 10.5, right). Each point Q of the reflected ellipse obeys the condition

$$2a = \widehat{S_1^*Q} + \widehat{S_2^*Q} = (\pi - \widehat{S_1Q}) + (\pi - \widehat{S_2Q}),$$

hence

$$\widehat{S_1Q} + \widehat{S_2Q} = 2\pi - 2a > \pi.$$

At the same time we have

$$\widehat{S_2Q} - \widehat{S_1^*Q} = \pi - 2a.$$

This is the second branch of our hyperbola. Hence, a spherical hyperbola consists of two mutually antipodal spherical ellipses.[1]

When both focal points are replaced by their antipodes, then

$$\widehat{S_1^* P} + \widehat{S_2^* P} = \pi - \widehat{S_1 P} + \pi - \widehat{S_2 P} = 2\pi - 2a.$$

We again come up with a constant sum; but this time for the same curve the sum is $> \pi$.

What is the locus of points P obeying $\widehat{S_1 P} + \widehat{S_2 P} = \pi$? Each such point P has the property $\widehat{S_2 P} = \widehat{S_1^* P}$. Hence, the locus is a great circle, the spherical bisector (note the definition on page 439) of S_2 and S_1^* and at the same time bisector of S_1 and S_2^*. Hence, we don't lose any proper spherical conic by the restriction $2a < \pi$.

Finally which curves are the spherical analogues of parabolas? What is the locus of points P with equal distances to a given point S_1 and a given great circle l which does not pass through S_1? Since the spherical distance to a great circle never exceeds $\frac{\pi}{2}$ the requested locus of points P lies within the hemisphere with center S_1. Let S_2 be the spherical center of the great circle l with $\widehat{S_1 S_2} < \frac{\pi}{2}$. Then, taking into account to which hemisphere the point P belongs to,

$$\widehat{PS_1} = \widehat{Pl} = \frac{\pi}{2} - \widehat{PS_2} \quad \text{or} \quad \widehat{PS_1} = \widehat{Pl} = \widehat{PS_2} - \frac{\pi}{2}.$$

The first equation is equivalent to $\widehat{PS_1} + \widehat{PS_2} = \frac{\pi}{2}$ and gives an ellipse with focal points S_1, S_2, and $2a = \frac{\pi}{2}$. The second equation is equivalent to $\widehat{PS_1} + \widehat{PS_2^*} = \frac{\pi}{2}$ with S_2^* denoting the antipode of S_2. This second condition can never be satisfied since, due to the spherical triangle inequality,

$$\widehat{PS_1} + \widehat{PS_2^*} \geq \widehat{S_1 S_2^*} = \pi - \widehat{S_1 S_2} > \frac{\pi}{2}.$$

Theorem 10.1.4 *On the sphere $\mathcal{S}^2$ we are given a point S_1 and a great circle l not passing through S_1. Let $S_2 \in \mathcal{S}^2$ be the spherical center of l*

---

[1] Sometimes, in the literature, a "spherical conic" is generally defined as a pair of antipodal curves, a spherical ellipse together with its mirror. Nevertheless, we distinguish between the two connected components.

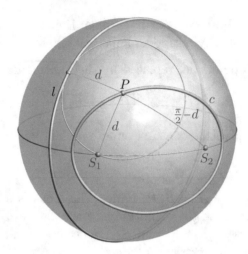

FIGURE 10.6. The spherical parabolas c are also spherical ellipses.

with $\overparen{S_1 S_2} < \frac{\pi}{2}$. Then the locus of points P with equal distances to S_1 and l is either an ellipse c with the major axis $2a = \frac{\pi}{2}$ and with focal points S_1 and S_2 (see Figure 10.6) or in the case $S_1 = S_2$ a circle with radius $\frac{\pi}{4}$.

Analogies to planar conics

Due to the fact that lines radiating from F_1 are reflected in a planar conic such that they all then pass through F_2 (Figure 10.7, left), these two points are called focal points.

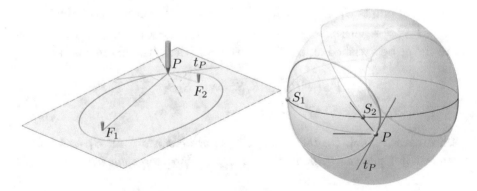

FIGURE 10.7. After reflection in the planar or spherical conic, all rays radiating from one focus pass through the other focus.

The same holds for spherical conics (Figure 10.7, right):

Theorem 10.1.5 *The tangent t_P at point P to the spherical ellipse with focal points S_1, S_2 bisects the exterior angle of the spherical triangle $S_1 P S_2$.*

Note that now tangents and normals are 'curved'. To be precise, they are great circles.

Proof: We can repeat the arguments used in the second proof of the planar version stated in Theorem 2.2.1 on page 31 (note Figure 2.11, right): With respect to each focal ray PS_i, $i = 1, 2$, the velocity vector $\mathbf{v}$ at P can be decomposed into two mutually orthogonal components. One is the radial component $\mathbf{v}_{r_i}$ tangent to $[P, S_i]$. It expresses the change of the focal distance $\widehat{PS_i}$. Because of the constant sum one distance increases while the other decreases by the same amount. Therefore, the radial components $\mathbf{v}_{r_1}$ and $\mathbf{v}_{r_2}$ of $\mathbf{v}$ have the same length but different directions w.r.t. the focal points S_1 and S_2. Therefore, the two triangles formed by $\mathbf{v}$ and the radial components are symmetric w.r.t. the tangent t_P (see Figure 2.11, right). This means that the tangent t_P bisects the exterior angle of the spherical triangle $S_1 P S_2$.

In terms of physical reflection on the sphere, this means that incoming and reflected rays build equal angles with the tangent and the normal, respectively. This is the *optical property* of spherical conics. ∎

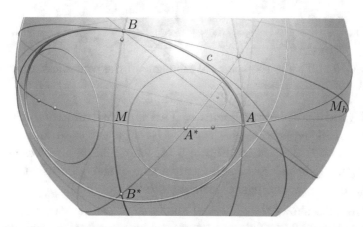

FIGURE 10.8. The points A^* and B^* are the centers of curvature at the vertices A, B of a spherical conic c. This construction is analogous to that for planar ellipses or hyperbolas.

At the end of this section, we mention two further properties which hold for spherical conics as well as for planar conics:

- Figure 10.8 reveals that the construction of the circles of curvature at the vertices A, B of an ellipse (Figure 3.11 or Figure 2.10, left) is

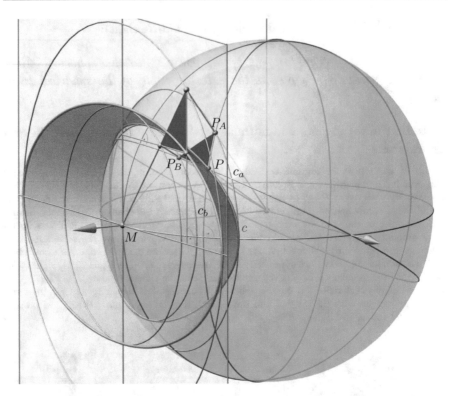

FIGURE 10.9. This is the spherical analogue of the construction named after
PROCLUS or DE LA HIRE.

still valid for spherical conics c. This is even true when c is seen as
a spherical hyperbola with center M_h, with vertex A and ideal point
B (Figure 2.10, right, or Figure 3.14, left). This can be verified either
analytically or using the spherical version of the quadratic transformation defined by conjugate normals (see Section 7.5).

- Figure 10.9 shows the spherical analogue of the construction of DE LA
  HIRE (Figure 2.6). This can be proved by a projection analogue to the
  one illustrated in Figure 10.11.

Analytic description of a spherical conic

Now we want to find an analytic representation of a spherical conic.
We assume that the foci are given by $S_1 = (\cos e, -\sin e, 0)$ and $S_2 =
(\cos e, \sin e, 0)$, where $e < \frac{\pi}{2}$. The point P traces the spherical conic if
$\widehat{S_1 P} + \widehat{P S_2} = 2a = $ const. Without loss of generality, we may assume

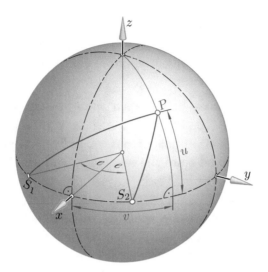

FIGURE 10.10. Deriving the equation of a spherical conic.

$0 < e < a < \frac{\pi}{2}$. We set $P = (\cos u \cos v, \cos u \sin v, \sin u)$. The parameters u and v can be interpreted as latitude and longitude on the unit sphere as shown in Figure 10.10.

By virtue of the spherical Pythagoras Theorem 10.1.2, we have

$$\cos \widehat{S_1 P} = \cos u \cos(v + e) \quad \text{and} \quad \cos \widehat{P S_2} = \cos u \cos(v - e). \tag{10.1}$$

From the definition of the spherical ellipse with major axis $2a$ we have

$$\cos \widehat{P S_2} = \cos \left(2a - \widehat{S_1 P}\right) = \cos 2a \cos \widehat{S_1 P} + \sin 2a \sin \widehat{S_1 P}. \tag{10.2}$$

Eqs. (10.1) and (10.2) together yield

$$\cos u \cos(v - e) = \cos 2a \cos u \cos(v + e) + \sin 2a \sin \widehat{S_1 P}. \tag{10.3}$$

We solve this equation for $\sin \widehat{S_1 P}$, square it, and insert $\sin^2 \widehat{S_1 P} = 1 - \cos^2 \widehat{S_1 P} = 1 - \cos^2 u \cdot \cos^2(v + e)$. Because of

$$\cos(v \pm e) = \cos v \cos e \mp \sin v \sin e$$

(10.2) simplifies to

$$\cos^2 u \left(\cos^2 v \frac{\cos^2 e}{\cos^2 a} + \sin^2 v \frac{\sin^2 e}{\sin^2 a} \right) = 1. \tag{10.4}$$

In order to give a geometric meaning to the coefficients in (10.4) we recall that the major and minor semiaxes are related to the linear eccentricity by $\cos a = \cos b \cdot \cos e$. Consequently, (10.4) can be rewritten as

$$\frac{\cos^2 u}{\sin^2 a \cos^2 b}\left(\cos^2 v \sin^2 a + \sin^2 v(\cos^2 b - \cos^2 a) \right) = 1.$$

The fact that this equation holds for $u = 0$ and $v = a$ and for $u = b$ and $v = 0$ confirms that a and b are the semimajor and semiminor axes.

Spherical conics on quadratic cones

In this section we prove that a spherical conic is not only contained in a sphere. It turns out that there are three quadratic cylinders and a quadratic cone that also pass through the spherical conic.

In the previous subsection we assumed that the point P moves along a spherical conic. Thus, $x = \cos u \cos v$, $y = \cos u \sin v$, and $z = \sin u$ are its coordinates with respect to the Cartesian standard coordinate system. Now, (10.4) reads

$$\mathcal{L}: \ \frac{\cos^2 e}{\cos^2 a} x^2 + \frac{\sin^2 e}{\sin^2 a} y^2 = 1 \qquad (10.5)$$

which is the equation of an elliptic cylinder with generators parallel to the z-axis. On the other hand, (10.5) can be viewed as the equation of an ellipse in the plane $z = 0$. The generators of $\mathcal{L}$ are lines of sight of the orthogonal projection onto the plane $z = 0$. Thus, we have

Theorem 10.1.6 *The orthogonal projection of a spherical conic onto the symmetry plane through its foci is located on an ellipse.*

Note that the coefficients in (10.5) are always positive, since they are squares of real numbers.

We can rephrase Theorem 10.1.6 by saying that the top view of any spherical conic with foci in the plane $z = 0$ is a part of an ellipse.

The equation of the unit sphere

$$\mathcal{S}^2: \ x^2 + y^2 + z^2 = 1 \qquad (10.6)$$

is also satisfied by the coordinates of the points of the spherical conic. Each linear combination of (10.5) and (10.6) gives an equation of a quadratic surface passing through the spherical conic. In particular, the difference of the equations leads to

$$\mathcal{C}: \ \left(\frac{\cos^2 e}{\cos^2 a} - 1\right)x^2 - \left(1 - \frac{\sin^2 e}{\sin^2 a}\right)y^2 - z^2 = 0, \qquad (10.7)$$

which is obviously free from linear and constant terms in x, y and z. Hence, it is the equation of a quadratic cone with its apex in the center of the sphere (compare with the standard equation (4.5) of quadratic conics).

Because of our assumption $0 < e < a < \frac{\pi}{2}$, the *signature* of the quadratic form on the left hand side, *i.e.*, the sequence of signs at the coefficients of x^2, y^2, and z^2 is $(+ - -)$.

Theorem 10.1.7 *A spherical conic is a connected component of the intersection between a quadratic cone and a sphere centered at its apex.*

Conversely, each quadratic cone centered at the origin and with signature $(+ - -)$ or $(+ + -)$ intersects the unit sphere in two antipodal conics or circles.

Proof: Following the above discussion, it only remains to show that for each quadratic cone satisfying the listed conditions, there is a Cartesian coordinate system such that its equation can be written in the form of (10.7). Here, we refer to (4.5): The cone's equation can be written in the form $u^2x^2 - v^2y^2 - z^2 = 0$. A comparison of coefficients with (10.7) results in

$$\cos^2 a = \frac{v^2}{u^2 + v^2}, \quad \cos^2 e = \frac{(1 + u^2)v^2}{u^2 + v^2}.$$

This defines the positive quantities a and e for the spherical conic uniquely, and they satisfy $a > e$. ∎

A parametrization of a spherical conic

In this subsection we shall derive a parametrization, where a uniform subdivision of the parameter interval serves for a fair distribution of the points on the curve. Thus, it is useful for computer drawings. The new parametrization allows us to prove in another way that spherical conics are the intersection curves of the sphere with quadratic cones.

Let us start with the projection of a suitably positioned ellipse e_0 with the semiaxes a_0 and b_0 onto the unit sphere $\mathcal{S}^2$, as depicted in the Figures 10.11 and 10.12. In a properly chosen Cartesian coordinate system, a point P on the ellipse e_0 has the coordinates $(x_0, a_0 \cos u, b_0 \sin u)^{\mathrm{T}}$, where

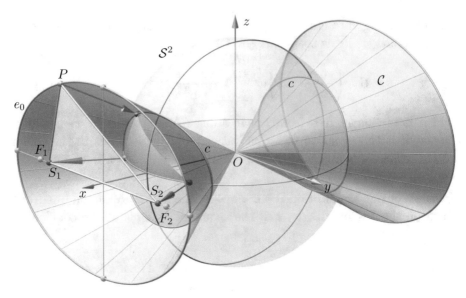

FIGURE 10.11. The planar ellipse e_0 is projected from the center O of a sphere S^2 onto the sphere (via a quadratic cone C). The result is a pair of antipodal spherical conics c. Note that the spherical foci S_1 and S_2 of c do *not* correspond to the foci F_1 and F_2 of the planar ellipse e_0.

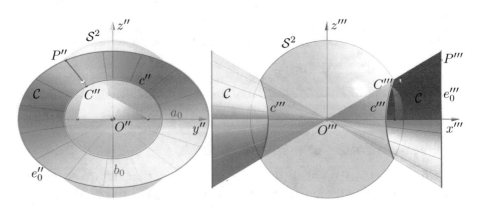

FIGURE 10.12. Front view (left) and side view (right) of the spherical conic c. The projections of c are portions of planar conics c'', c'''.

$u \in [0, 2\pi]$. Now we just have to normalize the vector $\overrightarrow{OE}$ in order to get the position vector $\mathbf{c}$ of the corresponding point C on the spherical curve

c (Figure 10.12)

$$\mathbf{c} = \begin{pmatrix} c_x \\ c_y \\ c_z \end{pmatrix} = \frac{1}{\sqrt{x_0{}^2 + a_0^2 \cos^2 u + b_0^2 \sin^2 u}} \begin{pmatrix} x_0 \\ a_0 \cos u \\ b_0 \sin u \end{pmatrix}.$$

In order to shorten the following calculations, we let $p = a_0/x_0$ and $q = b_0/x_0$. The parametrization of $\mathbf{c}$ gives

$$\mathbf{c} = \frac{1}{\sqrt{1 + p^2 \cos^2 u + q^2 \sin^2 u}} \begin{pmatrix} 1 \\ p \cos u \\ q \sin u \end{pmatrix}. \tag{10.8}$$

In spherical coordinates, $i.e.$, with longitude λ and latitude φ, the point C is described by

$$\lambda = \arctan \frac{c_y}{c_x} = \arctan(p \cos u)$$

and

$$\varphi = \arctan \frac{c_z}{\sqrt{c_x^2 + c_y^2}} = \arctan \frac{q \sin u}{\sqrt{1 + p^2 \cos^2 u}}.$$

We want to prove that the curve c is a spherical conic. Therefore, we need to show that for all $C \in c$ the sum of the distances of C to two points S_1 and S_2 inside the curve is constant. Then, S_1 and S_2 will automatically be the foci of the curve.

It would be a mistake to think that S_1 and S_2 are the projections of the foci of the ellipse – we will show below that this is not the case. We rather calculate the values of a and b of the projected curve c (the presumptive spherical conic) and from them the spherical distance e.

In order to get the values a and b we project the vertices of the ellipse e_0. The corresponding angles a and b are the latitude angle at $u = 0$ and the elevation angle at $u = \pi/2$:

$$a = \lambda(0) = \arctan p, \quad b = \varphi\left(\frac{\pi}{2}\right) = \arctan q.$$

Together with the spherical Pythagoras theorem (page 439) and the well-known formula $\cos(\arctan x) = 1/\sqrt{1 + x^2}$, this leads to

$$\cos e = \frac{\cos a}{\cos b} = \frac{\cos(\arctan p)}{\cos(\arctan q)} = \sqrt{\frac{1 + q^2}{1 + p^2}}$$

and further to

$$\sin e = \sqrt{1 - \cos^2 e} = \sqrt{\frac{p^2 - q^2}{1 + p^2}}.$$

Thus, we have found $S_1 = (\cos e, -\sin e, 0)$ and $S_2 = (\cos e, \sin e, 0)$. Their position vectors are

$$\mathbf{s}_{1,2} = \frac{1}{\sqrt{1 + p^2}} \begin{pmatrix} \sqrt{1 + q^2} \\ \mp\sqrt{p^2 - q^2} \\ 0 \end{pmatrix}.$$

The spherical distances $\widehat{S_1 C}$ and $\widehat{S_2 C}$, i.e., the angles δ_1 and δ_2, can be determined via the dot product of the corresponding position vectors. We have $\cos \delta_1 = \langle \mathbf{s}_1, \mathbf{c} \rangle$ and $\cos \delta_2 = \langle \mathbf{s}_2, \mathbf{c} \rangle$. Using the identity

$$\arccos x + \arccos y = \arccos \left(xy - \sqrt{1 - x^2}\sqrt{1 - y^2} \right),$$

we find, after some calculations,

$$\arccos \langle \mathbf{s}_1, \mathbf{c} \rangle + \arccos \langle \mathbf{s}_2, \mathbf{c} \rangle = \cdots = \arccos \frac{1 - p^2}{1 + p^2},$$

which equals $2 \arctan p = 2a$, provided that $p > 0$.

Thus, we know that the intersection of our quadratic projection cone $\mathcal{C}$ leads to a spherical conic. Any oblique section of $\mathcal{C}$ is again a conic, so that the choice of the carrier plane of the ellipse involved no loss of generality, and Theorem 10.1.7 is proven in a second way.

Now, it is obvious that the projections of the foci of the ellipse are *not* the foci of the spherical conic. Otherwise we would obtain $e = \arctan \sqrt{p^2 - q^2}$.

The principal views of a spherical conic

We have seen that the orthogonal projection of a spherical conic onto the symmetry plane σ_1 through its foci is a planar conic. The symmetry plane σ_2 of the foci S_1 and S_2 is another symmetry plane. The plane σ_3 perpendicular to both σ_1 and σ_2 that passes through O completes the set of symmetry planes of the spherical conic. We call the orthogonal projections onto these planes the *principal views* of the conic, and show:

Theorem 10.1.8 *The principal views of a spherical conic are located on ordinary (planar) conics.*

Proof: The symmetry planes have the following equations

$$\sigma_1 : z = 0, \quad \sigma_2 : y = 0, \quad \sigma_3 : x = 0.$$

Equations of the principal views of the spherical conic are obtained by eliminating z, y, and x from the equation (10.6) of the sphere $\mathcal{S}^2$, from eq. (10.5) of the elliptic cylinder $\mathcal{L}$ (or likewise (10.7) of the quadratic cone $\mathcal{C}$). The shape of the top view, *i.e.*, the curve in σ_1, is given by Theorem 10.1.6, and its equation is given by (10.5). Elimination of y and x from (10.5) and (10.6) results in

$$\left(1 - \frac{\cot^2 e}{\cot^2 a}\right) x^2 + z^2 = 1 - \frac{\sin^2 a}{\sin^2 e}, \quad \left(1 - \frac{\tan^2 e}{\tan^2 a}\right) y^2 + z^2 = 1 - \frac{\cos^2 a}{\cos^2 e}. \qquad (10.9)$$

These equations can be interpreted as equations of quadratic cylinders through the spherical conic or of conics in the planes σ_2 and σ_3. Because of $e < a$, in the latter case we obtain a hyperbola in $y = 0$ and an ellipse in $x = 0$. Of course, the principal views of the spherical conic are the restrictions to the closed unit disks.

Of course, the views of the spherical conic must lie inside the closed unit disk. ∎

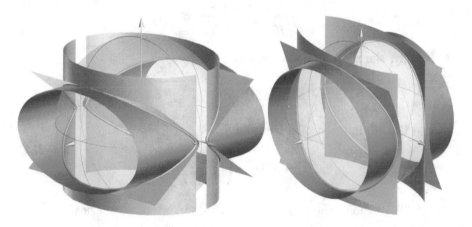

FIGURE 10.13. The principal projections of a spherical conic c are placed on planar conics. Hence, the curve c is the intersection of three quadratic cylinders. On the left $a : b = 3 : 1$ and a almost $\pi/2$, on the right $a : b = 5 : 4$.

We can summarize the results (Figure 10.13):

Theorem 10.1.9 *There are three quadratic cylinders and one quadratic cone (concentric with the sphere) passing through any spherical conic.*

Just to be complete, we do the same with the parametrization of our spherical conic given in (10.8):

1. The top view is given by

$$x = \frac{1}{r}, \quad y = \frac{p}{r} \cos u \quad \text{with} \quad r = \sqrt{1 + p^2 \cos^2 u + q^2 \sin^2 u}.$$

Elimination of the parameter u leads to the equation

$$(1 + q^2)x^2 + \frac{p^2 - q^2}{p^2} y^2 = 1.$$

2. The front view is described by

$$y = \frac{p}{r} \cos u, \quad z = \frac{q}{r} \sin u \quad \text{with} \quad r = \sqrt{1 + p^2 \cos^2 u + q^2 \sin^2 u}.$$

Elimination of the parameter u leads to the equation

$$\frac{1 + p^2}{p^2} y^2 + \frac{1 + q^2}{q^2} z^2 = 1.$$

3. Finally, the right side view is given by

$$x = \frac{1}{r}, \quad z = \frac{q}{r} \sin u \quad \text{with} \quad r = \sqrt{1 + p^2 \cos^2 u + q^2 \sin^2 u},$$

and the elimination of u leads to

$$(1 + p^2)x^2 + \left(1 - \frac{p^2}{q^2}\right) z^2 = 1.$$

Figure 10.12 shows the front and right side views of a spherical conic.

Ivory's theorem on the sphere

According to Theorem 10.1.7, each spherical conic is located on a quadratic cone $\mathcal{C}$ of signature $(+ - -)$. By (10.7), we have

$$\mathcal{C}: \quad x^2 \left(\frac{\cos^2 e}{\cos^2 a} - 1\right) - y^2 \left(1 - \frac{\sin^2 e}{\sin^2 a}\right) - z^2 = 0, \quad 0 < e < a < \frac{\pi}{2}.$$

The intersection $\mathcal{C} \cap \mathcal{S}^2$ consists of two mutually antipodal spherical conics. In the half-space $x > 0$, we get the conic with focal points $S_1 = (\cos e, \sin e, 0)$, $S_2 = (\cos e, -\sin e, 0)$ and principal vertices $(\cos a, \pm \sin a, 0)$.

A comparison of (10.7) with the ansatz

$$\frac{x^2}{u^2} - \frac{y^2}{v^2} - \frac{z^2}{w^2} = 0, \quad uvw \neq 0, \tag{10.10}$$

yields

$$u^2 : v^2 : w^2 = \frac{\cos^2 a}{\cos^2 e - \cos^2 a} : \frac{\sin^2 a}{\sin^2 a - \sin^2 e} : 1 = \cos^2 a : \sin^2 a : (\sin^2 a - \sin^2 e).$$

Under the assumption that $v^2 > w^2$, we can solve this for the quantities e, a, and b associated with the conic (see Figure 10.4, right):

$$\sin^2 e = \frac{v^2 - w^2}{u^2 + v^2}, \quad \sin^2 a = \frac{v^2}{u^2 + v^2}, \quad \sin^2 b = \frac{w^2}{w^2 + u^2}. \tag{10.11}$$

Therefore, all conics located on quadratic cones with equations

$$\frac{x^2}{u^2 + k} - \frac{y^2}{v^2 - k} - \frac{z^2}{w^2 - k} = 0, \quad k \in \mathbb{R} \setminus \{-u^2, v^2, w^2\} \tag{10.12}$$

have focal points in the set $\{S_1, S_2, S_1^*, S_2^*\}$. Therefore, we speak of a family of *confocal spherical conics*.

Under our assumption that $w^2 < v^2$, there are two *types* of conics to distinguish in this family:

- For $-u^2 < k < w^2$ the focal points are (S_1, S_2) or (S_1^*, S_2^*). In view of Figure 10.5, left, these are 'spherical ellipses'.

- For $w^2 < k < v^2$ we obtain branches of 'spherical hyperbolas' with respect to the focal points (S_1, S_2) or (S_1^*, S_2^*).

For $k < -u^2$ or $k > v^2$ the cones have the signature $(+ + +)$, they don't contain any real point except $\mathbf{0}$. The limit for $k \to -u^2$ is the great circle in $x = 0$ (as $a, b \to \frac{\pi}{2}$). The great circle in $y = 0$ occurs as limit for $k \to v^2$ (as $a \to 0$). The limits for $k \to w^2$ (as $b \to 0$) from $k < w^2$ are the segments $S_1 S_2$ or $S_1^* S_2^*$. As limits from $k > w^2$, we obtain the complementary segments $S_1^* S_2$ or $S_1 S_2^*$ on the great circle in $z = 0$.

The family of confocal conics together with these limiting curves forms an *orthogonal net* on the sphere (see Figure 10.14). Two mutually orthogonal net curves pass through each point $P \in \mathcal{S}^2$, and if P lies outside of any symmetry plane there is precisely one such conic of each type. This follows from Theorem 10.1.5 since the tangent line of an ellipse bisects the exterior angle between the focal rays, *i.e.*, the segments PS_1 and PS_2. Below we prove the spherical analogue of IVORY's Theorem 2.2.3.

Theorem 10.1.10 Ivory's theorem on $\mathcal{S}^2$. *The orthogonal net of confocal conics on the sphere has the property that in each curvilinear quadrangle the diagonals are of equal length.*

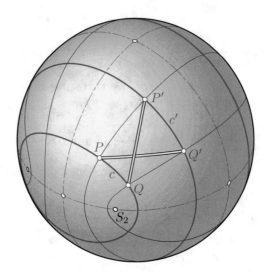

FIGURE 10.14. Ivory's theorem on the sphere states that $\widehat{P'Q} = \widehat{PQ'}$.

Proof: For the sake of simplicity, we identify points P with their position vectors $\mathbf{p}$. Therefore, we briefly speak of points $\mathbf{p}, \mathbf{q} \in \mathcal{S}^2$.

Let l be a linear map

$$l: \mathbb{R}^3 \to \mathbb{R}^3 \text{ with } \mathbf{p} \mapsto l(\mathbf{p}) = \mathbf{A}\mathbf{p}, \quad \mathbf{A} = \begin{pmatrix} \kappa & 0 & 0 \\ 0 & \lambda & 0 \\ 0 & 0 & \mu \end{pmatrix},$$

where $\kappa, \lambda, \mu \in \mathbb{R}$ and $(\kappa : \lambda : \mu) \neq (1 : 1 : 1)$. We look for points $\mathbf{p} = (x, y, z)^{\mathrm{T}} \in \mathcal{S}^2$ whose image points are again on $\mathcal{S}^2$. This holds if, and only if,

$$x^2 + y^2 + z^2 = \kappa^2 x^2 + \lambda^2 y^2 + \mu^2 z^2 = 1,$$

or equivalently

$$\|\mathbf{p}\| = 1 \text{ and } (\kappa^2 - 1)x^2 + (\lambda^2 - 1)y^2 + (\mu^2 - 1)z^2 = 0.$$

A comparison with the equation of the cone $\mathcal{C}$ in (10.10) shows that if

$$\kappa^2 - 1 > 0 > \lambda^2 - 1 > \mu^2 - 1, \text{ i.e., } \kappa^2 > 1 > \lambda^2 > \mu^2 \geq 0$$

these points $\mathbf{p}$ lie on the pair of antipodal conics given by

$$\mathcal{C}: \frac{x^2}{u^2} - \frac{y^2}{v^2} - \frac{z^2}{w^2} = 0 \text{ with } \frac{1}{u^2} = \kappa^2 - 1, \quad \frac{1}{v^2} = 1 - \lambda^2, \quad \frac{1}{w^2} = 1 - \mu^2,$$

after adjusting u, v, and w by an appropriate factor.

As a consequence, when $\mu^2 > 0$, the image points $l(\mathbf{p}) = (x', y', z')^{\mathrm{T}} = (\kappa x, \lambda y, \mu z)^{\mathrm{T}}$ are located on the pair of antipodal conics of

$$\mathcal{C}': \frac{x'^2}{\kappa^2 u^2} - \frac{y'^2}{\lambda^2 v^2} - \frac{z'^2}{\mu^2 w^2} = \frac{x'^2}{u^2 + 1} - \frac{y'^2}{v^2 - 1} - \frac{z'^2}{w^2 - 1} = 0.$$

We learn from (10.12) that the conics on $\mathcal{C}$ and $\mathcal{C}'$ are confocal (with $k = 1$) and of the same type (Figure 10.14), since

$$w^2 > 1 > -u^2.$$

Our linear mapping $l \colon \mathbf{p} \mapsto \mathbf{Ap}$ is *selfadjoint*, since due to the symmetry of the matrix $\mathbf{A}$, the adjoint mapping $l^* \colon \mathbf{q} \mapsto \mathbf{A}^{\mathrm{T}}\mathbf{q}$ equals l. Now, we use a well-known rule from Linear Algebra. It is a consequence of the following sequence of equations with switches between the dot product $\langle \cdot, \cdot \rangle$ and its matrix representation:

$$\langle l(\mathbf{p}), \mathbf{q} \rangle = l(\mathbf{p})^{\mathrm{T}}\mathbf{q} = (\mathbf{Ap})^{\mathrm{T}}\mathbf{q} = \mathbf{p}^{\mathrm{T}}\mathbf{A}^{\mathrm{T}}\mathbf{q} = \mathbf{p}^{\mathrm{T}}l^*(\mathbf{q}) = \langle \mathbf{p}, l^*(\mathbf{q}) \rangle.$$

This expresses directly the statement of IVORY's theorem $\overline{\mathbf{p}'\mathbf{q}} = \overline{\mathbf{p}\mathbf{q}'}$. It only remains to confirm that corresponding points $\mathbf{p} \in \mathcal{C}$ and $\mathbf{p}' = l(\mathbf{p}) \in \mathcal{C}'$ are always located on the same conic from the other type of the confocal family, in our case on conics (10.12) with k between v^2 and w^2:

Suppose that simultaneously

$$\frac{x^2}{u^2} - \frac{y^2}{v^2} - \frac{z^2}{w^2} = 0 \quad \text{and} \quad \frac{x^2}{u^2 + k} - \frac{y^2}{v^2 - k} - \frac{z^2}{w^2 - k} = 0,$$

and $k \neq 0$. Then, also

$$\frac{x'^2}{u^2 + k} - \frac{y'^2}{v^2 - k} - \frac{z'^2}{w^2 - k} = \frac{\kappa^2 x^2}{u^2 + k} - \frac{\lambda^2 y^2}{v^2 - k} - \frac{\mu^2 z^2}{w^2 - k} = 0,$$

because

$$\frac{\kappa^2}{u^2 + k} = \frac{u^2 + 1}{u^2(u^2 + k)} = \frac{1}{k}\frac{1}{u^2} + \frac{k-1}{k}\frac{1}{u^2 + k},$$

which finally yields an affine combination, *i.e.*, a linear combination with coefficients summing up to 1,

$$\frac{x'^2}{u^2 + k} - \frac{y'^2}{v^2 - k} - \frac{z'^2}{w^2 - k} = \frac{1}{k}\left(\frac{x^2}{u^2} - \frac{y^2}{v^2} - \frac{z^2}{w^2}\right) + \frac{k-1}{k}\left(\frac{x^2}{u^2 + k} - \frac{y^2}{v^2 - k} - \frac{z^2}{w^2 - k}\right).$$

When the second conic through $\mathbf{p}$ degenerates (*i.e.*, when $k \to v^2$ or $k \to w^2$), then it is still true that $l(\mathbf{p})$ lies on the same degenerate conic, since l maps each plane of symmetry onto itself. ∎

Remark 10.1.1 Ivory's statement holds also for degenerate conics in the family of confocal conics since the linear map l in the proof above is singular if $\mu = 0$. Then, one conic on $\mathcal{C}'$ is the segment $S_1 S_2$ between the focal points of $\mathcal{C}$. Since the endpoints S_1 and S_2 are the l-images of the vertices $A, B \in \mathcal{C}$, we obtain

$$\overline{PS_1} + \overline{PS_2} = \overline{PA'} + \overline{PB'} = \overline{P'A} + \overline{P'B} = \overline{AB} = 2a.$$

In this way, IVORY's theorem can be used to prove directly that the intersection curves between a quadratic cone and the unit sphere satisfy the properties of a spherical ellipse.

After inverting the coefficient matrices of the confocal conics in (10.12), we obtain their tangential equations (see Section 7.1)

$$(u^2 + k)x^2 - (v^2 - k)y^2 - (w^2 - k)z^2 = (u^2 x^2 - v^2 y^2 - w^2 z^2) + k(x^2 + y^2 + z^2) = 0.$$

We notice: The duals of confocal conics belong to a pencil which is spanned by one of the conics and the conic defined by the *isotropic cone* satisfying the equation $x^2 + y^2 + z^2 = 0$. The previously excluded parameter values $k = \{-u^2, v^2, w^2\}$ define the singular conics in this pencil, among them the pencils of great circles through the focal points.

Two particular spherical conics

A bundle in the projective three-space, *i.e.*, the set of lines and planes through a fixed point O, is a projective plane if lines and planes are seen as 'points' and 'lines' (cf. Section 5.1). In this sense quadratic cones are the 'conics' of the bundle. Hence, all projective properties of conics occur again at quadratic cones and at their spherical visualization in the form of antipodal pairs of spherical conics. For example, all theorems concerning porisms for planar conics are also true for spherical conics (Figure 10.15).

FIGURE 10.15. A spherical Poncelet porism with heptagons together with conics passing through the intersections between the sides' extensions.

However, the metric properties of spherical conics often differ from those of their planar counterparts. We have already seen this in the previous sections. A deeper reason for this difference lies in the fact that the isotropic lines (see Section 6.4, especially Example 6.4.7 on page 253) in the complex extension of the Euclidean plane form a singular dual line-conic while the quadratic cone of isotropic lines in the bundle is regular.

In Cartesian coordinates, the *isotropic cone* with the apex at the origin O obeys the equation

$$x^2 + y^2 + z^2 = 0.$$

The corresponding symmetric coefficient matrix is the unit matrix $\mathbf{I}_3$. Two vectors $\mathbf{v}_1 = (x_1, y_1, z_1)^{\mathrm{T}}$ and $\mathbf{v}_2 = (x_2, y_2, z_2)^{\mathrm{T}}$ are conjugate with respect to this cone if, and only if,

$$0 = x_1 x_2 + y_1 y_2 + z_1 z_2 = \langle \mathbf{v}_1, \mathbf{v}_2 \rangle,$$

i.e., the vectors are orthogonal. The polarity in the isotropic cone is also called *absolute polarity* of the bundle.

We set up the Cartesian equation of any quadratic cone with apex O as

$$\mathcal{C}:\ c_{00}x^2 + c_{11}y^2 + c_{22}z^2 = 0 \quad \text{with} \quad c_{00} > 0 > c_{11} \geq c_{22}. \tag{10.13}$$

The intersections of this quadratic cone with the planes $z = 0$ and $y = 0$ show that the semiaxes a and b of the corresponding spherical conic in the halfspace $x > 0$ obey

$$\tan^2 a = -\frac{c_{00}}{c_{11}}, \quad \sin^2 a = \frac{c_{00}}{c_{00} - c_{11}}, \quad \tan^2 b = -\frac{c_{00}}{c_{22}}, \quad \sin^2 b = \frac{c_{00}}{c_{00} - c_{22}}. \tag{10.14}$$

This follows also upon comparison with (10.7).

Among the quadratic cones we pick out the following two with special metrical properties:

- *equilateral* cones have a vanishing trace, *i.e.*, $c_{00} + c_{11} + c_{22} = 0$;
- *normal* cones are characterized by $c_{00} + c_{22} = c_{11}$.

We use these names also for the corresponding spherical conics.

Equilateral spherical conics

As a consequence of (10.14), equilateral spherical conics are characterized by

$$\sin^2 b = \frac{\sin^2 a}{3 \sin^2 a - 1}. \tag{10.15}$$

The condition $b < a$ for proper conics implies the limits

$$\arcsin \sqrt{\frac{2}{3}} < a < \frac{\pi}{2}, \quad \arcsin \sqrt{\frac{2}{3}} > b < \frac{\pi}{4}.$$

In one limiting case, our spherical conic becomes a circle with radius $a = b = \arcsin \sqrt{2/3}$. At the other limit, with $a = \pi/2$ and $b = \pi/4$, i.e., $c_{11} = 0$, the quadratic cone splits into two perpendicular planes obeying $x^2 - z^2 = 0$.

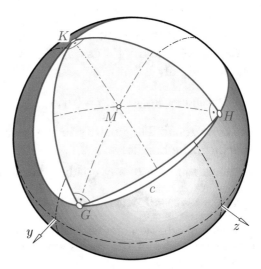

FIGURE 10.16. The regular right triangle GHK can move around in the equilateral spherical conic c while all three vertices G, H, and K run along c.

Theorem 10.1.11 *Within any equilateral spherical conic c we can inscribe an infinite number of octants, i.e., of regular triangles all of whose sides are of length $\pi/2$ and whose angles are all $\pi/2$.*

Proof: For the matrix $\mathbf{C} = \mathrm{diag}(c_{00}, c_{11}, c_{22})$ of c we obtain

$$\det(\sigma \mathbf{I}_3 + \tau \mathbf{C}) = \det \begin{pmatrix} \sigma + \tau c_{00} & 0 & 0 \\ 0 & \sigma + \tau c_{11} & 0 \\ 0 & 0 & \sigma + \tau c_{22} \end{pmatrix}.$$

The coefficient of $\sigma^2 \tau$ vanishes. Hence, by Lemma 9.5.2 (page 420), the isotropic cone is inpolar to c. In the bundle of lines and planes through O, each triple of lines which is self-polar with respect to the isotropic cone consists of three pairwise orthogonal lines. Therefore, by Lemma 9.5.1 (page 419), an infinite set of octants can be inscribed in the equilateral conic (Figure 10.16).   ■

The octant performs a periodic *continuous spherical motion* while the three vertices GHK simultaneously traverse the same equilateral conic c (Figure 10.17). During one full turn, the moving triangle GHK returns twice to its initial position, however rotated under $120°$ and $240°$.[2] The sides of the moving octant envelope again a spherical conic i. It is located on the cone which corresponds to the cone of c in the absolute polarity (compare also Lemma 9.5.4).

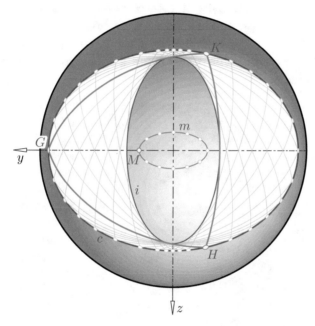

FIGURE 10.17. Top view of the motion of the octant GHK turning around in the equilateral spherical conic c, together with the path m of the triangle's center M and the envelope i of the three sides.

Thus, we obtain again a spherical Poncelet porism with a family of triangles which in this particular case are mutually congruent. One can traverse these triangles by a spherical motion which is overconstrained as three points trace the same curve. This motion is very special. There exists no counterpart in the plane. In the limiting case $a = b$, the octant's motion is a pure rotation about its axis of threefold symmetry. At the

---

[2]The generic point paths are of *spherical degree* 24, *i.e.*, they are projected from the origin by cones of degree 24. For details see [59].

other limit, with the splitting cone, the continuous motion consists of rotations which keep one of the three vertices fixed. There are poses which allow a bifurcation between these rotations.

Another spherical motion with the three vertices of a regular triangle running along the same curve t occurs in the spherical version of a *Wankel engine* (Figure 10.18): The curve t is a spherical trochoid generated as a point path when the rotor rotates about its center M with angular velocity -2 while additionally it rotates about the fixed center O with angular velocity 3. During this spherical motion the circle m rolls along the fixed circle o. In this case the rotor needs not be a spherical octant. However, the common point path t is not a spherical conic but a curve of spherical degree 6.

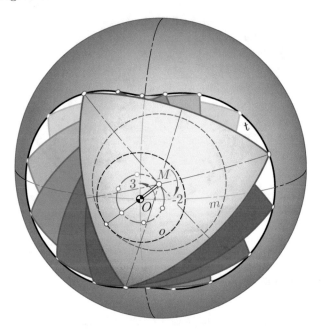

FIGURE 10.18. At the *'spherical Wankel engine'* the common path t of the three vertices of the rotor is a trochoid. It can be generated when the circle m which is attached to the rotor, is rolling along the fixed circle o. The angular velocities of the rotor's rotations about M and additionally about the fixed center O have the constant ratio $-3 : 2$.

Let us return to the equilateral conic c: Suppose, we project the moving octant from the origin into a plane ε which intersects the cone of c along a circle c_0 (Figure 4.23). Then, we obtain a one-parameter set of triangles with a common circumcircle c_0 and a common orthocenter O, which is the pedal point of ε w.r.t. the center $\mathbf{0}$ (Figure 10.19)[3]. Due to the properties of the Euler line, also the centroid G is common to these triangles in ε which envelope again a conic i_0. Conversely, these triangles can serve for an elementary approach to Theorem 10.1.11.

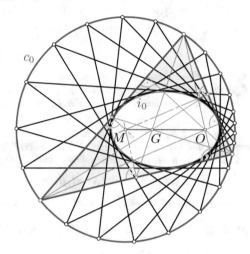

FIGURE 10.19. Triangles sharing the circumcircle c_0, the centroid G and the orthocenter O are circumscribed to the conic i_0.

Normal spherical conics

By (10.14), normal spherical conics are characterized by

$$\sin a = \tan b \quad \text{or} \quad \sin^2 b = \frac{\sin^2 a}{\sin^2 a + 1}. \tag{10.16}$$

On the other hand, the condition $c_{00} + c_{22} = c_{11}$ is equivalent to the fact that the corresponding cone $\mathcal{C}$ from (10.13) intersects the plane of symmetry $y = 0$ in lines which are perpendicular to planes

$$x\sqrt{c_{00} - c_{11}} \pm z\sqrt{c_{11} - c_{22}} = 0,$$

which cut $\mathcal{C}$ along circles.

---

[3] Figure 9.35 on page 422 shows a version with the orthocenter O outside of c. In this case the envelope of the triangles is a hyperbola.

We conclude this section with spherical analogues of wellknown theorems from plane geometry concerning the Thales circle and the angle of circumference (compare with Figure 6.9 on page 226).

Theorem 10.1.12 *Let A, B be two fixed points on the sphere $\mathcal{S}^2$ with $\overline{AB} = 2\lambda < \pi/2$. The set of points $P \in \mathcal{S}^2$ with $\sphericalangle APB = \pi/2$ is a normal spherical conic c with A and B as auxiliary vertices. Hence, the semiminor axis of the conic c is $b = \lambda$; the semimajor axis a satisfies $\sin a = \tan b$ (Figure 10.20a).*

Proof: We start with the generalized problem: Given two points $A, B \in \mathcal{S}^2$ with $\overline{AB} = 2\lambda \leq \pi/2$, what is the locus c of points $P \in \mathcal{S}^2$ such that $\sphericalangle APB = \varphi$ or $\sphericalangle APB = \pi - \varphi$ with constant φ and $0 < \varphi \leq \pi/2$? The curve c is called *isoptic curve* of the segment AB.

We denote the position vectors of A, B, and P with $\mathbf{a}, \mathbf{b}$ and $\mathbf{p}$, respectively, and we assume that $\mathcal{S}^2$ is centered at $\mathbf{0}$. For $P \neq A, B$ the great circles $[A, P]$ and $[B, P]$ have axes in the direction of the vectors $\mathbf{p} \times \mathbf{a}$ and $\mathbf{p} \times \mathbf{b}$, respectively. The angle $\sphericalangle APB$ or $\pi - \sphericalangle APB$ occurs also as angle between these two vectors. Hence, their scalar product must satisfy

$$\langle \mathbf{p} \times \mathbf{a}, \mathbf{p} \times \mathbf{b} \rangle = \pm \cos \varphi \, \|\mathbf{p} \times \mathbf{a}\| \|\mathbf{p} \times \mathbf{b}\|.$$

After squaring this basic equation, we substitute according to the formula $\|\mathbf{u}\|^2 \|\mathbf{v}\|^2 = \|\mathbf{u} \times \mathbf{v}\|^2 + \langle \mathbf{u}, \mathbf{v} \rangle^2$

$$\|\mathbf{p} \times \mathbf{a}\|^2 \|\mathbf{p} \times \mathbf{b}\|^2 = [(\mathbf{p} \times \mathbf{a}) \times (\mathbf{p} \times \mathbf{b})]^2 + \langle \mathbf{p} \times \mathbf{a}, \mathbf{p} \times \mathbf{b} \rangle^2$$

and obtain

$$\langle \mathbf{p} \times \mathbf{a}, \mathbf{p} \times \mathbf{b} \rangle^2 (1 - \cos^2 \varphi) = \cos^2 \varphi \, [(\mathbf{p} \times \mathbf{a}) \times (\mathbf{p} \times \mathbf{b})]^2.$$

Standard formulas for mixed products of vectors of $\mathbb{R}^3$ yield

$$[\langle \mathbf{p}, \mathbf{p} \rangle \langle \mathbf{a}, \mathbf{b} \rangle - \langle \mathbf{p}, \mathbf{a} \rangle \langle \mathbf{p}, \mathbf{b} \rangle]^2 = \cot^2 \varphi \, [\mathbf{a} \det(\mathbf{p}, \mathbf{p}, \mathbf{b}) - \mathbf{p} \det(\mathbf{a}, \mathbf{p}, \mathbf{b})]^2$$

and finally

$$[\langle \mathbf{p}, \mathbf{p} \rangle \langle \mathbf{a}, \mathbf{b} \rangle - \langle \mathbf{p}, \mathbf{a} \rangle \langle \mathbf{p}, \mathbf{b} \rangle]^2 = \cot^2 \varphi \langle \mathbf{p}, \mathbf{p} \rangle [\det(\mathbf{a}, \mathbf{p}, \mathbf{b})]^2.$$

We set (note the coordinate frame displayed in Figure 10.21)

$$\mathbf{p} = \begin{pmatrix} x \\ y \\ z \end{pmatrix}, \quad \mathbf{a} = \begin{pmatrix} \cos \lambda \\ -\sin \lambda \\ 0 \end{pmatrix}, \quad \mathbf{b} = \begin{pmatrix} \cos \lambda \\ \sin \lambda \\ 0 \end{pmatrix}$$

and obtain

$$[(x^2 + y^2 + z^2)(\cos^2 \lambda - \sin^2 \lambda) - (x^2 \cos^2 \lambda - y^2 \sin^2 \lambda)]^2$$
$$= \cot^2 \varphi (x^2 + y^2 + z^2)[2z \sin \lambda \cos \lambda]^2.$$

Finally, the cone connecting the center of $\mathcal{S}^2$ with the locus c of the requested points P satisfies the equation

$$[-x^2 \sin^2 \lambda + y^2 \cos^2 \lambda + z^2 (\cos^2 \lambda - \sin^2 \lambda)]^2 - 4z^2 \sin^2 \lambda \cos^2 \lambda \cot^2 \varphi (x^2 + y^2 + z^2) = 0. \quad (10.17)$$

This cone is symmetric with respect to all three coordinate planes (Figure 10.20b–d).

In the case $\varphi = \pi/2$, hence $\cot \varphi = 0$, the cone reduces to a quadratic one (Figure 10.20a). A comparison with (10.13) shows

$$c_{00} = \sin^2 \lambda, \quad c_{11} = \sin^2 \lambda - \cos^2 \lambda, \quad c_{22} = -\cos^2 \lambda.$$

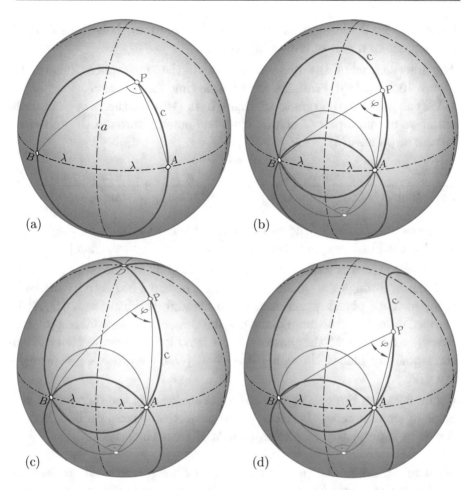

FIGURE 10.20. These are versions of spherical isoptic curves c for the segment AB, *i.e.*, sets of points $P \in \mathcal{S}^2$ with $\sphericalangle APB = \varphi$ or $\pi - \varphi$. (a) The spherical analogue of the Thales circle ($\varphi = \pi/2$) is a normal conic c with A and B as auxiliary vertices. In all other cases, c is a spherical quartic, *i.e.*, it is projected from the center of the sphere by a cone of degree 4. (b) the case $\overset{\frown}{AB} = 2\lambda < \varphi$; (c) $2\lambda = \varphi$, the point D is a singularity of the quartic c; (d) $2\lambda > \varphi$.

Obviously, $c_{00} + c_{22} = c_{11}$. By (10.14) the semiaxes a, b satisfy

$$\tan^2 b = -\frac{c_{00}}{c_{22}} = \tan^2 \lambda = \frac{c_{00}}{c_{00} - c_{11}} = \sin^2 a.$$

In the limiting case $\lambda = \pi/4$ and $\cot \varphi = 0$, the cone splits into two orthogonal planes.   ∎

Under $\cot\varphi \neq 0$, the cone in (10.17) is of degree four. It is the spherical counterpart of a pair of circles in the plane.

The intersection of the quartic cone with the unit sphere (shown in Figure 10.20, (b)–(d)) satisfies at the same time the equation $x^2 + y^2 + z^2 = 1$. We can use this equation to eliminate from (10.17) either x or y. In both cases, we obtain a pair of elliptic cylinders passing through the spherical locus c of points P:

$$\frac{x^2}{\cos^2\lambda/\sin^2\varphi} + \frac{(z \pm \cot\lambda\cot\varphi)^2}{\cot^2\lambda/\sin^2\varphi} = 1, \qquad \frac{y^2}{\sin^2\lambda/\sin^2\varphi} + \frac{(z \pm \tan\lambda\cot\varphi)^2}{\tan^2\lambda/\sin^2\varphi} = 1.$$

In the case $\lambda = \frac{\varphi}{2}$, all cylinders touch the sphere at a point on the z-axis. The locus c has two symmetric singularities (see Figure 10.20c).

As a consequence of Theorem 10.1.12, we can state for the bundle of lines and planes through the sphere center O: We obtain a projectivity between two pencils of planes when each plane through the fixed axis $[O, A]$ is mapped onto the orthogonal plane passing through the second axis $[O, B]$. However, contrary to the planar case, it is no longer a projectivity when corresponding planes through $[O, A]$ and $[O, B]$ enclose a given angle $\varphi \neq \pi/2$.

Figure 6.9 gives rise to a second spherical analogue of the theorem on the angle of circumference. It refers to the projective generation of spherical conics by a projectivity between pencils of great circles:

Theorem 10.1.13 *Let us assume that on $\mathcal{S}^2$ the projectivity α between the pencils of great circles through the points P and Q with $0 < \overline{PQ} < \pi$ is an orientation preserving congruence but no perspectivity. Then, the generated pair of antipodal conics consist of normal conics. The spherical bisector of P and Q is the minor axis of these conics (Figure 10.21).*

Remark 10.1.2 If on $\mathcal{S}^2$ the congruence α between P and Q is orientation reversing, then we can replace P by its antipode $\overline{P}$. Thus, we again can apply Theorem 10.1.13 because the assignment between the pencils $\overline{P}$ and Q is orientation preserving.

Proof: We confine our attention to the conic c passing through P and Q (cf. Figure 10.21). According to Figure 6.1, right, on page 218 the pre-image $1 = \alpha^{-1}([P,Q])$ and the image $2' = \alpha([P,Q])$ are tangent to c at the points P and Q. These tangents must build equal angles with the chord $[P,Q]$ since α is supposed to be an orientation preserving congruence. Hence, due to properties of the polarity with respect to c, the bisector of P and Q is an axis of symmetry of c.

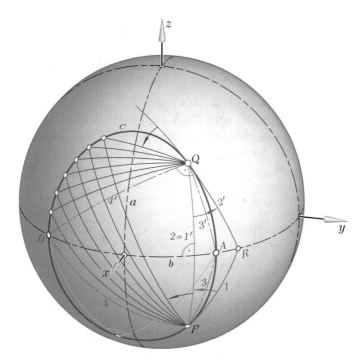

FIGURE 10.21. On any sphere, two congruent pencils of great circles generate a normal conic c.

Our congruence α must map the angle bisectors $3, 4$ between great circles 1 and 2 onto the angle bisectors $3', 4'$ of $1'$ and $2'$ (Figure 10.21). Corresponding angle bisectors intersect at the points $A, B \in c$ on our axis of symmetry. Because of the orthogonality of $(3, 4)$ and of $(3', 4')$, and by Theorem 10.1.12, the points P and Q are located on a normal conic with the auxiliary vertices A and B. The auxiliary vertices A and B together with the points P and Q define the conic c uniquely. Hence, the spherical conic c must be normal.

Conversely, any normal conic is already defined by the two auxiliary vertices A and B because of (10.16). Therefore, each normal conic c can be generated by a congruence, provided the basic points $P, Q \in c$ are symmetric with respect to the minor axis $[A, B]$. ∎

■ **Example 10.1.1** Spherical analogue of the Thales circle. Give a proof of Theorem 10.1.12 by confirming that the projection of our spherical conic c from the sphere center O into a fixed plane orthogonal to the diameter $[O, A]$ gives a circle. Note that corresponding orthogonal planes through $[O, A]$ and $[O, B]$ intersect our fixed plane along lines which are mutually orthogonal.

10.2 Conics in non-Euclidean planes

We are going to report on some Euclidean properties of conics which remain valid even in non-Euclidean planes. However, we restrict ourselves to two planes, the pseudo-Euclidean plane $\mathbb{M}^2$ and the hyperbolic plane $\mathbb{H}^2$. We focus in particular on IVORY's theorem which surprisingly is a connecting link between these geometries. Many other common properties are omitted, *e.g.*, the optical property of conics which is a consequence of the dual version of Desargues's involution theorem (Section 7.4).

The pseudo-Euclidean plane $\mathbb{M}^2$

The *pseudo-Euclidean plane* or *Minkowski plane*[4] $\mathbb{M}^2$ is defined as a real affine plane where the underlying vector space $\mathbb{R}^2$ is endowed with a non-degenerate indefinite symmetric bilinear form, called *pseudo-Euclidean scalar product* $\langle\ ,\ \rangle_m$. The *pseudo-Euclidean distance* $d_m(P,Q)$ of points P,Q with coordinate vectors $\mathbf{p},\mathbf{q} \in \mathbb{R}^2$ is defined as

$$d_m(P,Q) = \sqrt{\langle(\mathbf{p}-\mathbf{q}),(\mathbf{p}-\mathbf{q})\rangle_m}\,.$$

This distance is either a nonnegative number or the product of a positive number with the imaginary unit i.

A line segment PQ as well as the spanned line $[P,Q]$ are called *lightlike*, *spacelike* or *timelike* if the length $d_m(P,Q)$ is zero, positive or imaginary, respectively. Two vectors with vanishing scalar product as well as the two spanned lines are called *orthogonal* in the pseudo-Euclidean sense. Hence, lightlike vectors are self-orthogonal.

We call an affine coordinate system with basis vectors $(\mathbf{b}_1,\mathbf{b}_2)$ *orthonormal* in the pseudo-Euclidean sense, when the scalar product of two vectors $\mathbf{u} = u_1\mathbf{b}_1 + u_2\mathbf{b}_2$ and $\mathbf{v} = v_1\mathbf{b}_1 + v_2\mathbf{b}_2$ can be written as

$$\langle\mathbf{u},\mathbf{v}\rangle_m = u_1v_1 - u_2v_2 \text{ for all } \mathbf{u},\mathbf{v} \in \mathbb{R}^2. \tag{10.18}$$

Consequently, the basis vectors are orthogonal and satisfy $\langle\mathbf{b}_1,\mathbf{b}_1\rangle_m = 1$ and $\langle\mathbf{b}_2,\mathbf{b}_2\rangle_m = -1$. The components (u_1,u_2) of any vector $\mathbf{u} \in \mathbb{R}^2$ with respect to $(\mathbf{b}_1,\mathbf{b}_2)$ are $u_1 = \langle\mathbf{u},\mathbf{b}_1\rangle_m$ and $u_2 = -\langle\mathbf{u},\mathbf{b}_2\rangle_m$. The vectors $\mathbf{b}_1 + \mathbf{b}_2$ and $\mathbf{b}_1 - \mathbf{b}_2$ are lightlike. By virtue of (5.6), any two orthogonal non-lightlike lines separate the two lightlike directions harmonically.

---

[4]Named after HERMANN MINKOWSKI (1864–1909), mathematician and physicist in Göttingen.

The figures in this section are based on the standard model of the pseudo-Euclidean plane $\mathbb{M}^2$. This is embedded into the Euclidean plane $\mathbb{E}^2$ such that the basis vectors of the Cartesian coordinates (x, y) are also orthonormal in the pseudo-Euclidean sense. Lightlike lines make an angle of $45°$ with the x-axis, spacelike lines have an inclination $< 45°$. In the case of ambiguities we use the prefix "m-" or "e-" at geometric terms in order to emphasize whether they are meant in the pseudo-Euclidean (Minkowskian) or in the Euclidean sense. In this sense we can, $e.g.$, state that two intersecting lines in the standard model are m-orthogonal if and only if their e-angle bisectors are lightlike.

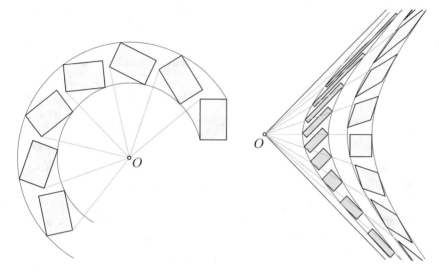

FIGURE 10.22. Euclidean (left) and pseudo-Euclidean rotations (right).

Affine transformations whose associated linear map $l \colon \mathbb{R}^2 \to \mathbb{R}^2$ preserves scalar products, $i.e.$,

$$\langle l(\mathbf{u}), l(\mathbf{v}) \rangle_m = \langle \mathbf{u}, \mathbf{v} \rangle_m \quad \text{for all} \quad \mathbf{u}, \mathbf{v} \in \mathbb{R}^2$$

are called *pseudo-Euclidean motions*. They preserve distances and orthogonality; the types of lines do not change, in particular, lightlike lines remain lightlike. When the matrix $\mathbf{A} = (a_{ik})$ represents the linear map l with respect to the orthonormal basis $(\mathbf{b}_1, \mathbf{b}_2)$, then the two column vectors $l(\mathbf{b}_1)$ and $l(\mathbf{b}_2)$ in $\mathbf{A}$ must be orthonormal too, hence

$$a_{11}^2 - a_{21}^2 = 1, \quad a_{12}^2 - a_{22}^2 = -1, \quad a_{11}a_{12} - a_{21}a_{22} = 0. \tag{10.19}$$

We obtain

$$
\begin{aligned}
(\det \mathbf{A})^2 &= a_{11}^2 a_{22}^2 - 2a_{11}a_{12}a_{21}a_{22} + a_{12}^2 a_{21}^2 \\
&= (1 + a_{21}^2)a_{22}^2 - 2a_{21}^2 a_{22}^2 + (a_{22}^2 - 1)a_{21}^2 = 1.
\end{aligned}
$$

Pseudo-Euclidean motions with $\det \mathbf{A} = +1$ are called *direct*, otherwise *indirect*. The lightlike directions remain fixed under direct motions, while they are exchanged in the indirect case (Exercise 10.2.1).

Translations are direct motions, as well as the motion

$$
\begin{pmatrix} x \\ y \end{pmatrix} \mapsto \begin{pmatrix} x' \\ y' \end{pmatrix} = \mathbf{R}_1(\varphi) \begin{pmatrix} x \\ y \end{pmatrix} \text{ with } \mathbf{R}_1(\varphi) = \begin{pmatrix} \cosh\varphi & \sinh\varphi \\ \sinh\varphi & \cosh\varphi \end{pmatrix} \tag{10.20}
$$

and $\varphi \in \mathbb{R}$ which keeps the origin O fixed. It is called a pseudo-Euclidean *rotation* about O. The parameter φ acts like an angle of rotation, since

$$
\mathbf{R}_1(\varphi_1)\mathbf{R}_1(\varphi_2) = \mathbf{R}_1(\varphi_1 + \varphi_2). \tag{10.21}
$$

Figure 10.22 (right) shows an iterated pseudo-Euclidean rotation of two parallelograms. The rotations preserve the e-area of the parallelograms.

For variable $\varphi \in \mathbb{R}$ in (10.20) we obtain a continuous rotation about O. Obviously, it is non-periodic. Points on the lightlike lines through O run along these lines, while the trajectories of all other points are *pseudo-Euclidean circles* with center O and satisfying the equation $\langle \mathbf{x}, \mathbf{x} \rangle_m = c = $ const., $c \neq 0$. In the Euclidean sense, they are branches of equilateral hyperbolas with the x- or y-axis as principal axis.

The motion

$$
\begin{pmatrix} x \\ y \end{pmatrix} \mapsto \begin{pmatrix} x' \\ y' \end{pmatrix} = \mathbf{R}_2(\tau) \begin{pmatrix} x \\ y \end{pmatrix} \text{ with } \mathbf{R}_2(\tau) = \begin{pmatrix} \cosh\tau & -\sinh\tau \\ \sinh\tau & -\cosh\tau \end{pmatrix} \tag{10.22}
$$

is indirect and involutory. It can be proved (Exercise 10.2.1) that this is the *pseudo-Euclidean reflection* in the line $y = \tanh \frac{\tau}{2} x$.

Remark 10.2.1 The Minkowski plane represents the geometry which is associated with the Einsteinian principle of relativity. The y-coordinate stands for the time t; points in $\mathbb{M}^2$ represent events (x, t) in space-time. Spacelike lines through the origin correspond to uniform motions along the x-axis. The scaling of t is such that the motion with maximum speed, *i.e.*, the speed of light, corresponds to lightlike lines with $45°$ inclination. The pseudo-Euclidean motions are socalled *Lorentz transformations* between inertial reference frames. The matrix equation (10.21) expresses the relativistic law of the 'velocity-addition' such that the speed of light is the same in all inertial reference frames. For further details see [68].

Pseudo-Euclidean conics

The pseudo-Euclidean plane $\mathbb{M}^2$ is an affine plane. Therefore, we still can distinguish between ellipses, hyperbolas and parabolas. Now the question arises: is the Apollonian definition of conics still valid in $\mathbb{M}^2$?

Given a focal point F, an associated directrix l with $F \not\in l$, and the numerical eccentricity $\varepsilon > 0$, what is the set of points P in $\mathbb{M}^2$ with proportional distances $d_m(P, F) = \varepsilon \cdot d_m(P, l)$, when $d_m(P, l)$ denotes the distance between P and its pseudo-Euclidean pedal point on l? Obviously, the directrix l must not be lightlike since otherwise all lines orthogonal to l would be parallel to l.

Up to translations and a rotation about O, we can confine ourselves to the cases that l is either the y-axis with $F = (p/\varepsilon, 0)$ or the x-axis with $F = (0, p/\varepsilon)$ for $p \in \mathbb{R} \backslash \{0\}$. Hence, for $P = (x, y)$ the condition $d_m(P, F) = \varepsilon \cdot d_m(P, l)$ is equivalent to either

$$x^2 - \frac{2p}{\varepsilon} x + \frac{p^2}{\varepsilon^2} - y^2 = \varepsilon^2 x^2 \quad \text{or} \quad x^2 - y^2 + \frac{2p}{\varepsilon} y - \frac{p^2}{\varepsilon^2} = -\varepsilon^2 y^2.$$

We can express these equations in the form

$$(\varepsilon^2 - 1)x^2 + y^2 + \frac{2p}{\varepsilon} x - \frac{p^2}{\varepsilon^2} = 0 \quad \text{or} \quad x^2 + (\varepsilon^2 - 1)y^2 + \frac{2p}{\varepsilon} y - \frac{p^2}{\varepsilon^2} = 0.$$

They reveal that the wanted geometric locus c of points P is a conic, namely for $\varepsilon > 1$ an ellipse, for $\varepsilon = 1$ a parabola and for $\varepsilon < 1$ a non-equilateral hyperbola which intersects the axis of symmetry through F in real points. (Note that in the Euclidean case we had $\varepsilon < 1$ for ellipses and $\varepsilon > 1$ for hyperbolas.) After applying pseudo-Euclidean motions, all obtained ellipses and hyperbolas have a center and a spacelike and a timelike axis of m-symmetry. The obtained parabolas have either a spacelike or a timelike axis of m-symmetry.

The lightlike lines through F intersect the directrix l at two points which satisfy the Apollonian condition, since their m-distances to F and l vanish. These lightlike lines cannot meet the conic c at another point since such a point must again have a zero-distance to l. Therefore, a point F is called a *focal point* or *focus* of a conic c in $\mathbb{M}^2$ if and only if the two lightlike lines through F are tangent to c (Figure 10.23).

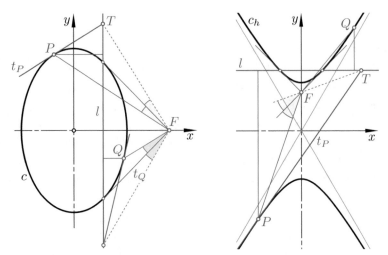

FIGURE 10.23. The ellipse c_e satisfies the Apollonian definition in $\mathbb{M}^2$ with $\varepsilon \sim 1.80$, the hyperbola c_h with $\varepsilon = 0.80$. The tangent t_P intersects the directrix l at T such that $[P,F]$ and $[T,F]$ are m-orthogonal.

Lemma 10.2.1 *In the pseudo-Euclidean plane $\mathbb{M}^2$, the locus c of points satisfying the Apollonian definition $d_m(P,F) = \varepsilon \cdot d_m(P,l)$ with a non-lightlike directrix l, $F \notin l$ and $\varepsilon > 0$ is a conic with an axis of symmetry. We obtain for $\varepsilon > 1$ an ellipse, for $\varepsilon = 1$ a parabola and for $\varepsilon < 1$ a hyperbola with non-lightlike asymptotes. Point F is a focus of c; the directrix l is polar to F with respect to c.*

Due to properties of the polarity with respect to c (Section 7.1), the tangent t_P to c at P can be constructed by virtue of the pseudo-Euclidean version of Theorem 2.1.4 via the pole T of the line $[P,F]$ (Figure 10.23).

According to Lemma 10.2.1, all conics in $\mathbb{M}^2$ which satisfy the Apollonian definition have an axis of symmetry. However conversely, not all conics c in $\mathbb{M}^2$ admit an axial symmetry. For ellipses and hyperbolas, the two axes of m-symmetry must be conjugate with respect to c and harmonic with respect to the lightlike directions. Such a pair always exists for ellipses; for hyperbolas it exists only, when the asymptotes do not separate the lightlike directions (note Figure 6.30). This gives rise to a classification of conics in $\mathbb{M}^2$ as given below:

Up to m-motions and a commutation of the coordinate axes, there are six types of conics to distinguish:

1. *Circles:* Their standard equation is $x^2 - y^2 = \sigma \neq 0$. Given two points A, B with $d_m(A, B) \neq 0$, e.g., $A = (a, 0)$ and $B = (-a, 0)$, the locus of points $P = (x, y)$, for which the lines $[P, A]$ and $[P, B]$ are orthogonal, is the circle $x^2 - a^2 - y^2 = 0$ with the diameter AB. This means, THALES's theorem is also valid in $\mathbb{M}^2$.

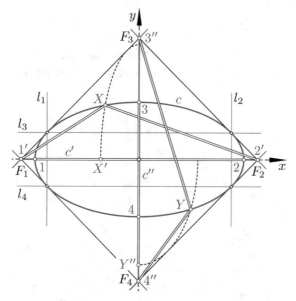

FIGURE 10.24. The ellipse c is of type 2 in $\mathbb{M}^2$. Points $X \in c$ satisfy either $d_m(X, F_1) + d_m(X, F_2) = d_m(1, 2)$ or $|d_m(X, F_1) - d_m(X, F_2)| = d_m(1, 2)$, as well as $d_m(X, F_3) + d_m(X, F_4) = d_m(3, 4)$ or $|d_m(X, F_3) - d_m(X, F_4)| = d_m(3, 4)$.

2. *Ellipses or hyperbolas with two axes of symmetry* and standard equation

$$\frac{x^2}{\sigma} + \frac{y^2}{\tau} = 1, \quad \text{where } \sigma\tau(\sigma + \tau) \neq 0.$$

Under $\sigma + \tau > 0$ these conics have four real focal points $(\pm e, 0)$ and $(0, \pm e)$ with $e := \sqrt{\sigma + \tau}$, like the ellipse c depicted in Figure 10.24. With respect to the two foci F_1 and F_2 on the x-axis and the associated directrices l_1 and l_2, points $X \in c$ satisfy

$$d_m(X, F_1) = \varepsilon \cdot d_m(X, l_1) \quad \text{and} \quad d_m(X, F_2) = \varepsilon \cdot d_m(X, l_2).$$

Consequently, points $X \in c$ between l_1 and l_2 have a constant sum of distances $d_m(X, F_1) + d_m(X, F_2) = \varepsilon \cdot d_m(l_1, l_2) = d_m(1, 2)$, while points outside the parallel strip bounded by l_1 and l_2 satisfy $|d_m(X, F_1) - d_m(X, F_2)| = \varepsilon \cdot d_m(l_1, l_2) = d_m(1, 2)$. Points on l_1 or l_2 satisfy both condititions.

3. *Hyperbolas with a spacelike and a timelike asympote:* These conics (see Figure 10.25) have a center, but no axis of symmetry. Their standard equation is

$$\sigma(y^2 - x^2) + (1 - \sigma^2)xy = \tau \quad \text{with} \quad \tau \neq 0.$$

Here, σ with $-1 < \sigma < 1$ is the *e*-slope of the spacelike asymptote. The pairwise complex conjugate *m*-focal points are located on the *e*-isotropic lines $y = \pm ix$.

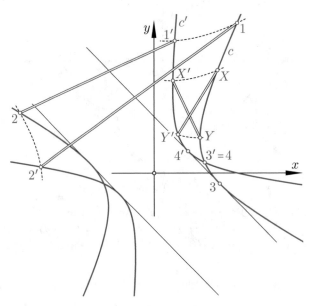

FIGURE 10.25. Confocal hyperbolas without axis of symmetry (type 3) in $\mathbb{M}^2$.

4. *Hyperbolas with one lightlike asymptote:* We specify the non-lightlike asymptote as x-axis and obtain the standard equation

$$xy - y^2 = \sigma \quad \text{with} \quad \sigma \neq 0.$$

In the case $\sigma > 0$ they have two real focal points $F_1 = (\pm\sqrt{2\sigma}, \pm\sqrt{2\sigma})$.

5. *Parabolas with an axis of symmetry:* We choose the axis of symmetry as x-axis and get the standard form

$$y_2^2 - 4\sigma x = 0, \quad \text{where} \quad \sigma > 0.$$

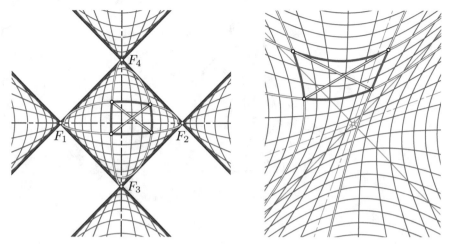

FIGURE 10.26. Nets of confocal conics of type 2 in $\mathbb{M}^2$ with real foci (left) or with imaginary foci (right).

with the focal point $(-\sigma, 0)$.

6. *Parabolas with lightlike diameters:* Their standard equation is

$$(x + y)^2 - 2\sigma(x - y) = 0 \text{ with } \sigma \neq 0.$$

They have no focal point.

A comparison with Lemma 10.2.1 reveals that in $\mathbb{M}^2$ only the conics of type 5 and those of type 2 with real foci satisfy the Apollonian definition. In the case of hyperbolas of type 2 (Figure 10.23, right) only foci on the principal axis are admitted, *i.e.*, on the axis which intersects the hyperbola in real points. Otherwise one of the distances $d_m(P, F)$ and $d_m(P, l)$ would be spacelike, the other timelike.

Confocal conics in $\mathbb{M}^2$

Two conics are called *confocal* in $\mathbb{M}^2$ if and only if their tangential equations span a linear system which contains the two pencils of lightlike lines as a singular curve. Therefore, confocal conics share the lightlike tangent lines and the foci. Also in the pseudo-Euclidean plane, a family of confocal conics forms an orthogonal net. We prove this below with the aid of DESARGUES's involution theorem.

Let P be a point of any conic c included in a confocal family. When P is not located on one of the common lightlike tangents of the family, the tangent to c at P is one of the two fixed lines of the induced Desargues

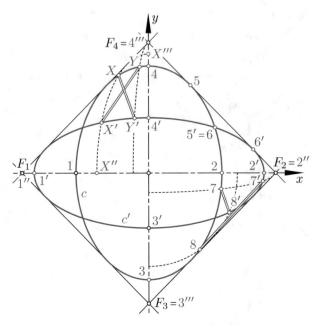

FIGURE 10.27. IVORY's theorem in $\mathbb{M}^2$, applied to confocal ellipses of type 2.

involution at P (Section 7.4). The fixed lines separate the lightlike lines through P harmonically. Therefore, there exists a second conic of the family which intersects c orthogonally at P.

Figure 10.26 shows a family of confocal ellipses and hyperbolas of type 2 with real focal points. When one semiaxis of an included ellipse tends to zero, the limit of the ellipse is the segment $F_1 F_2$ and $F_3 F_4$. The analogous limits of the included hyperbolas consist of pairs of disjoint aligned half-lines, each terminated by a focus. In the standard model of $\mathbb{M}^2$ which is used in all figures, the pseudo-Euclidean foci belong to the e-director circle or e-orthoptic circle (Theorem 2.2.6) of all ellipses and hyperbolas included in the confocal family.

In the Euclidean plane $\mathbb{E}^2$, an important property of confocal conics has been stated in Theorem 2.2.3 (see Figure 2.23): according to IVORY's theorem, in each curvilinear quadrangle formed by two pairs of conics the two diagonals have the same length. Another formulation of this theorem uses the fact that for any two confocal conics c, c' of the same type an affine transformation $\alpha\colon c \to c'$ can be defined such that curves of the confocal family which intersect c and c' orthogonally, pass through corresponding points $X \in c$ and $\alpha(X) \in c'$. Then IVORY's theorem in $\mathbb{E}^2$ states

$$\overline{X\,\alpha(Y)} = \overline{\alpha(X)\,Y} \text{ for all } X, Y \in c.$$

This statement holds even for singular α, when $c' = \alpha(c)$ degenerates into a point-set on an axis of symmetry (Figure 2.24).

The analogous results are valid for confocal spherical conics, as stated in Theorem 10.1.10 (see Figure 10.14). It has been proved in [60] that even in the pseudo-Euclidean plane $\mathbb{M}^2$ IVORY's theorem is valid for all types of confocal sets. However, one cannot generally define the term 'of the same type'. Sometimes the definition works that confocal conics c and c' are of the same type, if within the confocal family there is a continuous transition from c to c' without passing a singular curve (Figure 10.26, right). But this definition fails in the case depicted in Figure 10.26, left.

In terms of an affine transformation $\alpha \colon c \to c'$ with $X \mapsto X'$ and $Y \mapsto Y'$, the pseudo-Euclidean version of IVORY's theorem claims

$$d_m(X, Y') = d_m(X', Y) \text{ for all } X, Y \in c.$$

Figure 10.27 shows IVORY's theorem in several cases. The distances $d_m(X, Y') = d_m(X', Y)$ and $d_m(7, 8') = d_m(7', 8)$ are timelike, while $5' = 6$ implies that $d_m(5, 6') = d_m(5', 6) = 0$. Hence, $[5, 6']$ must be lightlike.

Remark 10.2.2 As in the Euclidean case, it can be proved that in each curvilinear quadrangle of a confocal net the diagonal lines are tangent to the same conic of the confocal family. This guarantees also that if one diagonal is lightlike the other must be lightlike, too.

In Figure 10.25 two confocal conics of type 3 are depicted. The distance $d_m(X, Y') = d_m(X', Y)$ is timelike and $d_m(1, 2') = d_m(1', 2)$ is spacelike.

Figure 10.24 shows singular cases with c' as x-axis and c'' as y-axis. The equations $d_m(X, 1') + d_m(X, 2') = d_m(X', 1) + d_m(X', 2) = d_m(1, 2)$ and $d_m(Y, 3'') - d_m(Y, 4'') = d_m(Y'', 3) - d_m(Y'', 4) = d_m(3, 4)$ reveal the constant sum or difference of the pseudo-Euclidean focal distances. Also in Figure 10.27 singular affine transformations $X \mapsto X''$ and $X \mapsto X'''$ between the ellipse c and its axes c'' and c''' are indicated.

● Exercise 10.2.1 (i) Verify with the aid of (10.19) that direct pseudo-Euclidean motions keep the directions of lightlike lines fixed, while in the indirect case these directions are exchanged.
(ii) Confirm that under the map (10.22) all points on the axis $y = \tanh \frac{\tau}{2} x$ are fixed, while other points remain on lines m-orthogonal to this axis.

Conics in the hyperbolic plane

The axioms of the hyperbolic plane $\mathbb{H}^2$ and the Euclidean plane $\mathbb{E}^2$ differ only in the parallel postulate. In $\mathbb{E}^2$ there is a unique parallel h to a given line g through any point P. In $\mathbb{H}^2$ the opposite is true: through each point $P \notin g$ there pass (at least) two lines which do not meet the line g.

We refer to the *projective* or CAYLEY-KLEIN *model* of $\mathbb{H}^2$: Given a conic ω, the *absolute conic* in a projective plane $\mathbb{P}^2$, points of $\mathbb{H}^2$ are the points in the interior of ω, and the lines of $\mathbb{H}^2$ are secants of ω. Hyperbolic motions are projective transformations in $\mathbb{P}^2$ which map the absolute conic ω onto itself — however restricted to $\mathbb{H}^2$. Hyperbolic reflections are the restrictions of harmonic homologies of ω. For further details see, *e.g.*, [17, 42, 44, 46].

We prefer the standard model which is embedded into the projective extension of the Euclidean plane $\mathbb{E}^2$. We use homogeneous Cartesian coordinates $(x_0 : x_1 : x_2) = (1 : x : y)$ and the absolute conic ω as the unit circle. Then, the polar form with respect to ω can be written as

$$\langle \mathbf{p}, \mathbf{q} \rangle_h = p_0 q_0 - p_1 q_1 - p_2 q_2.$$

Points in $\mathbb{H}^2$ have coordinate vectors $\mathbf{p}$ where $\langle \mathbf{p}, \mathbf{q} \rangle_h > 0$. The hyperbolic distance $d_h(P, Q)$ can be computed by virtue of

$$\cosh d_h(P, Q) = \left| \frac{\langle \mathbf{p}, \mathbf{q} \rangle_h}{\sqrt{\langle \mathbf{p}, \mathbf{p} \rangle_h \langle \mathbf{q}, \mathbf{q} \rangle_h}} \right|.$$

This distance formula satisfies the axioms of a metric space.

Conics c in $\mathbb{H}^2$ are projective conics which contain interior points of $\mathbb{H}^2$. One can classify all conics c in the hyperbolic plane $\mathbb{H}^2$ by classifying in $\mathbb{P}^2(\mathbb{R})$ the pencil of conics spanned by ω and c and, in addition, by considering the different possibilities concerning the reality of the base points.

A point $M \in \mathbb{H}^2$ is a *center* of c if the polars of M with respect to c and ω coincide. A line in $\mathbb{H}^2$ is an *axis* of c if its poles with respect to ω and c are the same. According to results in Section 7.1, conics c in $\mathbb{H}^2$ need not have a center or axis. Figure 10.28 shows a conic c without center and axis; c osculates the absolute conic c.

A point $F \in \mathbb{H}^2$ is a *focal point* of the conic c, if F is the point of intersection of two common (complex conjugate) tangents of c and ω. Two conics c, c'

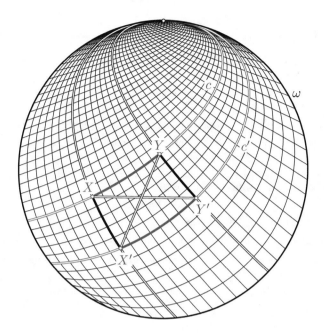

FIGURE 10.28. IVORY's theorem in the hyperbolic plane $\mathbb{H}^2$ in the case of confocal conics c and c' which osculate the absolute conic ω.

in $\mathbb{H}^2$ are called *confocal* if and only if their tangential equations span a linear system which contains the absolute quadric. In this case, common tangents of c and ω are also tangent to c'; therefore confocal conics share the focal points. Also in the hyperbolic plane a family of confocal conics forms an orthogonal net. Similarily to the pseudo-Euclidean plane, this can be concluded from DESARGUES's involution theorem.

Just as in the spherical case, the Apollonian definition of conics does not work in $\mathbb{H}^2$. In general, the geometric locus defined by $d_h(P, F) = \varepsilon \cdot d_h(P, l)$ is not even algebraic.

In [61] it is proved that IVORY's theorem is valid in the hyperbolic plane $\mathbb{H}^2$. This is illustrated in Figure 10.28: $d_h(X, Y') = d_h(X', Y)$. When a conic c has an axis s of symmetry with two real focal points, then similarly to previous cases (see, *e.g.*, Figure 10.24), one can use IVORY's theorem for proving that the sum or difference of distances to the focal points on the axis s is constant in the hyperbolic plane.

References

[1] S. ABU-SAYMEH, M. HAJJA, H. STACHEL: *Equicevian Points and Cubics of a Triangle.* J. Geometry Graphics **18**/2 (2014), 133–157.

[2] A.V. AKOPYAN, A.A. ZASLAVSKY: *Geometry of Conics.* Mathematical World - Volume 26, AMS, 2007.

[3] H.F. BAKER: *Principles of Geometry,* Vol. II, 2nd ed., Cambridge University Press, 1930.

[4] W.L. Black, H.C. Howland, B. Howland: *A theorem about zig-zags between two circles.* Amer. Math. Monthly 81 (1974), 754–757.

[5] M. BERGER: *Geometry I, II.* Translated from the French by M. Cole and S. Levy. Springer-Verlag, Berlin-Heidelberg, 1987.

[6] R. BIX: *Topics in Geometry.* Academic Press, 1993.

[7] R. BIX: *Conics and Cubics.* Springer, 2006.

[8] W. BLASCHKE: *Projektive Geometrie,* 3. Aufl., Verlag Birkhäuser, Basel, 1954.

[9] D.A. BRANNAN: *A First Course in Mathematical Analysis.* Cambridge University Press, Cambridge, England, 2006.

[10] D.A. BRANNAN: *Geometry.* Cambridge University Press, 2012.

[11] H. BRAUNER: *Geometrie Projektiver Räume I, II.* Bibliographisches Institut, Wien/Zürich, 1976.

[12] H. BRAUNER: *Lehrbuch der Konstruktiven Geometrie.* Springer, Wien, 1986.

[13] E. BRIESKORN, H. KNÖRRER: *Plane Algebraic Curves.* Birkhäuser, 1986.

[14] J. CILLERUELO, A. CÓRDOBA: *Lattice Points on Ellipses.* Duke Math. J. **76**, No. 3 (1994), 741–750.

[15] J. CILLERUELO, J. JIMÉNEZ-URROZ: *Lattice Points on Hyperbolas.* J. Number Theory **63** (1997), 267–274.

[16] J.L.A. COOLIDGE: *A Treatise on Algebraic Plane Curves.* Dover, New York, 1959.

[17] H.S.M. COXETER: *Non-Euclidean Geometry.* 5th ed., University of Toronto Press, 1965.

[18] H.S.M. COXETER *The real projective plane,* 3rd ed., Springer-Verlag, New York, 1993.

[19] H.S.M. COXETER & S.L. GREITZER: *Geometry Revisited.* Math. Assoc. Amer., Washington DC, 1967.

[20] D. COX, J. LITTLE, D. O'SHEA: *Ideals, Varieties, and Algorithms: An Introduction to Algebraic Geometry and Commutative Algebra.* 2nd ed. New York: Springer-Verlag, 1996.

[21] E.A. DIJKSMAN: *Motion Geometry of Mechanisms.* Cambridge University Press, 1976.

[22] F. DINGELDEY: *Kegelschnitte und Kegelschnittsysteme.* In: *Encyklopädie der math. Wiss.* III.2.1, no. C1, 1–160, B.G. Teubner, Leipzig, 1903.

[23] M.P. do Carmo: *Differential Geometry of Curves and Surfaces.* Prentice-Hall, 1976.

[24] V. DRAGOVIĆ, M. RADNOVIĆ: *Poncelet Porisms and Beyond: Integrable Billiards, Hyperelliptic Jacobians and Pencils of Quadrics.* Birkhäuser, 2011.

[25] V. DRAGOVIĆ, M. RADNOVIĆ: *Bicentennial of the great Poncelet Theorem (1813–2013): Current Advances.* Bull. Amer. Math. Soc. **51**/3 (2014), 373–445.

[26] L. FLATTO: *Poncelet's theorem.* American Mathematical Society, Providence, Rhode Island, 2009.

[27] O. Giering: Vorlesungen über höhere Geometrie. Vieweg, 1982.

[28] G. Glaeser, H. Stachel: *OpenGeometry + Advanced Geometry*. Springer, New York, 1999.

[29] G. Glaeser, H.-P. Schröcker: *Handbook of geometric programming using OpenGeometry and GL*. Springer, New York, 2001.

[30] G. Glaeser: *Geometry and its Applications in Arts, Nature and Technology*. Springer, Wien - New York, 2012.

[31] E. Ghys, S. Tabachnikov, V. Tomorin: *Osculating curves: around the Tasit-Kneser theorem*. Math. Intelligencer **35**/1 (2013), 61–66.

[32] D.L. Goodstein, J.R. Goodstein: *Feynman's lost lecture*. Vintage Books, London 1997.

[33] P. Griffiths, J. Harris: *On Cayley's Explicit Solution to Poncelet's Porism*. Enseign. Math. **24**, 31–40, 1978.

[34] W.R. Hamilton: *The Hodograph, or a new method of expressing in symbolical language the Newtonian Law of Attraction*. Proc. Royal Irish Academy **3** (1847), 344–353.

[35] D. Hilbert, S. Cohn-Vossen: *Geometry and the Imagination*. Chelsea, New York, 1999.

[36] W.P. Hirschfeld: *Projective Geometries over finite Fields*. Oxford University Press, 1998.

[37] F. Hohenberg: *Konstruktive Geometrie in der Technik*. Springer, Wien, 1961.

[38] F. Hohenberg: *Ein Schließungssatz über gleichseitige Polygone, deren Ecken abwechselnd auf zwei Kreisen liegen*. Sitzungsber., Abt. II, österr. Akad. Wiss., Math.-Naturw. Kl. **186** (1977), 281–300.

[39] J. Ivory: *On the Attractions of homogeneous Ellipsoids*. Phil. Trans. of the Royal Society of London **99** (1809), 345–372.

[40] C. Kimberling: *Triangle Centers and Central Triangles*. Congr. Numer. 129, Winnipeg, 1998.

[41] C. Kimberling: *Encyclopedia of Triangle Centers - ETC*. http://faculty.evansville.edu/ck6/encyclopedia/ETC.html

[42] F. Klein: *Vorlesungen über höhere Geometrie*. 3. Aufl., Springer, Berlin 1926.

[43] G. Kohn, G. Loria: *Spezielle algebraische Kurven*. In: *Encyklopädie der math. Wiss.* III C 5, B.G. Teubner, Leipzig 1903–1915.

[44] G. Kowol: *Projektive Geometrie und Cayley-Klein Geometrien in der Ebene*. Birkhäuser, Basel, 2009.

[45] E. Kruppa: *Analytische und konstruktive Differentialgeometrie*. Springer, Wien, 1957.

[46] G.E. Martin: *The Foundations of Geometry and the Non-Euclidean Plane*. Springer, New York, 1982.

[47] G. Meurant: *Handbook of Convex Geometry*. J.M. Wills, P.M. Gruber (eds.), Elsevier (ISBN-13 9780080934396), North-Holland, Amsterdam, 2014.

[48] A.F. Möbius: *Der barycentrische Calcul*. Verlag J.A. Barth, Leipzig, 1827.

[49] E. Müller, J.L. Krames: *Vorlesungen über Darstellende Geometrie II: Die Zyklographie*. Deuticke, Leipzig & Wien, 1929.

[50] B. Odehnal: *Generalized Gergonne and Nagel points.* Contr. Alg. Geom. **51**/2 (2010), 477-491.

[51] J. Richter-Gebert, Th. Orendt: *Geometriekalküle.* Springer, 2009.

[52] J. Richter-Gebert: *Perspectives on Projective Geometry: A Guided Tour Through Real and Complex Geometry.* Springer, 2011.

[53] R.E. Schwartz: The Poncelet Grid. Available at:
www.math.brown.edu/~res/Papers/grid.pdf

[54] W.R. Scott: *Group Theory.* Dover Publications Inc., revised, 2000.

[55] C. Segre: *Mehrdimensionale Räume.* In *Encyklopädie der math. Wiss.* B.G. Teubner, Leipzig, 1928. III.2.2A, no. C7, 779-972.

[56] J.E. Shigley, J.J. Uicker: *Theory of Machines and Mechanisms.* McGraw-Hill, New York, 1980.

[57] W.M. Smart: *Text-Book on Spherical Astronomy* 6th ed., Cambridge University Press, Cambridge, 1960.

[58] M. Spivak: *A Comprehensive Introduction to Differential Geometry.* Vols. I – V, Publish or Perish, 1979.

[59] H. Stachel: *A Remarkable Overconstrained Spherical Motion.* In: J. Lenarčič, M. Husty (eds.) Advances in Robot Kinematics: Analysis and Control, Kluwer Academic Publ. (ISBN 0-7923-5169-X), Dordrecht 1998, 287-296.

[60] H. Stachel: *Ivory's Theorem in the Minkowski Plane.* Math. Pannonica **13** (2002), 11-22.

[61] H. Stachel, J. Wallner: *Ivory's Theorem in Hyperbolic Spaces.* Sib. Math. J. **45** (2004), no. 4, 785-794.

[62] O.J. Staude: *Flächen 2. Ordnung und ihre Systeme und Durchdringungskurven.* In *Encyklopädie der math. Wiss.* III.2.1, no. C2, 161-256, B.G. Teubner, Leipzig, 1915.

[63] D. Wells: *The Penguin Dictionary of Curious and Interesting Numbers.* Penguin Books, 1986.

[64] W. Wunderlich: *Darstellend-geometrischer Beweis eines merkwürdigen Schließungssatzes.* Anz. österr. Akad. Wiss., Math.-Naturwiss. Kl. **115** (1978), 150-152.

[65] W. Wunderlich: *Darstellende Geometrie I, II.* Bibliographisches Institut, Wien/Zürich, 1967.

[66] W. Wunderlich: *Ebene Kinematik.* Bibliographisches Institut, Wien/Zürich, 1970.

[67] W. Wunderlich: *Mechanisms related to Poncelet's closure theorem.* Mech. Mach. Theory **16** (1981), 611-620.

[68] I.M. Yaglom: *A Simple Non-Euclidean Geometry and Its Physical Basis.* Springer-Verlag, New York, 1979.

[69] P. Yiu: *Introduction to the Geometry of the Triangle.* http://math.fau.edu/Yiu/GeometryNotes020402.pdf

[70] D. Zwillinger: *Spherical Geometry and Trigonometry,* §6.4 in: CRC Standard Mathematical Tables and Formulae, 468-471, CRC Press, Boca Raton, FL, 1995.

Index

Printed in the United States
By Bookmasters